GENERAL POWER MECHANICS

McGRAW-HILL PUBLICATIONS IN INDUSTRIAL EDUCATION
Chris H. Groneman, CONSULTING EDITOR

BOOKS IN SERIES

DRAWING AND BLUEPRINT READING   Coover

GENERAL DRAFTING: A COMPREHENSIVE EXAMINATION   Blum

GENERAL INDUSTRIAL EDUCATION   Groneman and Feirer

GENERAL METALS   Feirer

GENERAL POWER MECHANICS   Crouse, Worthington, Margules, and Anglin

GENERAL WOODWORKING   Groneman

TECHNICAL ELECTRICITY AND ELECTRONICS   Buban and Schmitt

TECHNICAL WOODWORKING   Groneman and Glazener

UNDERSTANDING ELECTRICITY AND ELECTRONICS   Buban and Schmitt

WILLIAM H. CROUSE
ROBERT M. WORTHINGTON
MORTON MARGULES
DONALD L. ANGLIN

GENERAL POWER MECHANICS

SECOND EDITION

McGRAW-HILL BOOK COMPANY

New York St. Louis San Francisco Dallas Auckland
Düsseldorf Johannesburg Kuala Lumpur London Mexico
Montreal New Delhi Panama Paris São Paulo
Singapore Sydney Tokyo Toronto

Library of Congress Cataloging in Publication Data

Main entry under title:

General power mechanics.

  (McGraw-Hill publications in industrial education)
  Robert M. Worthington's name appeared 1st on title
page of 1st ed.
  Includes index.
  1. Power (Mechanics) 2. Engines. I. Crouse,
William Harry, date
TJ153.W75 1976   621   75-22410
ISBN 0-07-014697-7

2345678910II2 MUMU 987

Editors: John Aliano and Gordon Rockmaker

Design Supervisor and Cover Designer: Joe Nicholosi

Editing Supervisors: Linda Richmond and Barbara Tannenbaum

Production Supervisor: Ted Agrillo

GENERAL POWER MECHANICS
Second Edition

PREFACE

It is the aim of the second edition of *General Power Mechanics* to give students an understanding of, and an ability to deal practically with, power and power devices. The book presents, in logical sequence, a course of instruction which will enable students to acquire a basic understanding of the operations and skills needed in the service and repair of power sources used today. Power, from whatever source it is developed, touches our lives in employment, medicine, recreation, research, and transportation. In view of the importance of power and its increased use and cost, knowing something about power is essential preparation for coping with the world of tomorrow, whether as a consumer, service technician, or engineer. *General Power Mechanics* has been written to provide instruction at the middle school and the junior and senior high school as well as for others who have special interest in this area of study.

Since the publication of the first edition, many things have happened in the important area of power. Overtaxing our electricity generating capacity has caused brownouts and blackouts. The use of atomic energy has failed to develop as rapidly as forecast a few years ago. Because only a few new nuclear-powered electric generating plants have been built and brought on line, our dependence on oil—much of it imported oil—has increased.

Current conditions brought about by the nation's energy crisis and the need for cleaner air were considered carefully in the revision of *General Power Mechanics.* In addition, the authors were concerned that the essential body of knowledge about power, and power sources and how to service them, be covered in a thorough manner of maximum benefit to the student. To do this most effectively, the text is no longer divided into units, but is now divided into chapters with topic sections. As a result, emphasis now is placed on power and power applications, instead of on the automobile.

General Power Mechanics has many unique features. Despite the technical nature of the subject matter, the book maintains an acceptable reading level. The text has been almost totally rewritten to simplify explanations and improve comprehension. The sentences are short and simple, and the language is direct and to the point. New terms are shown in boldface. A second color is used functionally throughout illustrations and text. Many new simple line drawings are used, and many new photographs are included. The pages have been designed to give maximum visual impact and easy comprehension of the textual material.

A new chapter early in the book covers energy sources. Other chapters cover fuels and fuel systems, new material on automotive emission controls, and alternative power sources to the conventional piston engine, such as the Wankel rotary engine. Motorcycle engines are also covered. In line with concern about career guidance, the last chapter discusses the future for employment in the power mechanics field.

A special feature of the new edition is that it has the metric equivalents of all United States customary measurements. The metric equivalent follows in brackets each USCS measurement given in the book.

There are still other important additions to the book—new material on safety in the shop, electronic ignition systems, and a host of other developments in automotive, marine, and aircraft engines and power sources.

To help the teacher and the student, the review questions at the end of each chapter have been completely redone. The study problems now contain many practical suggestions for student activities that will both interest and expand the student's knowledge and skills.

During the planning and preparation of the second edition of *General Power Mechanics,* the authors and publishers had the advice and assistance of many people—educators, researchers, artists, editors, and automotive service and power industry specialists. It would take pages to acknowledge them all individually, but the authors gratefully acknowledge their indebtedness and offer their sincere thanks to those many people.

WILLIAM H. CROUSE
ROBERT M. WORTHINGTON
MORTON MARGULES
DONALD L. ANGLIN

ABOUT THE AUTHORS

WILLIAM H. CROUSE

Behind William H. Crouse's clear technical writing is a background of sound mechanical engineering training as well as a variety of practical industrial experiences. After finishing high school, he spent a year working in a tin-plate mill. Summers, while still in school, he worked in General Motors plants, and for three years he worked in the Delco-Remy Division shops. Later he became Director of Field Education in the Delco-Remy Division of General Motors Corporation, which gave him an opportunity to develop and use his natural writing talent in the preparation of service bulletins and educational literature.

During the war years, he wrote a number of technical manuals for the Armed Forces. After the war, he became editor of technical education books for the McGraw-Hill Book Company. He has contributed numerous articles to automotive and engineering magazines and has written many outstanding books. He was the first Editor-in-Chief of the 15-volume McGraw-Hill Encyclopedia of Science and Technology.

William H. Crouse's outstanding work in the automotive field has earned him membership in the Society of Automotive Engineers and in the American Society of Engineering Education.

ROBERT M. WORTHINGTON

Dr. Robert M. Worthington is currently an educational consultant. He previously served as Associate U.S. Commissioner of Education for Adult, Vocational and Technical Education. From 1965 to 1971, he served as Assistant State Commissioner of Education and State Director of Vocational, Technical and Adult Education in the state of New Jersey. He was supervisor of veterans' training and industrial education for two years in the Minnesota State Department of Education. Dr. Worthington was for seven years Professor and Chairman of the Department of Industrial Education and Technology at Trenton State College. He was a Visiting Professor of Education in the Graduate School of Education at Rutgers University and was a member of the faculty in the School of Engineering at Purdue University and at the University of Minnesota.

Dr. Worthington received his Master of Arts and Doctor of Philosophy degrees from the University of Minnesota.

MORTON MARGULES

Morton Margules is Superintendent of the Hudson County Area (New Jersey) Vocational-Technical Schools. He has held the following posts in the New Jersey State Department of Education: Associate Director of Vocational-Technical Education, Director of the Research and Development Bureau, Director of Pilot and Demonstration Programs, and Director of Introduction to Vocations. In addition to having taught vocational subjects on the secondary level, Dr. Margules has served as an adjunct professor at Trenton State College and as a consultant on career education both in this country and abroad. Dr. Margules has a B.S. degree from Kean State College, an M.Ed. degree from Rutgers University, and a Ph.D. degree from the University of Ottawa (Canada). He has served as a high school administrator and a board of education member. In addition to his academic background, Dr. Margules has had wide trade and industrial experience.

DONALD L. ANGLIN

Trained in the automotive and diesel service field, Donald L. Anglin has worked both as a mechanic and as a service manager. He has taught automotive courses and has also worked as curriculum supervisor and school administrator for an automotive trade school. He has been an author and technical advisor in the development of independent media programs for automotive training. Interested in all types of vehicle performance, he has served as a race-car mechanic and as a consultant to truck fleets on maintenance problems.

Currently, he serves as editorial assistant to William H. Crouse and visits automotive instructors and service shops. Together they have coauthored magazine articles on automotive education and several books in the McGraw-Hill Automotive Technology Series.

Donald L. Anglin is a Certified General Automotive Mechanic and holds many other licenses and certificates in automotive education, service, and related areas. His work in the automotive service field has earned him membership in the American Society of Mechanical Engineers and the Society of Automotive Engineers. In addition, he is an automotive instructor at Piedmont Virginia Community College, Charlottesville, Virginia.

CONTENTS

1 | HISTORY OF POWER | 1

2 | ENERGY SOURCES | 17

3 | STEAM POWER | 23

4 | ELECTRICAL GENERATING PLANTS | 36

5 | FUNDAMENTALS OF INTERNAL-COMBUSTION ENGINES | 43

6 | TWO-CYCLE ENGINES | 52

7 | TWO-CYCLE ENGINE FUEL SYSTEMS | 65

8 | TWO-CYCLE ENGINE IGNITION AND STARTING SYSTEMS | 73

9 | FOUR-CYCLE ENGINES—OPERATION AND TYPES | 84

10 | AUTOMOTIVE ENGINE CONSTRUCTION—CYLINDER BLOCK, HEAD, CRANKSHAFT, AND BEARINGS | 101

11 | AUTOMOTIVE ENGINE CONSTRUCTION—PISTONS, CONNECTING RODS, VALVES, AND VALVE TRAINS | 117

12 | ENGINE MEASUREMENTS | 135

13 | INTRODUCTION TO AUTOMOTIVE ELECTRICITY | 142

14 | AUTOMOTIVE BATTERY AND STARTING MOTOR | 150

15 | AUTOMOTIVE CHARGING SYSTEMS | 163

16 | AUTOMOTIVE IGNITION SYSTEMS | 175

17 | AUTOMOTIVE FUEL SYSTEMS | 189

18 | ENGINE COOLING SYSTEMS | 207

19 ENGINE LUBRICATING SYSTEMS 214

20 AUTOMOTIVE EMISSION CONTROLS 226

21 OTHER AUTOMOTIVE POWER SOURCES 235

22 ROTARY ENGINES 244

23 DIESEL ENGINES 254

24 MOTORCYCLE ENGINES 267

25 SMALL MARINE ENGINES 278

26 AIRCRAFT ENGINES 305

27 ROCKETS AND SPACE TRAVEL 334

28 MECHANICAL POWER TRANSMISSION 354

29 AUTOMATIC TRANSMISSIONS 381

30 FLUID POWER TRANSMISSION 401

31 SHOP SAFETY 414

32 HAND AND POWER TOOLS 421

33 MEASUREMENTS AND MEASURING TOOLS 439

34 USING THE METRIC SYSTEM 452

35 SMALL-ENGINE MAINTENANCE AND SERVICE 467

36 AUTOMOTIVE-ENGINE TROUBLESHOOTING 488

37 AUTOMOTIVE-ENGINE TUNE-UP AND SERVICE 497

38 SERVICING THE AUTOMOTIVE-ENGINE SYSTEMS 511

39 YOUR FUTURE IN POWER MECHANICS 522

INDEX 525

EDITOR'S FOREWORD

General Power Mechanics is a well-accepted title in the McGraw-Hill Publications in Industrial Education, covering one of the newer, more progressive areas within the field of industrial education.

This revision's reorganized format presents thirty-nine individual chapters subdivided by sections. In addition to updating the vastly changing technology of transportation and power, new information is given on the metric system of measurement, pollution, and the selection of careers relating to transportation and power. A second color emphasizes important learning activities. This accent captures attention and reinforces the teaching-learning process. End-of-chapter problems and questions challenge and stimulate further learner interest and response.

In this age of spectacular achievement in all forms of transportation and power, *General Power Mechanics* provides information on the latest technologies which are helping solve the challenges of speed and space.

This edition was revised under the leadership of William H. Crouse. Mr. Crouse has contributed much professional literature and text materials, and he is a recognized authority in the broad areas of transportation and power.

The editor, authors, and publisher are confident that *General Power Mechanics* will continue to make a contribution to the concept of modern technologies in our socio-economic growth and development.

CHRIS H. GRONEMAN

1 HISTORY OF POWER

§1-1 INTRODUCTION TO POWER

Can you imagine a world where the only power was the power of human or animal muscles? It would be a world without homes as we know them, without factories, electric lights, modern medicine, automobiles, and adequate food and clothing. In fact, if you took away power, you'd have a world you could hardly recognize.

In this modern world of ours, there are countless wheels spinning and working for us. Wheels and gears turn in automobiles, in factories and in offices, on farms and in the cities, in the air and under the oceans, and on land and sea. There is hardly any human activity that does not depend upon the rotation of wheels and gears, upon the use of power (Fig. 1-1).

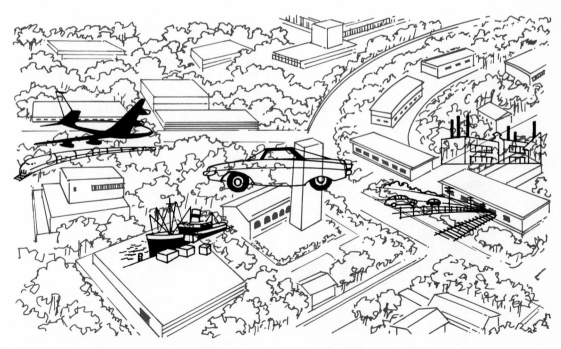

Fig. 1-1. There are few human activities that do not depend on the use of power.

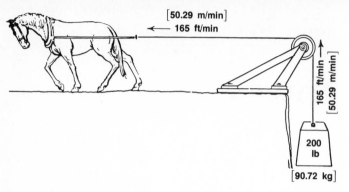

[50.29 m/min]
← 165 ft/min

165 ft/min →
[50.29 m/min]

200 lb

[90.72 kg]

Fig. 1-2. Power is measured in horsepower.

Power is measured in horsepower and in kilowatts (Fig. 1-2). One horsepower is the power that a theoretical horse can produce when it is working. In the United States, each one of us has the equivalent of 100 horses working for us. This is the power of automobile and airplane engines; of boats and industrial engines; of electric motors in homes, offices, and factories; of steam engines; and of water pouring through hydroelectric power plants.

The ability of the human race to produce and utilize power has helped to shape our modern civilization. A better understanding of power, and of the machines that produce and use it, will give you a greater insight into the world of today.

§ 1-2 WATER MILLS

Years ago, all power came from muscles, human and animal. But muscles get tired, and bodies must be fed to keep the muscles working. Over the centuries, many people must have observed the power in moving water. But they continued to use their own muscles to do the work. For hundreds of years, people ground grain by hand. Later, oxen were used to rotate a heavy millstone by walking in a circle and pulling a heavy beam attached to the millstone. The grain was ground to flour between the rotating stone and a stationary, or bed, stone. Then, perhaps 2,500 years ago, some clever person figured out a way to put moving water to work grinding grain. This early inventor made a wheel that turned by water power and caused the millstone to rotate.

There are three general types of water mills (Fig. 1-3): the **undershot wheel,** the **overshot wheel,** and the **breast wheel.** The undershot wheel was probably the first, because it was simplest to conceive and build. The overshot- and breast-wheel types, which are more effi-

Fig. 1-3. Three types of water wheels: *Left,* overshot; *center,* breast; *right,* undershot.

cient, came later. By then, people had learned to build dams to raise the water high enough to flow over or down into the wheel.

§ 1-3 WINDMILLS

The windmill came much later than the water mill. Its place of origin is not known, but by 640 A.D. a Persian claimed to have built "a mill that is rotated by the wind" (Fig. 1-4). The development of windmills proceeded slowly during the next several centuries, but they finally began to appear in Europe.

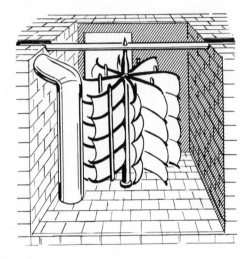

Fig. 1-4. An early horizontal windmill used by the Persians.

During the middle ages (500 A.D. to 1500 A.D.), windmills and water mills were used for a variety of purposes. One of the first uses was to grind grain into flour. They were also used to pump water to irrigate crops, to drain low-lying land, and to clear mines of water. Many other uses were found. By the twelfth century, there were water mills for turning wood, making cloth and paper, pressing oil from olives, crushing ore, operating iron forges, and driving sawmills. A census taken in 1086 A.D. listed 5,624 water mills in southern England alone.

Water mills were common in rather hilly country, where the streams ran swift and strong and the wind was apt to be uncertain. Windmills were favored in flat country, where the streams were slow and sluggish but the winds were more apt to be steady (Fig. 1-5).

About 1335 A.D., an Italian physician proposed a form of war vehicle somewhat like a tank. In it, soldiers could ride completely enclosed. The power to drive it was to have been supplied by a windmill mounted on top. (There is no record, however, of such a vehicle ever having been built.) The development of the powered vehicle had to wait many centuries until the invention of the internal-combustion and the external-combustion engines.

§ 1-4 A TRY THAT FAILED

In the first century A.D., in Alexandria, Egypt, on the Mediterranean sea, there lived a man called Hero. He wrote extensively of many mechanical marvels that had been developed before his time. Hero himself is credited with inventing many remarkable devices.

One device made use of the expansive force of air when it is heated. This device could be used to open and close the doors to a temple (Fig. 1-6). When a fire was built on the altar beside the doors, it heated the air in the chamber in the bottom of the altar. Since heated air expands, pressure was produced in the container below the altar. The water in the container was thus forced to flow through a pipe into the bucket hanging from ropes. The ropes were attached through pulleys to two shafts. When the bucket became sufficiently full of water, its weight caused the shafts to rotate. As they rotated, the doors, fastened to the upper ends of the shafts, swung open. Imagine the effect of this device on the average person of that time. In those days, most people knew nothing of such things as air pressure.

§ 1-5 HERO'S STEAM TURBINE

Hero also reported many other remarkable mechanical devices of that time. His crowning achievement was the invention of a crude steam turbine. This device (Fig. 1-7) had a

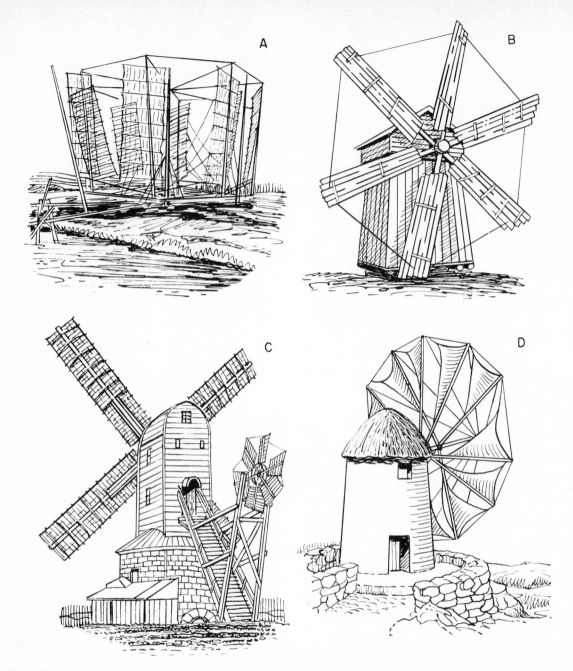

Fig. 1-5. Types of windmills used in various locations. (*A*) Chinese; (*B*) Russian; (*C*) English, showing fantail used to face mill to the wind; (*D*) Aegean Sea.

hollow metal ball suspended on a pair of tubes, one of which was hollow. The ball had two other tubes which were bent at right angles. Below the metal ball was an enclosed container partly filled with water. When a fire was lighted under the container, the water boiled. The steam from the boiling water passed up the hollow supporting tube and into the hollow ball. From there it spurted forth from the two bent tubes with sufficient force to cause the

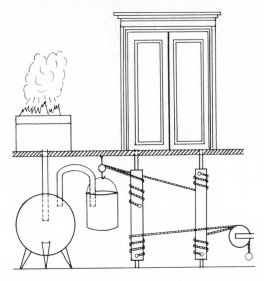

Fig. 1-6. Hero's device for opening the temple doors. A fire is built on the altar to provide the necessary heat.

Fig. 1-7. Hero's steam turbine.

ball to spin. As far as is known, this early steam turbine was not put to work and it remained essentially a toy.

Nearly 2,000 years ago, Hero invented a steam turbine. Yet, 1,600 years were to pass before steam was put to work! No one then could visualize a method of using the force of steam to turn wheels, to operate a mill, or to drive great ships.

If steam power had actually been developed many centuries ago, we probably would have had our steam age 1,000 years earlier. Our electrical age would have started perhaps 800 years ago, and our atomic and space age 600 years ago. Possibly, we would now have flourishing colonies on the moon, Mars, Venus, and other planets. Possibly, humans would already have ventured out into deep space in huge spaceships.

But the opportunity was missed. For many centuries, muscles, and later wind and moving water, were the only sources of power that people used.

This is not the only example of a try that failed, of a discovery or an invention that was overlooked for ages. When people do not keep their eyes and minds open, they fail to see the possibilities of new discoveries. This is unfortunate. People depend on new developments and better ways of doing things to continue to improve their way of life.

§1-6 HOW STEAM CHANGED THE WORLD

Steam finally came into its own about 1,600 years after Hero had devised his steam turbine. The first practical use of steam was to pump water from flooding mines. Pumps powered by horses had been used for many years. But such pumps were not powerful enough for the larger and deeper mines being worked at that time.

Then, about 1700, an Englishman, Thomas Savery, invented a steam pump (Fig. 1-8). Other people before him had proposed somewhat similar devices, but Savery apparently was the first to work out a practical device. Steam engines are described in detail in Chapter 3.

Steam engines were steadily improved after 1700 and were applied first to pumping water from mines. Later, they were used to run manufacturing and milling machinery. In 1769, James Watt obtained his first patent for an improved model of an earlier engine. It was a relatively crude affair. During the next few

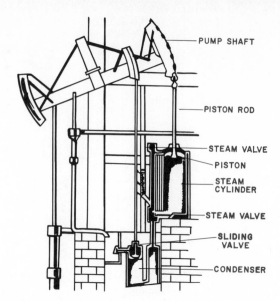

Fig. 1-9. A partial cutaway of Watt's first steam engine.

years he improved it in many ways (Fig. 1-9). By 1800, there were at least 500 of Watt's steam engines in service.

§ 1-7 STEAMBOATS

Not only were steam engines used in stationary applications for mines, mills, and factories, but before 1800, several more or less successful steamboats had been made. In 1790, John Fitch was operating a steamboat between Philadelphia, Pennsylvania, and Bordentown, New Jersey, 28 miles [45.1 kilometers] away (Fig. 1-10). The first really practical steamboat was built in 1802. Later, Robert Fulton and John Stevens built successful steamboats. Within the next 30 years, hundreds of steamboats of larger and more advanced designs were built. It was not until about 1840, however, that steamships regularly crossed the Atlantic Ocean between the United States and Europe.

§ 1-8 LOCOMOTIVES

The invention of the horse collar and the horseshoe enabled horses to pull heavy wagons. For thousands of years, draft animals

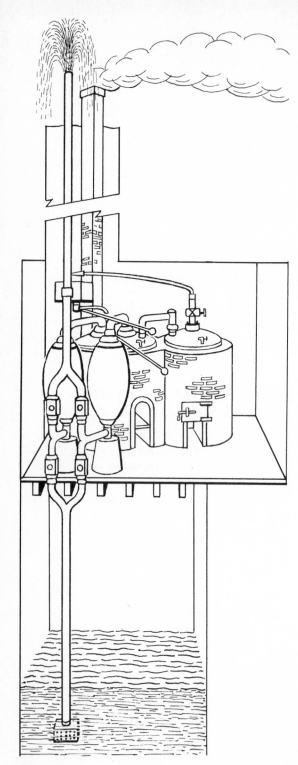

Fig. 1-8. Savery's engine.

Fig. 1-10. Fitch's steamboat.

were the only known source of power to transport people and goods overland. Then, around 1670 or 1680, a Flemish Jesuit priest in China, Ferdinand Verbiest, built a toy cart. It was about 2 feet [0.61 meter] long and was propelled by a steam turbine modeled after Hero's invention. This was only a toy, and no one carried the idea further for many years. In 1769, a French engineer, N. J. Cugnot, put the first self-propelled vehicle on the road (Fig. 1-11). It used a crude form of steam engine which was too weak to make the vehicle practical. A beginning, however, had been made.

About 1803, Richard Trevithick had a steam locomotive (Fig. 1-12) operating in the streets of London. But it was considered more a novelty than a practical machine. As the years passed, however, other people worked at im-

Fig. 1-11. Cugnot's steam carriage.

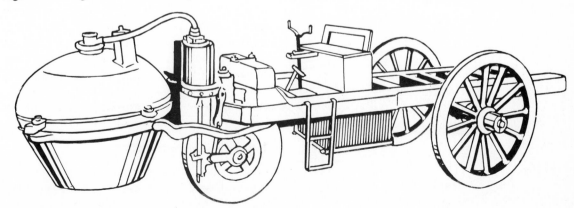

proving the steam-powered vehicle. By 1832, a steam carriage made by Goldsworthy Gurney was running regularly in England between Gloucester and a town 7 miles [11.3 kilometers] distant. That year, it made 396 trips, carrying more than 3,000 passengers without trouble.

§1-9 RAILROADS

Steam carriages did not really come into their own until the idea of running them on rails was adopted. The iron industry was ready, in the early 1800s, to supply iron rails so that railroads could be developed. Railroads were such a tremendous success that they spread rapidly over Europe and the eastern part of the United States. Thousands of miles of rails were laid down, and hundreds of locomotives and freight and passenger cars were built. In 1893, the famous locomotive No. 999 of the New York Central reached a speed of 112.5 miles per hour [181 kilometers per hour]. This was the first time in history that a vehicle had ever traveled more than 100 miles per hour [161 kilometers per hour]. In the heyday of the

steam locomotive, some very large, powerful engines were built. One of the largest weighed more than 440 tons [399 metric tons] and was more than 100 feet [30.5 meters] long, including the coal tender. Figure 1-13 shows a size comparison between an early steam locomotive and a later design.

For many years the steam locomotive reigned supreme on the railroads of the world. But by the mid-1930s, the diesel-powered locomotive was beginning to replace it. Today, the steam locomotive has been almost completely replaced by electric- and diesel-powered locomotives in commercial service, at least in the United States.

§1-10 EFFECTS OF STEAM POWER

The steam engine gave people, for the first time, a source of power that was steady and reliable. It could be used to greatly increase the productivity of mines. It was applied to factories and mills. It almost completely changed the patterns of transportation as steamships and railroads were developed. It

Fig. 1-12. Trevithick's steam locomotive. This is the 1804 model.

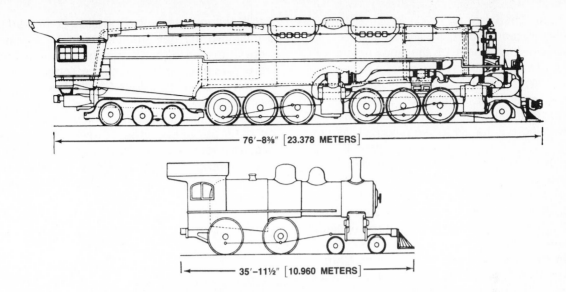

Fig. 1-13. A size comparison of steam locomotives. The locomotive shown at the top is an Allegheny type; it stood 16′7″ [5.1 meters] and weighed 771,300 pounds [349,085 kilograms]. The locomotive shown at the bottom is an Old Hocking Valley Class P-6; it stood 14′8″ [4.5 meters] and weighed 141,400 pounds [63,630 kilograms]. The Alleghenys of the type shown were built from 1941 through 1944. The engines of the Old Hocking Valley Class were built in 1907.

was applied to the farm as well as to the factory, and gradually, farm productivity went up. This meant that fewer workers could produce more food and more basic materials for clothing. This freed more people to work in offices and in factories to produce manufactured goods.

In factories, too, the productivity of each person went up as steam power was applied to drive the machinery. Engineers and inventors, with greatly increased power at their disposal, were able to develop new ways of doing things. Powered machinery enabled one person to turn out more work than a dozen persons working by the old methods.

New devices and inventions followed rapidly. With greatly increased production from mines and factories, transportation had to be improved in order to move these goods to markets. Better communication was necessary, and the telegraph and telephone came into being.

§ 1-11 EARLY DISCOVERIES IN ELECTRICITY

In a sense, electricity took up where steam left off. The steam engine was a remarkable device and made possible the beginnings and development of the modern industrial system. But it had disadvantages, one of the major ones being its inability to transmit its power efficiently over long distances.

The early Greeks knew that amber (a hard brownish fossil resin) that had been rubbed would attract bits of straw. They also knew about natural magnets. Rubbing amber caused an electric charge to develop on the amber, and this attracted the straw. Natural magnets are nuggets of iron ore, naturally magnetized from the effects of the earth's magnetic field. For a more detailed description of magnetism and electrical charges, see Chapter 13.

For many centuries there was very little progress in the study of electricity and magnetism. Then, about 1600, an English physician, William Gilbert, published a book. In it, he described some of the known effects of these two phenomena and certain experiments he had carried out. Other people began to experiment, and gradually over the next 200 years a great deal was discovered. By 1800, Alessandro Volta, an Italian physicist, had built the first electric battery (Fig. 1-14). It was a crude affair, made up of copper and zinc disks piled one on top of another and having cloth moistened with salt water between them. This device was called a **Voltaic pile,** and it gave scientists, for the first time, a source of electric current.

A professor in Copenhagen, Hans Christian Oersted, announced in 1820 that electric current flowing in a wire would produce magnetism (magnetism from electricity is called **electromagnetism**). Shortly after that, a Frenchman, André Ampère, observed that coils of wire with electricity flowing through them would attract and repel each other, according to the direction in which the current flowed in the coils.

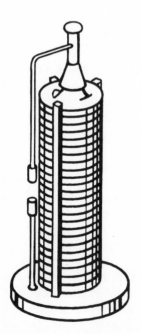

Fig. 1-14. Voltaic pile.

A German professor of physics, Georg Ohm, discovered in 1826 the relationship between electric pressure (**volts**), electric current (**amperes**), and electric resistance (**ohms**). This basic relationship, called **Ohm's law,** will be discussed later.

In 1831, an Englishman, Michael Faraday, became convinced that, if electricity could produce magnetism, magnetism could produce electricity. He showed how this could be done by moving a coil of wire through a magnetic field.

§ 1-12 THE BASIC DISCOVERIES

The basic discoveries had now been made. Different metals separated by certain chemicals could produce electric current—the principle of the battery. Coils of wire with current flowing through them would attract or repel each other—the principle of the electric motor. A wire moved through a magnetic field would have a flow of current induced in it—the principle of the electric generator. But many years were to pass before anything resembling the modern electric generator or motor appeared.

§ 1-13 THE FIRST GENERATORS

In 1856, an Englishman, Frederick Holmes, patented an electric generator that produced electricity by spinning a series of wire coils past a bank of magnets. Possibly the first practical application of his invention was in a lighthouse. In 1860, one of Holmes' machines, driven by a steam engine, was installed in a lighthouse on the east coast of England. It produced enough current to operate a powerful arc light that could be seen far out at sea.

Many early designs of generators were very inefficient and gave much trouble when operated. But gradually the designs were improved and were made more powerful and reliable. The hit of the 1878 season in Paris was the lighting of the avenue leading to the Paris Opera House with 62 powerful arc lights (Fig. 1-15). People came from afar to marvel at these lights that made the avenue "as bright as day."

Fig. 1-15. (*Above*) Avenue de l'Opéra (Avenue of the Opera) in Paris. It was lighted with arc lights.

Fig. 1-16. (*Right*) Edison's first successful bulb.

§1-14 EDISON'S POWER PLANTS

The American inventor Thomas Alva Edison produced the first successful incandescent light bulb (Fig. 1-16). Its filament glowed brilliantly (incandesced) when heated by an electric current. In 1880, he arranged a public exhibition of 500 of these in and around his laboratories at Menlo Park, New Jersey.

Edison's electric-light bulb was an instant success. By 1882 he had installed more than 150 generating plants for lighting purposes in hotels, steamships, factories, and homes. His famous Pearl Street Station, in New York City, went into operation in September, 1882. The Pearl Street Station had six generators totaling 900 horsepower [671 kilowatts] and produced enough electricity to light 7,200 incandescent lamps.

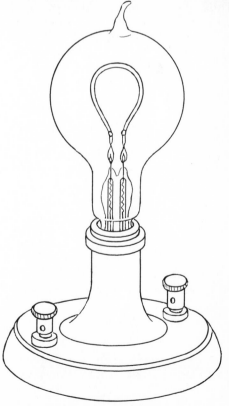

§ 1-15 THE GROWTH OF ELECTRIC POWER

Within the next few years, hundreds of electric power plants were built and put into operation. These plants supplied electricity not only for lighting but also for operating the electric motors that were then beginning to come into common use. Today, it would be almost impossible even to estimate the number of electric lights and electric motors in the world.

§ 1-16 THE INTERNAL-COMBUSTION ENGINE

The earliest powered vehicles were run by steam engines (Figs. 1-11 and 1-12). Steam engines are called **external-combustion engines** because the combustion, or fire, takes place outside the engine itself, in a boiler. Many engineers had the idea that a better engine could be made for highway vehicles if the combustion could somehow be made to take place inside the engine.

Thus, these engineers started on a somewhat different line of development. They had the idea that if an explosive charge could be concentrated in a cylinder and then ignited, the force of the resulting explosion could be carried through a connecting rod to turn a crankshaft. Although some experimental work had been carried on for centuries, much more was done early in the nineteenth century. By 1860, people began to operate **internal-combustion engines** with reasonable success (Fig. 1-17). The first of these was made by the French inventor Jean Joseph E. Lenoir. Shortly after, in 1867, Nickolaus Otto, a German engineer, started production of his gasoline engine. Others also built engines, and by 1885, what is generally considered to be the first gasoline-engine automobile had been built.

Developmental work increased rapidly after that. The first pneumatic (air-filled) tires came out in 1888. The French Panhard car of 1894 had its engine in front under a hood. The car also had a slanting steering column with a steering wheel and had floor pedals in front of the driver. The first geared transmission was introduced in 1899.

One of the most famous of the automobile designers was, of course, Henry Ford. An early Ford automobile is shown in Fig. 1-18. Most of the automobile designers and manufacturers at the turn of the century were building bigger and more luxurious cars. Henry Ford had a different idea. Instead of making the cars larger and more expensive, he wanted to make them as cheaply as possible so that millions could afford to own them. By 1908, Ford had his Model T in production. During the next 20

Fig. 1-17. The first successful internal-combustion engine.

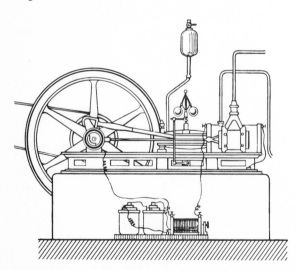

Fig. 1-18. The 1896 Ford.

12

Fig. 1-19. The first assembly-line production of the automobile was achieved by Henry Ford. The Ford Motor Company's 1914 Model T assembly line at Highland Park. (*Ford Motor Company*)

years, he sold 15 million Model T Fords, manufacturing them on the first modern automobile assembly line (Fig. 1-19).

Today, the automotive business is big—one of the biggest businesses in the world. One of every six persons working in the United States is employed in the automotive business! That amounts to more than 12 million people. Included among these 12 million are the people needed to service the more than 100 million cars and trucks rolling on our highways. Right now, about 1 million men and women are working in the automotive service business. Quite a pattern of growth since the days of Henry Ford and his Model T!

§ 1-17 OFF-THE-ROAD ENGINES

In addition to the millions of vehicles on the roads today, there are at least 6 million farm tractors, more than 7 million outboard engines for boats, and additional millions of other specialized vehicles for off-the-road operation. Internal-combustion engines are also used for transportation inside factories, for material handling, and for producing industrial power. There are also tens of millions of small machines powered by engines, including lawn mowers, power saws, snow removers, golf carts, tillers, and lawn sweepers.

§ 1-18 THE NUCLEAR AGE

The vast increase of power from external- and internal-combustion engines has changed our way of life. Now, the world is in the nuclear age, with new sources of power from the heart of the atom becoming available (Fig. 1-20).

What does the future hold in store, as the power within the nucleus of the atom is put to

Fig. 1-20. (*Top*) The nuclear power plant of the Yankee Atomic Electric Company at Rowe, Massachusetts. (*Wide World Photos*)

Fig. 1-21. (*Bottom*) The nuclear-powered ship *Savannah.* (*New York Shipbuilding Corporation*)

work in a major way? The future now promises remarkable new developments in science and technology (Fig. 1-21), which we shall discuss later.

§ 1-19 FUEL FOR POWER

In the last third of the twentieth century, people began worrying about the rapid rate at which our supplies of gas and oil were being used. In highly industrialized nations, such as the United States and Great Britain, oil and natural gas are used for heating homes, operating factories, and making electricity. Gasoline, which is made from oil, is used for operating automobiles. As we shall discuss further in Chapter 2, it is obvious that sooner or later we will begin to run out of oil and gas. In an earlier age—in the eighteenth and nineteenth centuries—coal was king. That is, coal was used for heating homes and for making steam and electricity for industry and transportation. Then, when oil became plentiful, everyone wanted to change to oil. The reason was that oil was cheaper and simpler to transport and burn.

With a future prospect of a much more limited supply of oil, industry and the power plants that make electricity must look for alternate sources of power and energy. One of the most promising of these sources is actually the basic source of all power—the sun. There is further discussion of the sun as an energy source in Chapter 2.

§ 1-20 THE SPACE AGE

It takes power, power in huge quantities, to lift a rocket (Fig. 1-22) from the earth and send it hurtling through space to the moon and beyond. To get this huge power thrust, engineers have made rockets that burn 1 ton [0.91 metric ton] or more of fuel per second. The terrific push that results can lift a spacecraft weighing hundreds of tons from the earth and send it blasting off into space.

We are now in the space age, which promises many interesting new developments, and probably many surprises, as space exploration unfolds.

Fig. 1-22. The mighty *Saturn V.*

To get ready for these new surprises and changes that are coming, we should have a good understanding of what power is, how the machines that produce power operate, and how power is transmitted. All this information is fundamental but essential. It helps us to better understand the changes that are coming in the world of general power mechanics.

Review questions

1. How has power affected the development of our civilization?
2. How is power measured?
3. Name several ways in which power is produced.
4. Describe the three general types of water wheels and explain how they work.

5. Windmills have served people since ancient times. Describe where and how windmills are used in the world today.
6. How is the basic idea of Hero's steam turbine used in today's power units?
7. Where has steam power found its greatest application?
8. For what purpose was electric power first developed?

9. What is the basic difference between an external-combustion engine and an internal-combustion engine?
10. Ford was responsible for producing automobiles so cheaply that millions could own them. How did he accomplish this?

Study problems

1. Get a piece of amber. Rub it, as you hold it near bits of straw or bird feathers. What effect will the rubbed amber have on the straw and feathers? What basic principle does this demonstrate?
2. Then go to the library and find out how a Voltaic cell is made. Build Volta's battery complete with two wires connecting a small flashlight bulb into the circuit. Place a magnetic compass near the wire. What basic principle is demonstrated?
3. List the tasks that electric motors perform in your home. Describe the final use of the work done by each of the motors (for example, the final use of a fan motor is to move air).

2 | ENERGY SOURCES

§ 2-1 THE LANGUAGE OF ENERGY

Our world runs on energy. It takes energy to do almost everything. You must have energy to do work, and an energetic person can do a lot of work. It is the same with machines. A machine that produces a lot of energy can do a lot of work. We often use the word "power" when talking about machines. For example, we say that a big automobile has a high-powered engine because it produces a lot of energy and can move the big car at a high speed.

Now we should stop and define the terms we are using: **energy, work,** and **power.** Of course, these are everyday words and we know in a general way what they mean. But let us look at them in the same way engineers and scientists do.

§ 2-2 WORK

You work a crossword puzzle or a math problem. Studying is sometimes hard work. It is work to push a lawn mower. We are using the word "work" in the common everyday way of talking about our activities. But engineers have a more specialized definition of work. They say that *work is the energy used to move an object against an opposing force.* The object is moved by a push, a pull, or a lift. When the object is moved, work has been done. For example, if you lift a weight, you lift it against the pull of gravity (Fig. 2-1). You have done work.

Work is measured in terms of force and distance. If you lift a 10-pound (lb) [4.53-kilogram] weight 5 feet (ft) [1.52 meters] off the ground,

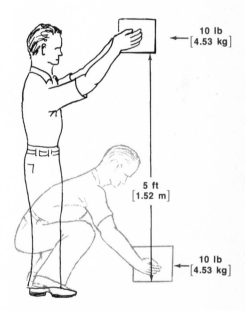

Fig. 2-1. When a weight is lifted, work has been done.

you have done 50 foot-pounds (ft-lb) of work. If you lift a 20-pound [9.06-kilogram] weight 4 feet [1.22 meters], you have done 80 foot-pounds of work. If you drop the 20-pound [9.06-kilogram] weight 4 feet [1.22 meters] onto a stake (Fig. 2-2), the weight does 80 foot-pounds of work as it drives the stake into the ground.

§ 2-3 ENERGY

It takes energy to do work. *Energy is the ability, or the capacity, to do work.* Energy comes in

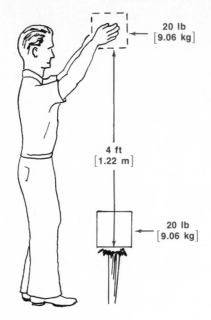

Fig. 2-2. Work is also done when a weight is dropped.

many forms. For example, gasoline, coal, and oil contain potential energy. That means they have the potential, or ability, to do work. Something must happen to this potential energy before it can be released. As you know, what happens is that the gasoline, coal, or oil are burned to release the energy.

As a simple example of energy, consider the lifting of a weight against the pull of gravity (Fig. 2-1). When you lift the weight, you put energy into the weight. If you lift a 10-pound [4.53-kilogram] weight 5 feet [1.52 meters], you have stored 50 foot-pounds of energy in that weight. That is, the weight has the ability to do 50 foot-pounds of work. Before you take a more careful look at energy, we should look at our third term—**power.**

§ 2-4 POWER

Work can be done slowly, or it can be done rapidly. The rate at which work is done it measured in terms of power. A machine that can do a great deal of work in a short time is called a **high-powered** machine. That is, it can use a lot of energy to do a lot of work in a short time. The definition of power is *the rate, or speed, at which work is done.*

By putting the three terms together, we have: **Energy** is the ability to do work. **Work** is the amount of energy used in moving an object against an opposing force. **Power** is the rate at which work is done.

§ 2-5 HORSEPOWER

Power (the rate at which work is done) is often measured in terms of horsepower. When steam engines were first developed, engineers realized that they had to have some way of measuring how powerful their engines were. They decided to compare the engine power with the power of an average horse. Thus, an engine with an output of 100 horses could match the power of 100 horses (Fig. 2-3). It was therefore called a 100-horsepower engine. As you know,

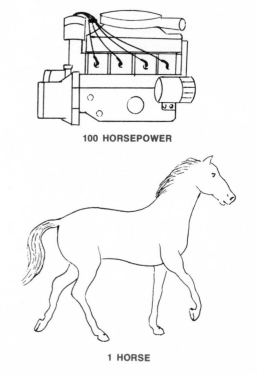

100 HORSEPOWER

1 HORSE

Fig. 2-3. A 100-horsepower engine can match the power of 100 horses.

18

automobile engines are rated in horsepower. A 300-horsepower engine is a big engine. You might say that the engine has 300 horses. That is, it can match 300 horses in the amount of power it can produce. Later in the book we will have more to say about horsepower and how it is measured.

§ 2-6 ENERGY SOURCES

Energy starts out in several forms. Coal and oil are examples. When they are burned, heat is produced. This heat boils water to produce steam. The steam is then used to run steam engines. Or oil is refined to produce gasoline, and the gasoline is burned in the automobile engine to produce power. The burning gasoline produces heat, and this is what makes the engine run.

Most of the energy used in the world today comes from coal or oil. Some energy comes from the wind, some from heat from the earth, some from atomic power plants, some from hydroelectric power plants, and some from the sun. But before we discuss these, let us look at coal and oil.

§ 2-7 COAL

Coal and oil are called **fossil fuels.** A fossil is the remains of a plant or animal that has been preserved in the earth's crust for millions of years. Millions of years ago, huge forests covered many parts of the earth. Gradually, over the ages, these forests were sometimes covered with up to thousands of feet of earth and rock. As centuries upon centuries passed, the wood of the trees gradually turned into coal. Sometimes, miners dig shafts deep into the earth to get to the coal. At other times, huge machines are used to strip the earth off the layers of coal by a process known as strip mining. People who want a clean earth object to either method of mining, but they object more to strip mining. Strip mining can leave huge areas of land devastated (Fig. 2-4). However, modern strip-mining methods can leave the land in good condition.

Fig. 2-4. Two recently mined seams near the top of Barrett Mountain in Boone County, W. Va., would appear to be nearly impossible to reclaim—except that a seam mined earlier shows it can be done, because it is almost invisible already. The previously mined seam may be traced faintly below the bench halfway up the hill at the left. Meanwhile, the benches and outslope of the newest operations have been seeded by helicopter, and vegetation is taking hold. (*National Coal Association*)

There are other objections to the use of coal. One is the atmospheric pollution that coal can cause when it is burned (Fig. 2-5). Ash (particulates), sulfur oxides, nitrogen oxides, carbon monoxide, and other gases come out of the chimneys of power and industrial plants. Most of these gases can be trapped by special and expensive devices, but it takes time and costs money to install and maintain these devices. Because of the desire to produce and sell electricity and goods at the lowest possible prices, many of our industrial and power plants have switched to oil.

§ 2-8 OIL

Oil was formed deep in the earth by a process similar to that which formed coal. Oil from organic matter was trapped in pools thousands

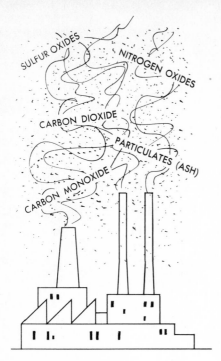

Fig. 2-5. Atmospheric pollution.

Fig. 2-6. An oil field with derricks.

of feet below the earth's surface. To get the oil, holes are drilled through the earth to the pools of oil. These holes, called oil wells, are lined with steel pipe. Usually, many wells are drilled in an area overlying a pool of oil. The area is called an oil field (Fig. 2-6).

The oil that flows from the wells is called crude oil. It is carried by pipe line, special ships, or tank cars to oil refineries (Fig. 2-7). In the oil refinery, the oil goes through processes that convert it into many products (Fig. 2-8), such as gasoline, fuel oil, lubricating oil, grease, and kerosene. As you can see, the oil is easier to obtain, easier to transport and easier to use than coal. Also, the land usually is not injured when oil is removed from the earth, although afterward the land sometimes subsides, or settles, to a lower level.

§2-9 OUR OIL SUPPLY

There is a limited amount of oil in the world. When all the oil fields in the world have given

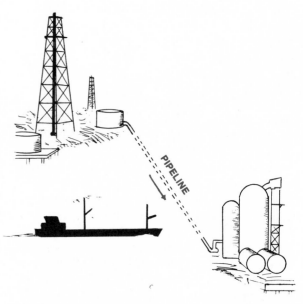

Fig. 2-7. Oil goes from the oil field (at left) to the refinery by such means as pipeline and ship.

up their oil, there will be no more. It takes millions of years for oil to form, and we have come close to using most of the oil in the world in just 100 years. Of course, new oil fields are being discovered all the time, but they are getting harder and harder to find. Some are miles deep in the earth. Some are off the ocean

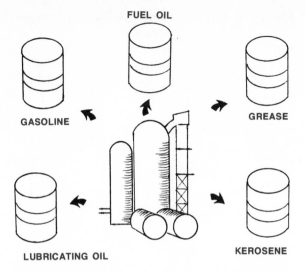

FUEL OIL

GASOLINE

GREASE

LUBRICATING OIL

KEROSENE

Fig. 2-8. An oil refinery produces many products.

shores of the continents, under hundreds of feet of water and thousands of feet of earth and rock.

Years ago, the United States had more than enough oil to supply its own needs. But as the population and the demand for more oil grew, oil fields began to run dry. New ones are being found, but not fast enough to meet the demand. So today, the United States must import millions of barrels of oil daily.

§ 2-10 OIL SHALE

In some areas of the world, there are huge beds of porous rock called shale. In many places, the pores of the shale are filled with a form of oil. This oil is generally rather gummy and is hard to get out of the shale. However, it is estimated that the known beds of oil shale contain billions of barrels of oil. The difficulty is getting the oil out of the shale. The shale must be mined, crushed, and heated to get the oil out. All this requires expensive machines. Until recently, there was little interest in making the investment of time, materials, and money needed to build the necessary machines and plants. Now, however, interest has grown, and

there are plans to build large shale-treating plants.

§ 2-11 OUR COAL SUPPLY

It is estimated that the United States has nearly half the unmined coal in the world. At the present rate of use, our coal could last us for hundreds of years. However, mining coal can damage the earth, and burning coal can pollute the atmosphere—the air we breathe. Of course, both these problems can be solved by modern technology, but it will be costly.

§ 2-12 NATURAL GAS

Quite often, the process that formed crude oil in the earth also formed natural gas. The gas is under tremendous pressure, and when a well is drilled down to a gas pocket in the earth, the gas rushes out. Pipe lines carry the gas to cities, where it is burned to warm homes and to run power and industrial plants. Natural gas is considered a good source of energy because it is easy to get, transports easily, and produces very little atmospheric pollution. However, we are also running out of natural gas. Most of the big natural gas fields have been discovered, and all are giving up their gas at a rate that cannot continue for many more years.

§ 2-13 OTHER ENERGY SOURCES

In Chapter 1, we described the power sources that changed the world and brought civilization to its present state of development. These were water, wind, steam engines, electric power plants, internal-combustion engines, and nuclear power plants. Some other sources of energy that could give us power in the future include the following power plants.

1 Geothermal power plants. These plants use the heat from the earth to produce steam that drives a turbine which produces electricity. There are hot spots in the earth where hot water and steam gush out of the surface.

Power plants for producing electricity can be located over these hot spots.

2 Wind power plants. In Chapter 1, we described various windmills. These were relatively small and nothing like those now being considered as possible energy sources for the future. Engineers now visualize hundreds of large windmills located in a broad belt across the country or off the East Coast. It is possible that a network of these windmills could supply most of the demand for electricity in the United States.

3 Water power plants. As far back as the fourteenth century, waterwheels existed that could produce as much as 60 horsepower [44.76 kilowatts]. Today, huge hydroelectric power plants, located below enormous artificial dams, are producing millions of horsepower—or as electric power is actually measured, millions of kilowatts of electricity.

4 Solar power plants. Energy streams out into space from the sun, and a tiny part of this energy strikes the earth. We see this energy as light and feel it as heat. Almost all the energy sources on earth today came from the sun. Coal, oil, and natural gas are, in effect, preserved sunshine, since the sun caused the growth of the plants from which these fuels were formed.

The sun also heats the atmosphere unevenly, and this causes winds. So energy from windmills is really energy from the sun.

The water that provides hydroelectric plants with energy is placed where it is by sun power. The sun evaporates water from our oceans and lakes, and the wind carries the evaporated water over land, where it then falls as rain. This rain runs into rivers and thus into the huge reservoirs behind the hydroelectric power-plant dams.

Solar energy can be utilized in other ways. One way is to use the heat to boil water and make steam. This requires reflectors to concentrate the sun's rays or a system of solar cells that make electricity directly from the sun's rays.

Review questions

1. Define the terms energy, work, and power. Explain the differences.
2. How is work measured?
3. Describe some examples of potential energy.
4. What is horsepower?
5. Name several energy sources.
6. Where does most of the energy used in the world come from?
7. What are fossil fuels, and how are they created?
8. What creates the pockets of oil in the earth?
9. Of what value is oil shale?
10. Describe the operation of a geothermal power plant and a solar power plant.

Study problems

1. Go on a field trip to an energy source. If you live close to the coal fields, visit a coal mine. Should you live close to an oil field, visit an oil well and a refinery. Decide for yourself how much more of that particular energy source remains to be used in the future.
2. Using references in your library, can you diagram how light is converted to power? Build a machine that uses light to rotate the vanes.

3 | STEAM POWER

§ 3-1 USES OF STEAM

Years ago, practically everyone in the country was familiar with the sight of steamships or of steam locomotives pulling railroad trains (Fig. 3-1). There were steam engines in nearly every factory and shop. Today, diesel engines are used almost exclusively for locomotives, and many ships also use diesel power. The machinery in many shops is now powered by electric motors.

Fig. 3-1. A steam locomotive of the Allegheny class. (*Chesapeake & Ohio Railroad*)

23

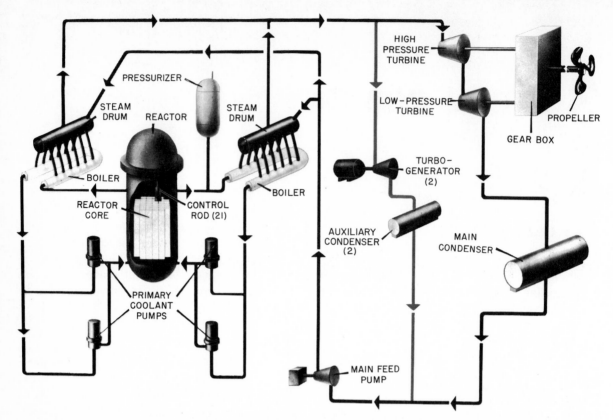

Fig. 3-2. A diagrammatic view of the powerplant of the *N.S. Savannah.* This powerplant combines a nuclear reactor, steam generating units, and steam turbines. (*New York Shipbuilding Corporation*)

But steam is by no means out of date. Nearly 80 percent of the electric power in the United States is produced in generators powered by steam turbines. Steam is also put to wide use in industry, particularly in the metal industries and the petroleum industry. Even nuclear-powered ships like the *NS Savannah* use steam to drive their electric generators (Fig. 3-2).

§3-2 PRODUCING STEAM POWER

The first step in producing steam power is to heat water until steam (water in a gaseous state) is formed. As the water is boiled, it vaporizes and, in so doing, expands. By the time steam is formed, it occupies a volume of as much as 1,670 times that of the water. In a closed container, the great expansion of the steam causes a large pressure rise within the container. This pressure is used to move pistons, as in the reciprocating steam engine (Fig. 3-3), or to spin the rotors (bladed wheels) of a huge steam turbine.

Steam engines are called **external-combustion engines** because the fire, or combustion, takes place outside the engine proper. Basically, the steam engine is a machine that changes heat energy (from the burning fuel) into mechanical work. The steam produced by the heat is delivered to the engine by the steam generating unit, which includes the boiler and furnace and their accessories.

§3-3 STEAM GENERATING UNITS

Steam generating units are primarily made up of the furnace and the boiler but may, and

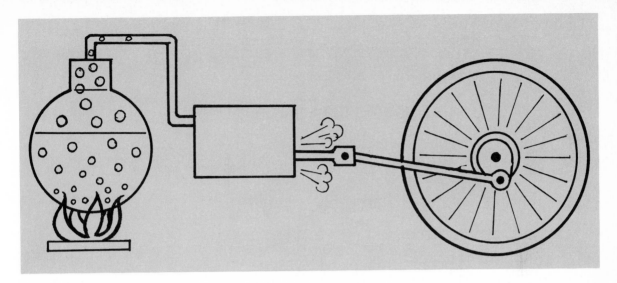

Fig. 3-3. A reciprocating steam engine.

usually do, include such devices as water puri-fiers, superheaters, and reheaters. The **fuels** used include coal, petroleum oils, lignites, al-cohol, and sometimes, combustible wastes or wood.

The water supplied to the boiler is called **feedwater.** One of the major problems in the use of steam is caused by impurities in this water. Not only does water pick up dirt and other particles, but it contains oxygen, carbon dioxide, etc.; and all these pollutants cause trouble in a steam generating unit. The steel of the boilers may corrode or become brittle, and the heat transmitted to the generating unit can be reduced by scale forming on its walls. Also, the efficiency of the steam engine or turbine can be affected through the use of dirty steam. If there are too many gases in the steam, it becomes more difficult to condense. Great care is taken, therefore, to purify the feedwater. The major effort in treating feed-water is aimed at removing these solids and gases. The purifying devices consist of filters, chemicals, and deaerators. In the deaerator, the water is first heated to form vapor and then is condensed. The gases are separated as the condensate drops to the bottom of the deaera-tor tank.

§ 3-4 BOILERS

There are two general types of steam boilers: water tube and fire tube. In the **water-tube boiler** (Fig. 3-4), water flows in metal pipes or tubes placed just above the fire. The heat ap-plied to the tubes causes the water to boil, and the resulting steam is then carried off.

In the **fire-tube boiler,** the pipes, or tubes, carry the heat from the fire, and the water is outside, and surrounding, the tubes (Fig. 3-5). The heat thus applied to the water causes it to boil. The tubes can be either vertical or horizontal. Small fire-tube boilers were those formerly used for concrete mixers and fire en-gines. Today they are primarily used for heat-ing purposes. Locomotive boilers were of the horizontal fire-tube type. Such boilers are now found in places like oil fields. They give reliable service, even under poor operating conditions. They also develop good power for their size.

§ 3-5 SAFETY AND CONTROL

Steam boilers are provided with several control and safety devices. The amount of pressure in the boiler must not be allowed to rise above a certain critical level. If the steam pressure

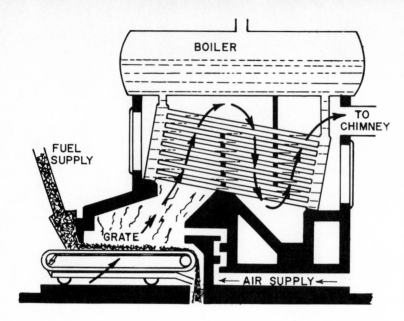

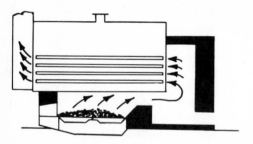

Fig. 3-4. Simplified sectional view of a water-tube boiler.

Fig. 3-5. Simplified sectional view of a fire-tube boiler.

rises above this level, valves open automatically to allow the steam to escape. Just as dangerous as excessive steam pressure is lack of water. If the water level becomes too low in the boiler, the heat of the furnace will melt the metal. To prevent this, boilers are provided with water columns. These columns are mounted on the boilers where they can be seen by the operators. When the water level is either too low or too high, a valve opens, and the escaping steam operates an alarm.

The amount of feedwater delivered to the boiler must correspond to the water loss (from steam production) in the boiler. To accomplish this, several methods of feedwater control have been introduced. One of them, the bal-

anced-flow feedwater control, is presented diagrammatically in Fig. 3-6. The balanced-flow control measures the amount of water in the boiler drum, the rate of steam flow out of the boiler, and the rate of feedwater flow into the boiler. These measurements control an air pilot valve, which in turn controls the boiler feed-pump switches.

§ 3-6 SUPERHEATERS

After the high-pressure steam leaves the boiler, it is further heated in a superheater. This is done to give the steam additional energy. There are two types of superheaters: the **convection** type, which receives its heat from the stack gases, and the **radiant** type, which receives its heat from the furnace flames. Often, the two are combined in one generating unit. The convection type develops a higher heat than the radiant type, and the inclusion of both in a single unit gives a steady heat in spite of changes in steam pressure. Sometimes, control of superheat is attained through the use of **attemperators.** These are devices that use boiler feedwater to modify the temperature. Sometimes, as in the generating unit in Fig. 3-7, the superheater is divided into two sections, a primary superheater and a second-

26

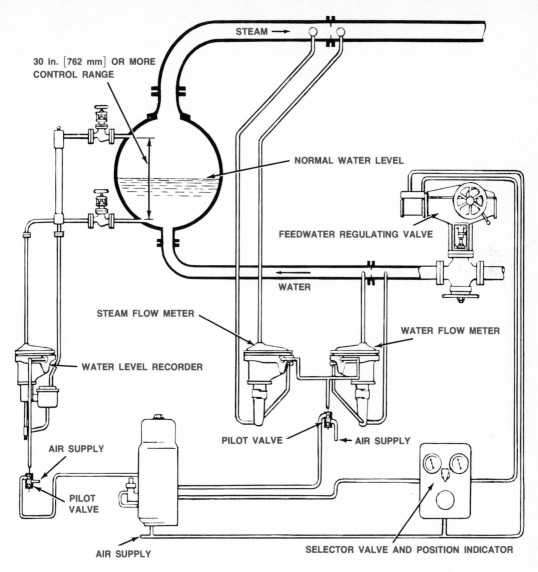

Fig. 3-6. A balanced-flow feedwater control.

ary superheater. Attemperators are then installed between the two.

In practically all modern industrial applications, both superheated and saturated
steam (steam at lower temperatures) are required. Therefore, a device to bring superheated steam down to saturated steam is
needed. Such a device is called a **desuper-
heater.** There are two types of desuperheaters:
the spray type and the surface type. Both use
feedwater to cool the superheated steam.

§ 3-7 THE RECIPROCATING STEAM ENGINE

The function of the steam engine is to change
the energy of the steam into mechanical energy. Formerly, the reciprocating steam engine
found wide application in the generation of
electricity. It was also used in propulsion units
such as the steam locomotive and the large
marine engine. Today, steam turbines have
replaced the reciprocating steam engine in

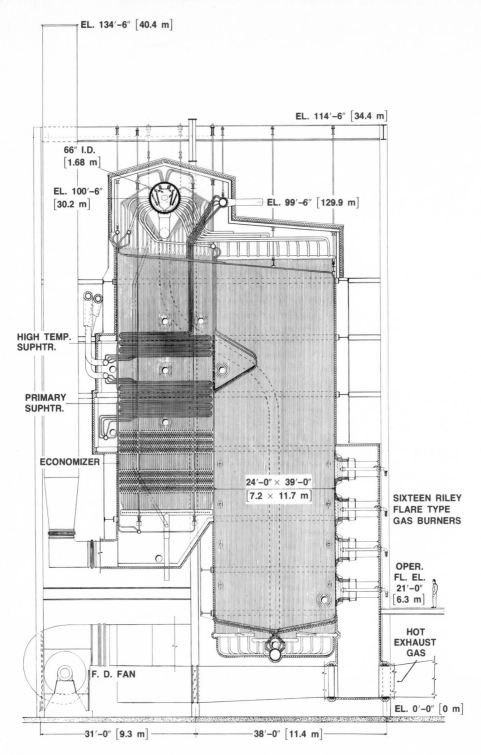

EL. 134'-6" [40.4 m]

EL. 114'-6" [34.4 m]

66" I.D.
[1.68 m]

EL. 100'-6"
[30.2 m]

EL. 99'-6" [129.9 m]

HIGH TEMP.
SUPHTR.

PRIMARY
SUPHTR.

ECONOMIZER

24'-0" × 39'-0"
[7.2 × 11.7 m]

SIXTEEN RILEY
FLARE TYPE
GAS BURNERS

OPER.
FL. EL.
21'-0"
[6.3 m]

HOT
EXHAUST
GAS

F. D. FAN

EL. 0'-0" [0 m]

31'-0" [9.3 m]

38'-0" [11.4 m]

Fig. 3-7. A large central-station boiler with primary and secondary superheaters. (*Riley Stoker Corporation*)

generator applications. Steam locomotives have largely been replaced by diesels (at least in the United States). And more and more ship-builders are coming to rely upon marine diesels. At the present time, reciprocating steam engines are used primarily to power fans, pumps, and small electric generators and to compress air and refrigerants.

§ 3-8 CYLINDERS

The position of the cylinder in the reciprocating steam engine may be either vertical or horizontal. Steam engines may also have more than one cylinder. If the steam is not expanded further as it passes from one cylinder to another, the engine is said to be a **simple engine. In a compound engine,** the steam is expanded twice, regardless of the number of additional cylinders. If the steam is expanded three times, the engine is called a **triple-expansion engine,** and if it is expanded four times, a **quadruple-expansion engine.** This further expansion of the steam in successively larger cylinders makes it possible (within certain limits) to utilize the steam pressure more economically.

The cylinder is essentially a length of iron pipe (Fig. 3-8). The piston (Fig. 3-9), which operates within the cylinder, is a round metal disk, designed to fit snugly within the cylinder without binding or sticking. The reciprocating

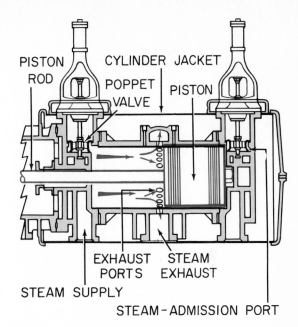

Fig. 3-9. A steam piston and rod, showing uniflow cylinder exhaust.

steam engine may be either single- or double-acting. That is, there may be two opposed pistons in each cylinder or one piston per cylinder. The engine described below is single-acting.

§ 3-9 VALVES

Steam from the boiler, after passing the engine governor, is stored in the **steam chest** (Fig. 3-10). The steam chest must be kept full at all times with high-pressure steam. This is accomplished with a throttle valve and a governor. Below the steam chest are two openings in the cylinder. The openings are controlled in many steam engines by a **D slide valve** (Figs. 3-10 and 3-11). The valve slides back and forth, admitting steam first to one end of the cylinder and then to the other.

The pressure of the steam on the piston causes it to move within the cylinder. As the piston moves, the new steam expands, and as the piston nears the other end of the cylinder, the other supply port is uncovered, and new

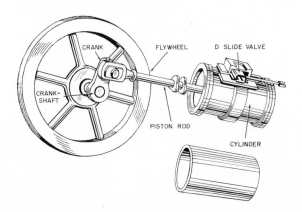

Fig. 3-8. A cylinder is basically an iron pipe in which the piston moves.

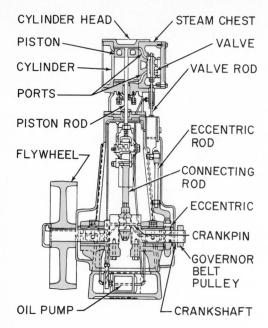

CYLINDER HEAD — STEAM CHEST
PISTON — VALVE
CYLINDER — VALVE ROD
PORTS
PISTON ROD
ECCENTRIC ROD
FLYWHEEL
CONNECTING ROD
ECCENTRIC
CRANKPIN
GOVERNOR BELT PULLEY
OIL PUMP — CRANKSHAFT

Fig. 3-10. A simple vertical steam engine.

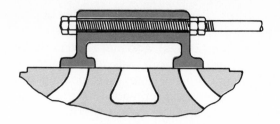

Fig. 3-11. A D slide valve. Note position of the valve in the vertical engine in Fig. 3-10.

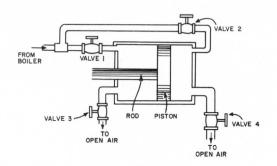

VALVE 2
FROM BOILER
VALVE 1
VALVE 3
ROD PISTON
VALVE 4
TO OPEN AIR
TO OPEN AIR

Fig. 3-12. A steam engine with four valves.

steam is admitted to that end. While the new steam is entering one end, the exhaust port is uncovered in the opposite end, and the newly expanded steam passes into the exhaust passage. The return of the piston on the next stroke helps to completely exhaust the used steam.

Multivalve steam engines increase the operating economy of steam engines by allowing for an earlier cutoff of new steam.

Figure 3-12 shows the action in a cylinder of a multivalve steam engine. The steam is supplied to the cylinder from the boiler, as shown. When valve 1 is opened, steam from the chest can flow into the left end of the cylinder. If valve 4 is open, the steam pressure can push the piston to the right. Any air in the right side of the cylinder can escape. Now, after the piston has moved to the right, valves 1 and 4 are closed and valves 2 and 3 are opened. Steam under pressure enters the right side of the cylinder and forces the piston to move to the left. Steam in the left end of the cylinder would escape through valve 3.

Note that the valves shown in Fig. 3-12 are

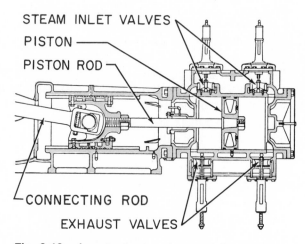

STEAM INLET VALVES
PISTON
PISTON ROD
CONNECTING ROD
EXHAUST VALVES

Fig. 3-13. A poppet-valve steam engine.

not slide valves. There are several other types of valves used in reciprocating steam engines: poppet valves, Corliss valves, and piston valves. Poppet valves are vertical shafts with a disk on top that lift to open. In most poppet-valve engines there are four valves for each cylinder, two steam valves and two exhaust valves. A poppet-valve engine is shown in Fig.

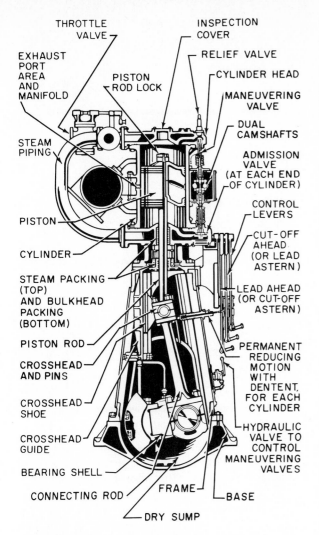

THROTTLE VALVE

EXHAUST PORT AREA AND MANIFOLD

PISTON ROD LOCK

INSPECTION COVER

RELIEF VALVE

CYLINDER HEAD

MANEUVERING VALVE

DUAL CAMSHAFTS

ADMISSION VALVE (AT EACH END OF CYLINDER)

STEAM PIPING

CONTROL LEVERS

CUT-OFF AHEAD (OR LEAD ASTERN)

PISTON

CYLINDER

STEAM PACKING (TOP) AND BULKHEAD PACKING (BOTTOM)

LEAD AHEAD (OR CUT-OFF ASTERN)

PISTON ROD

PERMANENT REDUCING MOTION WITH DENTENT, FOR EACH CYLINDER

CROSSHEAD AND PINS

CROSSHEAD SHOE

CROSSHEAD GUIDE

HYDRAULIC VALVE TO CONTROL MANEUVERING VALVES

BEARING SHELL

CONNECTING ROD

FRAME

BASE

DRY SUMP

Fig. 3-14. Cross section of a two- to eight-cylinder vertical marine steam engine. Steam enters the cylinder from the end and is exhausted through ports arranged around the center of the cylinder that are uncovered by the piston at the end of the power stroke. Steam therefore flows in one direction only; such an engine is called a **uniflow engine.** The admission valves in this engine are solid double-seat poppet valves. Maneuvering valves are single-seat poppet valves; they are used to relieve combustion pressure in the cylinder when the engine is reversed and to drain the cylinder when the engine is warmed up. Admission valves and maneuvering valves are operated off the main camshafts. (*Skinner Engine Co.*)

3-13. Poppet valves are often used with multi-cylinder uniflow engines.

The friction wear of slide valves is very great when the steam pressure of the particular engine is high, because the great pressure forces the valve against its seat. In such applications, the piston valve is used instead of the slide valve. Steam locomotives, for example, use piston valves. The piston valve is somewhat similar in operation to the slide valve but is cylindrical in shape. While steam pressure does not cause much friction between the piston valve and the surrounding surfaces, this type of valve tends to leak steam.

Figure 3-14 is a diagram of a marine steam engine.

§ 3-10 RECIPROCATING TO ROTARY MOTION

In order to change the reciprocating motion of the piston to the rotary motion needed to turn wheels, a connecting rod and crank are used. (For a further, more detailed explanation of how a crank works, see Chapter 5.)

Figure 3-15 shows how the piston rod is connected by a connecting rod to a crank mounted on the crankshaft. The crank is the offset section of the crankshaft that swings in a circle as the crankshaft rotates. A flywheel mounted on the crankshaft turns with it.

Figure 3-16 shows how the piston, in moving back and forth, causes the crank to swing around as the connecting rod pushes and pulls on it. In order to make the slide valve operate in proper synchronism with the piston, it is connected to an eccentric crank on the flywheel. This eccentric crank rotates as shown in Fig. 3-17, causing the slide valve to shift back and forth in time with the piston movement. The offsetting effect of the eccentric crank causes the slide valve to shift very quickly from one position to the other so that the openings to the steam chest or to the exhaust pipe are opened quickly. This fast movement permits the engine to run rapidly.

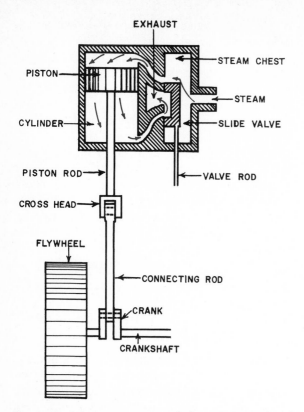

Fig. 3-15. How a piston is connected by a connecting rod to a crank on the flywheel.

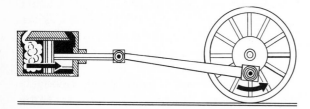

Fig. 3-16. As the piston moves back and forth, the connecting rod and crank move to cause the flywheel shaft to rotate.

§ 3-11 ENGINE SPEED CONTROL

Either of two types of mechanical governors is used to regulate the speed of reciprocating steam engines. The first is the **cutoff governor,** which operates by limiting the amount of steam delivered to the cylinder. This is accomplished by the timing of the opening of the valve (angle

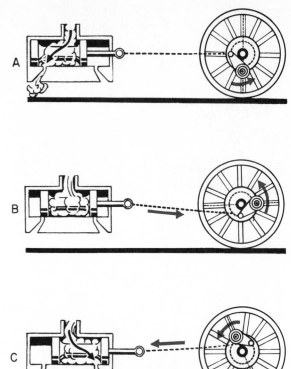

Fig. 3-17. (*A*) The valve is in its middle position, and the crank is on the head-end dead center. (*B*) The valve has moved forward, and the crank is on head-end center. (*C*) The valve moved forward.

of advance) or the valve travel. **Throttling control** simply changes the initial steam pressure in the cylinder.

§ 3-12 STEAM TURBINES

Over 80 percent of the electric power generated in the United States is generated in plants using steam turbines (Fig. 3-18). These turbines are huge steady-flow machines that transmit the mechanical energy produced by steam pressure to a rotating shaft. There are two basic kinds of steam turbine: the **impulse turbine** and the **reaction turbine.** In the impulse turbine, a nozzle mounted on the turbine

Fig. 3-18. The interior of a power plant, showing a condensing steam turbine.

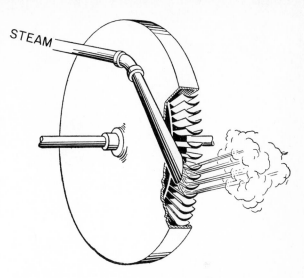

Fig. 3-19. A simple steam turbine wheel of the impulse type.

frame sends a jet of steam against the turbine blades, causing the blade wheels to rotate (Fig. 3-19) and turn a shaft. In the reaction turbine, on the other hand, the nozzles are mounted on the wheels themselves (these nozzles usually take the form of blades), and the reaction of the nozzle to its own jet of steam turns the wheels.

Steam turbines, and turbines in general, are examples of the operation of Newton's third law of motion, which says that *for every action there must be an equal and opposite reaction.* Jet engines and rocket motors also illustrate this principle.

The steam turbine consists of a casing, the stationary blades, the moving blades (mounted on rotors), the shaft, the shaft-supporting bearings, and a control system. A partial cutaway view of a large steam turbine and generator is shown in Fig. 3-20. Note that there is more than one section (stage) of blades on the turbine illustrated. Each stage of the turbine is larger than the preceding one, since the steam must be expanded to utilize its energy.

§ 3-13 TURBINE CYCLES

There are several different types of turbine steam-utilization plans. A turbine that only

partly expands the steam and then reheats it and passes it through again is said to operate on a **reheat** cycle. A turbine that uses a condenser to condense its used steam operates on a **regenerative** cycle. In a few small plants, especially where fuel is inexpensive, the steam is exhausted and does not return to the boiler at all. Such units are therefore said to be **non-condensing.** If, instead of wasting its used steam, a turbine sends it to another unit (which operates necessarily at a lower pressure), it is called a **back-pressure** turbine. **Extraction** turbines are arranged so that some steam can be removed (bled) at different points to operate various other machines.

§ 3-14 SPEED GOVERNORS

The governors on steam turbines may be of either the nozzle or the throttle type. In the **nozzle type,** poppet valves are used. The valves uncover only enough banks of nozzles to meet the immediate requirements of the turbine. The **throttle-type** governor is shown in Fig. 3-21. In this type, the amount of steam entering the turbine is increased or decreased

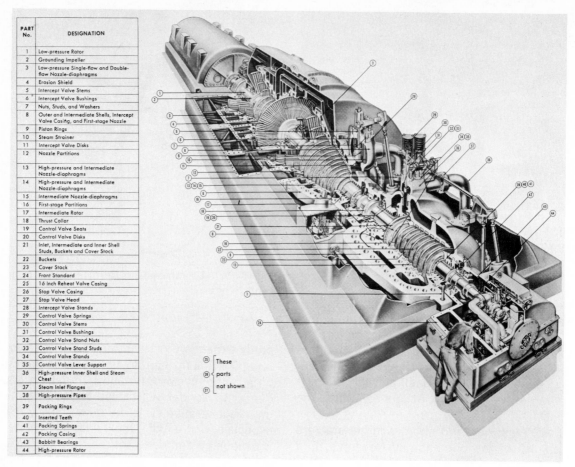

PART No.	DESIGNATION
1	Low-pressure Rotor
2	Grounding Impeller
3	Low-pressure Single-flow and Double-flow Nozzle-diaphragms
4	Erosion Shield
5	Intercept Valve Stems
6	Intercept Valve Bushings
7	Nuts, Studs, and Washers
8	Outer and Intermediate Shells, Intercept Valve Casing, and First-stage Nozzle
9	Piston Rings
10	Steam Strainer
11	Intercept Valve Disks
12	Nozzle Partitions
13	High-pressure and Intermediate Nozzle-diaphragms
14	High-pressure and Intermediate Nozzle-diaphragms
15	Intermediate Nozzle-diaphragms
16	First-stage Partitions
17	Intermediate Rotor
18	Thrust Collar
19	Control Valve Seats
20	Control Valve Disks
21	Inlet, Intermediate and Inner Shell Studs, Buckets and Cover Stock
22	Buckets
23	Cover Stock
24	Front Standard
25	16 Inch Reheat Valve Casing
26	Stop Valve Casing
27	Stop Valve Head
28	Intercept Valve Stands
29	Control Valve Springs
30	Control Valve Stems
31	Control Valve Bushings
32	Control Valve Stand Nuts
33	Control Valve Stand Studs
34	Control Valve Stands
35	Control Valve Lever Support
36	High-pressure Inner Shell and Steam Chest
37	Steam Inlet Flanges
38	High-pressure Pipes
39	Packing Rings
40	Inserted Teeth
41	Packing Springs
42	Packing Casing
43	Babbitt Bearings
44	High-pressure Rotor

These parts not shown

Fig. 3-20. Cutaway view of a steam turbine directly connected to an electric generator.

by the position of a valve, which is itself controlled by centrifugal governor weights. The greater the turbine speed, the further the weights move, adjusting the valve through a linkage to decrease the amount of steam entering the turbine.

It is necessary to provide steam generators with **emergency overspeed governors,** since a sudden decrease in load can cause disastrous overspeeding. Great care is taken to ensure reliability for such governors. A valve-type overspeed governor is illustrated in Fig. 3-22. The nozzle-type governor has the advantage of greater flexibility of control, but the throttle-type governor is mechanically simple and is therefore used on many turbines, particularly the smaller units. Some turbines have speed governors designed into them. Such turbines are used in the United States to turn large power generators that produce electrical energy.

§3-15 TURBINE EXHAUST

The used steam is exhausted from a steam turbine through the exhaust end stages. The exhaust system is designed to increase the space occupied by the steam and, by so doing,

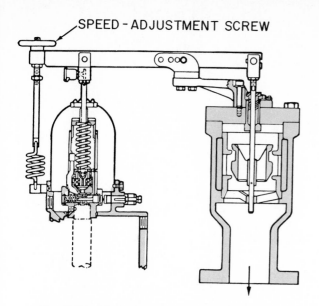

SPEED - ADJUSTMENT SCREW

Fig. 3-21. A shaft governor and throttle valve of the constant-speed throttling type.

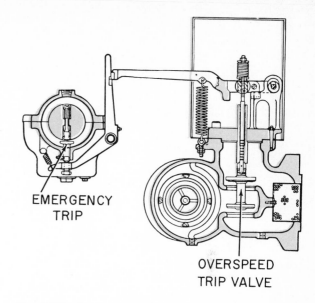

EMERGENCY TRIP

OVERSPEED TRIP VALVE

Fig. 3-22. An overspeed governor and valve of the constant-speed throttling type.

to increase the exit speed of the steam. This vacuum-creating action helps to draw more steam through the turbine. (The condenser in condensing turbines does somewhat the same job, because when the condenser condenses the steam, the volume the steam occupies is greatly decreased.)

Review questions

1. For what main purpose is steam used as a source of power today?
2. List five different kinds of fuel used to supply heat for steam.
3. Name and describe the differences between the two general types of steam boilers.
4. Describe the operation of a reciprocating steam engine using a sliding D valve.
5. Explain how the steam turbine operates.
6. Why are governors necessary in steam turbines?

Study problems

1. Sketch the sliding D valve mechanism of a simple reciprocating steam engine and explain its operation.
2. Construct a simple model of a steam turbine. Label all main parts.
3. Obtain a map of your community. Locate on the map the places, other than homes, using steam for power and heat.

4 ELECTRICAL GENERATING PLANTS

§4-1 A SIMPLE GENERATOR

Generators are machines that force electrons to move along a wire so that a flow of electricity (electric current) results. Electrons are present in fantastic numbers in everything around us. We explain more about electrons in Chapter 8. For now, just remember that when electrons flow in a wire, a current of electricity is produced.

You can make a simple generator with a horseshoe magnet and a loop of wire (Fig. 4-1). If you connect the ends of the wire to an electric meter of the proper type, and if you then move the wire between the ends of the magnet, the meter will indicate that electricity is flowing.

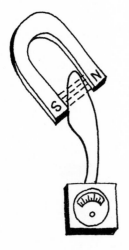

Fig. 4-1. A simple generator.

Generators contain many wires and can produce a heavy current. Each wire supplies some electrons. The many wires together can supply a heavy current of electrons, or electricity. The amount of electric current supplied is measured in amperes. A small generator might supply only 1 ampere. However, the actual number of electrons is enormous. A 1-ampere generator supplies more than 6 billion billion electrons each second!

An ordinary electric-light bulb may require 0.5 ampere of electric current. Therefore, a generator in a power plant that supplies electricity to many homes and factories must be large enough to produce thousands of amperes of electric current.

Electricity must have a complete path, or circuit, outside the generator. For example, an electric light in your home is connected by two wires to the generator in the power plant (Fig. 4-2). Electricity flows from the generator through one of the wires. After the electricity passes through the light bulb, it flows back to the generator through the other wire. The complete circuit includes both wires and the light bulb.

§4-2 ELECTRIC LIGHTS

If you could open an electric-light bulb, you would see two heavy metal wires sticking up from the base, with a coil of very fine wire stretching between their ends (Fig. 4-3). This

Fig. 4-2. An electric light connected to a generator in the power plant.

fine wire, which is called a **filament,** is made of the metal tungsten. Tungsten is a conductor of electricity because there are many free electrons in the tungsten wire. All chemical elements, including tungsten, are made up of tiny particles called **atoms,** which we will discuss later. The free electrons move around between the atoms of tungsten.

When the light filament is connected to a generator, electrons try to rush through the filament. The filament is hardly larger than a human hair, however, and so not very many electrons are able to get through easily. In spite of that, many electrons try to push

TUNGSTEN FILAMENT

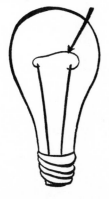

Fig. 4-3. A tungsten filament in a light bulb.

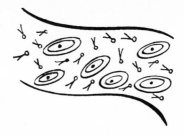

Fig. 4-4. In the tungsten wire, electrons constantly "jostle" tungsten atoms.

through, because electric pressure, or voltage, is pushing them. As a result, the filament becomes crowded with electrons, and these electrons are constantly "jostling" the atoms of tungsten (Fig. 4-4). Now, this introduces another new idea: Atoms in fast motion produce heat. We will go into this in more detail later. For now, just remember that when something gets hot, its atoms are moving fast. In fact, the tungsten atoms move so fast and the tungsten wire gets so hot that it turns white hot—it glows and gives off light.

The filament is enclosed in the glass bulb to keep oxygen in the air away from the filament. If oxygen were present, it would combine with the hot tungsten, and the tungsten would burn up. The bulb keeps the air out and prevents this from happening.

§ 4-3 ALTERNATING-CURRENT ELECTRICITY

It takes a very high voltage, or strong electric push, to make electricity flow long distances. Electric power lines that stretch long distances between cities may require as much as 230,000 volts pressure (Fig. 4-5). After electricity at this high voltage arrives in a city, it must be cut down to a pressure that is usable in homes and factories. It would be very dangerous to try to use such high-voltage electricity in a house. The high electric pressure would push electricity through ordinary insulation and make it jump through the air for as much as a foot. People would be electrocuted, and fires would

Fig. 4-5. Electric power lines.

start. Such high-voltage electricity would not be a servant in our homes, but a dangerous enemy.

To bring the voltage down to a safe range, transformers are used (Fig. 4-6). But in order for transformers to work, they must be fed a special form of electricity. This special form of electricity, called **alternating current,** or ac, may seem very strange to you. In a wire carrying alternating current, the electricity alternates in its direction of flow at rapid intervals. That is, it flows in one direction for a fraction of a second, and then it reverses and flows in the other direction (Fig. 4-7). The change from one direction to the other and back again is called a **cycle.** The alternating current we normally use does this 60 times a second and is thus known as 60-cycle alternating current. If you will look at the manufacturer's nameplate on an ac motor, you may find ''60-cycle'' stamped on it.

Why go to all the trouble of pushing the electrons in a wire in one direction and then in the opposite direction? Wouldn't it be simpler merely to start the electrons moving in one direction and then keep them moving?

Those were questions that advocates of alternating current had to answer, some 70 years ago, when they proposed the use of alternating current in place of the one-direction **direct current** (dc). Their answer was simple. High voltage was needed to transport electricity

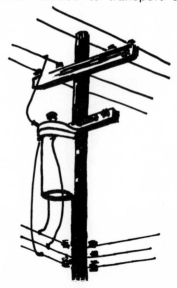

Fig. 4-6. The metal boxes on electric-light poles are transformers.

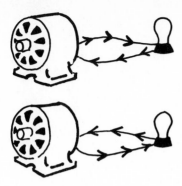

Fig. 4-7. With alternating-current electricity, the electricity flows one way in the wires and then reverses and flows in the opposite direction.

Fig. 4-8. As magnetism moves out from the first coil, current is induced in the secondary coil.

Fig. 4-9. When magnetism moves back toward the first coil, current is again induced in the secondary coil.

long distances. Low voltage was needed for homes and factories. With alternating current, transformers could step up or step down the voltage as needed. Transformers have no moving parts, and they can handle much higher voltages than generators can efficiently produce.

With direct current, however, no such simple device as a transformer would work to increase or decrease the voltage. To change the voltage of direct current, a motor-generator unit is required.

Another advantage of alternating current is that ac generators can be built in much larger sizes than dc generators. For example, a single ac generator might be able to supply enough electricity for a city of half a million people (Fig. 3-18). It would take at least a hundred dc generators to supply the same amount of electricity.

§4-4 HOW THE TRANSFORMER WORKS

The transformer works on the principle of magnetism. We can make a simple transformer by putting two coils of wire side by side. If we connect one coil to a source of alternating current, we will find that current will flow in the second coil, even though there is no connection between the two. This is the reason why: There is a magnetic field around the first coil

of wire only as long as current is flowing in the wire. But since alternating current is constantly changing direction, it is continually stopping and then starting again. This means that the magnetic field disappears and then reappears every time the current alternates, or changes direction. Consequently, the magnetic field is in constant motion. As the magnetic field appears around the first coil, it moves outward from the coil (Fig. 4-8). As it disappears, it moves in toward the first coil (Fig. 4-9). Magnetism is thus constantly moving through the second, or secondary, coil. Essentially, the action is no different from that in the generator, except that in the transformer, the wires are stationary and the magnetism moves. A transformer can be made that will change 110-volt ac to 550-volt ac. Or it can change the 550 to 110. Almost any combination of voltage can be arranged for merely by changing the number of turns of wire in the two transformer windings.

§4-5 THE IGNITION COIL

The ignition coil in an automobile (Fig. 4-10) is a small transformer. It transforms the 6 or 12 volts of the car battery to a high voltage,

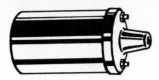

Fig. 4-10. The ignition coil is a small transformer.

which may be as much as 35,000 volts. This high voltage then causes a spark at the spark plug inside the engine cylinder. The spark ignites the gasoline vapor in the cylinder.

§ 4-6 AIR POLLUTION FROM POWER PLANTS

Now, we will look at the air pollutants that industry, home heating, burning of garbage and trash, and power plants produce (Fig. 4-11). When you add these all up, you find that they put into the air something like 120 million tons of pollutants a year in the United States. These pollutants consist of carbon monoxide from incomplete combustion, relatively small amounts of hydrocarbons and nitrogen oxides, and significant amounts of sulfur oxides. The sulfur oxides come from the burning of fuels that have sulfur in them. Coal and oil, for example, have sulfur in them. Some coal and oil have relatively large amounts; others have only small amounts. The sulfur oxides form very unpleasant compounds after they leave the smokestacks. For example, they form acid that stings the eyes, makes breathing painful, and can damage the lungs. And they thicken the smog in smoggy areas.

In addition to these pollutants, ash and fine particles float up into the air from combustion processes (Fig. 4-11). These are called **particulates,** and they amount to as much as 30 million tons a year in the United States. When you see smoke pouring out of a chimney or smokestack, you know that it is made up of particulates—that is, small particles of soot, ash, and other substances. These particles are what give the smoke its color.

We should mention that nature itself produces a great deal of atmospheric pollution and has been doing so for countless ages. Salt

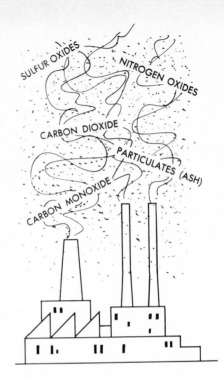

Fig. 4-11. Air pollutants.

particles from the oceans, dust and gases from volcanoes, dust from dust storms, pollen, and spores from various molds are some of the pollutants nature puts into the air. These pollutants settle out after a time and cause no great trouble, as a rule. But when people add their share of pollution to the air, it can become so loaded with pollutants that it becomes difficult to breathe. This can happen particularly in big cities and around industrial areas. We can't do much about nature polluting the air, but we can do something about pollution caused by people.

§ 4-7 POLLUTION CAUSED BY GENERATING PLANTS

First, let us look at the problem of sulfur oxides. If low-sulfur fuels are used, there will be much less sulfur dioxide coming from the chimneys and smokestacks. This is easier said than done, however. Much of our coal and oil contains considerable sulfur. Coal and oil can be treated to remove sulfur, but this is a rather expensive procedure. If high-sulfur fuels are burned, special devices can be installed in the smokestacks to capture the sulfur. Here again, it costs money to install such devices. However, with the application of our present-day technical knowledge, and money, the sulfur problem can be largely eliminated.

The carbon monoxide problem can be improved by careful operation of existing furnaces and boilers or by installing newer, more efficient equipment that will produce more complete combustion and less carbon monoxide (Fig. 3-20).

Other gases that escape from industrial processes can be reduced by careful design and use of equipment to prevent their escape. Particulates are a major problem. They cover everything with fine ash and soot as they settle out of the air. Many of the particles will rise up into the sky so that the smoke gradually disappears from sight. The particles are still there, however. Later, during rain or snow, they will be trapped and will fall to earth. As a matter of fact, particles in the upper atmosphere are largely responsible for rain.

Most of the particulates that come out of the stacks of power plants and industry can be trapped by special devices that either wash the air or take the particulates out by electrostatic means. Washing devices are called "scrubbers." They work by spraying water through the exhaust gases coming from the combustion process. The electrostatic device charges the particles electrically and then attracts them electrically to charged screens. Both systems work. Both add to the cost of the electricity or the goods that are coming out of the plant. Industry and the power companies are acting to add particulate-removing equipment to their systems so that the exhaust gases coming from their stacks will be cleaner.

§ 4-8 THERMAL POLLUTION

There is another kind of pollution that we have not mentioned previously but which is becoming increasingly important. This is thermal pollution—pollution of our waters and air by heat from power plants and industry (Fig. 4-12).

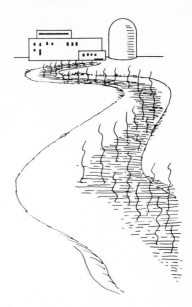

Fig. 4-12. Thermal pollution.

When fuel is burned in a power plant, a good part of the heat is used to produce steam that drives the generators which produce the electric power. Some of the heat has to be thrown away, however, to make the whole process work. And that is where the trouble starts. The most usual method of getting rid of this unwanted heat is to put it into a nearby river, lake, or ocean. This is done by taking the relatively cool water from the nearby body of water, pumping it through the plant, where it picks up the unwanted heat, and then returning the heated water to the river, lake, or ocean.

The trouble here is that if the temperature of the water goes up any great amount, all fish and plant life in the water is killed. This upsets the whole balance of the ecology in the vicinity of the plant. For this reason, new regulations and new ways of trying to get rid of the heat are being considered. One way is to use huge cooling towers that evaporate water to provide the cooling effect (Fig. 4-13). It takes much more heat to evaporate water than it does to heat the water a few degrees. These cooling towers are very effective and require relatively little water. They are widely used in England. However, they do have one bad effect: They raise the humidity of the atmos-

Fig. 4-13. Cooling towers for an electric power plant.

phere in the neighborhood and may even cause heavy fogs.

A still different method of losing the heat works somewhat like an automobile-engine cooling system. That is, the cooling water or other liquid is circulated through a huge radiator. Air passing through the radiator picks up the heat without any loss of water. The water continues to circulate between the power plant and the radiator, carrying heat from the plant to the air passing through the radiator. This method costs more than the cooling towers.

§ 4-9 ATOMIC ENERGY

More and more power plants are being built. Many of these power plants use the splitting of atoms—atomic energy—to produce heat (Fig. 1-20). This heat, in turn, boils water to produce steam. The steam then runs turbines that drive generators to produce electricity (Fig. 3-20). This type of power plant is clean from the standpoint of air pollution. There is no smoke, no fumes, from the atomic process. However, atomic power plants still produce unwanted heat that must be thrown away. So they have the same problem with thermal pollution that other kinds of power plants have. In addition, they have a special sort of problem with the "ash" that is left over after the atomic reactions are finished. This material is highly radioactive. That is, it gives off radiation that can kill. This material must be disposed of by burial deep in the earth or by some similar means.

Despite the drawbacks of atomic power plant, more are being built all the time. By the year 2000, many scientists say, more than half the electricity in the United States will be produced by atomic power plants. Aside from the pollution problem, atomic power plants are claimed to be a cleaner and cheaper source of electricity than other power plants.

Review questions

1. Describe how a simple generator works.
2. Why is the filament of an electric-light bulb enclosed in glass?
3. Why are transformers used in electrical systems?

4. Explain the basic difference between alternating-current electricity and direct-current electricity.
5. Upon what principle does the transformer work?
6. What is thermal pollution?

Study problems

1. Visit your local electrical generating plant. Find out what fuel is used and why that fuel is used. Ask what steps have been taken in recent years to control pollution from the plant.
2. Get up high, either on a hill or in a tall building, and look over the area in which you live. See

if you can identify any smokestacks or tall chimneys. Look for any source of pollution in the area. Then try to identify what that source is and what pollutants, such as smoke or steam, are being emitted.

5 FUNDAMENTALS OF INTERNAL-COMBUSTION ENGINES

§ 5-1 TYPES AND USES OF INTERNAL-COMBUSTION ENGINES

Internal-combustion engines are used in thousands of different kinds of machines. These include automobiles, trucks, buses, power mowers, airplanes, agricultural equipment, construction and mining equipment, and so on. The basic difference between external-combustion engines and internal-combustion engines is that the external-combustion engine has the combustion taking place outside the engine. The steam engine is an external-combustion engine. Combustion outside the engine produces the steam that makes the engine run. In the internal-combustion engine, the combustion process takes place inside the engine.

The automobile engine and the power-mower engine are two examples of the internal-combustion engine. These are reciprocating engines (Fig. 5-1). Reciprocating means moving up and down or back and forth. In the reciprocating engine, pistons move up and down in cylinders. Rotary engines, such as the Wankel engine, have their force developed in a circular manner. This force spins, or rotates, a rotor. We will discuss these engines in Chapter 22.

There are other kinds of internal-combustion engines. Jet engines, turbojets, and turbo-props are all internal-combustion engines. So also are the various types of rocket engines

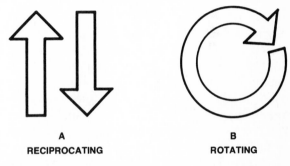

A	B
RECIPROCATING	ROTATING

Fig. 5-1. Reciprocating motion is up and down or back and forth as contrasted with rotary (rotating) motion.

used in guided missiles and space vehicles. All these are described in later chapters.

The internal-combustion engines that use cylinders and pistons differ greatly in size and shape. Automobile engines are much larger than power-mower engines, of course, because they have to produce more power. The small power-mower engine usually has only one piston, moving in one cylinder. Automobile piston engines usually have four, six, or eight cylinders and pistons. There are a few engines that have fewer than four cylinders, and some have more than eight.

§ 5-2 ATOMS AND MOLECULES

Before we take a close look at engine operation, let us review some physical principles that make possible the operation of the internal-combustion engine.

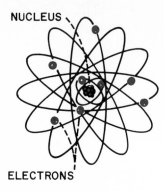

Fig. 5-2. An atom of oxygen.

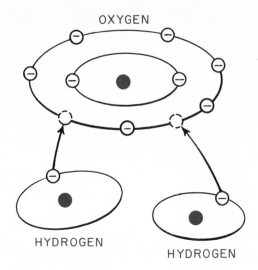

Fig. 5-3. A molecule of water.

There are more than 100 varieties of atoms. Each kind has a special structure and a special name, such as iron, lead, tin, oxygen, carbon, and so on. A piece of iron, for example, is made up of a great number of one particular kind of atom. A quantity of the gas oxygen is made up of another kind of atom (Fig. 5-2). Any substance made up of only one kind of atom is called an **element.**

The atoms of more than one element can be put together to form a great variety of **compounds.** Salt, water, wood, glass, the very blood and bones of the human body, are made up of different elements.

Compounds are made up of two or more elements that have combined. When one or more atoms of one element combine with one or more atoms of other elements, **molecules** are formed. Thus, one atom of oxygen and two atoms of hydrogen can combine to form a molecule of water (Fig. 5-3). During the combustion process in an engine, when gasoline is burned, the carbon and hydrogen atoms in the gasoline combine with oxygen in the air to form carbon dioxide and water. Because gasoline is made up largely of hydrogen and carbon atoms, it is called a **hydrocarbon** (abbreviated HC).

§5-3 HEAT

When fuel burns in an engine, high temperatures of as much as 4500°F [2482°C] result. From a scientific view, **heat** is the rapid motion of the atoms or molecules of a substance. The atoms and molecules of a substance are constantly in motion. Even though these substances seem solid and motionless, their atoms and molecules are moving rapidly. The more rapidly they move, the hotter the substance.

§5-4 CHANGE OF STATE

Look, for example, at what happens when ice cubes are melted in a pan (Fig. 5-4). If the pan is held over a fire, the ice cubes will melt and the water will boil away, or turn to vapor. These changes are called **changes of state,** the three states being **solid, liquid,** and **gas.**

Fig. 5-4. Melting ice cubes turn to water. The water boils, or turns to vapor.

A change in the speed of molecular motion results in the change of state. For example, in ice, the water molecules are moving relatively slowly and in restricted paths. But as the temperature increases, the molecules move faster and faster. A temperature increase, or heat, is an increase in the speed of molecular motion. Presently, the water molecules are moving so fast that they begin to break out of their restricted paths. The ice turns to water at 32°F [0°C]. As molecular speed is increased still further, by additional heating of the water, the boiling point is reached at 212°F [100°C]. Now, the molecules are moving so fast that great numbers of them fly clear out of the water. The water therefore boils, or turns to vapor.

§ 5-5 COMBUSTION

We have already mentioned that, during the combustion, or burning, of fuel in the engine, the atoms of carbon and hydrogen in the fuel unite with oxygen atoms in the air to form carbon dioxide and water.

The chemical formula for water is H_2O, with the H standing for hydrogen and the O standing for oxygen. Each atom of oxygen unites with two atoms of hydrogen to form a molecule of water. There are billions of molecules in even a drop of water.

The hydrogen and oxygen unite to form H_2O, or water (Fig. 5-3). The carbon in the fuel unites with oxygen to form CO_2, or carbon dioxide—each carbon atom has two oxygen atoms tied to it (Fig. 5-5).

All this activity is accompanied by heat. That is, the resulting molecules are set into extremely rapid motion. It is this rapid motion of molecules that causes the temperature to go up as high as 4500°F [2482°C].

§ 5-6 PRESSURE

We know that the burning of fuel in the engine cylinder causes high pressure. We also know that the resulting molecules are moving at very high speeds—that is, the burned gases are very hot. Let us relate these two conditions.

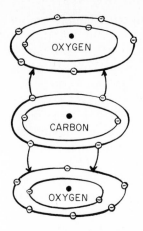

Fig. 5-5. One atom of carbon unites with two atoms of oxygen to form a molecule of carbon dioxide (CO_2).

Gas pressure in any container is the result of the ceaseless bombardment of the inner sides of the container by molecules (Fig. 5-6). Of course, a few molecules banging against the sides of the container would not increase the pressure very much. But there are billions of molecules in any container of gas. The more molecules there are, the more hits and the higher the pressure against the sides of the container.

It is also true that the faster the molecules are moving, the harder they will hit and thus the higher will be the pressure. Thus, there are two ways to raise the pressure in a container of gas. First, more molecules can be pushed into the container. Second, the gas can be heated—that is, the molecules can be made to move faster.

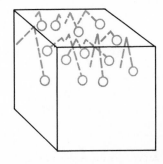

Fig. 5-6. Gas molecules bombarding the inner sides of a container cause gas pressure.

In the engine cylinder, both these things happen. First, as the piston moves up to the top, the mixture of air and fuel is compressed above the piston. That is, the piston pushes the molecules closer together so that the mixture occupies a smaller space. The pressure has increased. Next, a spark takes place at the spark plug so that the compressed mixture is ignited and burns. This combustion process sets the molecules of burned gases moving at very high speeds so that the pressure goes still higher. In a modern automotive engine, the pressure may go as high as 600 pounds per square inch (psi) [42.2 kilograms per square centimeter (kg/cm^2)]. This means that the hot gases are pushing down on every square inch [6.45 square centimeters] of the piston head with a weight of 600 pounds [272 kilograms]. A piston 3 inches [76.2 millimeters] in diameter, for example, would have a pressure on it of more than 4,000 pounds [1,814 kilograms].

§ 5-7 EXPANSION OF SOLIDS DUE TO HEAT

When such solid materials as iron are heated, they expand. For instance, suppose a steel rod measures 10 feet [3.05 meters] in length at 100°F [37.8°C]. If it is heated to 1000°F [538°C], it would then measure 10.07 feet [3.07 meters] in length (Fig. 5-7). This increase in length takes place because the molecules in the rod move faster and faster as the rod is heated. These molecules require more room and push against adjacent molecules, causing the rod to grow longer.

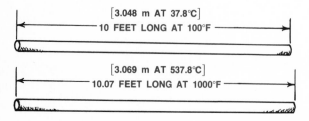

[3.048 m AT 37.8°C]
10 FEET LONG AT 100°F

[3.069 m AT 537.8°C]
10.07 FEET LONG AT 1000°F

Fig. 5-7. A steel rod expands and gets longer as it is heated.

Fig. 5-8. A coil-type thermostat. The coil winds up and unwinds as the temperature goes up and down. The resulting motion can be used to operate a control.

This property of solids is used in **thermostats.** These are devices that cause a part to move with changing temperature. For instance, the thermostat in a home heating system moves to close electrical contact points when the temperature drops. Then, when the house is warmed up enough, the thermostat has moved far enough to open the contact points, so the heater shuts off. Several thermostats are used in automobiles and are described in later pages.

A coil-type thermostat is shown in Fig. 5-8. The coil is made of two strips of different metals. Each metal expands at a different rate so that the thermostat coils up tighter or uncoils with changing temperature.

§ 5-8 EXPANSION OF LIQUIDS AND GASES

Liquids and gases also expand with increasing temperature. A cubic foot [0.0283 cubic meter] of water at 39°F [3.89°C], when heated to 100°F [37.8°C], will become 1.01 cubic feet [0.0286 cubic meter]. A cubic foot [0.0283 cubic meter] of air at 32°F [0°C] heated to 100°F [37.8°C] without a change of pressure will become 1.14 cubic feet [0.0323 cubic meter]. These expansions result from the more rapid molecular motion at the higher temperatures. This causes the molecules to push apart so that the water or gas expands.

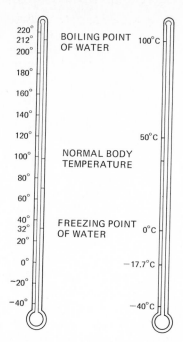

Thermometer markings (Fahrenheit, left):
220°
212° BOILING POINT
200° OF WATER
180°
160°
140°
120°
100° NORMAL BODY
80° TEMPERATURE
60°
40°
32° FREEZING POINT
20° OF WATER
0°
−20°
−40°

Thermometer markings (Celsius, right):
100°C
50°C
0°C
−17.7°C
−40°C

Fig. 5-9. Fahrenheit (*left*) and Celsius thermometers.

The expansive effect of liquids is used in ordinary thermometers (Fig. 5-9). The liquid, usually mercury or a special form of alcohol, is largely contained in a glass bulb at the bottom of a long glass tube. The tube is sealed at the top and has no air in it. As the temperature increases, the liquid expands, and part of it is pushed up into the stem. The higher the temperature, the more the liquid expands and the farther up the liquid is pushed. The stem is marked off, or **calibrated,** to indicate the degrees of temperature.

§ 5-9 GRAVITY

Gravity is the attractive force exerted by the mass of the earth. Gravity attracts objects on or near the earth's surface. If you release a stone from your hand, it will fall to the earth. When a car is at the top of a hill, gravity helps it to coast down.

Gravitational attraction is usually measured in terms of **weight.** An object weighs 10 pounds [4.54 kilograms] on a scale because

that is the amount of pull exerted by gravity on that object. It is the gravitational pull of the earth that gives an object its weight. Since the pull of gravity is much weaker on the moon, the same object will **weigh** less when measured on the moon.

§ 5-10 ATMOSPHERIC PRESSURE

We do not usually think of air as having weight. But it is composed of countless billions of molecules, and the earth pulls it downward. At sea level and average temperature, 1 cubic foot [0.0283 cubic meter] of air weighs about eight-hundredths (0.08) of a pound, or a little more than $1\frac{1}{4}$ ounces [36.3 grams]. Remember that the blanket of air surrounding the earth, the **atmosphere,** extends many miles [kilometers] upward. Thus, the total weight of this blanket of air amounts to about 2,160 pounds per square foot [10,550 kg/cm²] at sea level. Usually, this atmospheric pressure is referred to in terms of **pounds per square inch** (psi). Atmospheric pressure at sea level is approximately 15 psi [1.05 kg/cm²].

§ 5-11 VACUUM

A **vacuum** is the absence of air or other gases in a given space. Astronauts journeying far beyond the outer limits of the atmosphere encounter a nearly perfect vacuum (Fig. 5-10). Out in space there are relatively few molecules of any gases.

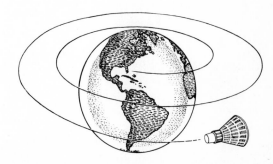

Fig. 5-10. Outer space, beyond the earth's atmosphere, is a vacuum.

A partial vacuum can be created here on earth. This is a matter of trying to remove the air from an enclosed space. For instance, movement of an engine piston causes a partial vacuum to form, and this vacuum then "pulls in" the air-fuel charge. Actually, a vacuum does not pull. Atmospheric pressure pushes the air-fuel mixture toward the vacuum in an effort to equalize the pressure. Now, let us relate these physical principles to the operation of an internal-combustion engine.

§ 5-12 PRINCIPLES OF ENGINE OPERATION

Imagine a tin can with one end cut out. Imagine a second tin can slightly smaller in size that will fit snugly into the first can (Fig. 5-11). Now, suppose you pushed the smaller can rapidly into the larger, trapping air ahead of it. This air would be pushed into a smaller space than it previously occupied. The air would be **compressed.** If the air contained a small amount of gasoline vapor, and if an electric spark were applied to this compressed air-fuel mixture, there would be an explosion, and the smaller can would be blown out of the larger one (Fig. 5-11).

This is about what happens in the internal-combustion engine, except that the smaller can is not blown all the way out. In the actual engine, the larger can is called the **cylinder** and the smaller can the **piston.** The piston slides up and down in the cylinder.

§ 5-13 PISTON RINGS

The piston must be a fairly loose fit in the cylinder. If the piston were a tight fit, it would expand as it became hot and might stick tight in the cylinder. If the piston sticks, it will ruin the engine. But if there is too much clearance between the piston and the cylinder wall, much of the pressure of the explosion will leak past the piston. This means that the push on the piston will be much less effective. It is the push on the piston that delivers the power from the engine.

To provide a good sealing fit between the piston and cylinder, pistons are equipped with piston rings (Fig. 5-12). The rings are made of cast iron or another metal. They are split at one point so that they can be expanded and slipped over the end of the piston and into the

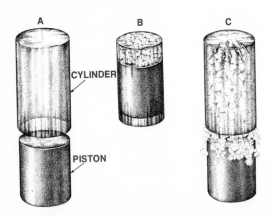

Fig. 5-11. Actions in an engine cylinder. (*A*) The piston is a second cylinder that fits snugly into the engine cylinder. (*B*) When the piston is pushed up into the cylinder, trapped air is compressed. (*C*) The pressure increase as gasoline vapor is ignited forces the piston out of the cylinder.

Fig. 5-12. A typical piston with connecting rod attached and piston rings installed. (*Chrysler Corporation*)

ring grooves that have been cut in the piston. When the piston is installed in the cylinder, the rings are compressed into the ring grooves so that the split ends come almost together. The rings fit tightly against the cylinder wall and against the sides of the ring grooves so that they make a good seal between the piston and the cylinder. The rings can expand or contract as they heat and cool and still make a good seal. Thus, they are free to slide up and down the cylinder wall. In the two-cycle engine, which we discuss in Chapter 6, oil mixed with the gasoline lubricates the rings and cylinder wall. This allows the rings and piston to move easily on the wall. In the four-cycle engine, a separate lubricating system lubricates the pistons, rings, and cylinder walls (Chapter 19).

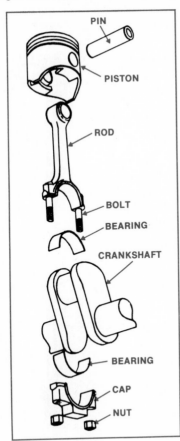

Fig. 5-13. A piston, connecting rod, piston pin, and crank, all in disassembled view. (*Chrysler Corporation*)

Fig. 5-14. A piston and connecting-rod assembly attached to the crankpin on the crankshaft.

§ 5-14 THE CRANK

The piston slides up and down in the cylinder, but it does not leave the cylinder. A connecting rod prevents this. The connecting rod joins the piston and a rotating crank mounted on the crankshaft (Figs. 5-13 and 5-14).

The pedal and its support on a bicycle form a crank (Fig. 5-15). There is also a crank on a pencil sharpener (Fig. 5-16). When you apply pressure to the foot pedal, or on the sharpener handle, so that it swings in a circle, you cause a shaft to turn. In the same way, when the piston is pushed down in the cylinder by the expanding gases, the push, carried through the connecting rod to the crank, causes the shaft to turn (Fig. 5-17).

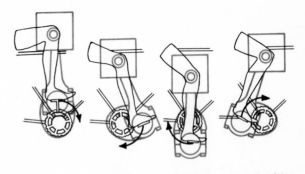

Fig. 5-15. The pedals on a bicycle are attached to cranks. The piston, connecting rod, and crank of the engine have been added in shadow to show the comparison between the connecting rod and the lower part of the leg.

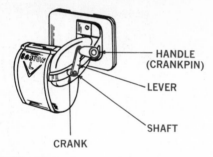

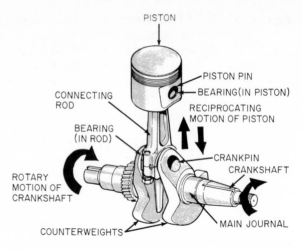

Fig. 5-16. There is a crank on a pencil sharpener.

Figure 5-18 shows the motions that the piston, connecting rod, and crank go through. As the piston moves up and down, the top end of the connecting rod moves up and down with it. At the same time, the bottom end of the rod moves in a circle with the crank.

Fig. 5-17. The crankpin moves in a circle around the crankshaft while the piston moves up and down.

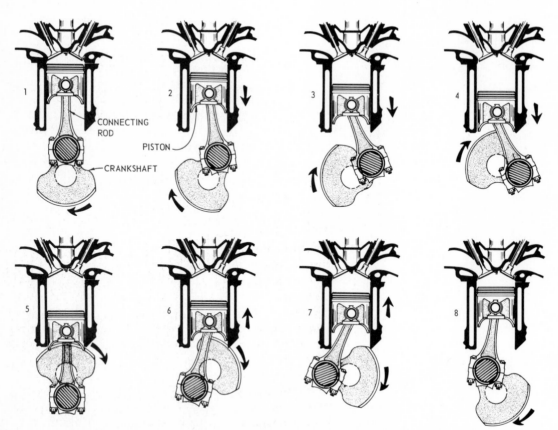

Fig. 5-18. The sequence of action as the crankshaft completes one revolution and the piston moves from top to bottom to top again.

§ 5-15 THE CRANKSHAFT

The crank is part of the crankshaft. It is an offset section, as shown in Fig. 5-17, to which the connecting-rod big end is attached by a bearing. The crankshaft is mounted in the engine on bearings that allow it to rotate. As it rotates, the crank swings in a circle, as shown in Fig. 5-18.

The crankshaft has counterweights, as you will note in Fig. 5-17. These counterweights balance the weights of the crankpin and connecting rod to reduce the tendency of the crankshaft to go out-of-round when it is rotating. This makes for a smoother running engine and for much less wear on the bearings that support the crankshaft.

§ 5-16 MAKING THE ENGINE RUN

In order to make the engine run, a charge of gasoline vapor and air must first get into the cylinder. As the piston moves upward, it compresses the air-fuel mixture. Next, a spark occurs as the piston nears the upper limit of its travel. The spark ignites the charge of gasoline vapor, and it burns very rapidly. The resulting expansion of gases pushes the piston down to convert the energy of the burning fuel into mechanical work.

After the piston has been pushed down, the gases that remain in the cylinder must be removed. This allows a fresh charge of air and gasoline vapor to enter. In the next chapter, we will put all this together and see how the two-cycle engine works.

Review questions

1. What is the basic difference between an internal-combustion engine and an external-combustion engine?
2. Explain what "reciprocating" means.
3. Name several kinds of internal-combustion engines.
4. Explain the difference between an atom and a molecule.
5. What is heat?
6. Describe the combustion process.
7. What causes gas pressure in a container?

8. Does the length of a solid piece of iron increase or decrease as it is heated? Why?
9. Explain what atmospheric pressure is, and then explain what a vacuum is.
10. What causes the air-fuel mixture to enter the cylinders of an engine?
11. Explain why piston rings are used in an engine.
12. Describe how the engine changes reciprocating motion to rotary motion.
13. What is the purpose of the counterweights on the crankshaft?

Study problems

1. Make a list of as many different kinds of internal-combustion engines as you know of. Alongside each, write the type of job usually done by each kind of engine.

2. After listing the different kinds of internal-combustion engines, note where each can be found in your area. Then try to visit and watch each type of engine in operation.

6 TWO-CYCLE ENGINES

§ 6-1 ENGINE DIFFERENCES

There are many differences among internal-combustion engines, but there are two that are basic. First, most of the larger engines are cooled by coolant circulating between the engine and a radiator. But most of the smaller engines, such as those used on power mowers and other garden equipment, are air cooled. That is, they are cooled by the flow of air moving around the engine cylinders.

The second difference is that many of the smaller engines are two-cycle and the larger engines are four-cycle.

§ 6-2 AIR-COOLED ENGINES

Figure 6-1 shows a one-cylinder, air-cooled engine. Note the fins that circle the cylinder block and head. These fins have large surface areas that allow heat to radiate from the cylinders. Many air-cooled engines are equipped with shrouds. These are metal shields that direct air from a fan attached to the crankshaft. The air circulates past the fins and helps to keep the engine cool. Figure 6-2 shows how the air flows over and around the fins. Figure 6-3 shows the temperature differences in the very hot combustion area, the cylinder wall, and the cooling fins. The heat radiates from the fins as fast as it arrives.

The multiple-cylinder, air-cooled engines, such as the Volkswagen and Corvair engines, use a similar air-circulating system along with shrouds to direct the air around the cooling fins on the cylinders and cylinder heads.

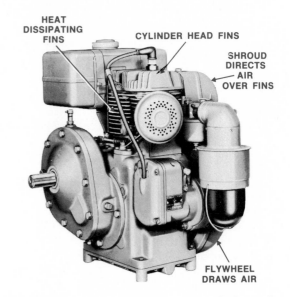

Fig. 6-1. A one-cylinder, air-cooled engine. (*Teledyne Wisconsin Motor*)

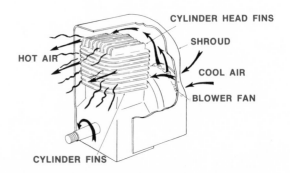

Fig. 6-2. The circulation of air around the cylinder fins.

52

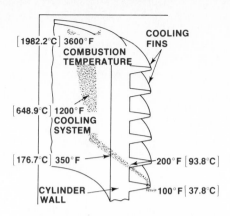

Fig. 6-3. Temperature differences in the cylinder of an air-cooled engine.

§ 6-3 ENGINE OPERATING CYCLES

As we stated earlier, the second difference among internal-combustion engines is that many of the smaller engines are two-cycle while the larger engines are four-cycle. These terms are explained more fully later. In the two-cycle engine, there is an explosion *every time* the piston moves up in the cylinder. In other words, the operating cycle is composed of two strokes (one up and one down) in the two-cycle engine and of four strokes in the four-cycle engine.

In the two-cycle engine, there is an explosion of air-fuel mixture (the mixture of gasoline vapor and air) every time the piston moves up to the top of the cylinder. Each downward movement, or stroke, of the piston produces power in the two-cycle engine.

§ 6-4 THE PISTON STROKE

In any piston engine, the movement of the piston from one limiting position to the other is called a **piston stroke.** The upper limiting position of the piston is called **top dead center** (TDC), and the lower limiting position is called **bottom dead center** (BDC). Thus, a piston stroke takes place when the piston moves from TDC to BDC or from BDC to TDC (Fig. 6-4). When the piston moves from TDC to BDC, after

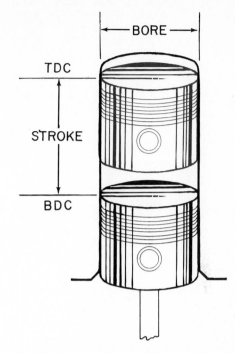

Fig. 6-4. The bore and stroke of an engine cylinder.

the explosion has taken place, the stroke is the **power stroke.** The high pressure of the explosion forces the piston to move during the power stroke, and this results in power from the engine.

§ 6-5 HOW THE TWO-CYCLE ENGINE GOT ITS NAME

The full name of the two-cycle engine is the two-stroke-cycle engine. The reason for this is that it takes two piston strokes, an up stroke and a down stroke, to complete a cycle of engine operation. In other words, everything that happens in the engine takes place in these two strokes, and these events continue to be repeated as the engine runs. This is the meaning of **cycle.** A cycle is simply a series of events that repeat themselves. For instance, the cycle of the seasons, spring, summer, fall, and winter, is repeated every year. In a similar way, the two piston strokes in the two-stroke-cycle engine form a cycle that is repeated continuously as the engine runs.

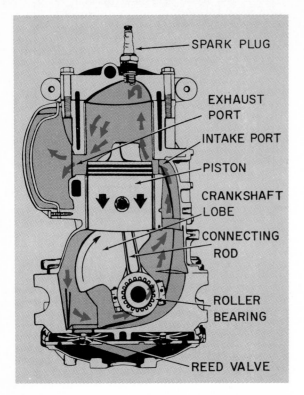

Fig. 6-5. When the piston moves down past the two ports, the air-fuel mixture can flow into the cylinder, and burned gases can flow out. (*Johnson Motors*)

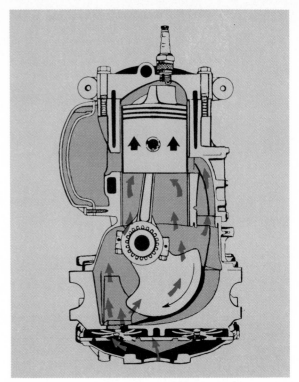

Fig. 6-6. Sectional view of a two-cycle engine with the piston nearing TDC. Ignition of the compressed air-fuel mixture occurs approximately at this point. (*Johnson Motors*)

As a rule, the word "stroke" is not used in the name of the engine, so that **two-stroke-cycle engine** has become, in common usage, **two-cycle engine.** Be very careful to remember, however, that there are two piston strokes in a single cycle of the two-cycle engine.

§6-6 ENGINE ACTIONS

The air-fuel mixture enters the cylinder, and burned gases leave the cylinder, through ports, or openings, in the cylinder wall. The port through which the air-fuel mixture enters is called the **intake port.** The port through which the burned gases leave is called the **exhaust port.** Figure 6-5 shows how this works. When the piston moves down on the power stroke, it clears, or moves past, the ports. Now, a fresh air-fuel charge can flow into the cylin-

der through the intake port. At the same time as the fresh charge is flowing in, burned gases can flow out from the cylinder through the exhaust port.

Let us follow the complete set of actions from the time that the fresh charge of air-fuel mixture is compressed and ignited.

As the piston nears TDC, ignition takes place (Fig. 6-6). The high combustion pressures drive the piston down, and the thrust through the connecting rod against the crank turns the crankshaft (see Fig. 5-17). As the piston nears BDC, it passes the intake and exhaust ports in the cylinder wall (Fig. 6-7). Burned gases, still under some pressure, begin to stream out through the exhaust port. At the same time, the intake port, now cleared by the piston, begins to deliver air-fuel mixture, under pressure, to

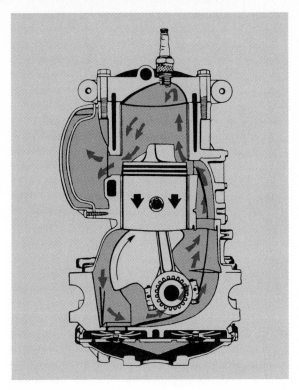

Fig. 6-7. As the piston approaches BDC, it uncovers the intake and exhaust ports. Burned gases stream out through the exhaust port, and a fresh charge of air-fuel mixture enters the cylinder, as shown by the arrows. (*Johnson Motors*)

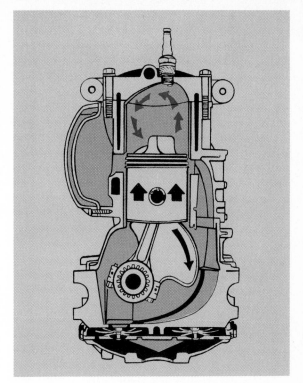

Fig. 6-8. After the piston passes BDC and moves up again, it covers the intake and exhaust ports. Further upward movement of the piston traps and compresses the air-fuel mixture. (*Johnson Motors*)

the cylinder. The top of the piston is shaped in such a way as to give the incoming mixture an upward movement. This helps to sweep the burned gases ahead and out through the exhaust port.

After the piston has passed through BDC and starts up again, it passes both ports, thus sealing them off (Fig. 6-8). Now, the fresh air-fuel charge above the piston is compressed and ignited. This series of events repeats again and continues as long as the engine runs.

§ 6-7 TWO-CYCLE ENGINES WITH EXHAUST VALVES

Another type of two-cycle engine uses a mechanical valve in the cylinder head to exhaust

the burned gases (Fig. 6-9). As the piston moves down past the intake ports (note that there is a ring of them around the cylinder), the mechanical valve opens. Now, the incoming air can efficiently **scavenge** (clear) the cylinder of exhaust gases. The air is delivered to the cylinder by a blower. The engine shown in Fig. 6-9 is a diesel engine, and only air enters the cylinder through the intake ports. The fuel is injected separately, as the piston nears TDC, by a high-pressure pump. Chapter 23 describes diesel engines.

§ 6-8 CRANKCASE PRESSURE

The air-fuel mixture is delivered to the cylinder under pressure. In most engines, this pressure is applied to the air-fuel mixture in the **crank-**

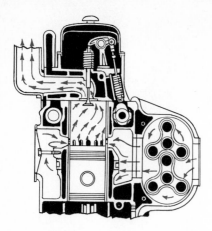

Fig. 6-9. Cutaway view of a two-cycle engine with exhaust valve and blower.

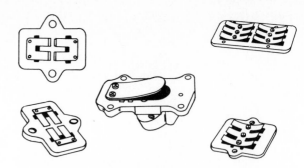

Fig. 6-10. Reed valves. The blades are flexible so that they can move away from the base to which they are attached or flatten down on the base so as to provide a seal. (*Tecumseh Products Company*)

case. (The crankcase is the lower part of the engine that contains the crankshaft.) The crankcase is sealed, except for a leaf (or reed) valve at the bottom. The **reed valve** is a flexible, flat metal plate that rests snugly against the floor of the crankcase (Fig. 6-10). There are holes under the reed valve that connect it with the engine carburetor. When the piston is moving up, a partial vacuum is produced in the sealed crankcase. Atmospheric pressure lifts the reed valve off the holes and pushes air-fuel mixture into the crankcase (Fig. 6-6). After the piston passes TDC and starts down again, pressure begins to build up in the crankcase. This pressure closes the reed valve so that further downward movement of the piston compresses the trapped air-fuel mixture in the crankcase. The pressure that is built up on the air-fuel mixture then causes it to flow up through the intake port into the engine cylinder when the piston moves down enough to clear the intake port (Fig. 6-7).

Instead of using a reed valve in the crankcase, some engines have a third, or transfer, port in the cylinder (Fig. 6-11). In this type of engine, the intake port is cleared by the piston as it approaches TDC. When this happens, the air-fuel mixture pours into the crankcase, filling the partial vacuum left by the upward movement of the piston. Then, as the piston moves down, the intake port is cut off by the piston. The air-fuel mixture in the crankcase is compressed, and the other actions then take place, as already described.

A disassembled view of a two-cycle engine is shown in Fig. 6-12. Study this illustration carefully; identify the reed valve assembly, cylinder, piston, connecting rod, and other parts. This is an air-cooled engine. Note that the cylinder and head have metal fins to help radiate

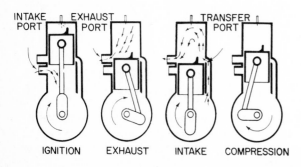

INTAKE PORT   EXHAUST PORT   TRANSFER PORT

IGNITION   EXHAUST   INTAKE   COMPRESSION

Fig. 6-11. Actions in a three-port, two-cycle engine. The third port is called the transfer port.

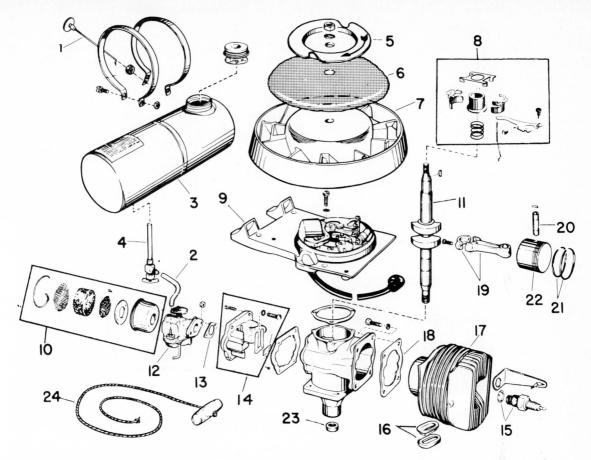

Fig. 6-12. Disassembled view of a one-cylinder, air-cooled engine used on a power lawn mower. (1) Choke-knob assembly; (2) gas line; (3) gas-tank assembly; (4) shut-off valve and screen assembly; (5) starter pulley; (6) flywheel screen; (7) flywheel assembly; (8) governor assembly; (9) magneto assembly; (10) air-filter assembly; (11) crankshaft; (12) carburetor assembly; (13) carburetor gasket; (14) reed-valve assembly; (15) spark plug; (16) exhaust sleeve; (17) cylinder and sleeve assembly; (18) gasket; (19) connecting-rod assembly; (20) connecting-rod pin; (21) piston rings; (22) piston; (23) crankcase; (24) starter rope. (*Lawn Boy Division of Outboard Marine Corporation*)

heat and thus prevent overheating of the engine.

This engine, installed in a lawn mower, is shown in cutaway view in Fig. 6-13. Note that the cylinder is placed horizontally (to the left in the illustration) and that the carburetor is on the opposite side of the crankshaft (to the right in the illustration). Chapter 7 describes the fuel system.

This engine has a built-in ignition system using a magneto. The magneto is described in Chapter 8. It also has a governor that controls engine speed so that the proper engine speed is maintained during engine operation. The governor is described in Chapter 7.

§6-9 TYPES OF BEARINGS

Wherever rotating shafts are supported, bearings must be provided. These bearings, which **bear** the load of the rotating shaft, greatly reduce friction and prevent undue wear of the

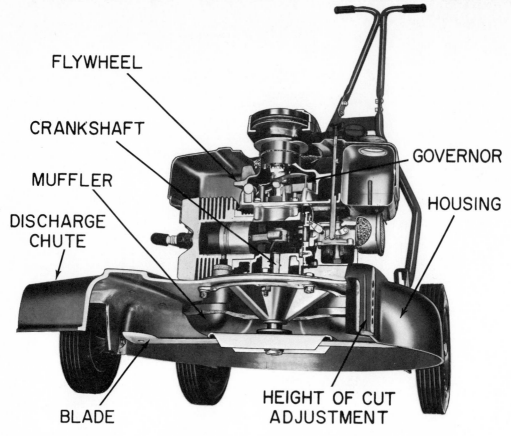

FLYWHEEL

CRANKSHAFT

MUFFLER

DISCHARGE
CHUTE

GOVERNOR

HOUSING

HEIGHT OF CUT
ADJUSTMENT

BLADE

Fig. 6-13. Cutaway view of a two-cycle engine used in a power lawn mower. (*Lawn Boy Division of Outboard Marine Corporation*)

moving parts. If wear takes place, the bearings themselves can be replaced rather than the much more expensive shaft and other parts.

There are two general types of bearings: the **sleeve** or **bushing** type and the **ball** or **roller** type (Fig. 6-14). The ball and roller type of bearings are often called **antifriction bearings** because they offer very little friction. Sleeve and bushing bearings are called **friction bearings;** they offer low friction also. Some two-cycle engines use ball or roller bearings. Others use sleeve bearings.

§ 6-10 FRICTION

Before we describe engine bearings, let us take a look at friction.

Friction is the resistance to motion between two objects in contact with each other. If you put this book on a table and push the book across the table top, you would have to use a certain amount of force. If you put a second book on top of the first, you will find more force is necessary to push them across the table. *Friction increases with load.* The higher the load, the greater the resistance to motion; that is, the greater the friction.

In the modern internal-combustion engine, bearings must take very heavy loads. In some automotive engines, for example, certain bearings will carry loads on their bearing surfaces of more than 4,000 pounds per square inch (psi) [281.23 kg/cm²]. With such loads, the friction would be too great for the metal to

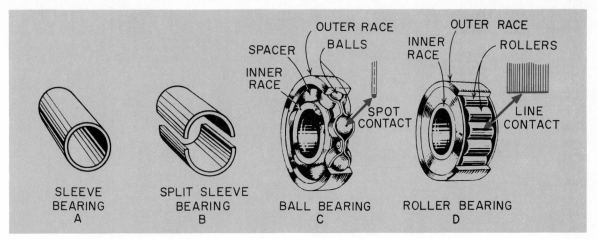

Fig. 6-14. Types of bearings. (*A*) One-piece sleeve bearing of the type used for piston pins. (*B*) Split-sleeve bearing of the type used to support the crankshaft and the big end of the connecting rod. (*C*) Ball bearing. (*D*) Roller bearing. The ball and roller bearings have been partially cut away to show their construction.

tolerate if it were not for the engine oil. The oil flows into the space between the rotating shaft journal and the bearing and prevents actual metal-to-metal contact. Therefore, the actual friction is low. There are three classes of friction: dry, greasy, and viscous.

§ 6-11 DRY FRICTION

Dry friction is the friction between two dry objects—a dry board being dragged across a dry floor, for example. If the board and the floor are rough, the friction is relatively high. If the board and floor are smooth, the friction is relatively low. Think of dry friction as the interference to relative motion between two objects caused by surface irregularities that tend to catch on each other. Even objects that are machined to a very smooth surface still have small irregularities that offer resistance to motion (Fig. 6-15).

§ 6-12 GREASY FRICTION

Greasy friction is the friction between two objects thinly coated with oil or grease. (It is

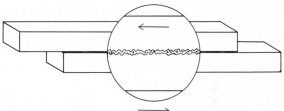

Fig. 6-15. Magnified view of metal-to-metal contact, showing chips shearing off as one piece slides over the other.

assumed that the thin coat of oil or grease tends to fill the low spots in the surfaces so that more nearly level surfaces result.) Thus, there is less tendency for the surface irregularities to catch on each other, and there is less friction than there is with dry friction (Fig. 6-16). Even so, high spots will still catch and wear as the surfaces move over each other.

When an automobile engine is first started, there may be greasy friction between the moving parts. If the engine has not been operated for a while, most of the lubricating oil will have drained away. Then, when the engine starts, the parts start to move against each other be-

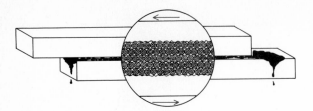

Fig. 6-16. Magnified view of metal-to-metal contact with oil between the two pieces. Note how the layer of oil keeps the two pieces of metal apart.

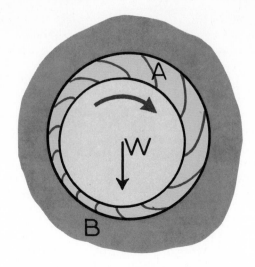

Fig. 6-18. Shaft rotation causes layers of clinging oil to be dragged around with it. The oil moves from the wide clearance *A* and is wedged into the narrow clearance *B*, thereby supporting the shaft weight *W* on an oil film.

fore the lubricating system can get oil to them. After a few moments, oil does flow and the moving parts are adequately lubricated. It is during these first few moments, however, that there is only greasy friction and the excessive wearing of parts takes place.

§ 6-13 VISCOUS FRICTION

The word "viscous" refers to the tendency of a liquid, such as oil or water, to resist flowing. A heavy oil is very viscous and flows slowly. A light oil, being less viscous, flows more easily. Water, still less viscous, flows very easily. Viscous friction is the friction, or resistance to motion, between layers of oil. For example, suppose an object (*W* in Fig. 6-17) is moving over a stationary surface. Between the two surfaces is a film of oil. This film of oil can be visualized as being in layers. The layer next to the stationary surface does not move at all. The layer next to the moving surface moves

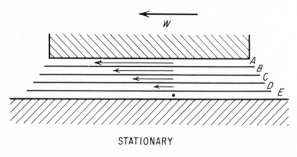

STATIONARY

Fig. 6-17. Viscous friction is the friction between layers of liquid moving at different speeds, or, in the illustration, between layers *A, B, C, D,* and *E*.

at the same speed as the moving surface. Between these two layers, the other layers of oil must be moving at varying speeds. Therefore, there must be slippage between the various layers of oil. There is resistance to this slippage, however, and this resistance is called **viscous friction.**

The resistance is very small compared with that of dry or greasy friction, but it becomes important in the internal-combustion engine, where there are high pressures between the moving surfaces. Of course, the heavier the oil, the higher the viscous friction and therefore the greater the resistance to relative movement between engine parts. During very cold weather, some types of oil can become so viscous in the engine that they act almost like glue and prevent cranking of the engine. This is especially important in automobile engines, as you will see when you read about these engines in later chapters.

In a shaft journal rotating in a sleeve bearing, the moving surface and the stationary surface are round, not flat as in Fig. 6-17. Such a journal and bearing are shown in Fig. 6-18, with the space between the two greatly exaggerated. In the actual engine, the **clearance,** or space, between the journal and bearing would be only one or two thousandths of an inch (0.001 or 0.002 inch) [0.025 or 0.051 millimeter]. As the shaft journal rotates, it carries layers of oil around with it, and these layers of oil wedge under the shaft journal, holding it away from the bearing so that metal-to-metal contact does not take place.

The ball and roller bearings provide rolling friction, rather than the sliding friction produced in the sleeve bearing as described above (Fig. 6-14). The action of the oil in ball and roller bearings is very similar to that in sleeve bearings. However, in the ball and roller bearings, the layers of oil are carried around with the balls and rollers and wedge between the balls or rollers and the inner and outer races. In effect, the balls or rollers are floating on layers of oil so that there is no actual metal-to-metal contact. This is much like what takes place in a sleeve bearing and rotating journal; that is, the journal floats in layers of oil so that there is no actual metal-to-metal contact between the journal and bearing.

Some ball and roller bearings are packed with grease when they are manufactured and then sealed with metal shields to hold the grease in place. The grease consists of oil mixed with a substance that makes it semisolid. These bearings never require additional lubrication. The built-in lubricant provides adequate lubrication throughout the life of the bearing.

§6-14 ENGINE BEARINGS

There are bearings at three places in the two-cycle engine (see Fig. 6-19):

1. At the piston, where it is fastened to the connecting rod by the piston pin
2. At the crankshaft end of the connecting rod, where the rod is fastened to the crankpin
3. At the two ends of the crankshaft, where it is mounted in the cylinder block

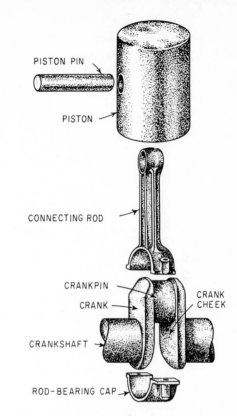

Fig. 6-19. Piston, connecting rod, piston pin, and crankpin on an engine crankshaft in disassembled view. The piston rings are not shown.

The bearing at the lower end of the connecting rod, in some engines, and the two bearings that support the crankshaft are of the split-sleeve type (Fig. 6-14). Figure 6-20 shows one half of one of this type of bearing. One half of the bearing is installed in the connecting rod, and the other half is installed in the rod cap. Then, bolts are used to attach the cap to the connecting rod, thus completing a full bearing around the crankpin (Fig. 6-21). The two split-sleeve bearings that support the crankshaft, in some engines, are installed in a similar manner. Bearing halves are installed in the cylinder block. The other halves are installed in bearing caps. Then the caps are bolted to the block to form complete bearing supports for the crankshaft. On many small engines, the bearing is the full-sleeve type or

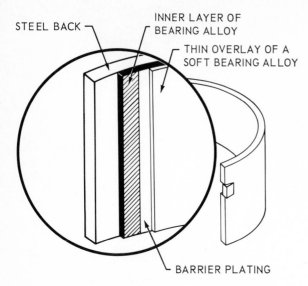

Fig. 6-20. Construction of a bearing half of the sleeve type. The softer bearing material is applied to a hard back. (*Federal-Mogul Corp.*)

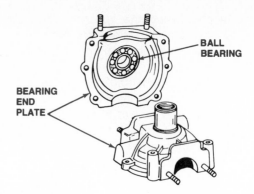

Fig. 6-22. Separate bearing end plates of a small engine using ball bearings to support the crankshaft. (*Tecumseh Products Company*)

the ball- or roller-bearing type (Fig. 6-14). On these, the cylinder block has separate bearing end plates (Fig. 6-22).

The piston pin rides in bearings in the piston (in most engines). Sometimes, there are no separate bearings; the pin rides on the piston metal itself.

The purpose of using bearings covered with oil is to provide a slippery surface on which the moving parts can ride. Also, if wear does take place, it will be the bearings that will wear, and not the more expensive engine parts. The bearings can be replaced with little trouble. If other parts wear, such as the crankshaft or cylinder block, then these major parts, or even the whole engine, must be replaced.

§ 6-15 LUBRICATING THE TWO-CYCLE ENGINE

When an automobile engine requires oil, it is poured into a filler tube, or opening, on the engine. From there, the oil runs down into the lower part of the engine—the crankcase. However, the oil for the two-cycle engines, those in power mowers, for example, is mixed with the gasoline. In the automobile engine, which is a four-cycle engine, the oil is pumped from the crankcase to the various moving engine parts to lubricate them. (These engines and their lubricating systems are described in later

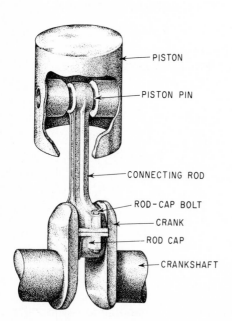

Fig. 6-21. Piston and connecting-rod assembly attached to the crankpin on a crankshaft. The piston rings are not shown. The piston has been cut away to show how it is attached to the connecting rod.

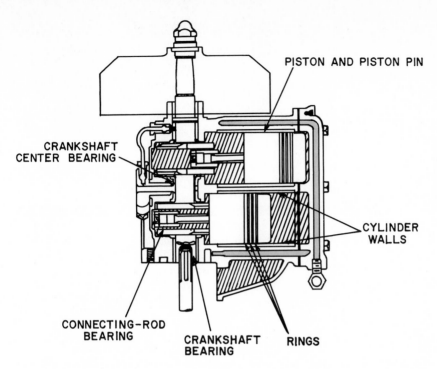

PISTON AND PISTON PIN

CRANKSHAFT
CENTER BEARING

CYLINDER
WALLS

CONNECTING-ROD
BEARING

CRANKSHAFT
BEARING

RINGS

Fig. 6-23. Oil mixed with gasoline is deposited upon all surfaces in the crankcase, upper cylinder walls and rings, and bearings.

chapters.) However, this system would not work in a two-cycle engine. The reason is this: If oil were kept in the crankcase, the incoming air-fuel mixture (which passes through the crankcase) would pick up some of the oil as it passed through on its way to the cylinder. Soon, all the oil would be picked up, carried into the cylinder, and burned. With no oil left, the engine would soon fail from lack of lubrication.

Another lubricating method is called for in the two-cycle engine. This method is to mix a little lubricating oil with the gasoline. The oil therefore enters the crankcase as a fine mist along with the air-fuel mixture. Some of it is carried on up to the engine cylinder, where it is burned. But part of it gets on the cylinder wall and engine bearings to provide adequate lubrication (Fig. 6-23). The amount of oil to be added to the gasoline varies with the engine. Some engines require only an ounce or so of oil per gallon of gasoline; other engines require more. The instructions of the engine manufacturer should always be followed when oil is mixed with gasoline for use in the engine.

§ 6-16 HOW THE OIL WORKS

As already mentioned, the oil clings to the moving engine parts and holds them apart so that metal does not touch metal. If metal is rubbed against metal, heating from friction and rapid wear will take place. But the films of oil on the metal parts hold them apart so that the parts slide over each other without actually touching. In effect, the moving parts are floating on films of oil. These films may be less than one-thousandth of an inch (0.001 inch) [0.025 millimeter] thick. But as long as the films are thick enough to prevent actual metal-to-metal contact, friction and wear are kept very low.

§ 6-17 ENGINE STARTING SYSTEMS

There are four general types of starters used with small two-cycle engines: rope-wind, rope-rewind, windup, and electric. These engine starting systems are discussed along with the engine ignition and charging systems in Chapter 8.

Review questions

1. Describe an air-cooled engine.
2. What is a two-cycle engine?
3. Explain the terms TDC and BDC.
4. Discuss the different ways that ports and valves can be used to get the air-fuel mixture into and out of the combustion chamber.
5. Explain why the air-fuel mixture is pressurized in the crankcase before being delivered to the combustion chamber.

6. What is a reed valve and how does it work?
7. Why are bearings used in an engine?
8. Describe two different types of bearings used in two-cycle engines.
9. What is friction? Name three classes of friction.
10. Where are the three places in a two-cycle engine that bearings are used?

Study problems

1. Examine various small engines as well as automobile engines to determine if each is an air-cooled or a liquid-cooled engine.

2. As you examine each small engine, determine how the air-fuel mixture gets into and out of the engine. Note whether the engine uses exhaust valves, a reed valve, or a transfer port.

7 TWO-CYCLE ENGINE FUEL SYSTEMS

§ 7-1 FUNCTION OF THE FUEL SYSTEM

The job of the fuel system is to supply the engine with a mixture of air and gasoline vapor. The proportions of air and gasoline must be correct for good engine operation. If the mixture does not have enough gasoline vapor in it (too lean), or if the mixture has too much gasoline vapor in it (too rich), the engine will not run properly. However, to start a cold engine, the mixture must be temporarily enriched by the choke.

§ 7-2 GASOLINE

It takes gasoline to run the internal-combustion engines we have been describing. Gasoline is a hydrocarbon. That is, it is a mixture of hydrogen and carbon compounds. These compounds split into hydrogen and carbon when the gasoline burns. Both the hydrogen and carbon unite with oxygen in the air during combustion, as we explained in Chapter 5. This produces heat and high pressure in the engine cylinders. The high pressure forces the pistons down in the cylinders so that the engine runs and produces power.

Gasoline is made from crude oil by a refining process. Actually, there is more to gasoline than just hydrocarbon compounds. The refiners add small quantities of other substances, called additives. Among these are chemicals to prevent formation of gum, antirust agents to protect metals the gasoline touches, detergents to keep the carburetor clean, and others.

§ 7-3 TYPICAL FUEL SYSTEM

A typical fuel system for a two-cycle engine used in a power mower is shown in Fig. 7-1. Figure 7-2 shows a similar system. These systems are called gravity-feed fuel systems because the fuel flows, or feeds, down into the carburetor due to the force of gravity acting on it. The engine used with the fuel system shown in Fig. 7-1 is described and illustrated in Chapter 6. The fuel system consists of the fuel tank with a gasoline filter and the carburetor with an air filter.

When the engine is running, air passes through the air filter on its way to the carburetor. In the carburetor, it picks up a charge of gasoline, and the mixture passes the reed valve as it enters the crankcase of the engine. This action has already been discussed in Chapter 6.

The carburetor itself is shown in sectional view in Fig. 7-3. Essential parts include the air filter, float bowl, choke valve, throttle valve, and fuel nozzle with adjustment needle.

§ 7-4 FUEL TANK

The fuel tank is made of sheet metal or plastic. It has a filler cap that can be removed to add gasoline. The filler cap has a small hole for air to enter the tank as gasoline is used.

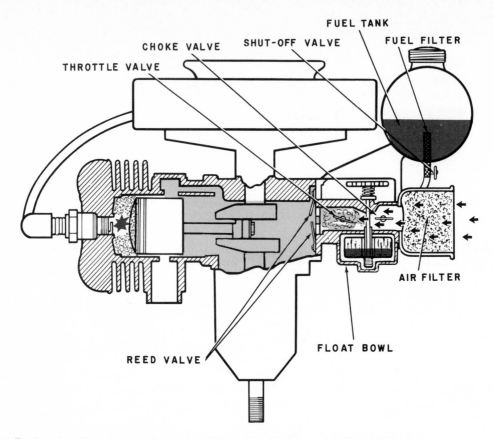

Fig. 7-1. Fuel system for a two-cycle engine. (*Lawn Boy Division of Outboard Marine Corporation*)

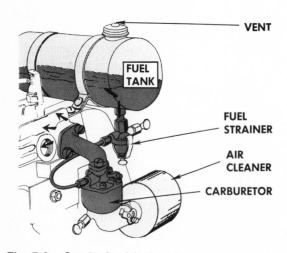

Fig. 7-2. Gravity-feed fuel system.

The fuel filter at the tank outlet filters out dirt that might have entered the tank, preventing it from entering the carburetor, where it could clog the fuel systems and stop the engine.

§ 7-5 AIR FILTER

The engine used with the fuel system shown in Fig. 7-1 uses a metal-mesh air filter. The mesh is packed into the filter case and is moistened with oil. It traps particles of dirt that enter with the air. Over a period of time, the filter can become so loaded with dirt that it restricts the flow of air. This would prevent normal operation of the engine. Before this happens, the filter should be removed and washed in clean gasoline. It should then be reoiled and reinstalled on the carburetor.

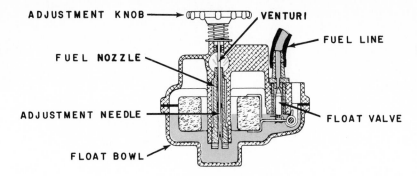

ADJUSTMENT KNOB → VENTURI

FUEL NOZZLE

FUEL LINE

ADJUSTMENT NEEDLE

FLOAT VALVE

FLOAT BOWL

Fig. 7-3. Sectional view of a carburetor for a two-cycle engine used on a lawn mower. (*Lawn Boy Division of Outboard Marine Corporation*)

Some carburetors use an oil-bath air filter. Different types of air filters are shown in Fig. 7-4. In the oil-bath air filter, there is a reservoir containing oil past which the incoming air must flow. As it does this, it picks up particles of oil and carries them up into the filter mesh. This tends to wash off dirt particles, which drain back into the oil reservoir with the oil. On these filters, the oil must be changed at the same time that the wire mesh is washed.

§7-6 FLOAT BOWL

The purpose of the float system is to prevent the delivery of too much gasoline to the carbu-

retor. Without the float system, all the fuel in the fuel tank would run down into the carburetor. The float system is made up of a small bowl, a float of metal or cork, and a needle valve that is operated by the float. Figure 7-5 is a simplified drawing of a float system. The sectional view of the carburetor (Fig. 7-3) shows how an actual float system looks. When gasoline from the fuel tank enters the float bowl, the float is raised. As the float moves upward, it lifts the needle valve into the inlet hole (called the **valve seat**). When the gasoline is at the proper height in the bowl, the needle valve is pressing tightly against its seat so that no more gasoline can enter. When the carbu-

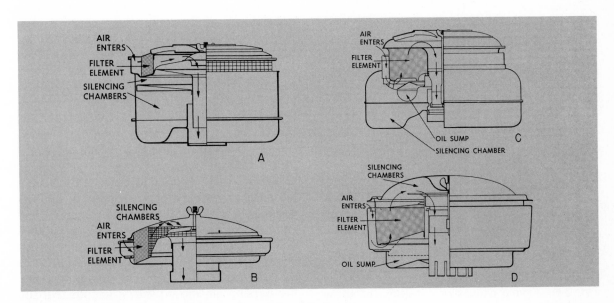

Fig. 7-4. Air filters. *A* and *B,* dry type; *C* and *D,* oil-bath type.

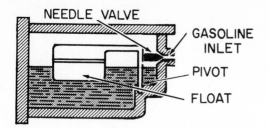

GASOLINE
INLET

PIVOT

FLOAT

Fig. 7-5. Simplified drawing of a carburetor float system.

retor withdraws gasoline to operate the engine, the gasoline level in the float bowl falls, the float and needle drop down, and more gasoline can enter. In operation, the needle valve holds a position that allows gasoline to enter at the same rate that the carburetor withdraws it. This keeps the level of gasoline in the float bowl at the same height.

§ 7-7 THE BASIC CARBURETOR

The carburetor has three basic parts besides the float system. These are the air horn, the fuel nozzle, and the throttle valve (Fig. 7-6). The **throttle valve** is a round plate fastened to a shaft. When the shaft is turned, the throttle valve is tilted more or less in the air horn (Fig.

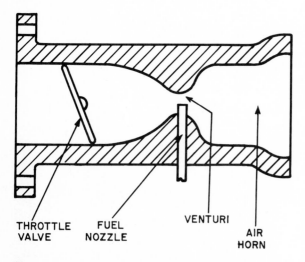

THROTTLE
VALVE   FUEL
NOZZLE   VENTURI   AIR
HORN

Fig. 7-6. Simple carburetor consisting of an air horn, a fuel nozzle, and a throttle valve.

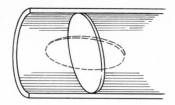

Fig. 7-7. Throttle valve in the air horn of a carburetor. When the throttle is closed, as shown, little air can pass through. When the throttle is opened, shown dashed, there is little throttling effect and the engine produces more power.

7-7). Thus, it can be tilted to allow air to flow through freely or turned to block the passage of air. This is the basic control, because turning the throttle allows more or less air-fuel mixture to feed to the engine so that the engine can produce more or less power.

The **air horn** has a restriction, or venturi, at the point where the fuel nozzle enters it (Fig. 7-8). The purpose of the venturi is to create a partial vacuum, or low-pressure area, when air is passing through the air horn.

The following is a simplified explanation of how the venturi can create a vacuum. As air moves into the air horn, all the air molecules are moving at the same speed and are about the same distance apart. But if all the molecules are to get through the venturi, they must begin to move faster as the air enters the venturi. As the first molecule enters the venturi, it speeds up, momentarily leaving the second molecule behind. The second molecule then enters the venturi and also speeds up. But the first molecule, in effect, has a head start and the second molecule cannot catch up. Therefore, in the venturi, the molecules of air are farther apart.

We know that, where molecules are relatively far apart, there is a partial vacuum, or low-pressure area. In the carburetor venturi, this vacuum is located around the end of the fuel nozzle. Then, air pressure on the fuel in the float bowl pushes fuel up the tube and out of the fuel nozzle (Fig. 7-8). The fuel sprays into the passing air, mixing with it to form the air-fuel mixture that the engine needs to operate.

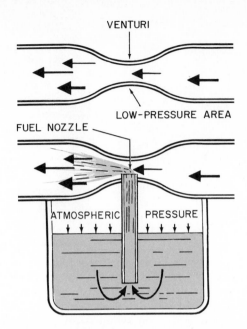

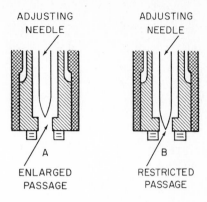

Fig. 7-9. Carburetor adjustment needle and seat, showing (*A*) enlarged passage allowing more fuel to flow and (*B*) restricted passage allowing less fuel to flow.

Fig. 7-8. *Top:* the venturi produces a vacuum, or low-pressure area, when air flows through it. *Bottom:* a nozzle leading up from the float bowl will pass gasoline upward into the air stream as atmospheric pressure pushes it up toward the vacuum. (*Lawn Boy Division of Outboard Marine Corporation*)

The more air that passes through the air horn, the higher the vacuum at the venturi and the greater the amount of fuel that feeds from the fuel nozzle. Thus, the proper proportions of air and fuel are maintained throughout the full range of throttle positions. When the throttle is partly opened, only a small amount of air flows through, and only a small amount of fuel feeds from the fuel nozzle. But when the throttle is wide open, a large amount of air flows through the venturi, and a large amount of fuel feeds from the fuel nozzle.

§7-8 ADJUSTING MIXTURE RICHNESS

The mixture richness can be varied by turning the adjustment knob to raise or lower the adjustment needle (Fig. 7-3). If the adjustment needle is lifted away from its seat, then the fuel passage around the needle tip is enlarged and

more fuel can flow. This means that there will be more fuel and that the air-fuel mixture will be richer. But if the adjustment knob is turned to move the needle tip toward its seat, then the passage is smaller and less fuel can flow (Fig. 7-9). In this case, the mixture will be leaner.

The proper richness is very important, because a mixture that is too lean or too rich will not burn well and engine performance will be poor. In addition, if the mixture is too rich, not all the gasoline will burn clean. Some carbon will be left, and this carbon will soon clog the piston rings, foul the spark plug, and fill the exhaust ports. All these cause engine troubles and a weak engine.

§7-9 CHOKE VALVE

The choke valve is located between the air filter and the carburetor venturi (Fig. 7-10). Its purpose is to help start the engine. During starting, especially when the engine is cold, only part of the gasoline will evaporate to form a combustible mixture. This means that the carburetor nozzle must deliver more gasoline to the air passing through. The choke valve has this job. The choke valve is a round plate, like the throttle valve. When the choke valve is closed, it partially blocks the air horn so that less air can get through. This produces a par-

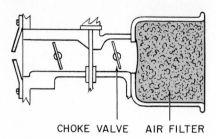

CHOKE VALVE   AIR FILTER

Fig. 7-10. The choke valve is located between the air filter and the venturi.

tial vacuum in the air horn when the engine is cranked and the piston pulls air from the air horn. This partial vacuum, added to the vacuum caused by the venturi, results in a greater vacuum at the fuel nozzle so that more fuel feeds from the fuel nozzle and the resulting mixture is enriched. After the engine starts, the choke valve must be opened to prevent delivery of an excessively rich mixture to the running engine.

§ 7-10 PRIMER

Many small-engine carburetors have a primer instead of a choke valve. One type is shown in Fig. 7-11. The primer, when operated, supplies extra fuel to the carburetor. Here is how it works. When you press down on the primer bulb, you shut off the vent hole and force air into the float bowl. This forces fuel up through

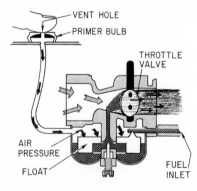

Fig. 7-11. Bulb-type primer. When the primer bulb is pressed down, it pushes air into the carburetor float bowl, causing the float bowl to discharge fuel into the air stream through the carburetor.

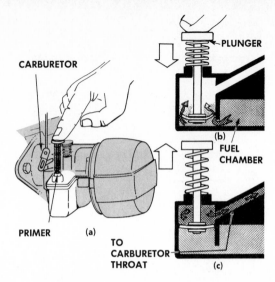

Fig. 7-12. Primer for a small engine. When the plunger is pushed down, as at (*b*), fuel passes by the cup-shaped disk. Then, when the plunger is released, as at (*c*), the spring raises the disk and lifts fuel up into the carburetor.

the fuel discharge hole. The extra fuel makes it easier to start the engine. The primer is also called a tickler.

Another type of primer is a small pump (Fig. 7-12). Pushing the plunger down and then releasing it allows the spring to push the plunger up. As it moves up, it carries fuel upward into the carburetor.

§ 7-11 GOVERNORS FOR TWO-CYCLE ENGINES

Where the load on the engine varies but a steady speed is required, as for instance on a lawn mower, an **engine governor** is needed to prevent the engine from bogging down when the going gets tough.

Basically, what the governor does is control the opening of the carburetor throttle valve. When the load is light, the engine starts to speed up. As this happens, the governor causes the throttle valve to move toward the closed position. This counterbalances the speed-up tendency so that engine speed remains constant. Likewise, if the going gets heavy, as for example when the mower meets

THROTTLE OPEN · THROTTLE CONTROL · AIR VANE · THROTTLE CLOSING · AIR · SPRING · FLYWHEEL

(a) (b)

Fig. 7-13. Details of an air-vane-type governor. (*a*) When the engine is not running, the spring holds the throttle open. (*b*) When the engine is running, air from the blades on the rotating flywheel causes the vane to move, thereby partly closing the throttle.

some high weeds or tough grass, then the engine tends to slow down. When this happens, the governor causes the throttle valve to open so that more air-fuel mixture can get to the engine. The engine will then develop more power to handle the heavier load without slowing down.

§ 7-12 TYPES OF GOVERNORS

There are two general types of governors. The air-vane type and the centrifugal type. The air-vane type works on the flow of air from the blades on the flywheel. The centrifugal type works off a centrifugal device that is operated by engine speed.

§ 7-13 AIR-VANE GOVERNOR

The air vane is located under the flywheel shroud close to the flywheel (Fig. 7-13). It is in the path of air coming from the flywheel blades. The air vane is connected by linkage to the throttle valve, as shown. The spring in the linkage tends to pull the throttle into the opened position. When the engine is stopped, the throttle is open.

When the engine runs, a flow of air from the flywheel blows against the governor vane. This pushes it to the right, as in Fig. 7-13. As the air vane moves, the linkage tends to close the throttle. The faster the engine runs, the stronger the air blast from the flywheel blades and the further the air valve moves. Thus, as the engine approaches rated maximum speed, the air vane has moved far enough to partly close the throttle so that no further engine speed can

result. However, if the engine should suddenly become loaded, as for instance when the mower hits a patch of tough grass, it will tend to slow down. When this happens, there is less air blowing from the flywheel. The air vane tends to relax, allowing the throttle to open wider so the engine can handle the greater load.

§ 7-14 CENTRIFUGAL GOVERNOR

A typical centrifugal governor used on a lawn-mower engine is shown in Fig. 7-14. In this unit, the lower collar is fastened to the crankshaft. The upper collar is attached to the lower by a pair of pivoted links. A spring holds the two

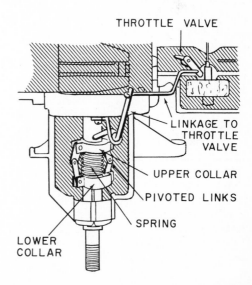

THROTTLE VALVE · LINKAGE TO THROTTLE VALVE · UPPER COLLAR · PIVOTED LINKS · SPRING · LOWER COLLAR

Fig. 7-14. Sectional view of the lower part of the engine, details of the governor, and the linkage to the throttle valve. (*Lawn Boy Division of Outboard Marine Corporation*)

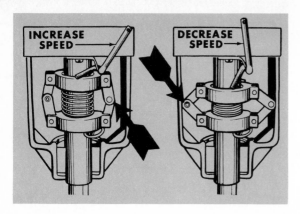

Fig. 7-15. As the engine speed increases, the pivoted links fly out, causing the control arm to move the throttle valve toward a closed position. (*Lawn Boy Division of Outboard Marine Corporation*)

collars apart. When the engine runs, the pivoted links move out due to centrifugal force. This action moves the upper collar down toward the lower, partly compressing the spring. The faster the engine runs, the greater the centrifugal force on the pivoted links and the farther down the upper collar moves by further compressing the spring. The upper collar is connected by linkage to the carburetor throttle valve. As it moves up or down, it opens or closes the throttle valve. If the engine slows down, for example, due to rough going, then the collar starts to move up. This causes the

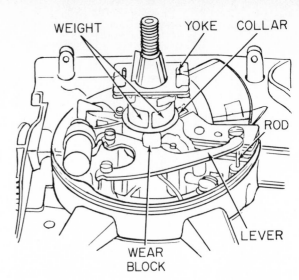

Fig. 7-16. Governor mounted under the flywheel, above the magneto. (*Lawn Boy Division of Outboard Marine Corporation*)

throttle valve to open and to supply additional air-fuel mixture to the engine so that it can produce added power. But if the engine starts to speed up, the collar moves down to cause the throttle valve to partly close and to reduce the amount of air-fuel mixture to the engine (Fig. 7-15).

There are other types of governors. Some are mounted at the upper end of the crankshaft, above the magneto (Fig. 7-16). All, however, work in a similar manner.

Review questions

1. What is gasoline? Where does it come from, and how is it made?
2. Describe the parts of a typical fuel system. Trace the fuel from where it enters the fuel tank to where it leaves the combustion chamber.
3. Name four types of air filters. Describe how each type works.
4. Besides the float system, what are the three basic parts of a carburetor?

Study problem

1. On a small engine, identify each part of the fuel system. Determine what type of air filter is used.

5. Discuss the operation of the choke valve. Why is it used?
6. What is a primer? On what type of engines is it used, and why?
7. Why do some two-cycle engines have governors?
8. Describe the basic operation of a governor.
9. What are the two general types of governors? Explain the operation of each.

Note whether the engine is equipped with a governor and with a primer.

8 TWO-CYCLE-ENGINE IGNITION AND STARTING SYSTEMS

§ 8-1 ATOMS, ELECTRONS, AND PROTONS

Before we study ignition and electric starting systems used in two-cycle engines, let us review the subject of electricity. An electric current is actually a movement of electrons. To see where the electrons come from, we will take another look at atoms.

Every substance is composed of atoms. Atoms are exceedingly small, far too small to be seen by even the most powerful microscope. But scientists have studied their actions and have constructed a picture of the atom.

Atoms, small as they are, are made up of still smaller units. For instance, an atom of the gas **hydrogen** consists of two parts. The center, or nucleus, of the atom is called the **proton,** and it has a charge of positive electricity. An outer particle, called an **electron,** spins around the nucleus. It has a charge of negative electricity (Fig. 8-1).

At 32° [0°C] and at atmospheric pressure, 1 cubic inch [16.39 cubic centimeters] of hydrogen gas would contain about 880 billion billion (880,000,000,000,000,000,000) atoms. If we were to expand this 1-inch [25.4-millimeter] cube until it was large enough to contain the earth—that is, to a cube 8,000 miles [12,875 kilometers] on a side—a single atom in this tremendously enlarged cube would measure about 10 inches [254 millimeters] in diameter.

The atom is tiny, but the particles that make up the atom are far smaller. For example, it would take something like 100,000 billion elec-

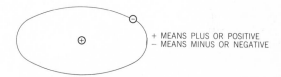

Fig. 8-1. The hydrogen atom consists of two particles: a proton with a positive electric charge and an electron with a negative electric charge.

trons, side by side, to measure 1 inch [25.4 millimeters]. It is impossible, of course, to see these particles with a microscope.

§ 8-2 ATTRACTION AND REPULSION

In the hydrogen atom (Fig. 8-1), a single electron whirls around a single proton, just as the earth orbits the sun. The electron is kept in its orbit by a combination of forces. One force is the attraction that positive and negative electrical charges have for each other (protons are positive, electrons are negative). *Unlike electrical charges attract each other.*

An opposing force is **centrifugal.** Centrifugal force attempts to get the electron to fly in a straight path. Compare this force with the force acting on a ball swung on a rubber band around your hand (Fig. 8-2). As the ball is swung, centrifugal force tries to make it fly away in a straight path. But the rubber band exerts an attractive force and keeps the ball moving in a circle.

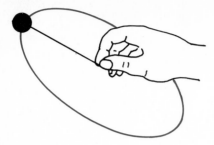

Fig. 8-2. The electron in a hydrogen atom circles the proton like a ball on a rubber band swung in a circle around the hand.

Fig. 8-3. In a copper wire, many free electrons move from atom to atom.

§ 8-3 ELECTRIC CURRENT

Certain atoms can lose some of their electrons. When these "lost" electrons gather in one area, that area may have an excess of electrons and therefore a charge of negative electricity. If there is a path along which these electrons can flow, they will move away from this concentration to any place in which there are atoms that are missing some of their electrons. This movement of electrons is called a **current of electricity,** or an **electric current.**

One reason the electron movement takes place is that *like electrical charges repel.* Negative repels negative; positive repels positive. Therefore, electrons repel each other. Remember, too, that *unlike charges attract.* That is, positive attracts negative.

Thus, the electron movement is caused by a combination of the two conditions: like charges repelling and unlike charges attracting. For example, when we connect an electric

circuit, or electron path, to a battery, electrons will flow from the negative terminal of the battery through the electron path to the positive terminal. There is a concentration of electrons at the negative battery terminal and a shortage of electrons at the positive battery terminal. The battery, by chemical action, produces this unbalanced condition. Therefore, if a path is provided, electrons will flow in an attempt to restore a balance of electrical charges.

§ 8-4 CONDUCTORS

A conductor provides an easy path through which electrons, that is, electricity, can flow. Copper is a good conductor. The reason for this is that copper atoms can very easily lose electrons. There are great numbers of electrons moving about in all directions among the atoms of copper (Fig. 8-3).

If the copper wire is not connected to a source of electrons, such as a battery, the electrons move about in all directions in the wire. But when the wire is connected between the terminals of a battery, electrons pour into the wire from the negative terminal. Electrons begin to move in the same direction all along the wire, moving from the negative terminal and pouring out of the wire into the positive terminal. The negative charge at the negative terminal pushes electrons into the wire. The positive charge at the positive terminal attracts, or "pulls," the electrons out of the other end of the wire. A flow of electrons, or an electric current, results.

§ 8-5 INSULATORS

Insulators are substances that are composed of atoms which hold on to their electrons. Thus, in such substances, there are few, if any, free electrons moving around. When an insulating material is placed around a negative terminal of a battery, for example, the electrons on the negative terminal cannot push into the insulating material. The insulating material stops the flow of current. Insulators are very important, and wires carrying electric current must be covered with insulation to prevent the current from leaking off through the supports holding the wire in place. Current-carrying parts in electrical devices such as alternators or motors either are supported so that they do not touch other parts or are covered with insulating material such as mica, rubber, Bakelite, varnish, fiber, or porcelain.

§ 8-6 ELECTROMAGNETISM

Electric batteries concentrate electrons at one terminal and take them from the other terminal. They produce this unbalanced condition by chemical means. An imbalance of electrical charge can also be produced by mechanical means, that is, by moving a magnet past a conductor. A magnet will attract bits of iron such as tacks, paper clips, and hairpins. This is called **magnetic attraction,** or **magnetism** (Fig. 8-4). Magnetism also has an effect on electrons. For instance, if a wire is moved past a magnet (Fig. 8-5), the free electrons in the wire are moved in one direction. In other words, they stop moving about in random directions and start moving in the same direction along the wire. This action is the basis of the ignition system for the two-cycle engine and of the generator in the automotive electrical system.

The magnet produces this effect by means of invisible magnetic lines of force that stretch in the air between the two poles of the magnet (Fig. 8-6). When a conductor moves across these lines of force, the magnetic action gives the electrons a push so that they will all move in the same direction. *Magnetism can produce current, and, conversely, current can produce magnetism.*

Conductors are surrounded by magnetic lines of force whenever current is flowing through them. The magnetism produced by electric current is called **electromagnetism.** Automotive generators, regulators, ignition coils, etc., contain conductors that are used to produce electromagnetism. We will learn more about this when we study automotive electrical systems.

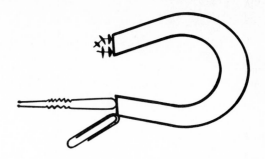

Fig. 8-4. A magnet will attract objects made of iron.

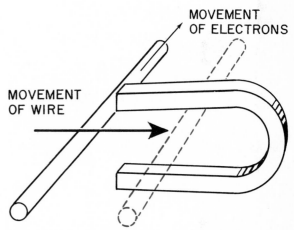

Fig. 8-5. Current flows in a conductor when it moves past a magnet.

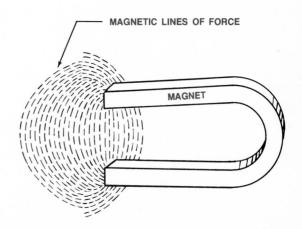

Fig. 8-6. A magnet has lines of force stretching between its poles.

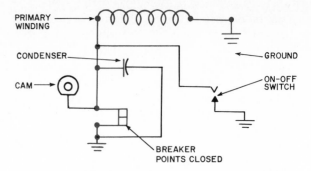

Fig. 8-8. Wiring diagram of a magneto ignition system with current flowing through the primary circuit. (*Lawn Boy Division of Outboard Marine Corporation*)

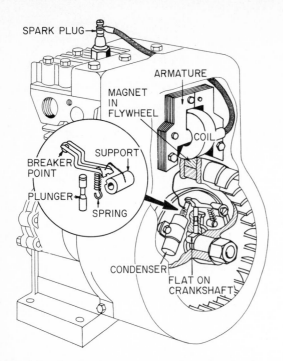

Fig. 8-7. Cutaway view of an engine, showing location and construction of the magneto. (*Briggs and Stratton Corporation*)

§ 8-7 IGNITION SYSTEM FOR A TWO-CYCLE ENGINE

If you ever shuffled your feet on a carpet on a dry day and then held your finger close to a friend's ear, you produced an electric spark that jumped from your finger with a sharp "crack." Under the right circumstances, such a spark would be strong enough to ignite the air-fuel mixture compressed in the engine cylinder. But such an "ignition system" would be highly impractical, to say nothing of being tiring!

The ignition system in the typical one-cylinder, two-cycle engine is built into the engine. This type of system is called a **magneto ignition system.** Usually, it is located at one end of the crankshaft. The principle of operation is simple. A series of magnets are whirled past a coil of wire. The magnets are mounted on the engine flywheel (see Fig. 8-7). When magnetic lines of force move through a conductor,

voltage is induced in the conductor. If the conductor is in a closed circuit, then current will flow.

Figure 8-8 is a wiring diagram of the primary circuit. It includes the coil of wire (primary winding) past which the magnets move. Also included in this circuit is a pair of breaker contact points. One of these points is on a lever, or arm. The other is stationary. One end of the arm rests on a cam on the crankshaft. This cam is round except for a high spot. When the crankshaft and cam rotate, the contact points remain closed until the high spot passes under the contact arm. When this happens, the contact points are separated.

Now, let us see how these actions can produce an electric spark. When the engine is running, the magnets are whirling past the coil primary winding. Voltage is induced and current flows when the contact points are closed. This current causes a strong magnetic field to build up around the winding. When the high spot on the cam comes around under the contact arm and the contact points separate, the current stops flowing. Now, the magnetic field rapidly collapses.

The **capacitor** (or condenser, as it is also called) aids this rapid collapse of the magnetic field. It contains two long strips of metal foil insulated from each other. When the points start to separate, the current would continue to flow, causing a momentary arc between the points, if it were not for the capacitor. But the

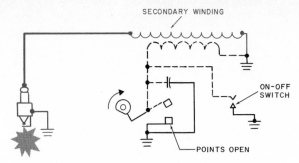

SECONDARY WINDING

ON-OFF
SWITCH

POINTS OPEN

Fig. 8-9. Wiring diagram of a magneto ignition system with breaker points open. Current has stopped flowing in the primary circuit, and a high-voltage surge has been induced in the secondary circuit to produce a spark at the spark-plug gap in the cylinder. (*Lawn Boy Division of Outboard Marine Corporation*)

capacitor provides a place for this current to flow. It acts somewhat like a check spring and brings the current to a quick stop. This produces the rapid magnetic-field collapse.

Surrounding the primary winding is a secondary winding (Fig. 8-9) made of many thousands of turns of a fine wire. The magnetic field from the primary winding, in collapsing, moves rapidly through the secondary winding. Since this is a movement of a magnetic field through a conductor, a voltage is induced in the secondary winding. And since there are many thousands of turns of wire in the secondary winding, a high voltage is induced.

The spark plug (Fig. 8-10) is connected to the two ends of the secondary winding (Fig. 8-9). One end is connected through the metal of the engine (called **ground**) and the other through a rubber-covered wire (called the **high-tension lead**). The voltage in the secondary winding quickly goes up high enough to cause a powerful spark to occur at the gap between the two spark-plug electrodes. One of these electrodes is connected to the metal shell of the plug that is screwed into the cylinder head of the engine. The other is insulated by a porcelain shell in which it is centered. The porcelain is breakable, just like glass, and this is the reason why the center electrode must never be bent when the spark-plug gap is adjusted. Only the outer electrode should be

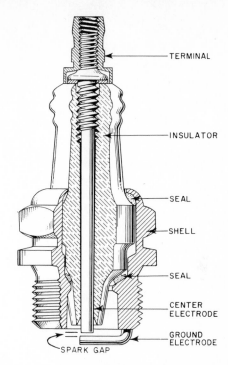

TERMINAL

INSULATOR

SEAL

SHELL

SEAL

CENTER
ELECTRODE

GROUND
ELECTRODE

SPARK GAP

Fig. 8-10. Spark plug partially cut away to show construction.

bent. If the center electrode is bent, it probably will break the porcelain and ruin the plug. This is also the reason why the plug must be removed and installed with care; improper handling will also break the porcelain and ruin the plug.

Figure 8-11 shows, in end view, what happens before and after the contact points separate. Figure 8-12 is a top view of a typical magneto. Notice the arrangement of the coil, the contact points, and the capacitor (condenser). Notice that the magnets are curved so that, as the flywheel rotates, they pass close to the coil.

An ON-OFF switch is used on many systems to turn the engine off (Fig. 8-9). When this switch is flipped so that it is closed, it grounds the contact-point end of the primary winding. Now, current continues to flow in the primary winding, and opening the points does not interrupt it. As a result, no sparks occur and the engine stops. The engine can also be stopped by a grounding blade (Fig. 8-13) near the spark plug that can be bent by hand to ground the

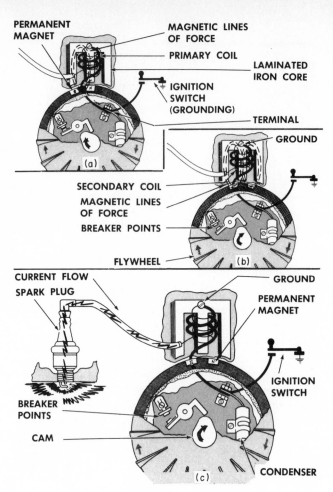

Fig. 8-11. Principles of the flywheel magneto ignition system. (*a*) Magnetic lines of force are built up around a coil of wire by the movement of a permanent magnet past the iron core on which the coil is wound. (*b*) As the permanent magnet passes the iron core and the contact points open, the lines of force collapse, thus producing a high voltage in the secondary winding of the coil. (*c*) The high voltage produces a spark at the spark-plug gap.

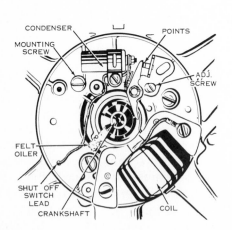

Fig. 8-12. Top view of a magneto. (*Lawn Boy Division of Outboard Marine Corporation*)

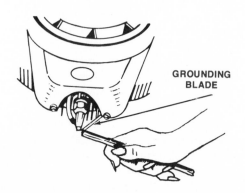

Fig. 8-13. Grounding blade near the spark plug used to stop the engine.

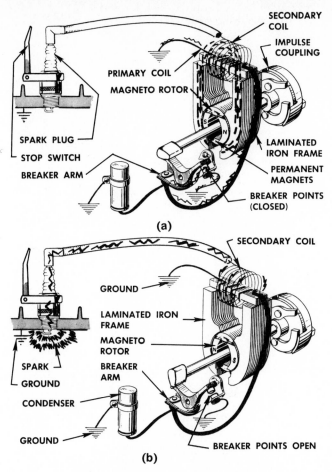

SECONDARY COIL

IMPULSE COUPLING

PRIMARY COIL

MAGNETO ROTOR

SPARK PLUG

STOP SWITCH

BREAKER ARM

LAMINATED IRON FRAME

PERMANENT MAGNETS

BREAKER POINTS (CLOSED)

(a)

SECONDARY COIL

GROUND

LAMINATED IRON FRAME

MAGNETO ROTOR

BREAKER ARM

SPARK

GROUND

CONDENSER

GROUND

BREAKER POINTS OPEN

(b)

Fig. 8-14. Schematic view of an external-type magneto ignition system. (*a*) Breaker points closed. (*b*) Breaker points open.

insulated terminal of the plug. When this happens, the high-voltage surges flow through the blade and no spark occurs.

§ 8-8 EXTERNAL MAGNETO

Some engines have an externally mounted magneto (Fig. 8-14). The magneto is driven through an impulse coupling, which will be explained later. As the rotor spins, it produces a magnetic field in the laminated iron frame on which the primary and secondary coils are wound. Each half-turn of the magnetic rotor causes a complete reversal of the magnetic field in the iron frame. This, in turn, causes the magnetic lines of force to build up and collapse through the two windings. Thus, a flow of current is induced in the primary winding all the time that the contact points are closed.

When the current flow is at its greatest, the breaker points are opened by the cam on the end of the rotor shaft. This stops the flow of current, and the magnetic lines of force collapse very rapidly. This induces a high voltage in the secondary winding. The voltage is high enough to produce a spark at the spark-plug gap.

The impulse coupling improves starting. It drives the rotor shaft. During starting, it repeatedly flips the rotor so that it spins very rapidly. This produces a strong spark for starting. The impulse coupling does this through a delayed spring action. That is, during cranking, spring tension builds up during a part turn of the coupling and then releases to spin the rotor ahead. The rotor thus turns part way, stops momentarily until spring tension builds up again, and then is flipped ahead once more.

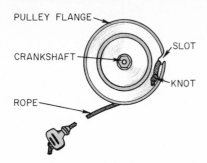

Fig. 8-15. The rope-wind starter is the simplest of all small-engine starters. You wind the rope on the pulley and pull it to spin the engine crankshaft.

After the engine starts, the impulse coupling unlocks due to centrifugal force so that it does not function. Now, the rotor turns steadily in time with the engine.

§ 8-9 ENGINE STARTING SYSTEMS

There are four general types of small-engine starters: rope-wind, rope-rewind, windup, and electric. In the rope-wind starter (Fig. 8-15), the engine has a pulley attached to the crankshaft. To use the rope, you hook the knot in the end of the rope into the slot in the pulley, as shown. Then you wind the rope around the pulley, adjust the choke, make sure the ignition is on, and give the rope a strong pull. This spins the crankshaft to start the engine. In most cases, it takes more than one pull to get the engine started.

§ 8-10 ROPE-REWIND STARTER

This starter rewinds the rope after each starting attempt. It has a spring inside (Fig. 8-16) that winds up as the rope is pulled out. Then, at the end of the starting attempt, the spring rewinds the rope on the pulley as the rope is released. The two pawls shown lock the pulley to the crankshaft for starting. They do this because centrifugal force pushes them out so that they lock on the grooves cut in the inside of the crankshaft adapter. After the engine

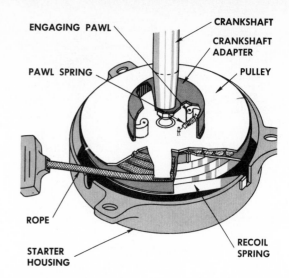

Fig. 8-16. A partially cutaway view of a typical rope-rewind starter.

starts, the pawl springs pull the pawls in out of the way.

§ 8-11 WINDUP STARTER

This starter has a spring that you wind up with a crank. Then, when you release the spring, the spring unwinds and spins the crankshaft to start the engine. Figure 8-17 shows how to use this starter.

§ 8-12 ELECTRIC STARTER

Electric starters for small engines are of two types. One is operated by connecting the extension cord from the starting motor to the 120-volt home wiring system (Fig. 8-18). The other uses a storage battery of the kind found in automobiles (Fig. 8-19). We will discuss batteries later. Both of these are electric motors that spin the engine crankshaft when they run. This starts the engine. Later chapters cover the operation of batteries and starting motors.

§ 8-13 STARTER-GENERATOR

In many small-engine installations using battery-type electric starting motors, the starter

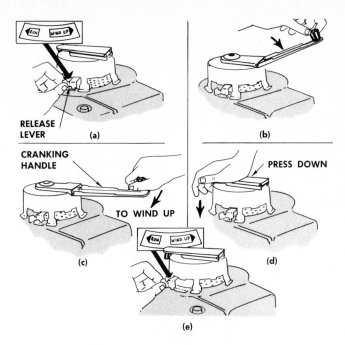

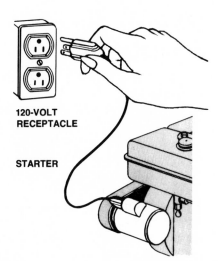

Fig. 8-17. Here are the steps in using a windup starter. (*a*) Lock the spring by moving the control lever to WIND UP. (*b*) Open the crank handle. (*c*) Wind up the recoil spring. (*d*) Fold the handle. (*e*) Release the spring by moving the control lever to RUN.

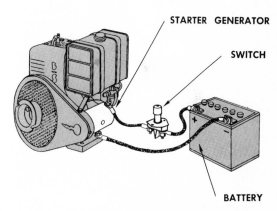

Fig. 8-19. Battery starter system. This system includes an electric starter that is also a generator to recharge the battery.

Fig. 8-18. Electric starter that uses house current to operate.

may also be a generator. That is, after the engine starts, the starter-generator begins to generate electric current. This puts back into the battery the current used to start. In addition, the generator can furnish current for electric lights or other electrical equipment.

Figure 8-20 shows the wiring circuit for this type of equipment. Figure 8-21 shows one system of mounting the starter-generator.

The voltage regulator is a device for preventing battery overcharge. It keeps the voltage from going up too high. By holding the voltage down, it cuts down the current going to the battery when the battery comes up to charge.

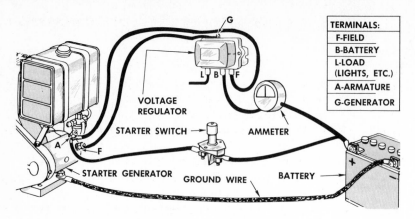

TERMINALS:
| F-FIELD |
| B-BATTERY |
| L-LOAD (LIGHTS, ETC.) |
| A-ARMATURE |
| G-GENERATOR |

Fig. 8-20. Wiring circuit of a typical starter-generator system. The starter-generator not only starts the engine but also generates current to charge the battery. The system includes a regulator to control the generator.

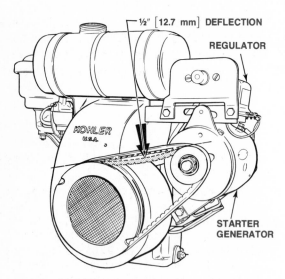

Fig. 8-21. Belt drive for a starter-generator. Proper belt tension is indicated as allowing ½-inch [12.7-millimeter] deflection and is adjusted by moving the starter-generator toward or away from the engine. (*Kohler Company*)

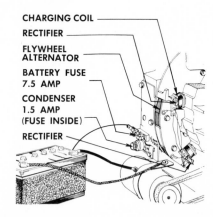

Fig. 8-22. An electric starter system using a battery-type starter and an alternator mounted on the engine under the flywheel to charge the battery.

§ 8-14 ALTERNATOR

The alternator is a special kind of electric generator. Except for some fairly old models, every automobile has one. Its purpose is to keep the battery charged and to supply electric current to handle any electrical equipment that is turned on. Some small engines have an alternator built into the flywheel housing (Fig. 8-22). The flywheel has a series of magnets. Charging coils are located close to the flywheel, as shown. As the magnets spin past the coils, current is produced in the coils. The current flows first in one direction and then in the other. That is, it alternates; it is alternating current (ac). This is also what happens in magneto windings, as you will recall from our description of magnetos. Alternating current is all right for magnetos but not for charging batteries. To charge batteries, the current must be changed to direct current (dc). That is, it must be rectified. Note that the alternator system has a rectifier (Fig. 8-22). Current flowing from the alternator is changed from ac to dc in the rectifier and then can flow to the battery and charge it.

Review questions

1. What is an electric current?
2. Name several good electrical conductors. Explain why these materials are good conductors of electricity.
3. What is an insulator? Name several.
4. Describe how a magnet and a piece of wire can be used to create electricity.
5. Describe a magneto ignition system for a one-cylinder, two-cycle engine.
6. What is the job of the condenser in the magneto ignition system?
7. Why do some small engines use an impulse coupling with the magneto? How does the impulse coupling work?
8. Name the four types of small-engine starters. Describe each type.
9. What is a starter-generator?
10. Explain the purpose of the alternator.

Study problems

1. Place a magnet under a piece of cardboard and sprinkle the cardboard with iron filings. Notice the patterns formed by the filings as they align themselves along the lines of force.
2. On a board, mount several small samples of good electrical conductors. Under these, make a row of several samples of materials that are good insulators.

9 FOUR-CYCLE ENGINES— OPERATION AND TYPES

§9-1 THE FOUR-CYCLE INTERNAL-COMBUSTION ENGINE

We will now look at the operation of four-cycle engines. Most automotive engines are of the four-cycle type, and many small engines also operate on the four-cycle principle. The basic difference between the two-cycle and four-cycle engines is in the manner in which the air-fuel mixture is introduced into the cylinder and the way in which the burned gases are removed from the cylinder after the power stroke of the piston.

The four-cycle internal-combustion engine is very similar in many ways to the two-cycle engine, which was described in Chapter 6. In both engines a piston moves up and down in each cylinder. The piston is attached by a connecting rod to a crank on the crankshaft. When ignition of the compressed air-fuel mixture takes place in the cylinder, the resulting high pressure drives the piston down. This push, carried through the connecting rod, forces the crankshaft to rotate.

To this point, the actions are similar in the two engines. However, in the two-cycle engine, the air-fuel mixture is admitted to the cylinder, and burned gases exit from the cylinder, through openings, or valve ports, in the cylinder wall (Figs. 6-5 to 6-8). Also, in the two-cycle engine, *every time* the piston nears top dead center (TDC), an explosion of air-fuel mixture takes place. Only two piston strokes are required to complete a cycle of engine operation.

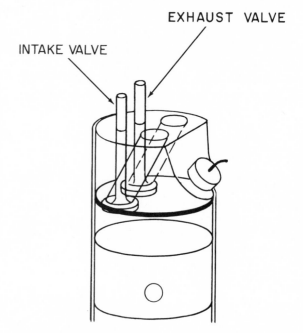

Fig. 9-1. Valves in the top of the cylinder admit the air-fuel mixture and allow the burned gases to escape.

But the four-cycle engine does not have valve ports in the cylinder wall. Instead, this engine has two metal plugs, called **valves,** in the top of the cylinder (Fig. 9-1), that move to admit the air-fuel mixture and to allow burned gases to exit. An explosion of air-fuel mixture occurs only *every other time* the piston nears TDC. That is, it requires four piston strokes (two up and two down) to complete a cycle of engine operation.

§9-2 MAKING THE ENGINE WORK

To make the four-cycle engine work, we must give it a charge of air and gasoline vapor to explode when the piston reaches top dead center on the compression stroke. Then, after the gasoline is burned, the cylinder must be cleared of the burned gases. We must continue to do this just as long as the engine runs.

In the four-cycle engine, the two valves at the top of the cylinder work in rhythm with the piston to do these jobs (Fig. 9-2). First, when the piston is moving down, one of the valves opens to allow air mixed with gasoline vapor to enter. The downward-moving piston in effect sucks the air-fuel mixture into the cylinder. (Actually, atmospheric pressure *pushes* the air-fuel mixture into the cylinder.) After the piston reaches the lower end of its travel and starts back up again, this valve closes so that the upper end of the cylinder is sealed. The piston therefore compresses the mixture. Then, when the piston reaches the top, this mixture is ignited and it explodes. The explosion drives the piston down. When it reaches its lower limit again, the second valve opens. This valve allows the burned gases to escape as the piston moves up again, pushing the burned gases ahead of it.

In automotive language, each time the piston moves from top to bottom, or from bottom to top, the movement is called a **piston stroke** (Fig. 9-3). A piston stroke takes place when the piston moves from TDC to BDC, or from BDC to TDC.

§9-3 THE FOUR PISTON STROKES

Four piston strokes are required to make a complete cycle of actions in the cylinder. First, the piston moves *down* to draw in the mixture of air and fuel. Second, the piston moves *up* to compress the mixture. Third, the piston is pushed *down* by the pressure produced when the mixture explodes. Fourth, the piston moves *up* to push out the burned gases.

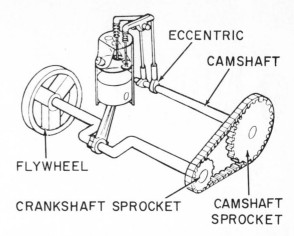

Fig. 9-2. The valves work in rhythm with the movement of the piston.

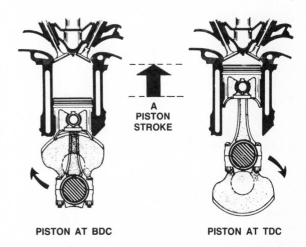

Fig. 9-3. A piston stroke takes place when the piston moves from TDC to BDC or from BDC to TDC.

The engine is therefore called a **four-cycle engine.** Actually, the full name is a four-*stroke*-cycle engine. There are four piston strokes in each cycle. As already mentioned in Chapter 6, a cycle is a series of events that repeats itself. The four piston strokes in the engine cylinder form a cycle that is repeated continuously as the engine runs.

§9-4 THE VALVES

The valves at the top of the engine cylinder are round metal plugs on stems. The plugs fit

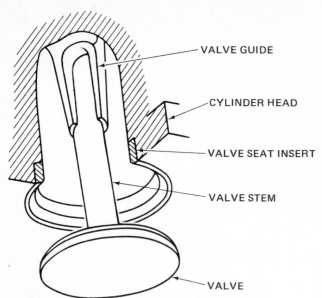

VALVE GUIDE

CYLINDER HEAD

VALVE SEAT INSERT

VALVE STEM

VALVE

Fig. 9-4. A valve and valve seat in the cylinder head.

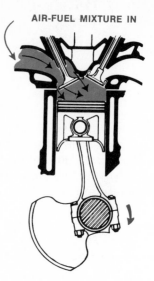

AIR-FUEL MIXTURE IN

Fig. 9-5. The intake stroke. The intake valve, at the left, has opened. The piston is moving downward, drawing air and gasoline vapor into the cylinder.

snugly into round holes cut in the top of the cylinder (Fig. 9-4). In a way, a valve is much like a cork fitted into the neck of a bottle. The valves are operated by a valve mechanism that is described later. Normally, the valves are held in the holes by springs. But when the engine is operating, the valve mechanism moves the valve off its seat so that an opening appears between the rim of the valve and the seat. In this opened position, air-fuel mixture or exhaust gas can pass through the opening and enter or leave the cylinder.

§ 9-5 HOW THE ENGINE OPERATES

Let's find out how the engine works. First, one of the valves opens to let the air-fuel mixture into the cylinder. Later, the other valve opens to let the burned gases out. Let's follow the cycle of events all the way through.

§ 9-6 THE INTAKE STROKE

First, the piston moves down, as shown in Fig. 9-5. As the piston moves down, it produces a

vacuum in the cylinder. A **vacuum** is the absence of air or any other substance. Outside air rushes in to fill the vacuum. The same thing happens, for example, when you drink a carbonated beverage through a straw. You move your jaw down to produce a vacuum in your mouth. Air tries to get in to fill the vacuum. But, instead, the beverage gets in the way, and the air pushes the beverage up the straw and into your mouth.

During this piston stroke, one of the valves—the intake valve—is open. The valve is down off its seat, and the air-fuel mixture can pass the valve and enter the cylinder. This air-fuel mixture fills the vacuum produced by the downward movement of the piston. This downward movement of the piston is called the **intake stroke** because the cylinder is taking in a mixture of fuel and air. As the air moves toward the cylinder, it first has to pass through the carburetor. The **carburetor** sits on top of the engine and has the job of mixing gasoline vapor with the air passing through. Carburetor actions are explained in Chapter 17.

The piston moves all the way down to BDC on the intake stroke. The intake valve is open and the air-fuel mixture pours into the cylinder. At the end of the intake stroke, the intake valve closes. We will find out later what makes the valve close.

§ 9-7 THE COMPRESSION STROKE

After the piston passes BDC at the end of the intake stroke, it starts to move up. Since both valves are closed, the air-fuel mixture has no place to go. It is pushed, or **compressed,** into a smaller volume (Fig. 9-6). The amount the mixture of air and gasoline vapor is compressed is called the **compression ratio.**

In a modern engine, the air-fuel mixture is compressed into one-eighth or one-ninth of its original volume. That is like taking a quart of air and squeezing it down to less than half a cup (Fig. 9-7).

If you squeezed the mixture from a quart down to half a cup, you would have a compression ratio of 8 to 1. There are 8 half-cups in a quart. When you compress the quart, which is 8 half-cups, into 1 half-cup, you have gone from 8 down to 1. That is, you have compressed the mixture down from 8 to 1. So, the

compression ratio is 8 to 1. Usually the ratio is shown as 8:1. You read this as "8 to 1." The piston movement from BDC to TDC while both valves are closed is called the **compression stroke.**

§ 9-8 THE POWER STROKE

As the piston nears TDC at the end of the compression stroke, a spark occurs in the top of the cylinder. The spark plug makes the spark (we will find out how this happens later). When a spark occurs in the compressed air-fuel mixture, there is an explosion. The pressure and temperature of the mixture go way up. Every square inch [645.16 square millimeters] of the piston head gets a push of up to 600 pounds [272.4 kilograms] or more. Do you know how much that amounts to overall? As much as 2 tons—4,000 pounds [1,816 kilograms]—are pushing down on that piston head! It is no wonder the piston moves!

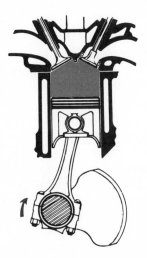

Fig. 9-6. The compression stroke. The intake valve has closed. The piston is moving upward, compressing the mixture.

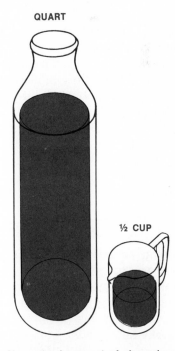

QUART

½ CUP

Fig. 9-7. If you had a quart of air and compressed it to half a cup, you would compress it to one-eighth of its original volume.

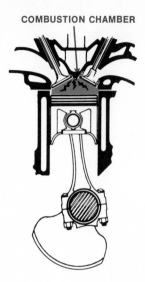

COMBUSTION CHAMBER

Fig. 9-8. The power stroke. The ignition system produces a spark that ignites the mixture. As the mixture burns, high pressure is created, pushing the piston down.

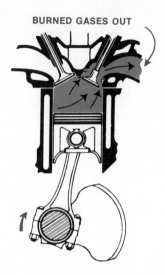

BURNED GASES OUT

Fig. 9-9. The exhaust stroke. The exhaust valve, at the right, has opened. The piston is moving upward, pushing the burned gases out of the cylinder.

The 2-ton pressure pushes the piston downward. This downward movement is called the **power stroke** (Fig. 9-8). The powerful push on the piston is carried through the connecting rod to the crank on the crankshaft. The crankshaft turns this downward movement into rotary motion, and the rotary motion is carried through the gears and shafts to the car wheels so that the car moves.

Note that the combustion chamber is the space at the top of the cylinder and above the piston in which the burning of the air-fuel mixture takes place (Fig. 9-8). Combustion chambers vary in shape.

§9-9 THE EXHAUST STROKE

As the piston reaches BDC on the power stroke, the exhaust valve opens. When the piston moves up again, it pushes out the burned gases through the exhaust port. This upward movement of the piston is called the **exhaust stroke** because the burned gases are pushed out, or exhausted, from the engine cylinder (see Fig. 9-9). Finally, as the piston reaches TDC on the exhaust stroke, the ex-

haust valve closes and the intake valve opens. The piston moves down once more on another intake stroke. The cycle of events in the cylinder is then repeated: intake stroke, compression stroke, power stroke, exhaust stroke. And this cycle continues as long as the engine runs.

§9-10 PISTON RINGS

Piston rings are essential to the operation of the engine. They are metal rings that fit into grooves in the pistons. You can see these piston grooves in Fig. 9-10. You can see piston rings installed on a piston in Fig. 5-12. The purpose of piston rings is to form a tight seal between the piston and the cylinder wall. As we have seen, there is great pressure above the piston during the compression stroke and the power stroke. The piston itself cannot be made to fit tightly enough in the cylinder to prevent this pressure from leaking past the piston. Such leakage could mean serious power loss. The rings press tightly against the cylinder wall and the sides of the piston grooves to provide the needed seal. The rings

Fig. 9-10. A piston for an engine.

are covered with oil by the engine lubricating system so that they can slide up and down easily on the cylinder wall. There is more on piston rings in Chapter 10.

§ 9-11 THE FOUR-STROKE CYCLE

Since it takes four piston strokes to complete one cycle, the engine is called a four-stroke-cycle engine. The name is usually shortened to four-cycle engine.

§ 9-12 TWO-CYCLE AND FOUR-CYCLE ENGINES COMPARED

We have explained that it takes two revolutions of the crankshaft to complete the four strokes in a four-stroke-cycle engine. The first half-revolution, or 180 degrees, the piston is moving down on the intake stroke (Fig. 9-11). The next half-revolution, the piston is moving up on the compression stroke (Fig. 9-12). The third half-revolution of the crankshaft, the piston is moving down on the power stroke. The fourth half-revolution, the piston is moving up on the exhaust stroke. Figures 9-5, 9-6, 9-8, and 9-9 show these four strokes.

Most larger internal-combustion engines, and practically all automotive engines, are of the four-stroke-cycle type. Every fourth piston stroke in each cylinder is a power stroke.

In the two-stroke-cycle engine (or two-cycle engine, as it is commonly called), a power

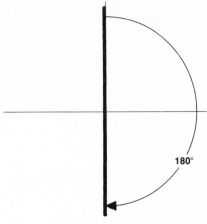

ONE PISTON STROKE

Fig. 9-11. The first half-revolution of the crankshaft. The piston is moving down on the intake stroke.

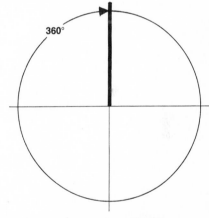

TWO PISTON STROKES

Fig. 9-12. The next half-revolution of the crankshaft. The piston is moving up on the compression stroke.

stroke occurs every two piston strokes. That is, every downward movement of the piston is a power stroke. In effect, the intake and compression strokes are combined. Also, the power and exhaust strokes are combined. We have described how the two-cycle engine works in Chapter 6.

You might think that, because the two-cycle engine has twice as many power strokes as a four-cycle engine (Fig. 9-13), it would produce twice as much horsepower as a four-cycle en-

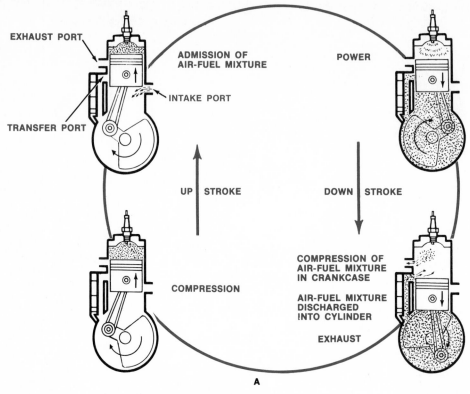

EXHAUST PORT

**ADMISSION OF
AIR-FUEL MIXTURE**

POWER

INTAKE PORT

TRANSFER PORT

UP STROKE

DOWN STROKE

COMPRESSION

**COMPRESSION OF
AIR-FUEL MIXTURE
IN CRANKCASE**

**AIR-FUEL MIXTURE
DISCHARGED
INTO CYLINDER**

EXHAUST

A

TWO STROKES—ONE REVOLUTION

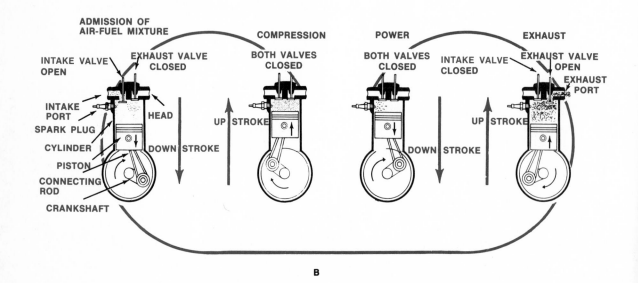

**ADMISSION OF
AIR-FUEL MIXTURE**

COMPRESSION

POWER

EXHAUST

**INTAKE VALVE
OPEN**

**EXHAUST VALVE
CLOSED**

**BOTH VALVES
CLOSED**

**BOTH VALVES
CLOSED**

**INTAKE VALVE
CLOSED**

**EXHAUST VALVE
OPEN**

**EXHAUST
PORT**

**INTAKE
PORT**

SPARK PLUG

HEAD

UP STROKE

UP STROKE

CYLINDER

PISTON

DOWN STROKE

DOWN STROKE

**CONNECTING
ROD**

CRANKSHAFT

B

FOUR STROKES—TWO REVOLUTIONS

Fig. 9-13. Two-stroke cycle and four-stroke cycle compared.

gine of the same size, running at the same speed. However, this is not true. In the two-cycle engine, when the intake and exhaust ports have been cleared by the piston, there is always some mixing of the fresh charge and the burned gases. Not all the burned gases get out, and this prevents a larger fresh charge from entering. Therefore, the power stroke that follows is not as powerful as it could be if all the burned gases were exhausted and a full charge of air-fuel mixture entered.

In the four-cycle engine, nearly all the burned gases are forced from the combustion chamber by the upward-moving piston. And a comparatively full charge of air-fuel mixture can enter, because a complete piston stroke is devoted to the intake of the mixture (contrasted with only part of a stroke on the two-cycle engine). Therefore, the power stroke of the four-cycle engine produces more power.

However, the two-cycle engine is widely used as a power plant for lawn mowers, motor-boats, snow removers, model airplanes, motor scooters, power saws, and other such equipment. These engines are often air cooled. Because they have no valve train or water cooling system, they are relatively simple in construction and light in weight. These are desirable characteristics for engines used on small, lightweight equipment that must be handled and moved around.

§ 9-13 MULTIPLE-CYLINDER ENGINES

Of course, a single cylinder, with a single piston working in it, would not produce enough power to run an automobile. Also, there would be only one power stroke for every two revolutions of the crankshaft, and this would make for a very rough ride. So almost all automobiles have engines with four, six, or eight cylinders. In these engines, one or more power strokes are going on all the time so that the engine produces a continuous flow of power and the automobile moves along smoothly.

§ 9-14 ENGINE CLASSIFICATIONS

We have already mentioned some of the different types of automotive engines—both reciprocating and rotary. Now we are going to look at other classifications of engines. Automotive piston, or reciprocating, engines can be classified in at least seven different ways. They are the following:

1. Number of cylinders
2. Arrangement of cylinders
3. Arrangement of valves
4. Types of cooling systems
5. Number of piston strokes per cycle
6. Type of fuel burned
7. Firing order

This is not a complete list, but it will give you an idea of the ways in which engines can be classified.

§ 9-15 NUMBER AND ARRANGEMENT OF CYLINDERS

Almost all piston engines have either four, six, or eight cylinders. Usually, the cylinders in four-cylinder and six-cylinder engines are arranged in a single row, or line. Some V-4 and V-6 engines have been made. In these, the cylinders are placed in two rows set at an angle to each other. Eight-cylinder engines are all of the V-8 type. The two rows of cylinders are set at an angle to each other. Figure 9-14 shows the various arrangements. Now, let us look at these different engines in detail.

§ 9-16 FOUR-CYLINDER, IN-LINE ENGINES

As already mentioned, the cylinders in a four-cylinder engine can be arranged in a single line, in pairs set at an angle to form a V, or in pairs set opposite each other. A typical four-cylinder, in-line engine is shown in Fig.

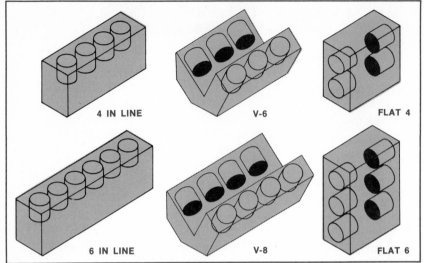

Fig. 9-14. Several cylinder arrangements.

9-15. Note that this engine has overhead valves operated by pushrods and rocker arms. Study this picture to locate the pistons, connecting rod (one shown), camshaft, camshaft drive gear, spark plugs, pushrods, rocker arms, and valves. If you look carefully, you will see a spiral gear on the camshaft that is meshed with another spiral gear on a nearly vertical shaft. The purpose of these gears is to drive the oil pump that lubricates the engine and also the ignition distributor, which is not shown in the picture. You can see part of the oil pump at the bottom of the picture. See Chapter 15 for a description of the oil pump and Chapter 12 for a look at the ignition distributor.

As you can see, you can learn a great deal from the pictures in this book by studying them carefully.

§ 9-17 V-4 ENGINES

Not very many V-4 engines have been built, but there are a few in this country. Figure 9-16 is a phantom view of a V-4 engine, showing the working parts of the engine. You can see the crankshaft, pistons, connecting rods, camshaft, and valve train. The extra shaft with a gear (near the bottom of the picture) is a special balance shaft that is needed to balance the engine. A balanced engine runs smoothly and

Fig. 9-15. A partial cutaway view of a four-cylinder, in-line, overhead-valve engine.

does not vibrate or run unevenly. The difficulty of balancing a V-4 engine is one reason why very few of them have been built.

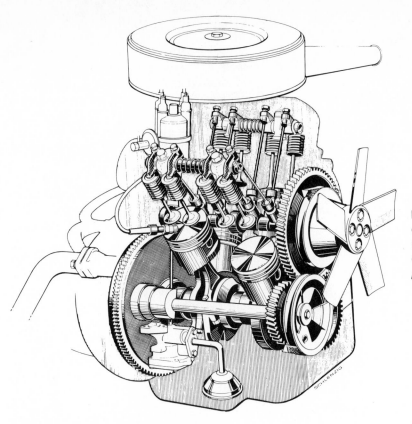

Fig. 9-16. A phantom view of a V-4 engine, showing major moving parts in the engine. (*Ford Motor Company of Germany*)

§ 9-18 FLAT-FOUR ENGINES

The best example of the flat-four engine is the Volkswagen engine, which is shown in sectional view in Fig. 9-17. The four cylinders are arranged in two opposing rows of two cylinders each. If you study this picture, you will find pistons, connecting rods, cylinders, rocker arms, valves, and pushrods. You will also notice something different about this engine. The cylinders are surrounded by flat metal rings, or metal fins. These fins provide large surfaces from which heat can radiate. The fins are

Fig. 9-17. A flat four-cylinder engine with two opposing rows of two cylinders each. This is an air-cooled engine. (*Volkswagen*)

needed because the engine is air-cooled. That is, heat from the combustion of the air-fuel mixture is radiated from the cylinder fins into the air. In most automobiles, the engines are liquid-cooled. We will look more closely at cooling systems in Chapter 18.

Another interesting thing about the Volkswagen engine is that the pushrods are carried in hollow tubes. Four tubes per cylinder hold together the cylinder heads, the cylinders, and the cylinder block.

§ 9-19 SIX-CYLINDER, IN-LINE ENGINES

Most six-cylinder engines are of the in-line type. In these engines, the six cylinders are arranged in a single row, or line. Figure 9-18 is a partial cutaway view of a six-cylinder, in-line engine. As you can see, it is an overhead-valve engine. The picture is especially good because you can see so many different parts. Find the crankshaft, camshaft, valves, rocker arms, pushrods, pistons, connecting rods, and

other parts. Notice that you can see the oil pump and the ignition distributor in this picture, as well as the drive gears. They are located front and center in the picture. The oil pump is located at the bottom in the oil pan. As previously mentioned, the oil pump supplies oil to the moving parts in the engine so that they are kept well lubricated. The ignition distributor works with the other parts in the ignition system to supply sparks to the spark plugs.

Another six-cylinder, in-line engine is shown partly cut away in Fig. 9-19. This engine is especially interesting because the cylinders are slanted to one side, as you can see. Thus, it is often called a "slant six." The engine has been supplied in two styles, one with a cast-iron cylinder block and the other with a cast-aluminum cylinder block. Cylinder blocks are discussed in more detail in Chapter 10. Figure 9-19 is also a very good picture because you can see so many of the internal engine parts. Study the picture and try to locate as many of the parts as you can.

Fig. 9-18. A six-cylinder, in-line engine with overhead valves, partly cut away to show internal parts. (*Ford Motor Company*)

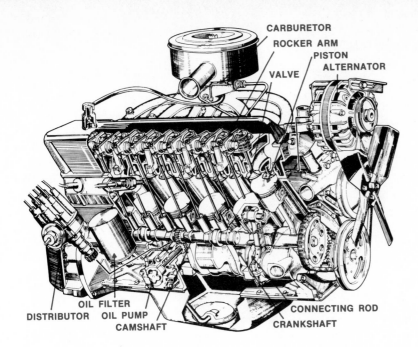

CARBURETOR
ROCKER ARM
PISTON
VALVE   ALTERNATOR

DISTRIBUTOR
OIL FILTER
OIL PUMP
CAMSHAFT
CONNECTING ROD
CRANKSHAFT

Fig. 9-19. A slant-six, in-line, overhead-valve engine, cut away to show internal parts. The cylinders are slanted to permit a lower hood line. (*Chrysler Corporation*)

§ 9-20 V-6 ENGINES

A few companies have built V-6 engines. These have two rows of three cylinders each, set at an angle to form a V. Figure 9-20 is a cutaway view of a V-6 engine. You can see many details of the parts and interior construction of the engine.

§ 9-21 FLAT-SIX ENGINES

The flat-six engine is very similar to the flat-four engine, except that one more cylinder has been added to each bank. Figure 9-21 is a cutaway view from the top of the flat-six engine used in the Chevrolet Corvair. This engine is air-cooled and is mounted at the rear of the car.

§ 9-22 V-8 ENGINES

All eight-cylinder automotive engines made today are of the V-8 type. They have two rows of cylinders with four cylinders in each row.

The two rows are set at an angle to form a V. Figure 9-22 is a cutaway view of one model of a V-8 engine. Study the picture for a moment, and then pick out the various interior parts of the engine.

At one time, eight-cylinder, in-line engines were common, but today V-8 engines have taken their place. The V-8 is a shorter and more rigid engine. The cylinders are closer together and have a better chance of getting their share of the air-fuel mixture from the carburetor. The more rigid engine permits higher running speeds and higher compression pressures, with less difficulty from bending of the block and crankshaft. The shorter engine also makes possible more passenger space on the same wheel base.

§ 9-23 ARRANGEMENT OF VALVES

Another way to classify engines is according to the arrangement of the valves and valve trains. Now, let's discuss the L-head (or flathead), the I-head (or overhead-valve), and the

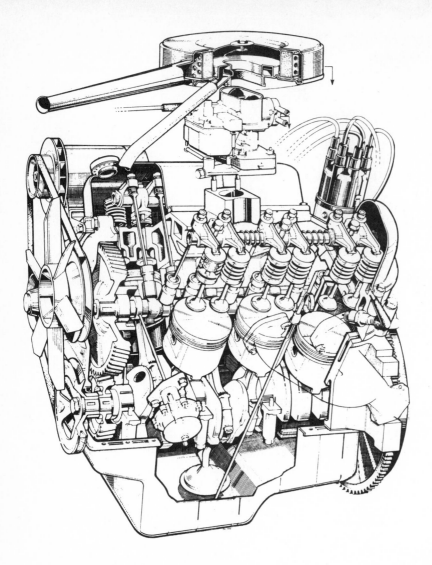

Fig. 9-20. A cutaway view of a V-6 overhead-valve engine. (*Ford Motor Company of Germany*)

overhead-camshaft arrangements. We take a good look at valve trains in Chapter 11.

§ 9-24 L-HEAD VALVE TRAIN

The L-head valve train is the simplest valve train. It is found in many of the one-cylinder engines used in lawn mowers, edgers, and so on. Years ago, nearly all automobile engines had L-head valve trains, but not anymore.

Figure 9-23 shows the valve train for an L-head engine. The valve moves up and down in a valve guide, which is installed in the cylinder block. The **cylinder block** is the large block

of cast iron that encloses the cylinders. All the other engine parts are installed in or on the cylinder block. See Chapter 10 for a detailed look at the cylinder block.

When the valve is down, its head seals off the valve port, so no air can get into or out of the cylinder. But when it is pushed up, the valve is raised off the valve seat, so air or gas can pass through. About three-quarters of the time, the valve is held down in the closed position by the valve spring.

Figure 9-24 shows the parts of the valve train for a one-cylinder, L-head engine in their proper order of assembly. Notice the cams on

Fig. 9-21. A sectional view from the top of a flat six-cylinder, overhead-valve, air-cooled engine. Flat engines are sometimes referred to as pancake engines. (*Chevrolet Motor Division of General Motors Corporation*)

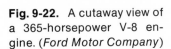

Fig. 9-22. A cutaway view of a 365-horsepower V-8 engine. (*Ford Motor Company*)

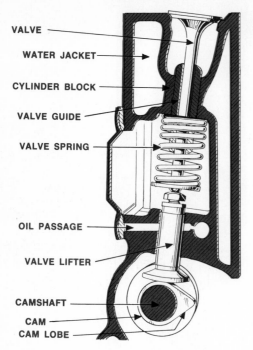

VALVE

WATER JACKET

CYLINDER BLOCK

VALVE GUIDE

VALVE SPRING

OIL PASSAGE

VALVE LIFTER

CAMSHAFT
CAM
CAM LOBE

Fig. 9-23. A valve train for an L-head engine. The valve is raised off the valve seat with every camshaft rotation.

on top. Figure 9-25 is a cutaway view of one cylinder of a flat-head engine. Study this picture to find the valves, the piston, and the combustion chamber. Look at Fig. 9-26 and you will see why the engine is called an L-head engine. Note that the combustion chamber and cylinder are in the shape of an upside-down L.

§ 9-25 I-HEAD

In recent years, automotive manufacturers have stopped making L-head engines and have switched to overhead-valve engines. There are several reasons for this change, which are explained later. First, let's look at the I-head engine, also called the overhead-valve engine or the valve-in-head engine. As the name implies, the valves are in the cylinder head instead of in the cylinder block.

Figure 9-27 shows the essential parts of the valve train for one cylinder in an overhead-valve engine. Do you see the parts that have been added? They are the pushrods and the rocker arms. The valve springs are not shown here, but they are used, just as in the L-head engine. When the cam lobe comes up under the valve lifter, it moves the valve lifter up, and this pushes up on the pushrod. The pushrod moves up, pushing up on one end of the rocker arm. The rocker arm is mounted on a shaft so

the camshaft. These are round collars with a high spot, called the cam lobe. The lobe opens the valve, as we will see in a moment.

The L-head engine is also called the flat-head engine, because the cylinder head is flat

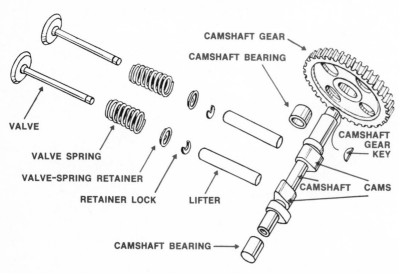

CAMSHAFT GEAR

CAMSHAFT BEARING

VALVE

VALVE SPRING

VALVE-SPRING RETAINER

RETAINER LOCK

LIFTER

CAMSHAFT BEARING

CAMSHAFT
GEAR
KEY

CAMSHAFT   CAMS

Fig. 9-24. Complete valve trains for a one-cylinder engine. (*Cushman Motor Works*)

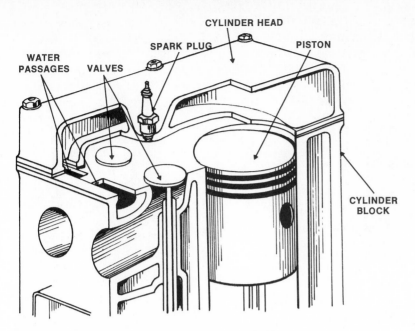

Fig. 9-25. A cutaway view of an L-head engine.

Fig. 9-26. The combustion chamber and cylinder of the L-head engine are shaped like an upside-down "L," and that is the reason for the name.

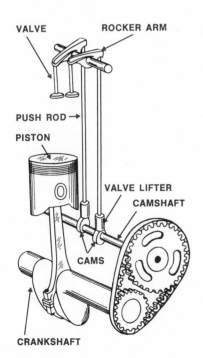

Fig. 9-27. Valve-operating mechanisms for one cylinder of an overhead-valve engine. Only the essential moving parts are shown.

that it can rock back and forth, just like a see-saw. When one end of the rocker arm is pushed up by the pushrod, the other end pushes down on the end of the valve stem. The result? The valve is pushed down off its seat and it opens.

§ 9-26 OVERHEAD-CAMSHAFT ENGINE

In the overhead-camshaft engine, the camshaft is located in the cylinder head. This eliminates the pushrod and puts the cams closer to the valves. With this setup, the valves respond quicker to the cam lobes, and the engine is more responsive. That is, the engine accelerates faster and can have a higher speed (in revolutions per minute, or rpm).

Figure 9-28 shows a simplified drawing of an overhead-camshaft engine and the drive belt. Note that the camshaft sprocket is twice as large as the crankshaft sprocket, just as in other kinds of engines. The drive belt also drives the ignition distributor. As we will explain later,

L-head and I-head engines have camshafts with gears that drive the ignition distributor. Many overhead-camshaft engines use the camshaft to drive the distributor. However, on the engine shown, in Fig. 9-28, the distributor is driven from the same belt that drives the camshaft. The distributor is, of course, essential to the running of the engine. It controls the ignition system and supplies the sparks to the spark plugs. Without these sparks the engine would not run.

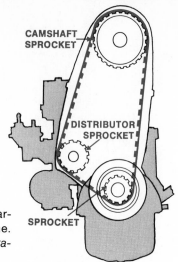

Fig. 9-28. A simplified drawing of the drive arrangement for an overhead-camshaft engine. (*Pontiac Motor Division of General Motors Corporation*)

Review questions

1. How many piston strokes are required to complete a cycle of engine operation in a four-cycle engine?
2. Name the four strokes of the four-stroke-cycle engine.
3. Explain what is meant by the term compression ratio.
4. Power is transmitted to the crankshaft during which piston stroke?
5. What are the differences in construction between a two-cycle engine and a four-cycle engine?
6. Explain why a two-cycle engine does not produce twice the power of a four-cycle engine of the same size.
7. Name seven different ways that can be used to classify automotive piston engines.
8. Describe the cylinder arrangement of an in-line engine.
9. What is a flat-four engine?
10. Where are the valves located in an L-head engine? In an overhead-camshaft engine?

Study problems

1. Make a list of all the different ways you can think of to classify an engine. Then make a quick inspection of 10 different automobiles, classifying each engine according to your list. When you are finished, use your findings to describe the typical automobile engine.
2. Using a cutaway engine, or an engine with the valve cover off and the spark plugs removed, turn the engine through the four strokes of engine operation. Note the position of the valves for each piston stroke.

10 AUTOMOTIVE ENGINE CONSTRUCTION—CYLINDER BLOCK, HEAD, CRANKSHAFT, AND BEARINGS

§10-1 ENGINE CONSTRUCTION

In this chapter we are going to take a close look at the various engine components. We will see how they are made and what they do in the engine. Let's start with the cylinder block.

§10-2 CYLINDER BLOCK

The cylinder block is the basic part—the foundation—of the engine. Every other engine part is put inside the block or attached to it. Figure 10-1 shows the main parts that are attached

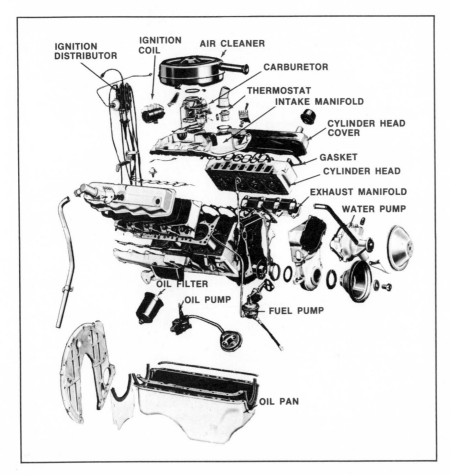

IGNITION DISTRIBUTOR

IGNITION COIL

AIR CLEANER

CARBURETOR

THERMOSTAT
INTAKE MANIFOLD

CYLINDER HEAD COVER

GASKET
CYLINDER HEAD

EXHAUST MANIFOLD

WATER PUMP

OIL FILTER

OIL PUMP

FUEL PUMP

OIL PAN

Fig. 10-1. External parts that are attached to the engine block when the engine is assembled. (*Chrysler Corporation*)

to the block when the engine is assembled. Figure 10-2 shows the parts that are put inside the cylinder block and hung from it. In Fig. 10-2, notice that only one piston and one connecting rod are shown, although the engine uses eight of each. This is done to make the picture easy to study.

The cylinder block is a very complicated piece of cast iron. (Some blocks are built of aluminum, and we talk about these later in the chapter.) Figure 10-3 shows one bank of a V-6 engine cylinder block partly cut away.

As you can see, the cylinder block contains not only the cylinders but also the water jackets that surround them.

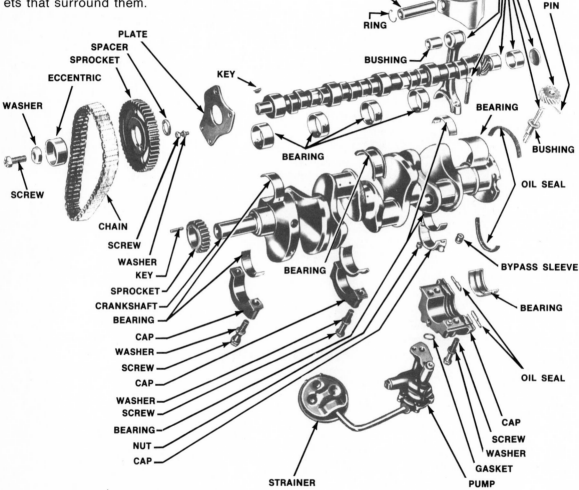

Fig. 10-2. Internal parts that are installed in the cylinder block of a V-8 engine. (*Chrysler Corporation*)

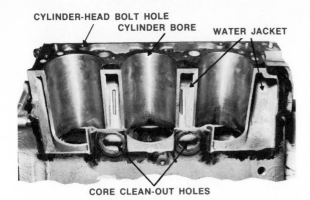

CYLINDER-HEAD BOLT HOLE
CYLINDER BORE   WATER JACKET

CORE CLEAN-OUT HOLES

Fig. 10-3. One bank of a V-6 engine block partly cut away to show the internal construction. (*Truck and Coach Division of General Motors Corporation*)

§ 10-3 WATER JACKETS

The water jackets are spaces formed by the inner shells of the cylinders and the outer shell of the cylinder block. You can see the water jackets in Fig. 10-3. Remember that the water jackets are spaces surrounding the cylinders through which coolant (water and antifreeze) can flow. The coolant then flows to the radiator. In the radiator, the coolant loses heat. It then returns, cooled, to the engine. It is this constant circulation of the coolant between the engine and the radiator that keeps the engine from overheating.

§ 10-4 MACHINING THE BLOCK

Figure 10-4 shows three views of a V-8 block with the various parts named. You can see that it takes a lot of work to finish a cylinder block. Also, remember that every machining operation must be done to within a thousandth of an inch (0.001 inch) [0.025 millimeter] or less. Notice that the block has water-jacket plugs and oil-gallery plugs. The water-jacket plugs are used to plug up the holes to the water jackets. You can see these holes in Fig. 10-3.

The oil-gallery plugs are used to plug up the oilholes that are drilled in the casting. These oilholes are called oil galleries. They carry the oil flow from the oil pump to various places in the engine where lubricating oil is needed.

§ 10-5 PARTS ATTACHED TO THE BLOCK

Parts that are attached to the block when the engine is assembled are shown in Fig. 10-1. The cylinder head or heads are mounted on top of the block. The water pump and the timing-chain cover are attached at the front. The timing chain, you remember, is mounted on the camshaft and crankshaft sprockets so that the crankshaft can drive the camshaft. (Some engines use a pair of gears to drive the camshaft instead of a timing chain and sprockets.) The clutch housing or automatic transmission is attached at the rear. The crankshaft is hung underneath the block by bearings and bearing caps. You can see the crankshaft and the crankshaft bearings and bearing caps in Fig. 10-2. You can get some idea of how the crankshaft is hung from the block by studying Fig. 9-22.

§ 10-6 OIL PAN

The oil pan is attached to the bottom of the cylinder block. It is formed from pressed steel, as shown in Fig. 10-5. The oil pan usually holds from 4 to 9 quarts [about 4 to $8\frac{1}{2}$ liters] of oil, depending on the engine design. Bigger engines require more oil. The oil pan and the lower part of the cylinder block form the **crankcase.** This is a box, or case, that encloses, or encases, the crankshaft; hence the name "crankcase."

The oil pump draws oil from the reserve of oil in the oil pan and sends it to all engine parts needing lubrication. The engine-lubricating system is described in detail in Chapter 19.

§ 10-7 ALUMINUM CYLINDER BLOCKS

Several engines have been made with aluminum cylinder blocks. Aluminum is a relatively light metal, weighing much less than cast iron,

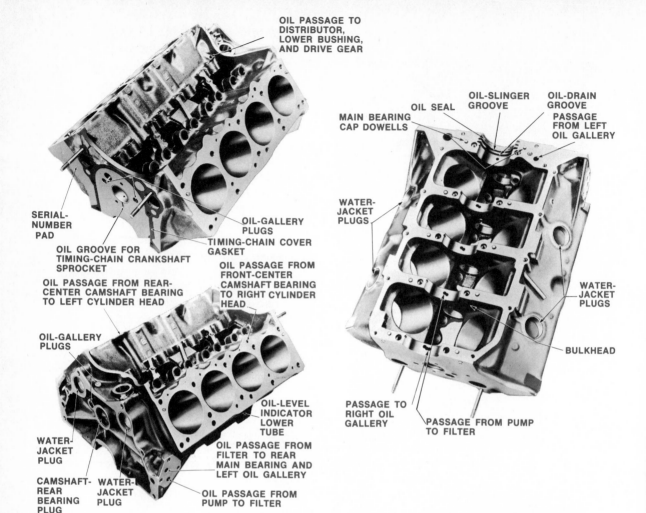

OIL PASSAGE TO
DISTRIBUTOR,
LOWER BUSHING,
AND DRIVE GEAR

MAIN BEARING
CAP DOWELLS

OIL SEAL

OIL-SLINGER
GROOVE

OIL-DRAIN
GROOVE

PASSAGE
FROM LEFT
OIL GALLERY

WATER-
JACKET
PLUGS

SERIAL-
NUMBER
PAD

OIL-GALLERY
PLUGS

TIMING-CHAIN COVER
GASKET

OIL GROOVE FOR
TIMING-CHAIN CRANKSHAFT
SPROCKET

OIL PASSAGE FROM
FRONT-CENTER
CAMSHAFT BEARING
TO RIGHT CYLINDER
HEAD

OIL PASSAGE FROM REAR-
CENTER CAMSHAFT BEARING
TO LEFT CYLINDER HEAD

WATER-
JACKET
PLUGS

BULKHEAD

OIL-GALLERY
PLUGS

OIL-LEVEL
INDICATOR
LOWER
TUBE

WATER-
JACKET
PLUG

OIL PASSAGE FROM
FILTER TO REAR
MAIN BEARING AND
LEFT OIL GALLERY

CAMSHAFT-
REAR
BEARING
PLUG

WATER-
JACKET
PLUG

OIL PASSAGE FROM
PUMP TO FILTER

PASSAGE TO
RIGHT OIL
GALLERY

PASSAGE FROM PUMP
TO FILTER

Fig. 10-4. Three views of a cylinder block from a V-8 engine.

and it conducts heat more rapidly than cast iron. Thus, there is less chance for hot spots to develop. However, aluminum is too soft to use as cylinder-wall material. It wears too rapidly. Therefore, aluminum cylinder blocks must have cast-iron cylinder liners. The one exception to this is the engine used in the Chevrolet Vega, which we will talk about later.

First, let's look at the cast-iron cylinder liners. These are sleeves that are either cast in the block or installed later. In the cast-in type, the cylinder liners are properly positioned, and the aluminum is poured around them. Thus, they are cast in the block.

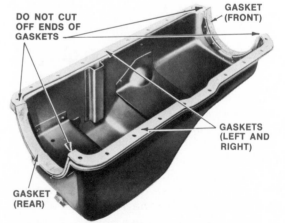

DO NOT CUT
OFF ENDS OF
GASKETS

GASKET
(FRONT)

GASKETS
(LEFT AND
RIGHT)

GASKET
(REAR)

Fig. 10-5. An oil pan with gaskets in place, ready for installation on the bottom of the cylinder block. (*Chrysler Corporation*)

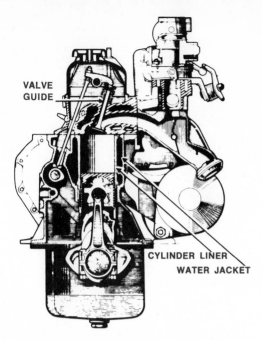

Fig. 10-6. A sectional view of an engine that uses removable wet cylinder liners. (*Renault*)

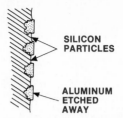

Fig. 10-7. Greatly enlarged view of the Vega cylinder-wall surface. (*Chevrolet Motor Division of General Motors Corporation*)

There are two types of cast-iron cylinder liners that are installed later: dry and wet. In the dry type, the liners are pressed in place with considerable pressure. They are in contact with the cylinder-bore hole for their full length. In the wet type, the liner is in contact with the engine coolant and is sealed at top and bottom. These liners are removable and can be replaced if they become worn or damaged. Figure 10-6 is a sectional view of an engine with wet cylinder liners.

Now, let's take a look at the Vega. The engine for this car has a cast-aluminum cylinder block without cylinder liners. The aluminum used is "loaded" with silicon particles (Fig. 10-7). Silicon is an extremely hard substance and has very good wearing qualities. In the final preparation of the cylinder block, the cylinders are honed, or smoothed down, and are then treated by a special process that etches away the surface aluminum. This leaves only the silicon particles exposed, and the pistons and rings ride on the silicon.

§ 10-8 CYLINDER HEAD

In the water-cooled engine, the cylinder head is cast in one piece from iron or aluminum. Aluminum is lighter and runs cooler than iron. However, most engines use the cast-iron head. In a V-8 engine, there are two cylinder heads.

A typical cylinder head for a V-8 engine is shown in Fig. 10-8. It is a complex casting. There are water jackets in the head, and there are also passages from the valve ports to the openings in the manifolds. The manifolds are described later. As you study Fig. 10-8, look for the following parts:

- Tapped holes for the spark plugs
- Water-passage holes
- Oil-passage holes
- Pushrod holes
- Tapped holes to attach the rocker-arm brackets
- Tapped holes to attach the manifolds
- Valve-guide holes to carry the valves
- Valve seats

§ 10-9 GASKETS

The joints between the cylinder block and the cylinder head or heads must be tight. They must hold in the high pressure of the burning air-fuel mixture. To ensure a good seal, gaskets are installed between the cylinder block and the cylinder head. Head gaskets are made

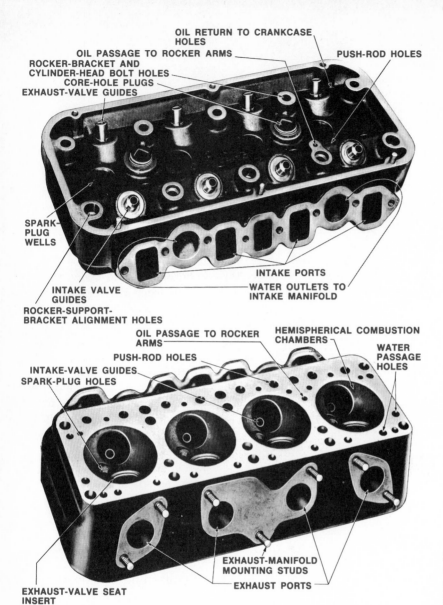

OIL RETURN TO CRANKCASE HOLES

OIL PASSAGE TO ROCKER ARMS

PUSH-ROD HOLES

ROCKER-BRACKET AND
CYLINDER-HEAD BOLT HOLES
CORE-HOLE PLUGS
EXHAUST-VALVE GUIDES

SPARK-
PLUG
WELLS

INTAKE PORTS

INTAKE VALVE
GUIDES
ROCKER-SUPPORT-
BRACKET ALIGNMENT HOLES

WATER OUTLETS TO
INTAKE MANIFOLD

OIL PASSAGE TO ROCKER
ARMS

HEMISPHERICAL COMBUSTION
CHAMBERS

PUSH-ROD HOLES

WATER
PASSAGE
HOLES

INTAKE-VALVE GUIDES
SPARK-PLUG HOLES

EXHAUST-MANIFOLD
MOUNTING STUDS

EXHAUST PORTS

EXHAUST-VALVE SEAT
INSERT

Fig. 10-8. Top and bottom views of a cylinder head for a V-8, overhead-valve engine. (*Chrysler Corporation*)

of thin sheets of soft metal or of asbestos and metal. They are shaped to fit the block, with all openings cut out.

When the engine is assembled, the head gasket is placed between the block and the head. Then the head bolts are installed and tightened. This squeezes the gasket between the head and the block, sealing the joint.

Gaskets are also used between the block and the other parts that are attached to the block. Figure 10-9 shows a set of gaskets for a six-cylinder engine. You can see the gaskets for the oil pan in Fig. 10-5. In this picture, the gaskets are shown in place on the oil pan, ready for oil-pan installation.

§ 10-10 INTAKE MANIFOLD

The intake manifold is an assembly of tubes through which the air-fuel mixture can flow

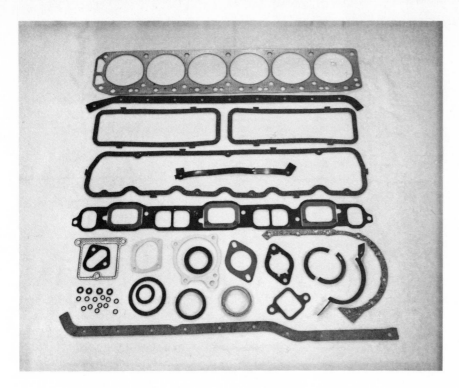

Fig. 10-9. Engine overhaul gasket set for a six-cylinder engine, showing gaskets and seals used in the engine. (*McCord Replacement Products Division of McCord Corporation*)

from the carburetor to the engine. The intake manifold for a six-cylinder engine is shown in Fig. 10-10. Note that it is fastened to the side of the cylinder head. You can see part of the intake manifold on the engine shown in Fig. 9-19. It is located just under the carburetor and has six pipes running from the carburetor to the cylinder head.

The intake manifold shown in Fig. 10-10 has only four passages. The outer two passages each serve two cylinders. Other intake manifolds have separate passages for each cylinder.

The intake manifolds for V-8 engines are more complicated. In these engines, the carburetor sits between the two banks of cylinders. The intake manifold is underneath the carburetor and is connected to both cylinder heads. The intake manifold must serve all eight cylinders; therefore, it has eight passages from the carburetor to the cylinder heads. Look at Fig. 10-8, which shows one cylinder head from a V-8 engine. Count the four intake ports in the top view.

Now look at Fig. 10-11, which shows a top view of an intake manifold for a V-8 engine. The flat surface to the left, with two holes in it, is where the carburetor is mounted. The carburetor used has two barrels. The two-barrel carburetor is like two separate carburetors. One barrel feeds half the cylinders and the other barrel feeds the other half. In this way, each cylinder gets the air-fuel mixture it needs. The arrows in Fig. 10-11 show how the air-fuel mixture flows from the two barrels to the eight cylinders.

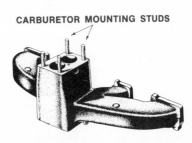

Fig. 10-10. The intake manifold for a six-cylinder, in-line engine.

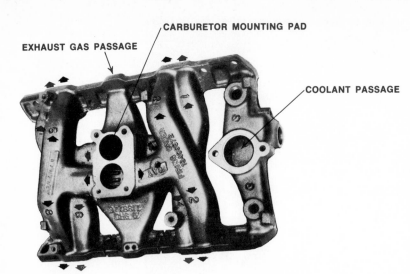

EXHAUST GAS PASSAGE

CARBURETOR MOUNTING PAD

COOLANT PASSAGE

Fig. 10-11. The intake manifold for a V-8 engine. The arrows show the directions in which the air-fuel mixture flows from the two barrels of the carburetor to the eight cylinders in the engine. The central passage connects the two exhaust manifolds. Exhaust gases flow through this passage during engine warm-up. (*Pontiac Division of General Motors Corporation*)

The center passage, marked "exhaust-gas passage," is open when the engine is cold. The hot exhaust gases, flowing through this passage, heat up the intake manifold, and this vaporizes the gasoline better. Thus the cold engine runs better. When the engine warms up, a heat-control valve closes to shut off the exhaust-gas passage. We take a more detailed look at the heat-control valve later.

There is one more thing we want to point out in Fig. 10-11. Look at the passage to the right. This is part of the cooling system that connects the cylinder-head water jackets and the hose to the radiator. When the engine is running, coolant circulates through this passage. The flat surface is for mounting the connection to the top of the radiator. We have more to say on this in Chapter 18.

§ 10-11 EXHAUST MANIFOLD

The exhaust manifold is an assembly of tubes carrying exhaust gases from the engine to the car's exhaust system. The exhaust manifold is attached to the side of the cylinder head. Figure 10-12 shows an exhaust manifold for a six-cylinder engine. This is mounted underneath the intake manifold, shown in Fig. 10-10. The heat-control valve for the six-cylinder engine is mounted in the exhaust manifold. You can see this valve in Fig. 10-12. When the en-

gine is cold, the heat-control valve sends the hot exhaust gases upward. They circulate around the base on which the carburetor is mounted. This puts heat in the air-fuel mixture coming from the carburetor. The heat causes the gasoline to evaporate and thus helps the engine to run better when it is cold. As the engine heats up, the heat-control valve automatically closes. When this happens, the hot exhaust gases go directly into the exhaust system without circulating around the mounting base of the carburetor. When the engine is hot, the extra heat is no longer needed to make the gasoline evaporate.

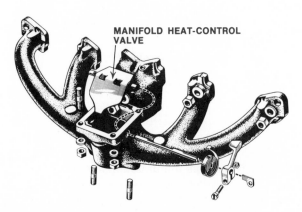

MANIFOLD HEAT-CONTROL VALVE

Fig. 10-12. The exhaust manifold for a six-cylinder, in-line engine. The heat-control valve and parts are shown in a disassembled view.

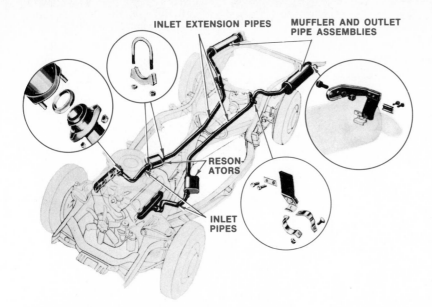

INLET EXTENSION PIPES

MUFFLER AND OUTLET PIPE ASSEMBLIES

RESONATORS

INLET PIPES

Fig. 10-13. A dual exhaust system for a V-8 engine. Each bank of cylinders has its own exhaust system. The circles show details of assembly and attachment. (*Ford Motor Company*)

§ 10-12 EXHAUST SYSTEM

The exhaust system carries the exhaust gases from the exhaust manifold to the tail pipe. The exhaust system is a system of pipes with a muffler. The muffler has a series of passages through which the exhaust gases must flow. This softens, or muffles, the sound of the engine.

Some V-8 engines have two separate exhaust systems, one for each bank, as shown in Fig. 10-13. The resonators shown in this picture are actually first-stage mufflers. They partly muffle the engine sounds. The other mufflers at the rear of the car complete the job.

Other V-8 engines have a single exhaust system. The two exhaust manifolds are connected by a crossover pipe, which is connected to the exhaust system.

§ 10-13 CRANKSHAFT AND ENGINE BEARINGS

The crankshaft is the backbone of the engine. It must take the hard downward thrusts of the pistons and connecting rods when the air-fuel mixture is burned in the cylinders. The crankshaft changes reciprocating action into rotary motion and sends it through gears and shafts

to the car wheels so that the car moves. The crankshaft hangs from the bottom of the cylinder block. It is supported in bearings that allow it to spin without excessive wear or friction. Now we will look at crankshafts and bearings.

§ 10-14 CRANKSHAFT

The crankshaft is made of high-strength steel (Fig. 10-14). It has to be strong enough to take the 2-ton thrust of each piston and connecting rod on every power stroke. Also, it must be balanced so that it will run smoothly without a lot of vibrations. The crankshaft also does another job—it distributes oil to the connecting-rod bearings.

Let's look at balance first. The crankshaft is composed of a series of cranks. These cranks, together with the connecting rods, change the reciprocating motion of the pistons into rotary motion. Now, look at a single crank. As it rotates, centrifugal action keeps pulling the crank away from the centerline of the crankshaft. This off-balance force could cause terrible vibration and very rapid wear of the bearings supporting the crankshaft. To provide balance and to avoid undue wear, the crankshaft has a series of counterweights. These weights balance the rotating forces of the

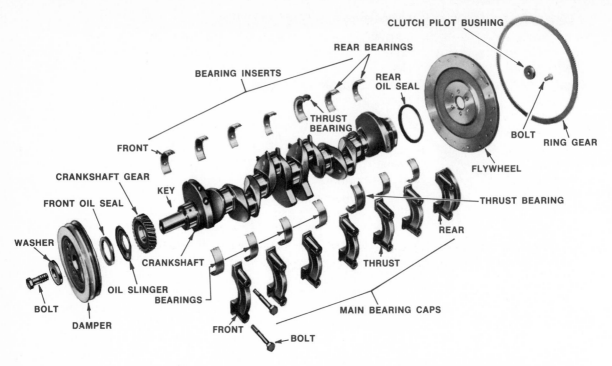

Fig. 10-14. A crankshaft and related parts for a six-cylinder engine. (*Ford Motor Company*)

cranks, so there is no force that tends to cause vibration. You can see the counterweights in Figs. 10-14 and 10-15.

§ 10-15 LUBRICATING THE CRANKSHAFT BEARINGS

As we explain in Chapter 19, the engine-lubricating system pumps oil to the moving parts in the engine. This oil covers all the moving parts so that they "float" on layers of oil. The

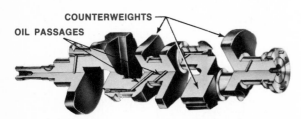

Fig. 10-15. A crankshaft cut away to show the oil passages drilled to the crankshaft journals.

layers of oil are very thin, but they are thick enough to prevent metal-to-metal contact. When metal-to-metal contact occurs in an engine, serious trouble can result. Metal rubbing on metal will scratch surfaces, waste power, overheat the metal, and can cause complete engine failure.

It is especially important to lubricate with oil the bearings in which the crankshaft turns. The **bearings** are the parts that bear the weight of the crankshaft. The bearings are replaceable. When they wear, they can be removed and new bearings can be installed. They are much less expensive to replace than the crankshaft or the cylinder block! Figure 10-14 shows a crankshaft and the bearings that support it. Note that the bearings are thin shells that fit into undercuts in the cylinder block and into the bearing caps.

The bearings supporting the crankshaft are called **main bearings.** Many mechanics just call them the "mains." The round sections of the crankshaft that rest on the bearings are called **journals.** They rotate, or "journey," in the bearings; thus the name "journal."

The cylinder block has oilholes—called oil galleries—drilled in it that carry the oil from the oil pump to the main bearings. You can see the locations of the oilholes in the pictures of the cylinder block in Fig. 10-4. Long holes, drilled from the front to the back of the block, are connected by short holes drilled in the half-round supports for the crankshaft. These short holes match the holes in the bearings that are installed in the support. Now you can see how the oil flows through the oil galleries to the holes in the bearings. The oil then flows through the bearing holes and onto the crankshaft journals rotating in the bearings. The crankshaft is therefore "floating" on a layer of oil. This allows the crankshaft to spin with relatively little effort. Little power is lost, and metal-to-metal contact is prevented.

§ 10-16 CRANKSHAFT OILHOLES

The crankshaft also has oilholes drilled in it. These oilholes have the job of keeping the connecting-rod bearings lubricated. The connecting rods are attached to the crankpins of the crankshaft by rod caps. Bearing shells are placed between the rods and the rod caps and the crankpins. These bearings must be kept well supplied with oil. The oilholes in the crankshaft do this job. You can see the oilholes drilled in the crankshaft in Fig. 10-16. These holes go from the main bearing journals to the

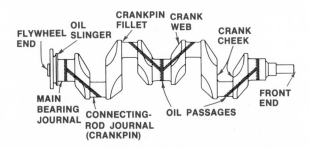

Fig. 10-16. This drawing shows how the oilholes drilled in the crankshaft feed oil to the main and connecting-rod journals.

crankpins. Some of the oil that is fed to the main bearings flows through these holes to the crankpins. It covers the crankpins and connecting-rod bearings, thus providing good lubrication.

§ 10-17 ENGINE BEARINGS

Now let's take a closer look at engine bearings. Bearings are needed at every point in the engine where there is rotary motion between engine parts. Take a look at Fig. 10-17, which shows the various places bearings are installed in an engine. These engine bearings are called **sleeve bearings.** They are shaped like a sleeve that fits around the rotating journal, or shaft. Some of the bearings are the split type; that is, the bearing is split into two halves. The main and connecting-rod bearings have to be the split type. Otherwise, they could not be placed on the crankshaft journals or crankpins. Figure 10-18 shows a typical bearing half with the various parts named. Some bearing halves have a groove—the annular groove—that helps distribute the oil around the bearing. The oilhole in the annular groove allows the oil from the holes in the cylinder block and crankshaft to feed through and onto the journal or crankpin.

As we have already mentioned, the crankshaft is hung from the bottom of the cylinder block by the main bearing caps. The upper bearing halves are installed in the cylinder-block supports. The lower bearing halves are put into the main bearing caps before the caps are bolted in place on the cylinder block.

The connecting rods are attached to the crankpins in a similar way. The upper bearing half is put into the rod. The lower bearing half is installed in the rod cap. Then the cap is bolted to the rod, with the crankpin between the cap and the rod.

§ 10-18 BEARING CONSTRUCTION

Each bearing half is made of a steel or bronze back to which layers of bearing material have

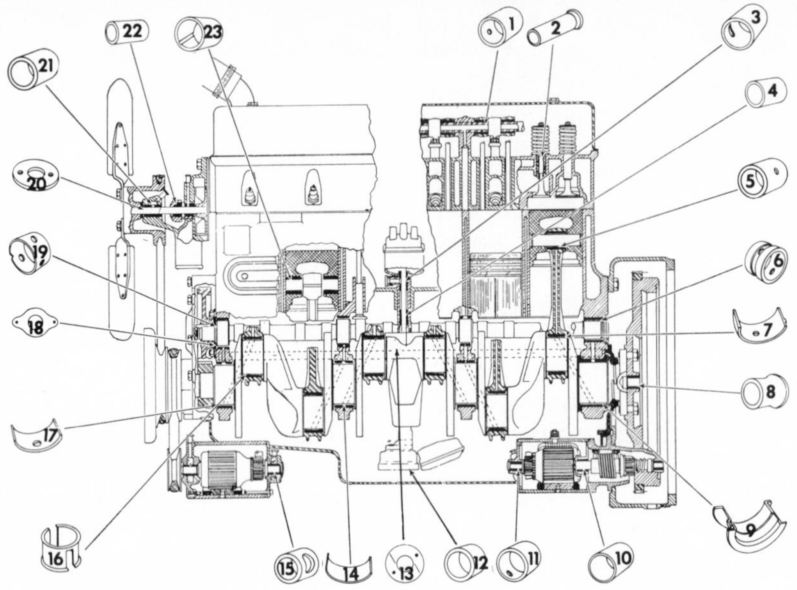

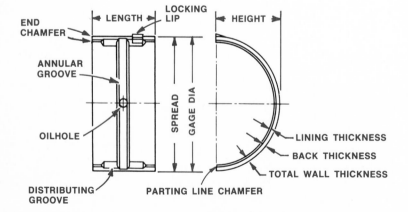

Fig. 10-17. Various bearings and bushings in a typical engine. (*Johnson Bronze Company*)

1 ROCKER-ARM BUSHING	13 DISTRIBUTOR THRUST PLATE
2 VALVE-GUIDE BUSHING	14 INTERMEDIATE MAIN BEARING
3 DISTRIBUTOR BUSHING, UPPER	15 ALTERNATOR BUSHING
4 DISTRIBUTOR BUSHING, LOWER	16 CONNECTING-ROD BEARING, FLOATING TYPE
5 PISTON-PIN BUSHING	17 FRONT MAIN BEARING
6 CAMSHAFT BUSHING	18 CAMSHAFT THRUST PLATE
7 CONNECTING-ROD BEARING	19 CAMSHAFT BUSHING
8 CLUTCH PILOT BUSHING	20 FAN THRUST PLATE
9 FLANGED MAIN BEARING	21 WATER-PUMP BUSHING, FRONT
10 STARTING-MOTOR BUSHING, DRIVE END	22 WATER-PUMP BUSHING, REAR
11 STARTING-MOTOR BUSHING, COMMUTATOR END	23 PISTON-PIN BUSHING
12 OIL-PUMP BUSHING	

Fig. 10-18. A typical bearing half with the parts named. Many bearings do not have annular and distributing grooves. (*Federal-Mogul Corp.*)

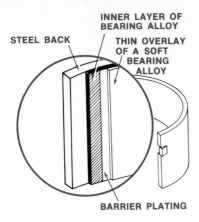

Fig. 10-19. The construction of a three-layer bearing. Some bearings have three layers, as shown; others have two layers. (*Federal-Mogul Corp.*)

been applied (see Fig. 10-19). The bearing material is relatively soft. If wear takes place, it will be the bearing that wears, rather than the much more expensive crankshaft or other moving part. It is relatively cheap and simple to replace worn bearings.

§ 10-19 BEARING LUBRICATION

We have already explained how the oil gets in between the bearing and the moving parts. Lubrication of the bearings allows the moving part to slide easily on layers of oil. The journal must be slightly smaller than the bearing to provide **oil clearance** (Fig. 10-20), which is about 0.001 inch [0.025 millimeter]. Oil feeds through the bearing oilhole into this clearance. The oil constantly flows across the bearing and then drops off the edges of the bearing. This movement of the oil not only cleans the bearings but also helps to keep them cool. As the oil moves across the bearing, it picks up heat as well as particles of dirt. The oil drops into the oil pan, where it loses heat. The oil pump continuously sends oil through the oil filter to the engine parts. The filter removes dirt from the oil. The oil then goes back up through the engine. This continuous circulation of oil allows the oil to both clean and cool the engine.

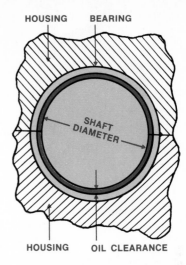

Fig. 10-20. The oil clearance between the bearing and shaft journal.

§ 10-20 THRUST BEARING

The crankshaft has to be prevented from moving endwise. Normal engine operation tends to push the crankshaft forward and backward. To prevent excessive forward and backward movement, one of the main bearings—called the **thrust bearing**—has flanges on its two sides, as shown in Fig. 10-21. In the engine assembly, these flanges are a close fit to the two sides of the crankpin. If the crankshaft tends to shift one way or the other, the crankpin sides come up against the flanges, preventing excessive endwise movement. You can

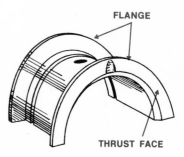

Fig. 10-21. A crankshaft thrust bearing. (*Federal-Mogul Corp.*)

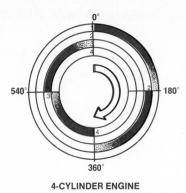

4-CYLINDER ENGINE

Fig. 10-22. Power impulses in a four-cylinder engine. The complete circle represents two full crankshaft revolutions, or 720 degrees. Less power is delivered toward the end of the power stroke because cylinder pressure falls off. This is shown by the lightening of the shaded areas that show the power impulses.

see a thrust bearing in the picture of a crankshaft and its related parts (Fig. 10-14).

§ 10-21 FLYWHEEL

In the engine assembly, a flywheel is attached to the back end of the crankshaft. The flywheel does two jobs (three jobs in some engines): It acts as a stabilizer, and it works with the starting motor to start the engine. A flywheel is shown to the right in Fig. 10-14.

First, the flywheel is a stabilizer. It smooths out the power impulses from the pistons. When a power stroke starts, the push is very strong. The pressure from the burning air-fuel mixture is at the maximum. However, as the piston moves down, the pressure drops. Toward the end of the power stroke, the pressure is only a small fraction of what it was at the beginning. This means that the push on the crankpin starts high but soon tapers off.

The power strokes in an engine follow one another in regular order. The whole sequence for a four-cylinder engine is shown in Fig. 10-22. Although the process is shown in a single circle, we know it takes two revolutions of the crankshaft to produce a complete cycle.

First, cylinder 1 fires and its piston moves down on its power stroke. This movement is shown by the colored space curving down to the right from the top of the circle numbered 1. Notice that the color thins out as it approaches the end of the power stroke (which is marked to the right as 180 degrees). It thins out because the power stroke weakens toward the end.

Then cylinder 2 fires and its piston moves down on the power stroke. The power stroke in cylinder 4 follows, and then cylinder 3 fires. The order in which the cylinders *fire* is called the **firing order.** In Fig. 10-22, the firing order is 1-2-4-3.

If you study Fig. 10-23, you will see that the flow of power is not steady. Each power stroke starts off strong and then fades. If there were no flywheel, you would feel this unsteadiness in the engine. In other words, you would feel a sort of vibration as the separate power strokes hit the crankshaft. The flywheel helps to smooth out these power surges. When a power stroke starts, the engine tends to speed up. Then the engine tends to slow down toward the end of a power stroke. The flywheel resists this speed-up–slow-down action.

Any rotating wheel, including the flywheel, resists any effort to change its speed of rotation. When the engine tends to speed up, the flywheel resists it. When the engine tends to slow down, the flywheel also resists it, and it is this resistance that serves to smooth out the power surges.

In the six-cylinder and eight-cylinder engines, the power impulses overlap, as shown in Fig. 10-23. Even in these engines, however, the power impulses do not add up to a perfectly smooth engine. The flywheel helps to smooth out the engine so that the power impulses are less noticeable.

The second job of the flywheel is to work with the starting motor to start the engine. The starting motor is an electric motor that cranks the engine to get it started. This action is described in Chapter 12. There is a ring gear on the flywheel (see Fig. 10-14). When the starting

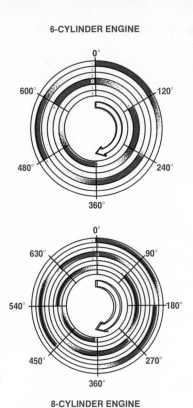

6-CYLINDER ENGINE

8-CYLINDER ENGINE

Fig. 10-23. Power impulses in six- and eight-cylinder engines. The complete circles represent two full crankshaft revolutions, or 720 degrees. Note the power overlap.

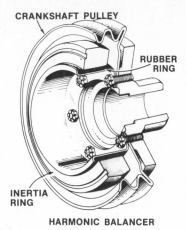

Fig. 10-24. A partial cutaway view of a torsional vibration damper.

motor works, it meshes a small driving gear with this ring gear.

A third job of the flywheel on some cars that do not have automatic transmissions is to form part of the clutch. In a car with a manual transmission, you have to shift the gears by hand. The clutch is the mechanism that allows you to shift. We take a more detailed look at the clutch in Chapter 28.

§ 10-22 VIBRATION DAMPER

The power impulses from the power strokes hit the crankshaft in one place and then another as the engine runs. Every time a power impulse hits a crankpin, that part of the crankshaft actually twists a little. Then, when the power stroke ends, that part of the crankshaft untwists. This action sets up an oscillating—or back and forth—motion. This motion is called **torsional vibration.** That is, it is a twist-untwist vibration. If it is not controlled, torsional vibration can build up enough at some speeds to break the crankshaft.

Torsional vibration is controlled by a **vibration damper** (Fig. 10-24), also called a **harmonic balancer.** The vibration damper has a small damper flywheel, which is bonded to the crankshaft pulley by a rubber ring. The damper flywheel is also called the inertia ring. The vibration damper works in the following way: When the twist-untwist action starts, the inertia ring has a dragging effect. It works through the rubber ring to hold the twist action down. Then, when the untwist motion starts, the inertia ring holds this action down too. The result is that torsional vibration is kept to a minimum.

The pulley part of the vibration damper is connected by belts to drive the water pump and the engine fan. The belts also drive various accessories, such as the alternator, the power-steering pump, and the air-conditioner compressor.

Review questions

1. What is the name of the spaces that surround the cylinders and through which the coolant flows?
2. What is attached to the top of the cylinder block? To the bottom?
3. What is the name of the part that is installed between the cylinder block and the cylinder head to seal the joint?
4. How is heat sent into the intake manifold when the engine is cold?

5. Why is it important to prevent the crankshaft journals from making metal-to-metal contact with the main bearings?
6. How does oil get to the crankshaft journals?
7. What is the purpose of the thrust bearing?
8. What is the name for the order in which the power strokes take place?
9. What is the purpose of the vibration damper?

Study problems

1. Locate a bare cylinder block that you can examine. Identify the water jackets and the core clean-out holes. Inspect each machined surface and determine what part should be attached to each surface. Find out if the cylinder block is made of cast iron or aluminum. Check cylinder-head surfaces to determine if the cylinders are sleeved.

2. Do you know a real master mechanic?—a person highly skilled in repairing automobiles and highly regarded by the community for having that skill? If so, ask the master mechanic to tell you what is the best engine and how it should be made. Ask if the cylinder block or head should be cast iron or aluminum. Ask what type of cylinder arrangement and valve arrangement should be used, and so on.

11 AUTOMOTIVE ENGINE CONSTRUCTION—PISTONS, CONNECTING RODS, VALVES, AND VALVE TRAINS

§ 11-1 PISTONS AND CONNECTING RODS

The job of the piston and connecting-rod assembly is to carry the high pressure of the burning air-fuel mixture through the connecting rod to the crankshaft. This pressure spins the crankshaft so that the car moves. The piston and the connecting rod are described in Chapter 5. In this chapter we take another, closer look at the piston, connecting rod, and valves.

§ 11-2 CONNECTING ROD

The connecting rod (Fig. 11-1) is made of high-strength steel. Its purpose is to connect the piston to the crankshaft. It is attached at one end to a crankpin on the crankshaft and at the other end to a piston. The end of the connecting rod that is connected to the crankpin is often called the **rod big end.** The other end, connected to the piston, is called the **rod small end.**

The connecting rod is attached to the crankpin with a rod cap, bolts, and nuts, as shown in Fig. 11-1. Two bearing halves are installed in the rod and cap before the assembly is completed. These bearing halves, and the crankpin they surround, are oiled through oilholes drilled in the crankshaft.

Some connecting rods have oilholes drilled up through them. Oil flows through these holes up to the bearing or bushing at the piston end

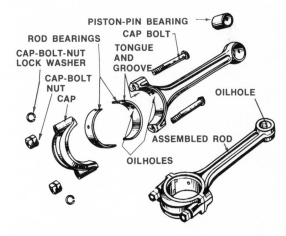

Fig. 11-1. Connecting rod with bearings and bearing cap in disassembled view (*top*) and assembled view (*bottom*). (*Chrysler Corporation*)

of the connecting rod. You can see the oilholes in Fig. 11-1.

The connecting rod is attached to the piston with a piston pin. The piston pin goes through two holes in the piston and one hole in the connecting rod. Three methods of attaching the piston and rod are shown in Fig. 11-2.

In one method, the piston pin is locked to the connecting rod. The most widely used version of this method is to **press fit** the pin in the rod (Fig. 11-2A). That is, the fit is so tight that you have to press the pin through the hole in the rod, using high pressure. The press fit holds the pin in place, centered in the con-

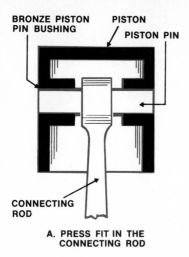

A. PRESS FIT IN THE CONNECTING ROD

BRONZE PISTON PIN BUSHING — PISTON — PISTON PIN — CONNECTING ROD

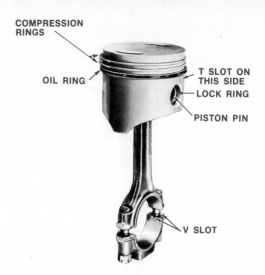

Fig. 11-3. A piston and connecting-rod assembly. This type has lock rings to hold the piston pin in position in the piston and connecting rod. (*Chrysler Corporation*)

COMPRESSION RINGS — OIL RING — T SLOT ON THIS SIDE — LOCK RING — PISTON PIN — V SLOT

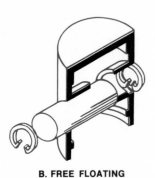

B. FREE FLOATING

C. LOCKED TO PISTON

Fig. 11-2. Three piston-pin arrangements.

necting rod. The two ends of the pin rest in the holes in the piston into which bushings are sometimes installed. When the connecting rod tilts one way or the other, the piston pin turns with it. A film of oil on the piston or bushings allows the turning action of the piston pin.

A second method of attaching the piston and rod with the piston pin is shown in Fig. 11-2B. Here, the piston pin is not locked to either the piston or the connecting rod. It is free to turn in both parts. A pair of lock rings is used, one at each end of the pin. They keep the pin centered. The lock rings fit into undercuts in the piston. An assembly with this arrangement is shown in Fig. 11-3.

A third method is rarely used. The piston pin is locked to the piston with a lock bolt (Fig. 11-2C). The connecting rod has a bushing, which allows the connecting rod to rock back and forth on the locked piston pin. The bushing is covered with oil so that the connecting rod can move easily on the piston pin. Another variation of this method is to use a lock bolt to lock the piston pin in the connecting rod.

§ 11-3 PISTON-PIN LUBRICATION

Whether or not the piston pin turns in the piston, in bushings, or in both, it must be lubricated. In some engines the connecting rod has a hole drilled through it so that oil can flow from the crankpin up to the piston pin (Fig. 11-1).

In most engines the piston pin and its bushings get lubricated from oil thrown off the rotating crankpin and rod big end. Some of this oil splashes onto the cylinder walls, where it lubricates the walls and also the piston and piston rings. This oil is scraped off the cylinder walls by the piston rings. Some of the oil that is scraped off gets to the piston pin.

§ 11-4 CONNECTING-ROD BALANCE

All connecting rods in an engine must be of equal weight. If one were heavier than the others, an out-of-balance condition would result and produce vibration. This vibration could soon damage the engine seriously.

When an engine is first assembled in a factory, rods and caps are matched. Therefore, if you tear down an engine for a service job, never mix rod caps. For example, never put the rod cap for cylinder 2 on the rod for cylinder 1. If you do, the chances are that you will get a poor bearing fit, and bearing failure will result after a few hundred miles of operation.

§ 11-5 PISTONS

Pistons for automobile engines are made of aluminum; they are about 4 inches [101.6 millimeters] in diameter and weigh about 1 pound [0.454 kilograms]. They are a loose, sliding fit in the engine cylinders. A piston is shown attached to the connecting rod in Fig. 11-3. Figure 11-4 shows all the parts separated. The assembly shown in Fig. 11-4 is the type in which the piston pin is a press fit in the connecting rod. The piston has two bearing surfaces in which the piston pin can rock as the connecting rod rocks back and forth.

Notice that the lower edge of the piston is cut away so that the piston skirt is short on the sides. This type of piston is called a **slipper** piston. The piston skirt is shortened on two sides for two reasons. One reason is that it lightens the weight of the piston. A light piston

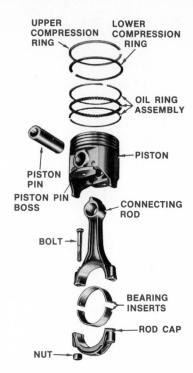

Fig. 11-4. A piston and connecting-rod assembly disassembled so that the various parts can be seen. The oil-control ring is of the three-piece type. The piston is the slipper type, which has the skirt partly cut away. (*Ford Motor Company*)

moves up and down in the cylinder more easily and wastes less power. A lighter piston is also easier on the bearings.

The second reason for the cutaway piston skirt is that it allows a shorter (from top to bottom) engine. Figure 11-5 shows why the cutaway piston skirt allows the shorter engine. The crankshaft has to have counterweights, as explained in Chapter 5. Cutting away the piston skirt allows the connecting rod to be made shorter but leaves enough room between the counterweights and the piston, as shown in Fig. 11-5.

§ 11-6 PISTON CLEARANCE

There must be some space between the piston and the cylinder wall. This space, called the **piston clearance,** allows the piston to move up

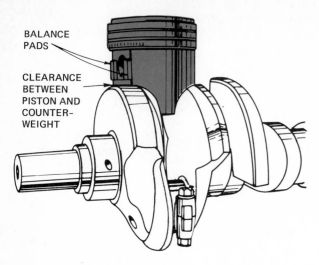

BALANCE
PADS

CLEARANCE
BETWEEN
PISTON AND
COUNTER-
WEIGHT

Fig. 11-5. Modern slipper piston and connecting rod assembled to the crankshaft. Note the small amount of clearance between the piston and the counterweights on the crankshaft. (*Chevrolet Motor Division of General Motors Corporation*)

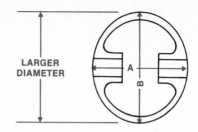

LARGER
DIAMETER

Fig. 11-6. A cam-ground piston viewed from the bottom. When the piston is cold, its diameter at A (at the piston-pin holes) may be from 0.002 to 0.003 inch [0.05 to 0.08 millimeter] less than at B. (*Chrysler Corporation*)

and down easily in the cylinder. The clearance is somewhere between 0.001 and 0.004 inch [0.025 and 0.102 millimeter] in most engines.

As cylinder walls wear, the clearance increases. When the clearance gets excessive, the piston fits too loosely, and **piston slap** occurs. In other words, there is so much clearance that the piston shifts from one side of the cylinder wall to the other when the power stroke starts. As the piston hits the other side of the cylinder wall, it gives off a hollow, bell-like sound. When you hear this sound, you know the engine is in need of repair.

§11-7 EXPANSION CONTROL IN PISTONS

A piston expands when it gets hot. Any metal expands with heat. The cylinder walls expand, too, but not nearly as much as the piston. If the piston were perfectly round, it could expand so much that it would "seize." This means that the piston would expand so much

that all clearance would be gone. The piston would jam in the cylinder and stop moving. The connecting rod and crankshaft would continue to move, and the connecting rod or rod bolts would break. The chances are that part of the rod would go through the cylinder block. The result would be a ruined engine.

To prevent this from happening, pistons are built with expansion control. One method of building pistons with expansion control is to make them slightly oval in shape. These slightly oval pistons are called **cam-ground** pistons because they are finished on a machine that uses a cam. The cam moves the piston toward and away from the grinding wheel as the piston is revolved. A cam-ground piston is shown in Fig. 11-6. The diameter of the piston at the piston-pin holes (A in Fig. 11-6) is less than the diameter 90 degrees from the holes (B in Fig. 11-6).

§11-8 PISTON-PIN OFFSET

In many engines, the piston pin is offset to one side (Fig. 11-7). It is offset to minimize the effect of the sudden heavy push on the piston at the start of the power stroke. This sudden heavy push that occurs as the piston starts down tends to slam the piston to one side of the cylinder wall. By offsetting the piston pin to one side a few thousandths of an inch, the sudden slamming effect is prevented.

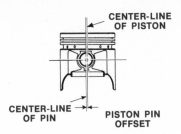

Fig. 11-7. The piston pin is offset to one side to minimize the effect of the sudden push on the piston at the start of the power stroke.

§ 11-9 PURPOSE OF PISTON RINGS

The piston rings provide a good seal between the piston and the cylinder wall. Do you remember what you read about piston clearance? There must be enough clearance so that the piston can slide up and down easily in the cylinder without danger of seizure.

If it were not for the piston rings, piston clearance would allow much of the pressure from the burning air-fuel mixture to blow by the piston and escape. **Blow-by** is the name given to the burned gases that pass the piston and flow down into the crankcase. When burned gases blow by the piston, power is lost because the burned gases cannot add to the push on the top of the piston. Blow-by can also cause engine troubles, as you will learn later.

In addition to providing a seal between the piston and cylinder wall, the piston rings do a second job. They scrape off excessive oil from the cylinder walls and return it to the oil pan. This action keeps oil from working its way up into the combustion chamber, where it would burn. Oil that is burned in the combustion chamber leaves a carbon deposit that can clog valves and piston rings as well as short out spark plugs. Both these conditions cause engine trouble.

Actually, there are two types of piston rings. One type fights blow-by. The other type scrapes oil off the cylinder walls. The blow-by fighters are called **compression rings.** The oil-scraper rings are called **oil-control rings.** We will look at both types of rings in detail.

§ 11-10 RING JOINT GAP

The piston rings are made a little larger than the diameter of the cylinder wall. They are cut at one point, as shown in Fig. 11-8. The cut allows the ring to expand slightly so that it can be slipped over the head of the piston. It can then be slid down into the ring groove that has been cut in the piston. Figure 11-3 shows the rings in place in the piston grooves.

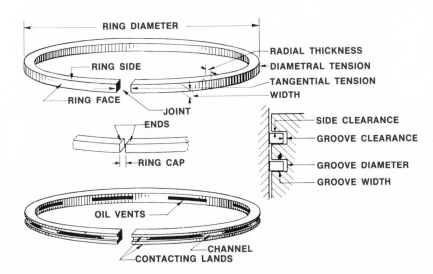

Fig. 11-8. A compression ring (*top*) and an oil-control ring (*bottom*). (*Sealed Power Corporation*)

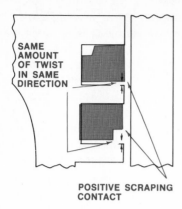

SAME AMOUNT OF TWIST IN SAME DIRECTION

POSITIVE SCRAPING CONTACT

Fig. 11-9. Action of counterbored and scraper compression rings during an intake stroke. Internal forces of the rings tend to twist them so that they put a sharp corner on the cylinder wall. This scrapes the wall clean of any excess oil that has worked up past the oil-control ring. (*Perfect Circle Division of Dana Corporation*)

§ 11-11 COMPRESSION RINGS

Compression rings are made of cast iron. They are not just plain rings, but have a corner cut out, as shown in Fig. 11-9. The purpose of this is to give the rings a slight twist. This causes the rings to put a sharp corner against the cylinder wall on the intake stroke. The sharp corner is more effective in scraping oil off the cylinder wall. The oil that is scraped off returns to the oil pan rather than working its way up into the combustion chamber to be burned.

You can see in Fig. 11-9 how the slight twist of the compression rings puts sharp corners against the cylinder wall.

Now, let's see how the compression rings hold in the high pressure of the burning air-fuel mixture and thus fight blow-by. When the high pressure occurs, the pressure works around behind the rings, as shown in Fig. 11-10. The pressure presses the rings down against the lower sides of the ring grooves. The pressure also presses the rings out against the cylinder wall. This provides a good seal at both the cylinder wall and at the lower side of the ring groove. Most of the pressure that gets by the upper ring is caught and contained by the lower ring. The two rings, working together,

hold in the combustion pressures and, in a good engine, hold blow-by to a minimum.

§ 11-12 OIL-CONTROL RINGS

The compression rings, as we have seen, help keep oil from getting into the combustion chamber. However, they can do only part of the job. The main job is done by the oil-control ring, which is below the compression rings (Fig. 11-3). The oil gets on the cylinder walls because oil is thrown off the rod-big-end bearings. Far more than enough oil gets on the cylinder walls. Most of the oil must be scraped off.

Figure 11-8 shows an oil-control ring. It is a cast-iron ring with a channel and oil vents cut in it. Oil that the ring scrapes off the cylinder wall passes through these vents and through holes in the back of the ring groove in the piston. The oil then returns to the oil pan. Some of the oil gets onto the piston-pin bearing surfaces and lubricates them.

A later type of oil-control ring consists of three parts. This ring is made up of two thin rings, or **rails,** and an expander ring between them, as shown in Fig. 11-4. You can see how they fit together in Fig. 11-11. The expander ring pushes the two rails up against the two

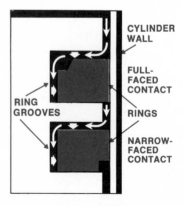

CYLINDER WALL

FULL-FACED CONTACT

RING GROOVES

RINGS

NARROW-FACED CONTACT

Fig. 11-10. Action of counterbored and scraper compression rings during a power stroke. The combustion pressure presses the rings against the cylinder wall with full-face contact, thus forming a good seal. (*Perfect Circle Division of Dana Corporation*)

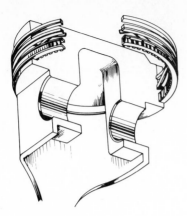

Fig. 11-11. Cutaway views of rings and piston to show construction. The second compression ring has an inner tension ring. The oil-control ring is a three-piece ring consisting of an expander spacer and two rails. (*TRW Inc.*)

sides of the ring groove. At the same time, the expander ring pushes the two rails out against the cylinder wall. The arrows in Fig. 11-12 show the direction of these pushes. The two rails do a better job of scraping the oil off the cylinder wall.

Some engines—especially older engines—have two oil-control rings per piston. Because of better design and better rings, most modern engines need only one oil-control ring per piston.

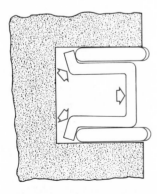

Fig. 11-12. The action of the expander spacer, shown by the arrows, forces the rails out against the cylinder wall and up and down against the sides of the ring groove. (*Perfect Circle Corporation*)

§ 11-13 EFFECT OF ENGINE SPEED ON OIL CONTROL

The higher the engine speed, the poorer the job the rings do in controlling the oil. At high speed, more oil is pumped through the lubricating system. More oil gets on the cylinder walls. The rings are moving faster and have less time to do their scraping job. The result is that more oil gets past the rings and into the combustion chamber.

§ 11-14 EFFECT OF ENGINE WEAR ON OIL CONTROL

As an engine gets older, its parts wear. The cylinder walls wear unevenly, with more wear at the top, where the combustion pressures are greatest. It is important to remember that the higher the combustion pressure, the harder the rings are pressed against the cylinder wall (see Fig. 11-10). Uneven wear of the cylinder wall makes it harder for the rings to maintain good contact against the wall. The rings have to change in size, expanding as they move up into the area of maximum wear. The result is that the rings do a poorer job of scraping off oil. Thus, more oil works up into the combustion chamber, where it burns.

Worn cylinders are not the only cause of burning oil. We will find out about the other causes in later chapters.

§ 11-15 OIL CONSUMPTION

The amount of oil that an engine uses depends on engine speed and engine wear, as we have said. All engines burn some oil. A new engine operated at moderate speed would probably not require any additional oil between oil changes. An old engine that is operated at high speed could require a quart of oil every few hundred miles.

§ 11-16 REPLACEMENT RINGS

When cylinder walls are worn and old rings are not doing a very good job, good performance can sometimes be restored by installing a set of special rings. These rings are sometimes called **severe** rings because they have more tension than regular rings. Their greater tension allows them to follow the changes in the size of the cylinder walls as they move up and down. Thus, they can better control the oil. Their ability to follow the changing size of the cylinder wall also reduces blow-by, which results from worn cylinder walls.

Installing severe rings is not the best solution to the problem of worn cylinder walls. A better solution is to hone or bore the cylinders to a larger size and then to install oversize pistons and rings. How this is done is explained in Chapter 37.

§ 11-17 VALVES AND VALVE TRAINS

Valves and valve trains are discussed briefly in Chapter 9. In the following sections, the I-head and the overhead-camshaft valve trains are discussed in greater detail. The L-head valve train is not used in today's automobiles.

§ 11-18 I-HEAD VALVE TRAIN

As is explained in Chapter 9, the I-head, or overhead-valve, valve train uses pushrods and rocker arms to operate the valves. The camshaft is driven by a chain or gears from the crankshaft. As the camshaft rotates, the cam lobes on the cams raise the valve lifters. The valve lifters push up on the pushrods. The rocker arms rock, and the valves are pushed down off their seats. In other words, the valves open when the cam lobes move up under the valve lifters.

§ 11-19 DRIVING THE CAMSHAFT

The camshaft is driven by gears or by sprockets and chain. Figure 11-13 shows the chain-and-sprocket arrangement for a six-cylinder engine. Many four-cylinder and six-cylinder engines use a pair of gears, as shown in Fig.

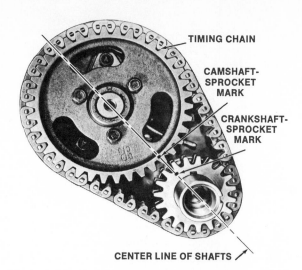

Fig. 11-13. Crankshaft and camshaft sprockets with chain drive for a six-cylinder engine, showing timing marks on sprockets. Note that the larger of the two sprockets is on the camshaft so that it turns at one-half crankshaft speed. (*Chrysler Corporation*)

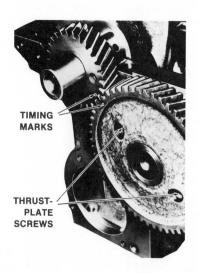

Fig. 11-14. Crankshaft and camshaft gears for a six-cylinder engine. Note the timing marks on the gears. (*Buick Motor Division of General Motors Corporation*)

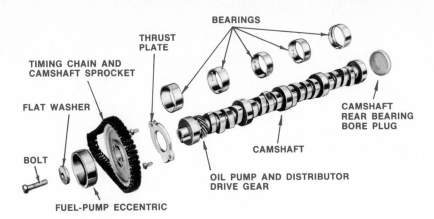

BEARINGS

THRUST PLATE

TIMING CHAIN AND CAMSHAFT SPROCKET

FLAT WASHER

BOLT

FUEL-PUMP ECCENTRIC

CAMSHAFT REAR BEARING BORE PLUG

CAMSHAFT

OIL PUMP AND DISTRIBUTOR DRIVE GEAR

Fig. 11-15. The camshaft and related parts for a V-8 engine. (*Ford Motor Company*)

11-14. In four-cycle engines the camshaft must turn only once for every two revolutions of the crankshaft. Each intake valve and each exhaust valve opens once every complete cycle. One complete cycle takes four piston strokes, or two revolutions of the crankshaft. The camshaft gear or sprocket is twice as big as the crankshaft gear or sprocket. Thus, it takes two crankshaft revolutions to turn the camshaft once.

The camshaft turns in sleeve bearings or bushings in the cylinder block. Figure 11-15 shows a camshaft with the bearings and other parts for a V-8 engine. The bearings are larger than the cams on the camshaft so that the cams can pass through the bearings when the camshaft is installed in the cylinder block.

The chain in Fig. 11-13 is called the **timing** chain. The timing chain and sprockets determine when the valves open and close; that is, they "time" the valve action.

§ 11-20 INTAKE-VALVE TIMING

Until now, we have been saying that the intake valve opens when the intake stroke starts and closes when the intake stroke ends. We have also been saying that the exhaust valve opens when the exhaust stroke starts and closes when the exhaust stroke ends.

This is not exactly true. Actually, the intake valve starts to open before the intake stroke starts. And it stays open for some time after

the intake stroke ends and the compression stroke starts. This gives the air-fuel mixture more time to enter the cylinder.

Valve timing depends on the shape of the lobe on the cam that operates the valve. It also depends on the relationship between the gears or chain and sprockets on the crankshaft and the camshaft.

§ 11-21 EXHAUST-VALVE TIMING

As we have mentioned, the exhaust valve starts to open before the power stroke ends and stays open after the exhaust stroke ends. This gives the exhaust gases more time to get out of the cylinder.

The early opening of the exhaust valve does not waste power, as you might think. By the time the exhaust valve opens, the pressure in the cylinder has already dropped. The purpose of getting the exhaust valve open early is to start the exhaust gases moving out.

Also, the exhaust valve does not completely close until a few degrees after TDC on the intake stroke. The upward push of the piston during the exhaust stroke has started the exhaust gases moving out. This upward movement toward the exhaust-valve port continues even after the piston has reached TDC and is starting down on the intake stroke. So, the reason for leaving the exhaust valve open just a little longer is to give most of the remaining exhaust gases a chance to get out also.

§ 11-22 VALVE OVERLAP

As we have explained, the intake valve opens before the piston reaches TDC on the exhaust stroke. Also, the exhaust valve does not close until after the piston starts moving down the cylinder on the intake stroke. This period, during which both valves are open, is called **valve overlap** and is measured in degrees. Changing the amount of valve overlap can be used to help reduce the amount of polluting gases in the exhaust.

§ 11-23 VALVE GUIDES

The valves slide up and down in valve guides in the cylinder head. In some engines, the valve guides are tubes made of special metal installed in the cylinder head. In other engines, the valve guides are integral with the cylinder head, which means that they are holes bored into the cylinder head. The cylinder head for a V-8 engine, shown in Fig. 10-8, has valve guides that have been installed separately. You can see these valve guides in the upper view of the cylinder head. The valve guides must provide a close fit with the valve stems but must also permit the valves to move up and down easily.

§ 11-24 VALVES

As we have mentioned, each cylinder has two valves: an intake valve and an exhaust valve (Fig. 11-16). The valve face must closely match the valve seat. In other words, the two must be in full contact all around the seat. This match is necessary for two reasons: to keep the combustion pressure from leaking out between the valve face and the valve seat and to help conduct heat away from the valve head.

Let's see why it is necessary to conduct heat away from the valve head. When the engine is running, the exhaust valve gets very hot. It is repeatedly passing the hot exhaust gases, and it can reach a temperature of more than 1000°F [538°C]. Some of the heat is passed down the valve stem to the valve guide. The

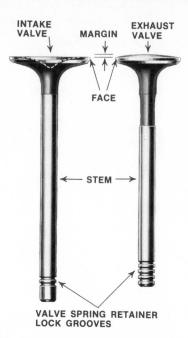

Fig. 11-16. Typical engine valves. (*Chrysler Corporation*)

valve guide is cooled by the coolant circulating in the cylinder head. But much more of the heat is passed from the valve head to the valve seat when the valve is closed. Remember, the exhaust valve is closed nearly three-quarters of the time. With a good match between the valve face and the valve seat, the maximum amount of heat can be passed to the seat. The seat, being in the cylinder head, is cooled by the engine's cooling system.

If the contact between the valve face and the valve seat is poor, less heat will be passed to the seat, and the valve will run hotter. Also, hot exhaust gases will leak out between the valve and the seat at the points of poor contact. The escaped hot exhaust gases will heat the valve still more and may even burn the valve and the seat. Poor contact, or poor seating, can be caused by dirt or other things.

§ 11-25 VALVE SEATS

In many engines, the valve seats are integral with the cylinder head. In other words, the valve seats are machined circles bored in the

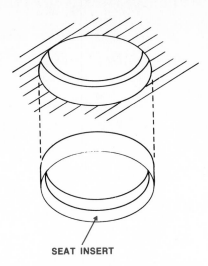

Fig. 11-17. An exploded view of a valve-seat insert in the cylinder head. The insert is indicated by the arrow.

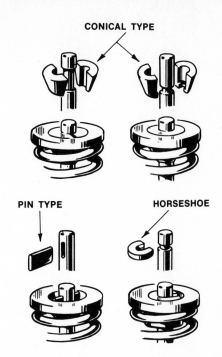

Fig. 11-18. Types of valve-spring-retainer locks, or keepers.

cylinder head. They are carefully ground to match exactly the angle of the valve face. Also, they are carefully centered with the valve guides so that the valves will seat evenly all around.

In some engines the exhaust-valve seats are metal rings, called valve-seat inserts, that are inserted in the cylinder head. Figure 11-17 shows a valve-seat insert in place in a cylinder head. The seat insert is made of a special metal that can hold up under the heat and pounding it gets in the engine.

In most engines the exhaust-valve seats are integral with the cylinder head. A special process called **induction hardening** is used to harden the valve seats in the cylinder head. This improves their wearing ability.

§ 11-26 VALVE-SPRING-RETAINER LOCKS

The valve spring is held onto the end of the valve stem by a spring retainer and a lock. The retainer is a large washer, and the lock is a small piece of steel that fits into or onto the valve stem. Three types of valve-spring-retainer locks are shown in Fig. 11-18. Once the lock

is put into place on the valve stem and the spring retainer is released, the retainer holds the lock in place. This maintains valve-spring pressure on the valve stem. The lock prevents the spring retainer from moving farther out toward the end of the valve stem.

§ 11-27 OIL SEALS AND SHIELDS

Oil flows up from the oil pump to lubricate the rocker arms and valve stems. Usually, there is so much oil in the cylinder head that the valve stems must be protected from the excessive oil. Without protection, oil could work its way down the valve guide, past the valve stem, and into the combustion chamber. Too much oil in the combustion chamber could foul the spark plugs and piston rings. The excessive oil on the valve stems, combined with the high temperature, could cause gum to form on the valve stems. If this happened, the valves would not open and close properly. They could stick in

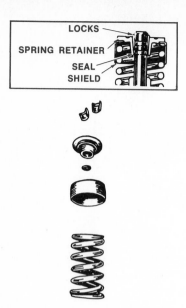

Fig. 11-19. Disassembled and sectional views of a valve-and-spring assembly with oil seal and shield.

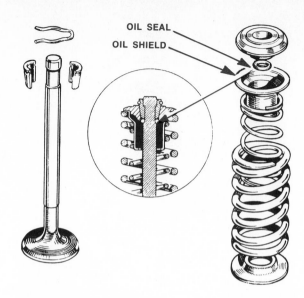

Fig. 11-20. In another arrangement, disassembled and sectional views of a valve-and-spring assembly with oil seal and shield. (*MG Car Company, Limited*)

a partly open position. Also, oil could get on the valve seats, and the oil would burn. This would prevent proper valve seating. With either condition, the valves would overheat and burn.

To protect the valve stems from excessive oil, oil seals and shields are used. Figure 11-19 shows one arrangement. The shield is a rubber cup that fits around the outside of the spring. It prevents excessive amounts of oil from being thrown through the spring and onto the valve stem. The seal is a small rubber ring that fits around the valve stem at the bottom of the spring retainer. The seal keeps oil from seeping down the valve stem. The seal-and-shield combination keeps excessive oil off the valve stem.

Figure 11-20 shows another type of oil seal and shield. In this arrangement, the oil shield is placed inside the valve springs.

§11-28 VALVE SPRINGS

In many engines, each valve uses only one spring. In other engines, two valve springs are used per valve; one spring is inside the other (see Fig. 11-20).

§11-29 ROCKER ARMS

There are several types of rocker arms. One type, shown in Fig. 11-21, is mounted on a shaft that is attached to the cylinder head. Figure 11-22 shows similar rocker arms mounted on a shaft on the cylinder head. The rocker arm in Fig. 11-21 has a screw and a lock nut. The purpose of the screw and lock nut is to allow for valve-clearance adjustments. The valve must close completely every time. Otherwise, hot combustion gases will flow between the valve seat and the valve face, burning both of them. The result: an expensive service job or a ruined engine.

Fig. 11-21. One type of rocker arm used in overhead-valve engines. (*Chevrolet Motor Division of General Motors Corporation*)

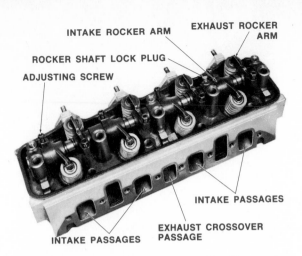

Fig. 11-22. A cylinder-head assembly. The rocker arms are mounted on a shaft. (*Chrysler Corporation*)

Fig. 11-23. Rocker arms of the type using ball pivots. (*Chevrolet Motor Division of General Motors Corporation*)

The adjustment screw in the end of the rocker arm can be turned in or out to adjust the clearance in the valve train, that is, to give the valve the clearance it needs.

Another kind of rocker arm is shown in Fig. 11-23. This rocker arm is mounted on a stud that is installed in the cylinder head, as shown. The rocker arm is held in place by a cup-shaped washer called a **ball pivot.** It resembles a steel ball that is cut in two and has a hole

in its center. The ball pivot is positioned on the stud by a self-locking nut. In operation, the rocker arm rocks back and forth on the ball pivot. The valve train is adjusted by turning the nut up or down on the stud. The ball pivot is lubricated by oil that flows up through the hollow stud from an oil gallery in the cylinder head.

A third kind of rocker arm is shown in Fig. 11-24. Here, the rocker arms are supported in

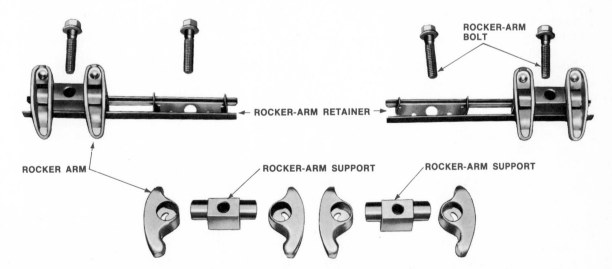
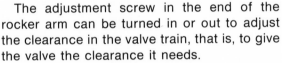

Fig. 11-24. Rocker arms, supports, and retainers. (*Cadillac Motor Car Division of General Motors Corporation*)

pairs by rocker-arm supports. At the top of Fig. 11-24, you can see how the rocker arms look when they are installed on the supports. At the bottom in the figure, the rocker arms have been removed from the supports and laid down on their sides.

§ 11-30 HYDRAULIC VALVE LIFTERS

Hydraulic valve lifters are used in engines to reduce noise and eliminate valve adjustments. In the past, before hydraulic valve lifters were used, valves had to be adjusted every so often. As normal wear took place in the engine, the clearances in the valve train increased. As we have noted, there must be some clearance to guarantee that the valves will close completely. Thus, every time the valve is opened, the valve-train parts must come together to eliminate this clearance. The clearance must be taken up before the lifter and pushrod movement can carry all the way through the rocker arm to the valve stem. Taking up of the clearance produces a noise, called **tappet noise.**

However, with the hydraulic valve lifter, all clearance in the valve train is taken up by the oil in the lifter every time the valve closes. Then, when the valve train starts to open the engine valve, there is no clearance between metal parts. Because there is no clearance to be taken up, the engine valve opens silently.

Figure 11-25 shows the installation of a hydraulic valve lifter in a V-8 engine. Figure 11-26 shows how the lifter works. As you can see, there are several parts mounted inside the lifter body. The principle of operation is that the lifter is kept filled with oil from the engine lubricating system, and this takes up all the clearance. Let's see how it all works.

The picture on the left in Fig. 11-26 shows the situation when the valve is closed. Notice that the lobe of the cam has passed out from under the valve lifter. Oil from the lubricating system is flowing through holes in the valve lifter body and plunger. This oil is under pressure from the engine oil pump. As it flows in-

Fig. 11-25. A sectional view of a V-8 engine, showing the location of the hydraulic valve lifter in the valve train. (*Cadillac Motor Car Division of General Motors Corporation*)

side the plunger, most of it flows up through the hollow pushrod to lubricate the rocker arms.

Some of the oil pushes down on the ball-check valve, which opens the valve. This oil can then flow under the plunger. The oil raises the plunger enough to take up any clearance—usually only a few thousandths of an inch—in the valve train.

Now let's see what happens when the cam lobe comes around, as shown on the right in Fig. 11-26, and pushes the valve lifter up. The increase in pressure under the ball-check valve closes the valve. The valve lifter then acts like a solid lifter and moves up to push the pushrod up. The pushrod operates the rocker arm and opens the engine valve. During this interval,

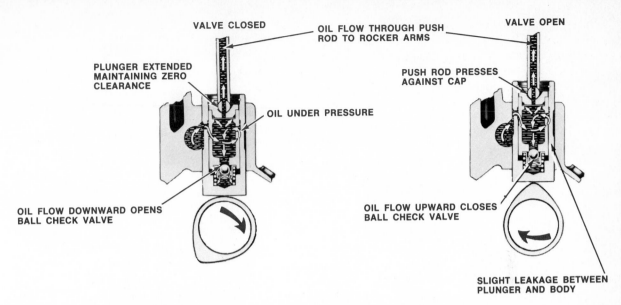

Fig. 11-26. The operation of a hydraulic valve lifter. (*Cadillac Motor Car Division of General Motors Corporation*)

some of the oil may leak out from under the plunger. This does not matter, however. The oil lost is replaced by the lubricating system as explained above. As soon as the cam lobe moves out from under the valve lifter and the engine valve closes, any clearance is taken up once again.

§11-31 OVERHEAD CAMSHAFTS

In all the engines we have looked at so far, the camshaft is located in the cylinder block. A pushrod and rocker arm are required to carry the motion of the valve lifter to the valve. When the camshaft is installed in the cylinder head, as shown in Fig. 11-27, the pushrod is no longer needed. And in some engines the rocker arm is also eliminated. Figure 11-28 shows, in an end-sectional view, an overhead-camshaft engine that uses neither pushrods nor rocker arms. The cams work directly on valve tappets that rest on the ends of the valve stems.

What is the advantage of an overhead camshaft? When the camshaft is in the overhead position, the path between the cams and the valves is much shorter. Thus the valves respond more quickly. Higher engine speeds are possible because there are fewer valve-train parts to move. Some domestic cars and many imported ones use engines with overhead camshafts.

Figure 11-29 is a partial cutaway view of an engine using an overhead camshaft. This engine is used in the Saab automobile, made in Sweden. The camshaft is driven by a chain and sprockets, which can be seen at the left in the picture.

Figure 11-30 is an end-sectional view of a V-8 engine using four overhead camshafts, two in each cylinder head. This engine has a rather complicated camshaft drive arrangement. The crankshaft gear drives a second gear, which then drives a third gear. The third gear is mounted on the same shaft as a chain sprocket. The chain then drives the two camshafts in the cylinder head. This arrangement

Fig. 11-27. (*Above*) A cutaway view of the Chevrolet Vega four-cylinder engine. This is an overhead-valve engine with the camshaft in the cylinder head. (*Chevrolet Motor Division of General Motors Corporation*)

Fig. 11-28. (*Right*) A sectional view of the Chevrolet Vega overhead-camshaft engine. (*Chevrolet Motor Division of General Motors Corporation*)

is duplicated for the other bank of cylinders.

An engine with a single overhead camshaft is known as an SOHC (single-overhead-camshaft) engine. An engine with two overhead camshafts is called a DOHC (double-overhead-camshaft) engine. The name is often shortened to overhead-cam engine.

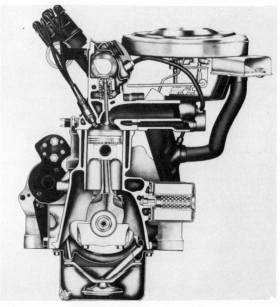

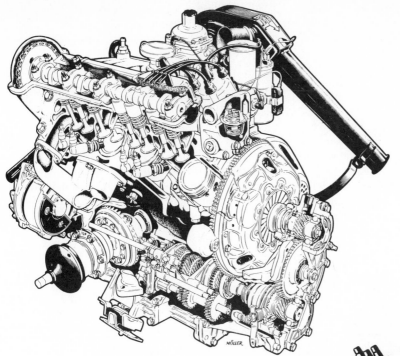

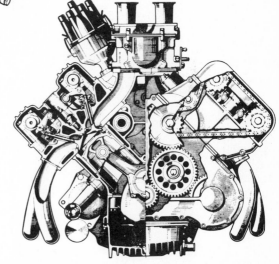

Fig. 11-29. A slant four-cylinder engine with an overhead camshaft. The clutch and transmission are part of the assembly. (*Saab of Sweden*)

Fig. 11-30. A sectional view from the end of a V-8 engine with four overhead camshafts (a DOHC engine). The right bank has been cut away to show the camshaft drive arrangement. The left bank has been cut away to show the internal construction of the engine and locations of the valves and other components. (*Renault*)

Review questions

1. What is the job of the piston and connecting-rod assembly?
2. Why is piston fit so important?
3. What happens to the oil that is not scraped off the cylinder walls by the rings? What happens to the oil that is scraped off the cylinder walls by the rings?
4. What effect does speed have on the piston rings' job of oil control?
5. What are the two main factors that determine how much oil an engine will burn? Which is more likely to burn oil, a new or an old engine?
6. What are the two types of valves? Which gets hotter? Why?

7. Why must excessive amounts of oil be kept off the valve stems?
8. What would happen if an exhaust valve failed to close completely?

9. How is the rocker-arm ball pivot lubricated?
10. When the valve is open, is the ball-check valve in the hydraulic valve lifter open or closed?

Study problems

1. Obtain a completely assembled piston and connecting-rod assembly. Determine how the piston pin is attached to the rod and if the piston pin is offset. Carefully remove the piston rings, and note the appearance and position of each of the rings. Notice the difference between the compression rings and the oil-control rings.

2. Locate a completely assembled cylinder head that has been removed from an overhead-valve engine. Name each part of the valve train. Determine how the rocker arm is lubricated. Then, using Fig. 10-8 as a guide, identify each passage and hole in the cylinder head.

12 | ENGINE MEASUREMENTS

§ 12-1 BORE AND STROKE

Before we describe the engine fuel system and other engine details, let us look at the various ways in which engines are measured. These measurements include cylinder diameter, length of piston movement, and piston and engine displacement.

The size of an engine cylinder is referred to in terms of **bore,** or diameter, and **stroke,** or distance the piston moves from BDC to TDC (Fig. 12-1). The bore is always mentioned first, as, for instance, 4 by $3\frac{1}{2}$. This means the bore is 4 inches [101.6 millimeters] and the stroke is $3\frac{1}{2}$ inches [88.9 millimeters].

§ 12-2 PISTON DISPLACEMENT

This is the volume of space that the piston displaces as it moves from BDC to TDC. It can be calculated if the bore and stroke are known. Piston displacement is the volume of a cylinder having the diameter of the bore and the length of the piston stroke. For instance, a cylinder with a bore and stroke of 3×4 inches [76.2×101.6 millimeters] would have a piston displacement of

$$\frac{\pi \times D^2 \times L}{4} = \frac{3.1416 \times 3^2 \times 4}{4}$$

$$= 28.27 \text{ cubic inches}$$

where D = diameter of the bore of the cylinder
L = length of the stroke

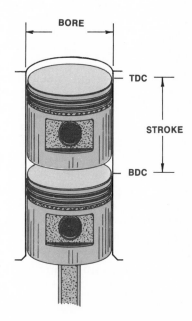

Fig. 12-1. The bore and stroke of an engine cylinder.

If the engine had six cylinders, then the total displacement would be 169.62 cubic inches (6×28.27). Piston displacement is an important specification of racing cars.

When you talk with the hot-rodders and with the people who build and race cars, they will tell you that limits are set for engine displacements in most races. In a recent race at the Indianapolis 500—the "Indy 500," as it is called—the maximum allowable displacement was set at 305.1 cubic inches [5,000 cubic centimeters (cu cm), or 5 liters] for nonsupercharged engines.

In many races, especially those in Europe, the displacement is given in **liters,** a metric measurement. One liter equals 61.02 cubic inches, or 1.057 quarts. Thus, the Indy 500 specification of 305.1 cubic inches equals 5 liters (305.1 divided by 61.02, or exactly 5 liters).

§ 12-3 COMPRESSION RATIO

This is a very important engine measurement. The higher the compression ratio, the more powerful an engine is. This ratio is a measure of the amount that the air-fuel mixture is squeezed, or compressed, on the compression stroke. It is calculated by dividing the volume in the engine cylinder with the piston at BDC by the volume with the piston at TDC. In other words, it is A divided by B in Fig. 12-2.

The volume above the piston with the piston at TDC is also called the **clearance volume** because it is the clearance that remains above the piston at TDC.

As an example of how to figure the compression ratio, suppose an engine had a cylinder volume of 45 cubic inches [738 cu cm] with the piston at BDC (A in Fig. 12-2) and a clearance volume of 5 cubic inches [82 cu cm] (B in Fig. 12-2). Then the compression ratio of the engine would be 45 divided by 5 [738 divided by 82], or 9 to 1 (written 9/1 or 9:1).

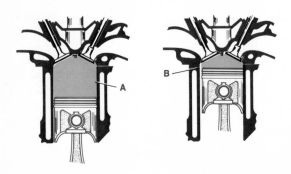

PISTON AT BDC   PISTON AT TDC

Fig. 12-2. Compression ratio is the volume in a cylinder with the piston at BDC divided by the volume with the piston at TDC. That is, A divided by B.

When the compression ratio of an engine is increased, its power and economy also increase. In effect, an engine with a higher compression ratio squeezes the air-fuel mixture harder by compressing it to a smaller volume. When this happens, the air-fuel mixture gives up more power on the power stroke. Here is why. The higher compression ratio means higher pressure at the end of the compression stroke. It also means that higher combustion pressures will be reached when ignition takes place. Thus, a harder push will be made on the piston. The result is more power.

§ 12-4 IMPORTANCE OF HIGH COMPRESSION

You may be asking, "What's so important about the compression ratio?" You should know two facts about compression ratio. First, when compression ratio goes up, power goes up. Thus, a car with a high-compression engine has more getaway, quicker passing on the highway, and higher speed. Second, the high-compression engine needs what we commonly call **high-octane** gasoline. If the engine does not get it, the engine will "knock." Feed a high-compression engine low-octane gas, and you can ruin it. It can knock so hard that engine parts will be broken! What's more, settings and adjustments on a high-compression engine must be just right in order to get top performance.

The high-compression engine has been blamed for a lot of the air pollution that comes from automobiles. Chapter 20 explains air pollution, how the automobile adds to this pollution, and what the automotive mechanic can do to reduce the air pollution that comes from automobiles.

§ 12-5 ENGINE PERFORMANCE MEASUREMENTS

Now, let us look at the various ways in which engine performance is measured. These include engine torque, horsepower, and volumetric efficiency.

§ 12-6 WORK

As we mentioned in Chapter 2, work is the moving of an object against an opposing force. The object is moved by a push, a pull, or a lift. Regardless of the direction in which the object is moved, it requires effort to move it. Work is measured in terms of distance and force. If a 5-pound weight is lifted off the ground 1 foot, the work done on the weight is 5 foot-pounds (ft-lb). If the 5-lb weight is lifted 3 feet, the work done is 15 ft-lb.

Work equals force times distance.

§ 12-7 POWER

Work can be done slowly, or it can be done fast. The rate at which work is done is measured in terms of power. A machine that can do a great deal of work in a short time is a **high-powered** machine.

Power is the rate, or speed, at which work is done.

§ 12-8 ENGINE POWER

Engine performance is measured in terms of power—horsepower. One horsepower is the power of one average horse, or a measure of the speed at which a horse can work. It may seem somewhat strange that in this modern day we still compare the power of engines with the speed at which a horse can work.

Years ago, when engines were first being developed, people realized that they had to use some unit of measurement so that they could compare the power of different engines. Since the horse was then the most common source of power, it was natural that the power of engines should be measured in terms of the power of horses, that is, in horsepower. If people had harnessed cats instead of horses to do their work for them, we would probably be referring to engines in terms of "catpower."

It was found that an average horse can raise a weight of 200 pounds [90.718 kilograms] a distance of 165 feet [50.292 meters] in 1 minute. Figure 12-3 shows a simplified version

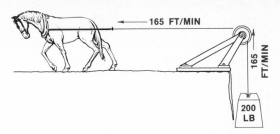

Fig. 12-3. One horse can do 33,000 ft-lb [4,562.25 m-kg] of work per minute.

of the method by which this test was made. The horse walks 165 feet [50.292 meters] in 1 minute, and the cable, running over the pulley, raises the 200-pound [90.718-kilogram] weight 165 feet [50.292 meters] in that minute. The amount of work done is 33,000 foot-pounds (ft-lb) [4,562.25 meter-kilograms (m-kg)]. The length of time it takes to do this amount of work is 1 minute. The amount of power required is 1 horsepower [0.746 kilowatt (kW)]. In other words, this is the amount of work one horse can do in 1 minute.

On the above basis, if 400 pounds [181.436 kilograms] were to be raised 165 feet [50.292 meters] in 1 minute, two horses would be required (one for each 200-pound [90.718-kilogram] weight). The amount of work required would be 66,000 foot-pounds (165 × 400) [9,127.8 m-kg (50.292 × 181.436)]. When this amount of work is done in 1 minute, then 2 horsepower [1.492 kilowatts] is required. Similarly, if 33,000 foot-pounds [4,562.25 m-kg] of work is to be done in 2 minutes, then only $\frac{1}{2}$ horsepower [0.373 kilowatt] is required. In this case, work is being done at the rate of 16,500 foot-pounds/minute [2,281.95 m-kg/minute]. The formula for horsepower is

$$HP = \frac{\text{ft-lb per min}}{33,000} = \frac{L \times W}{33,000 \times t}$$

where HP = horsepower

L = length, in feet, through which W is forced

W = push, in pounds, exerted through distance L

t = time, in minutes, required to force W through L

Note that in the metric system, power output from an engine is often measured in kilowatts (kW). That is, the power output of an engine is measured in terms of the amount of electricity the engine could produce if it were driving an electric generator. The equivalent of 1.34 horsepower is 1 kilowatt, and 1 horsepower is equivalent to 0.746 kilowatt. Thus, a 200-horsepower engine would be equivalent to a 149-kilowatt engine.

§ 12-9 VOLUMETRIC EFFICIENCY

Take a look at these two words—"volumetric" and "efficiency." You know that "volumetric" has something to do with volume and that "efficiency" has something to do with how well a job is done. The two words used together refer to how well the engine cylinder fills up on the intake stroke. Remember, the piston moves down on the intake stroke, and the air-fuel mixture pours in to fill the vacuum left by the downward movement of the piston. This mixture does not pour in instantly. It takes time for the air-fuel mixture to flow through the carburetor and past the intake valve.

The same thing happens when you drink a carbonated beverage through a straw. The beverage does not come up into your mouth the instant you start sucking. It takes a fraction of a second to reach your mouth.

Likewise, it takes time for the air-fuel mixture to get into the engine cylinder on the intake stroke. And there is not much time! At highway speed, the intake stroke takes less than one-hundredth of a second. The piston is really moving! So, before the cylinder really fills up, the intake valve closes, and the compression stroke starts. If the cylinder fills up almost completely, then the volumetric efficiency is high. If the cylinder fills up only partly, then the volumetric efficiency is low.

The key idea here is that the shorter the time it takes to fill up the cylinder, the lower the volumetric efficiency. At low speeds, when the intake stroke takes as long as one-tenth of a second, the cylinder can almost fill up. This means that there is more air-fuel mixture to be compressed and burned, and as a result, the power stroke is stronger.

At high speeds, the intake stroke takes such a short time that the cylinder does not get nearly as much air-fuel mixture. There is less air-fuel mixture to be compressed and burned, and as a result, the power stroke is weaker.

That is the reason you cannot go on increasing engine speed indefinitely. As speed increases, the intake-stroke time gets shorter and shorter. Volumetric efficiency gets lower and lower. Less and less air-fuel mixture gets in, and the power output of the engine falls lower and lower. Finally, a speed is reached at which the power strokes are so weak that they can no longer produce any further speed increase.

Here is the formula for volumetric efficiency:

$$VE = \frac{\text{actual amount of air-fuel mixture entering cylinder}}{\text{amount that could enter under ideal conditions}}$$

Suppose that at a certain speed, 40 cubic inches [656 cu cm] enter the cylinder, but under ideal conditions 50 cubic inches [820 cu cm] could enter the cylinder. The volumetric efficiency is 40 divided by 50 [656 divided by 820], or 80 percent.

§ 12-10 IMPROVING VOLUMETRIC EFFICIENCY

You can readily see that volumetric efficiency can be improved by increasing the size of the passages through which the air-fuel mixture travels. If the intake valve, the valve port, and the passages through the carburetor are made larger, the air-fuel mixture can get into the cylinder more easily. This is done in modern engines. Engineers called it "improving the engine's breathing." They point out that an engine "breathes" just as you do. It takes in air and then blows it out.

With improved engine breathing, the engine can put out more power at higher speeds. When engineers are assembling a racing engine, for example, they open up the air-fuel passages as much as possible so that the engine can breathe better at high speed. That is, they improve the volumetric efficiency of the engine. For example, suppose an engine has a volumetric efficiency of 80 percent. Modifying the engine raises the actual amount of air-fuel mixture taken in from 40 cubic inches [656 cu cm] to 45 cubic inches [738 cu cm]. The VE would then be 90 percent.

You have possibly heard about "two-barrel" and "four-barrel" carburetors. These are the carburetors you find on the high-horsepower engines. The extra barrels are additional air passages that let the engine breathe easier—they give the engine higher volumetric efficiency. The higher volumetric efficiency allows the engine to put out more horsepower, especially at high speeds.

§12-11 TORQUE

Torque is twisting or turning effort. You apply torque to the top of a screw-top jar when you loosen it (Fig. 12-4). You apply torque to the steering wheel when you take a car around a curve. The engine applies torque to the wheels to make them rotate.

Do not confuse torque with power. Torque is turning effort that may or may not result in motion. Power is the rate at which work is being done, and this means that something must be moving.

Torque is measured in pound-feet (lb-ft, not to be confused with ft-lb of work). For example, suppose you pushed on a crank with a 20-pound [9.072-kilogram] push, and suppose the crank handle were $1\frac{1}{2}$ feet [0.457 meter] from the center of the shaft (Fig. 12-5). You would be applying 30 lb-ft (20 × $1\frac{1}{2}$) [4.146 kg-m (9.072 × 0.457)] of torque to the crank. You would be applying this torque regardless of whether or not the crank was turning, just as long as you continued to apply the 20-pound [9.072-kilogram] push to the handle.

Notice that in the metric system, torque is measured in kilogram-meters (kg-m), not pound-feet.

Fig. 12-4. Torque, or twisting effort, must be applied to loosen and remove the top from a screw-top jar.

Fig. 12-5. Torque is measured in pound-feet (lb-ft) or kilogram-meters (kg-m). It is calculated by multiplying the push by the crank offset, that is, the distance of the push from the rotating shaft.

§12-12 ENGINE TORQUE

Engine torque comes from the pressure of the burning gases in the cylinders. This pressure pushes down on the pistons and causes the crankshaft to turn; the harder the push on the piston, the greater the torque. *Engine torque is not engine power.* Torque is the twisting effort that the engine applies to the crankshaft. Power is the rate at which the engine is working.

Engine torque varies with the speed of the engine. An engine develops more torque at intermediate speed (with open throttle) than at high speed. Here is the reason: At intermediate speed, there is more time for the air-fuel mixture to enter the cylinder. In other words, the volumetric efficiency is high. This means that more air-fuel mixture enters the cylinder, the combustion pressure goes up, and there is a stronger push on the piston. In other words, more torque is applied to the crankshaft.

At higher speeds, there is less time for the air-fuel mixture to get into the cylinder. Volumetric efficiency drops off. There will be less air-fuel mixture to burn, and the combustion pressure will be lower. The push on the piston will be less, and the engine torque will be lower.

The fact that torque drops at high speed is shown in a graph (Fig. 12-6). At low speed, about 500 rpm (revolutions per minute), the torque is about 180 lb-ft [24.9 kg-m]. Find this point on the graph by moving your finger up the 500-rpm line until it crosses the curved line. From this point, move your finger to the left. Your finger will come to a spot slightly above the 175 lb-ft [24.2 kg-m] line, at about 180 lb-ft [24.9 kg-m]. As engine speed increases, the torque goes up. Torque reaches a peak in the 1,500- to 2,000-rpm range. Then it begins to drop, and at about 4,000 rpm, the torque is less than 125 lb-ft [17.29 kg-m].

This explains why at high speed the engine won't give you the performance that it does at intermediate speed. It is the torque of the engine that gives acceleration.

Here is another formula for measuring horsepower, using the torque of the engine:

$$HP = \frac{torque \times rpm}{5,252}$$

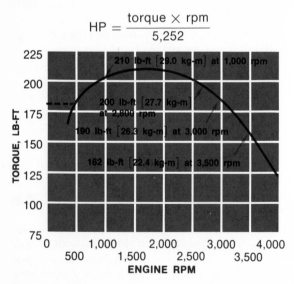

Fig. 12-6. The torque curve of an engine, showing the relationship between torque and speed.

When you work in the shop and use the shop dynamometer to measure engine output, you will see why this formula is more convenient than the one presented in § 12-8. You can measure torque and rpm, as well as horsepower, with some dynamometers.

§ 12-13 BRAKE HORSEPOWER

The **brake horsepower** of an engine is the actual horsepower that the engine is putting out. It is called brake horsepower because when engineers first started measuring engine performance, they used a type of brake to measure this power. Today, engine power is measured by a dynamometer. This word is from "dynamo" and "meter." Originally, the dynamometer was an electric generator with a meter to measure the amount of electricity the dynamometer produced. Today, the term "dynamometer" is applied to several types of power-measuring devices. When a dynamometer is used, an engine drives it, and the amount of power the engine produces is measured. The more power the engine produces, the more power is absorbed by the dynamometer and the higher the meter reads. Thus, engine output is measured as the amount of power the dynamometer is absorbing.

The amount of horsepower that an engine puts out depends on torque and speed in rpm. As speed goes up, horsepower goes up. As torque goes up, horsepower goes up. See the formula in § 12-12.

The graph of horsepower output of an engine is shown in Fig. 12-7. Note that the horsepower starts out low at low speed and builds up to about 110 hp at 3,500 rpm. After that, as speed increases further, the horsepower output drops off.

The drop-off of horsepower output is due not only to the decrease in torque but also to the increase in friction horsepower at higher speeds. Friction horsepower is the power required to overcome the friction of the moving parts in the engine. Therefore, the decrease in torque and the increase of friction horsepower

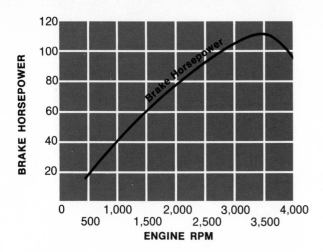

Fig. 12-7. A curve showing the relationship between brake horsepower and engine speed.

are two things working against continued increases of engine horsepower with further increases of rpm.

Note that the two graphs (Figs. 12-6 and 12-7) are for only one particular engine. Different engines have different torque and brake-horsepower curves. Peaks may be at higher or lower speeds, and the relationships may not be the same as in these graphs.

§ 12-14 ENGINE POWER LOSS

Unfortunately, engine power loss is considerable. Power is lost because of friction and also because only part of the energy in gasoline is converted into power. A good deal of the potential power in gasoline is lost as heat. The engine cooling system, which is described in Chapter 18, takes about one-third of the heat energy of the gasoline. Another one-third is lost because the exhaust gases leave the engine very hot. More power is lost because of friction. The result is that only about 15 to 20 percent of the energy in the fuel is actually used to move the car.

Review questions

1. What is cylinder bore? Piston stroke?
2. What is the formula for figuring cylinder displacement?
3. Does power increase or decrease when compression ratio is raised?
4. How is engine power measured? How is it measured in the metric system?
5. Is volumetric efficiency better at low speed or at high speed?

6. What is torque? How is it measured? Does torque require motion?
7. With what piece of equipment is engine power measured?
8. What are two causes of a drop-off of engine horsepower at higher engine speed?
9. About how much of the energy in the fuel is lost through the cooling system?

Study problems

1. Select an engine with which you are familiar. Write the year, make, displacement, and horsepower rating on a sheet of paper. Then, list every change that you could make to raise the volumetric efficiency of that engine.

2. On an engine having the cylinder head removed, measure the bore and stroke of one cylinder. Using the formula given in **§** 12-2, figure the piston displacement and the engine displacement in cubic inches. Then, convert this measurement to liters.

13 INTRODUCTION TO AUTOMOTIVE ELECTRICITY

§ 13-1 HOW WE USE ELECTRICITY

Electricity does all sorts of things for us. It gives us light; it runs much of the machinery around us (refrigerators, factory equipment, television sets, subway trains); and it heats our homes. In the car, electricity does several jobs. It starts the engine when we turn the ignition switch on. It makes the sparks that ignite the compressed air-fuel mixture. It operates the radio, the electric gages, and the lights. Figure 13-1 shows the major components of the automotive electrical system. In this chapter, we briefly describe all these electrical devices, and then we find out what electricity is all about.

§ 13-2 AUTOMOTIVE ELECTRICAL EQUIPMENT

The typical internal-combustion engine, for example, the automobile engine, requires an electrical system to start it and to keep it running. The electrical system supplies electric sparks to the cylinder spark plugs. These sparks set fire to, or ignite, the compressed air-fuel charges in the engine cylinders. The electrical system also cranks the engine (turns the crankshaft) for starting. It has a battery to operate the starting motor and an alternator to keep the battery charged and to furnish electric current for other electrical units. The elec-

trical system also includes lights for driving at night as well as the horns, radio, heater, and other electrical accessories.

In Chapter 8, the two-cycle-engine ignition system is described. It uses a magneto (Fig. 8-7) to create current for ignition. This same type of magneto is used in single-cylinder, four-cycle engines.

Let us review briefly the purpose of the various units in the electrical system of a vehicle (Fig. 13-1). These units are connected to each other by wires. The wires form circuits, or paths, through which the electricity can move. And as the electricity moves through the circuits formed by the wires and the units, it causes the units to do their jobs. In later pages, these units will be described in more detail.

§ 13-3 BATTERY

The battery (center in Fig. 13-1) is a source of electricity that is available when other electrical units are not operating. For instance, the battery will supply electricity to light the headlights when the engine is not running. The battery also supplies electricity to the starting motor for starting the engine. The battery cannot, however, serve as a source of electricity indefinitely. It can supply only so much electricity before it is run-down, or dead. To keep the battery from running down, it is supplied with electricity, or recharged, by the car alternator when the engine is running.

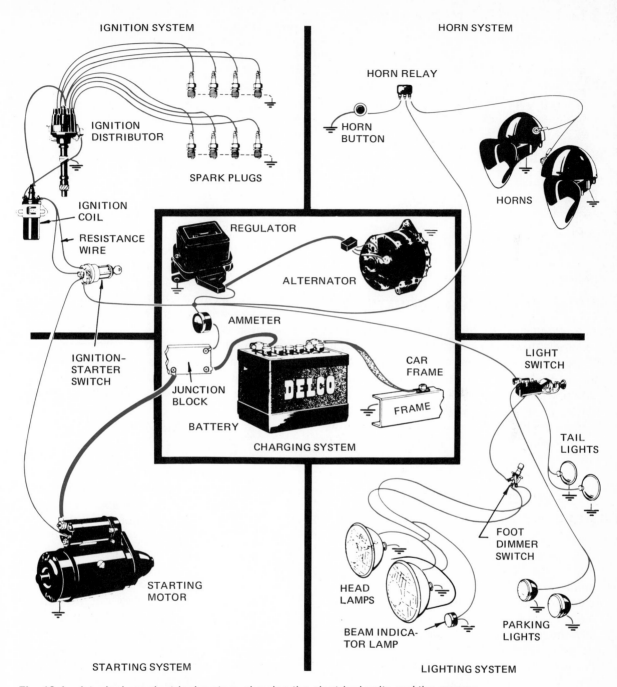

IGNITION SYSTEM

IGNITION DISTRIBUTOR

SPARK PLUGS

IGNITION COIL

RESISTANCE WIRE

IGNITION-STARTER SWITCH

HORN SYSTEM

HORN RELAY

HORN BUTTON

HORNS

REGULATOR

ALTERNATOR

AMMETER

JUNCTION BLOCK

BATTERY

DELCO

CAR FRAME

FRAME

CHARGING SYSTEM

LIGHT SWITCH

TAIL LIGHTS

FOOT DIMMER SWITCH

HEAD LAMPS

BEAM INDICATOR LAMP

PARKING LIGHTS

STARTING MOTOR

STARTING SYSTEM

LIGHTING SYSTEM

Fig. 13-1. A typical car electrical system, showing the electrical units and the connections between them. The symbol ⏚ means ground, or the car frame. Using the car frame as the return circuit requires only half as much wiring. (*Delco-Remy Division of General Motors Corporation*)

§ 13-4 STARTING MOTOR

The starting motor (lower left in Fig. 13-1) is a special electric motor that turns the engine crankshaft when the driver operates the starting control. As the crankshaft turns, the engine begins to run. The starting motor thus starts the engine.

§ 13-5 ALTERNATOR

The alternator (center in Fig. 13-1) is driven by the engine when it is running. It supplies electricity to operate different electrical devices such as the ignition system, lights, radio, heater, and so on. It also restores to the battery the electricity used in starting and keeps the battery in a charged condition.

§ 13-6 ALTERNATOR REGULATOR

To prevent the alternator from producing too much electricity (as it might at high speed, for instance), an alternator regulator is used (center in Fig. 13-1). The regulator automatically controls the amount of electricity the alternator produces and prevents it from producing too much. In many modern alternators, the regulator is built-in; that is, it is an **integral part** of the alternator.

§ 13-7 IGNITION SYSTEM

The ignition system (upper left in Fig. 13-1) produces electric sparks that ignite the compressed air-fuel mixture, or **charge,** in the cylinder. The ignition system produces these sparks in rapid succession so that each charge is ignited at the right time toward the end of each compression stroke.

In later pages, we will look at each of these various electrical units in detail. But first, let's add to what we learned in Chapter 8 and find out more about electricity.

§ 13-8 WHAT IS ELECTRICITY?

As we mentioned in Chapter 8, nobody has ever seen what electricity is made of, and so we have to rely on the description of the experts who have studied electricity. Electricity is composed of tiny particles—particles so tiny that it would take billions upon billions of them, all piled together, to make a spot big enough to be seen through a microscope. These particles are called electrons. Electrons are all around us in fantastic numbers. In an ounce of iron, for example, there are about 22 million billion billion electrons. Electrons are normally locked into the elements that form everything in our world.

§ 13-9 ELECTRIC CURRENT

If many, many electrons are forced to move together in the same direction, in a wire, for example, we have a flow, or current, of electrons. We call this flow an electric current. The job of the battery and the generator or the alternator is to get the electrons to move together in the same direction. When many electrons are moving, we say the current is high. When relatively few electrons are moving, we say the current is low.

§ 13-10 MEASURING ELECTRIC CURRENT

The movement of electrons, or electric current, is measured in **amperes,** or **amps.** One ampere of electric current is a very small amount of current. A battery can put out 200 or 300 amps as it operates the starting motor. Headlights draw 10 or more amps. A single ampere is the flow of 6 billion billion electrons per second.

Now, obviously, nobody is going to sit there and count electrons to find out how many amperes are flowing in a wire. You have to use an ammeter (Fig. 13-2) to find this out. The ammeter uses a strange effect of electron flow. This effect is that a flow of electrons, or electricity, produces magnetism.

§ 13-11 MAGNETISM

There are two forms of magnetism: natural and electrical. Natural magnets are made of iron or

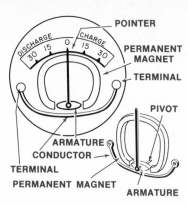

Fig. 13-2. A simplified drawing of a car ammeter, or charge indicator.

other metals. Electrically produced magnets are called **electromagnets.** Natural and electromagnets act in the same way: they attract iron objects. Here are two important facts about magnets:

- Magnets can produce electricity.
- Electricity can produce magnets.

We have more to say about magnets and electromagnets and how they act later in this chapter.

§13-12 THE AMMETER

Now, let's look at how the ammeter measures electric current. The simplest kind of ammeter is shown in Fig. 13-2. This kind of ammeter is found on the instrument panel in many cars. Its purpose is to tell the driver whether the alternator is charging the battery. If the alternator doesn't charge the battery when it should, the battery will run down. If the battery runs down, the car won't start.

Let's see how the ammeter works. The conductor is connected at one end to the battery. The pointer is mounted on a pivot. There is a small piece of oval-shaped iron mounted on the same pivot. This oval-shaped piece of iron is called the **armature.** A permanent magnet, almost circular in shape, is positioned so that

its two ends are close to the armature. The permanent magnet attracts the armature and tends to hold it in a horizontal position. In this position, the pointer or needle points to zero. Nothing is happening. Now, suppose the alternator starts sending current to the battery. This current passes through the conductor. The current produces magnetism. This magnetism attracts the armature and causes it to swing clockwise. This moves the pointer to the "charge" side. The more current that flows, the stronger the magnetism and the farther the pointer moves. The meter face is marked off to show the number of amperes flowing.

Now, suppose the alternator is not working and you turn on the lights of your car. Current will flow from the battery to the lights. In this situation, current will flow in the reverse direction through the conductor in the ammeter. Therefore, the armature is attracted in the opposite direction. It swings in a counterclockwise direction. This moves the pointer to the "discharge" side of the ammeter. The more current that is taken out of the battery, the farther the pointer moves across the discharge side of the ammeter.

§13-13 WHAT MAKES THE ELECTRONS MOVE?

As we said, electrons on the move make up electric current. But what makes the electrons move? Simply this: too many electrons in one spot. When electrons are concentrated in one place, they try to move. The battery and the alternator are devices that concentrate electrons at one terminal and take them away from the other. Therefore, if the two terminals are connected by a conductor, electrons will flow from the terminal that has too many electrons to the terminal that has too few.

§13-14 VOLTAGE

Suppose there are a great many electrons concentrated at one terminal. And suppose there is a great shortage of electrons at the other terminal. When there is a great excess and a great shortage, we say that the electrical pressure is high. We mean that the pressure on the electrons to move from the "too many" terminal to the "too few" terminal is high.

We measure electrical pressure in **volts.** High pressure is high voltage; low pressure is low voltage. Car batteries are 12-volt units. Twelve volts is considered low pressure. The spark at the spark-plug gap is a flow of electrons at high pressure, or voltage. The voltage at the spark-plug gap can be 20,000 volts or more. That's high, but not nearly as high as the voltage on the power lines that carry electricity from power plants to your home and to factories. The voltage on power lines is in the hundreds of thousands of volts.

§ 13-15 INSULATION

If electrons escape from the wire in which they are flowing, electric power is lost. That's the reason wires are covered with insulation. That's also the reason power lines are hung from long insulators on the power poles or towers. In addition to power loss, electrons on the loose can be dangerous. For instance, if the insulation on the wire to a household appliance or a lamp is damaged, a fire could result, or a person who touches the wire or appliance would get an electric shock.

In the car, the wires between the battery, the alternator, and other electrical devices are all covered with insulation. The insulation is a nonconductor, which means that electrons, or electric current, cannot flow through it. But if the insulation goes bad, electric current will go where it is not supposed to. It could take a shortcut through the metal of the car frame and the engine. Such a shortcut is called a **short circuit,** and it can cause all sorts of trouble, as you will discover in later sections of the book.

Just remember that the insulation has the job of keeping the electrons, or the electric current, moving in the proper path, or circuit. Circuits include the wires and the electrical devices in the car.

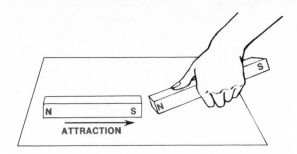

Fig. 13-3. Unlike magnetic poles attract. As the north pole is moved near the south pole, the south pole will pull toward the north pole.

§ 13-16 MAGNETS—ANOTHER LOOK

Let's take another look at magnets. Magnets act through **lines of force.** These lines of force stretch between the ends of the magnet. The two ends of the magnet are called the **magnetic poles,** or the **poles.** One pole is called the north pole; the other, the south pole. The area surrounding the poles is called a **magnetic field.**

§ 13-17 LINES OF FORCE

The lines of force have two characteristics. First, the lines of force try to shorten up. For example, if you hold the north pole of one magnet close to the south pole of another magnet, the two magnets pull together (Fig. 13-3). If we drew the lines of force between the two poles, the picture would look something like Fig. 13-4. The lines of force, stretching between the two poles, try to shorten up and thus pull the two poles together.

The second characteristic is that the lines of force run more or less parallel to each other and try to push away from each other. We can understand this if we bring like poles together—two north poles, for example (Fig. 13-5). The lines of force run parallel to each other and try to push away (see Fig. 13-6). The magnet that is free to move will actually move away as the like pole of the other magnet is brought closer.

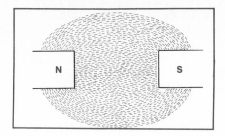

Fig. 13-4. Magnetic lines of force stretching between two unlike magnetic poles try to shorten up. This pulls the two unlike poles together.

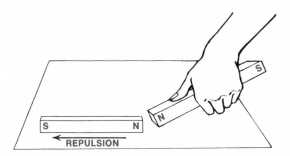

Fig. 13-5. Like magnetic poles repel each other. When a north pole is brought near another north pole, they push away from each other.

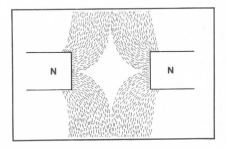

Fig. 13-6. Magnetic lines of force between two like magnetic poles. These magnetic lines of force run parallel to each other. This forces the two like poles apart.

We can draw these conclusions:

- Like magnetic poles repel each other. North repels north; south repels south.
- Unlike magnetic poles attract each other. North attracts south; south attracts north.

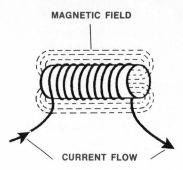

Fig. 13-7. An electromagnet can be made by wrapping a wire around a rod.

§ 13-18 ELECTROMAGNETS

Electromagnets act just like natural magnets. An electromagnet can be made by wrapping a wire around a rod (Fig. 13-7). We saw what happened in the ammeter when current flowed one way or another through the conductor. That is, the current produced magnetism, or magnetic lines of force.

Current flowing through a single wire, or conductor, will not produce much magnetism. But suppose you wind a conductor, or wire, around a rod and connect the ends of the wire to a source of electric current (or electrons). Then, the wire winding will produce strong magnetism. In other words, a strong magnetic field will develop around the coil of wire.

With current flowing through the winding, the winding acts just like a bar magnet. You can point one end of the winding toward a pole of a bar magnet, and it will either attract or repel the bar-magnet pole. One end of the winding is a north pole, and the other end is a south pole. You can change the poles by reversing the leads to the source of current. Thus, you can see that when the electrons flow through in one direction, one of the poles becomes north. But when the electrons flow through in the opposite direction, the poles reverse. The north pole becomes the south pole, and the south pole becomes the north pole.

An electromagnet such as the one you made by winding wire around a rod is also called a

solenoid. It is used in several places in the electrical system of the automobile. Solenoids are explained in more detail in Chapter 14.

§ 13-19 RESISTANCE

An insulator has a high resistance to the movement of electrons through it. A conductor, such as a copper wire, has a very low resistance. Resistance is found in all electrical circuits. In some circuits we want a high resistance in order to keep down the amount of current flow. In other circuits we want as little resistance as possible so that a high current can flow.

Resistance is measured in **ohms.** For instance, a 1,000-foot [304.8-meter] length of wire that is about 0.1 inch [2.54 millimeters] in diameter has a resistance of 1 ohm. A 2,000-foot [609.6-meter] length of the same wire has a resistance of 2 ohms. In other words, the longer the path, the greater the resistance.

On the other hand, a 1,000-foot [304.8-meter] length of wire that is about 0.2 inch [5.08 millimeters] in diameter has a resistance of only $\frac{1}{4}$ ohm. Thus, the heavier the wire, the lower the resistance.

The explanation is simple. The longer the path, or circuit, the farther the electrons have to travel and the higher the resistance to the electric current. With the heavier wire, the path is larger and the resistance is lower.

§ 13-20 OHM'S LAW

There is a definite relationship between amperes (electron flow), voltage (electrical pressure), and resistance. As the electrical pressure goes up, more electrons flow. Thus, increasing the voltage increases the number of amperes of current. However, increasing the resistance decreases the number of amperes of current. These relationships can be summed up in a formula known as Ohm's law:

Voltage is equal to amperage times resistance

The main thing to remember about Ohm's law is that increasing the ohms, or resistance, cuts down on the current flowing. We will talk about this again when we discuss the electrical systems in a car.

§ 13-21 ONE-WIRE SYSTEMS

For electricity to flow, there must be a complete path, or circuit. The electrons must flow from one terminal of the battery or the alternator, through the circuit, and back to the other terminal. In the automobile, the engine and car frame are used to carry the electrons back to the other terminal. Therefore, no separate wires are required for the return circuit from the electrical device to the battery or the alternator. The return circuit is called the **ground,** and it is indicated in wiring diagrams by the symbol shown in Fig. 13-8. Look at Fig. 13-1 and you will see many of these symbols. Just remember that the ground—the engine and car frame—is the other half of the circuit that runs between the source of electricity (battery or alternator) and the electrical device.

§ 13-22 ALTERNATING CURRENT AND DIRECT CURRENT

Most of the electricity generated and used is alternating current (ac), as we discussed in Chapter 4. The current flows first in one direction, then in the opposite direction; that is, it alternates (Fig. 13-9). The current you use in your home is ac. It alternates 60 times a second and is therefore called 60-cycle ac.

Ground,
or
Return Circuit

Fig. 13-8. The symbol used for ground, or return circuit, in wiring diagrams.

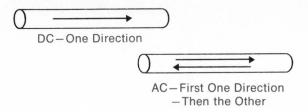

DC—One Direction

AC—First One Direction
—Then the Other

Fig. 13-9. *Top:* direct current. *Bottom:* alternating current.

The automobile cannot use ac. The battery is a direct-current (dc) unit. When you discharge the battery, that is, connect electrical devices to it, you take current out in one direction only. The current does not alternate, or change directions. Likewise, the other electrical devices in the car operate on dc only. In the following chapters, we will discuss the dc electrical devices used in the automobile.

Review questions

1. What is the job of the electrical system in an automobile?
2. How is the battery kept from running down?
3. What is the function of the alternator regulator?
4. What is the purpose of the ammeter?
5. What makes electrons move in a wire? What is a flow of electrons in a wire called?
6. What is the name given to the material used to cover wires and prevent electrons from escaping?
7. Do like magnetic poles attract or repel each other?
8. What is the name given to a winding that has current flowing through it?
9. Does increasing the resistance in a circuit increase or decrease the amount of current flowing?
10. In the one-wire system used in automobiles, what forms the return circuit?
11. Is the battery a dc or an ac unit?

Study problem

1. Raise the hood of an automobile. Using Fig. 13-1 as a guide, identify each component of the automotive electrical system. Try to locate, identify, and trace each of the wires shown in Fig. 13-1 on the automobile. Note how the car engine and frame are used as ground, thus enabling the automobile to have a one-wire system.

14 | AUTOMOTIVE BATTERY AND STARTING MOTOR

§ 14-1 PURPOSE OF THE BATTERY

The battery supplies current to operate the starting motor and the ignition system when the engine is being started. It also supplies current for lights, radio, and other electrical accessories when the alternator is not handling the electric load. The amount of current the battery can supply is limited by the capacity of the battery. This, in turn, depends on the size of the battery and the amount of chemicals it contains.

§ 14-2 BATTERY CONSTRUCTION

The 12-volt automobile battery has a series of six cells in the battery case (Fig. 14-1). Each cell has a number of battery plates with separators between them. When battery liquid, called **electrolyte,** is put into each cell, the cell produces an electrical pressure of 2 volts. The six cells in the battery produce a total of 12 volts.

Each cell contains two groups of battery plates: one group is negative, the other positive. Figure 14-2A shows a battery plate. Notice that a plate group (Fig. 14-2B) consists of several plates attached in one corner to a heavy strap. The terminal post is part of this strap.

Two plate groups (one negative, one positive) are nested together, as shown in Fig. 14-2C, to make an element. Separators made of insulating material such as rubber, spun glass, or plastic are placed between adjacent

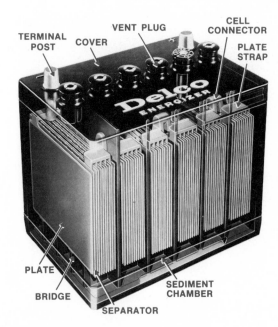

Fig. 14-1. A phantom view of a lead-acid storage battery. The case has been drawn as though it were transparent so that you can see the internal construction of the battery. (*Delco-Remy Division of General Motors Corporation*)

negative and positive plates. This prevents electrical contact between the plates.

The plates are made by applying pastes prepared from lead compounds to plate grids. These lead compounds change in composition as the battery is charged or discharged.

In the battery assembly, the two groups of plates are installed in a cell compartment and

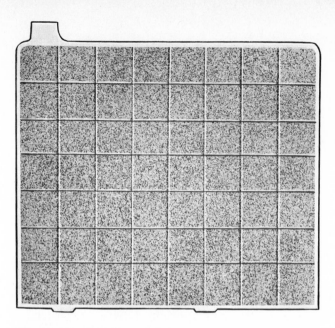

Fig. 14-2A. A battery plate.

Fig. 14-2B. A battery-plate group.

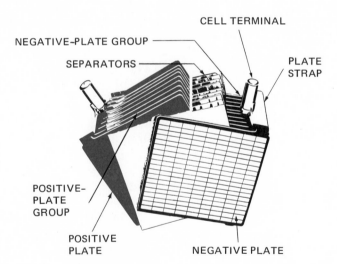

CELL TERMINAL

NEGATIVE-PLATE GROUP

SEPARATORS

PLATE STRAP

POSITIVE-PLATE GROUP

POSITIVE PLATE

NEGATIVE PLATE

Fig. 14-2C. A battery element partly assembled.

a cover is put on, as shown in Fig. 14-1. Then, the cell is filled nearly to the top with a mixture of water and sulfuric acid, the electrolyte.

Many batteries have vent plugs on the cover of each cell. These vent plugs let gas escape from the battery when the battery is being charged. The vent plugs can be removed so that water can be added when the electrolyte level is low.

Some batteries are sealed (Fig. 14-3). Because sealed batteries never need water, they do not need vent plugs. Delco-Remy calls its late-model batteries "Energizers." But don't let this confuse you. Energizers are still batteries.

§ 14-3 BATTERY ELECTROLYTE

The battery electrolyte is made up of water and sulfuric acid. Here is what happens when electric current is taken out of a battery: The sulfuric acid gradually goes into the battery plates, which means that the electrolyte gets weaker.

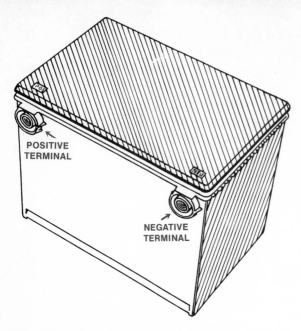

Fig. 14-3. A sealed battery of the type that never requires water.

As this happens, the battery runs down—goes dead—and then has to be recharged. The recharging job requires a battery charger. The charger pushes a current of electricity through the battery, thus restoring the battery to a charged condition. The current is pushed into the battery in a direction opposite to the direction in which it was taken out. There's more on this later in the book.

§14-4 HOW THE BATTERY SUPPLIES CURRENT

During discharge, the electrolyte chemically reacts with the lead compounds in the battery plates to produce voltage and, when a continuous electron path is present, a flow of current.

As the chemical reactions go on, the lead compounds change. In effect, the acid in the electrolyte (sulfuric acid) "goes into" the plates during the chemical reactions. After a while, most of the sulfuric acid has disappeared from the electrolyte (it is replaced by water during the chemical action). With this condition, the battery is discharged, and it

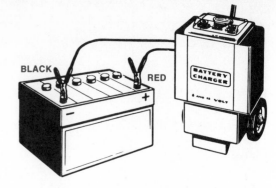

Fig. 14-4. A quick-charger is used to charge a battery. (*Chrysler Corporation*)

cannot supply very much additional current. However, if the battery is given a charge (Fig. 14-4), or is supplied with a flow of current, then the chemical actions in the battery will reverse. Sulfuric acid will reappear in the electrolyte, and the battery will be restored to a charged condition.

§14-5 BATTERY RATINGS

The amount of current that a battery can deliver depends on the area and volume of the active plate material. It also depends on the amount and strength of electrolyte. Batteries are rated in several ways.

1 Reserve capacity. This is measured in the length of time in minutes that a fully charged battery at 80°F [26.7°C] can deliver 25 amperes. A typical rating would be 125 minutes. This figure indicates the ability of a battery to carry the electrical operating load when the alternator is not operating.

2 Cold cranking rate. One of the two cold cranking rates is the number of amperes that a battery can deliver for 30 seconds when it is at 0°F [−17.8°C] without the cell voltages falling below 1.2 volts. A typical rating for a battery with a reserve capacity of 125 minutes would be 430 amperes. This figure indicates the ability of the battery to start the engine at low temperatures. The second cold cranking rate is measured at $-20°F$ $[-28.9°C]$. In this, the final voltage is allowed to drop to 1.0 volt per cell. A typical rating for a battery with a reserve capacity of 125 minutes would be 320 amperes.

3 Overcharge life units. This is a measure of how well the battery will stand up when it is overcharged.

4 Charge acceptance. This is a measure of how well the battery will accept a charge under normal operating conditions with a voltage-regulated automotive charging system.

5 Watts. Delco is using an additional rating—watts. This is roughly equivalent to the battery cold cranking rate.

§ 14-6 BATTERY PERFORMANCE

We can sum up battery performance as follows:

1. Battery voltage decreases when the battery is being discharged. The higher the discharge rate (the more amperes being drawn), the more the voltage is reduced. Also, lower battery temperature results in lower battery voltage when the battery is being discharged. In addition, as the battery is run down, its voltage drops off.
2. Battery voltage tends to increase when the battery is being charged. To increase the charging rate (to force more amperes into the battery), the charging voltage must go up. Also, lower battery temperatures require a high charging voltage to force current through the battery. In addition, as the battery approaches a fully charged condition (most of the sulfuric acid is restored to the electrolyte), the charging voltage must go higher to force current through the battery.

These fundamentals of battery performance are important since they help to explain why it is harder to start a car when the battery is cold, why the alternator voltage goes up as the battery nears a charged condition (thus making a voltage limiter or regulator necessary), and so on. These battery fundamentals will be referred to again when the ignition system, starting motor, alternator, and regulator are described.

§ 14-7 STARTING MOTORS

Starting motors—also called cranking motors and starters—are small but powerful electrical devices that spin the crankshaft to get the engine started. Starting motors are not very complicated and they do not require much service. Now, let's find out how they work and what service they need.

§ 14-8 PRINCIPLES OF STARTING MOTORS

The electric motor (Fig. 14-5) is a device that uses a flow of electric current to cause a shaft to rotate. The starting motor makes use of this simple principle: As the shaft rotates, gearing causes the engine crankshaft to rotate so that the engine is started. Of course, the starting motor is not as simple as the description just given, but the principle is the same.

The fact that magnetism can cause motion can be demonstrated with two bar magnets. A magnet will attract pieces of iron, or iron filings, at both ends, or poles (Fig. 14-6).

§ 14-9 LINES OF FORCE— ANOTHER LOOK

In order to explain these various actions of magnets, scientists think of magnets as acting through **lines of force** that can pull the magnets together or push them apart. These lines of magnetic force can be demonstrated, as shown in Fig. 14-7. Lay a pane of glass over a bar magnet and sprinkle iron filings on the glass. Then, tap the glass and watch the iron filings move into a pattern of lines, as shown. Note that the lines of force stretch between the two poles of the magnet. If we think of these

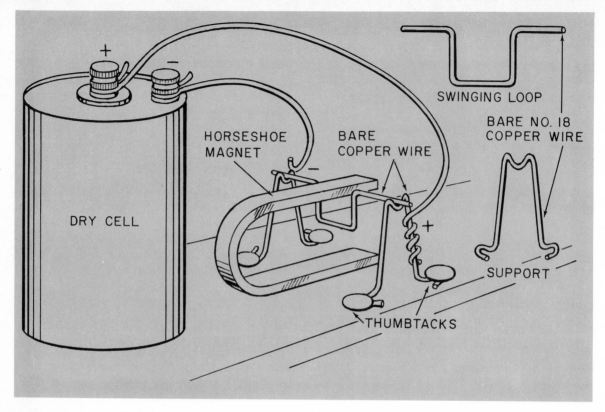

Fig. 14-5. (*Above*) Simple arrangement to demonstrate motor principles. When the circuit is closed,
the wire loop swings out from between the magnetic
poles.

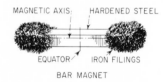

Fig. 14-6. (*Right*) Bar magnet, showing how the
ends, or poles, attract iron filings.

Fig. 14-7. (*Below*) *Left:* demonstrating magnetic lines of force. *Right:* magnetic lines
of force of a bar magnet as shown by iron filings.

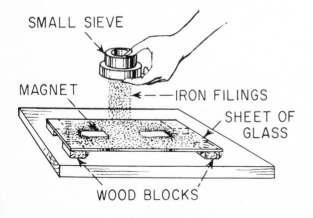

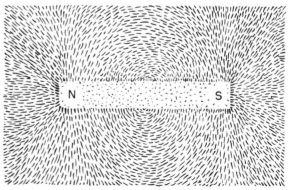

lines of force as having two basic characteristics, we can explain why like magnetic poles repel and unlike poles attract. These characteristics are:

- Lines of force act like rubber bands and attempt to shorten.
- Lines of force do not cross, but attempt to push away from each other.

With these two characteristics in mind, look at the bar magnets again. Lines of force stretch between unlike poles and try to shorten up. Therefore, when two unlike poles are brought near each other (Fig. 13-4), the lines of force between the poles tend to shorten and pull the two poles together.

On the other hand, when we bring two like poles together (Fig. 13-6), the lines of force run parallel to each other. The lines of force, as they run alongside each other, tend to push apart. This produces the repulsion between like poles.

§14-10 ELECTROMAGNETS— ANOTHER LOOK

Magnets have magnetism; they produce magnetic lines of force. Magnetic lines of force can also be produced by electric current. This is called electromagnetism (or magnetism by electricity).

If a wire were wrapped around in the form of a coil, and the coil connected to a battery, the coil would act magnetically, just like a bar magnet (Fig. 14-8). The coil would have a north and a south pole. Which end is north and which end is south would depend upon the direction in which the current flows through the coil.

§14-11 THE LEFT-HAND RULE

The left-hand rule can be used to determine the north pole. Put your left hand around the coil with your fingers pointing in the direction of current flow through the turns of wire (current flows from negative to positive). Your

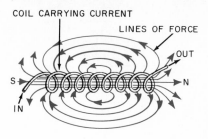

Fig. 14-8. An electromagnet is a coil of wire that is carrying electric current.

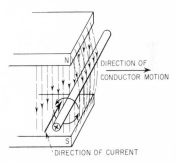

Fig. 14-9. A conductor held in the magnetic field of a magnet. The direction of current flow and encircling magnetic field around the conductor are shown by arrows.

thumb will point to the north pole of the electromagnet.

With these principles in mind, remember that there are magnetic lines of force between the two poles of the magnet (Fig. 14-5). When current flows through the loop, magnetic lines of force will circle the wire in the loop. These magnetic lines of force stretch from the north to the south magnetic pole (Fig. 14-9). The lines of force circle the conductor (wire), as shown by the circular arrow. The direction in which lines of force circle a conductor is known. The X in the cross section of the conductor means that the current is moving away from you (in the direction of the arrow). If it were coming toward you, it could be shown as a dot.

With the current moving away, as shown in Fig. 14-9, the lines of force would circle the conductor in a counterclockwise direction

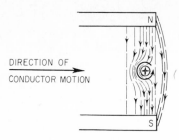

Fig. 14-10. An end view of the conductor shown in Fig. 14-9.

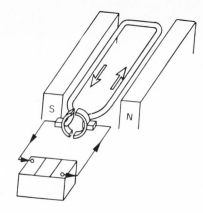

Fig. 14-11. A simple electric motor with a two-segment commutator.

(opposite to the direction of clock-hand movement). In an end view, this would look like Fig. 14-10. Notice that the lines of force to the left of the conductor tend to bunch up. The lines of force from the magnet and the circular lines of force add up on this side. But on the other side, they oppose. Opposing lines of force tend to cancel. Since lines of force try to shorten up, and also push apart, the lines of force to the left of the conductor (Fig. 14-10) tend to push the conductor to the right.

The left-hand rule can be used to show the direction of magnetic lines of force around a conductor. Wrap the fingers of your left hand around the conductor with your thumb pointing in the direction of current movement. Your fingers will point in the direction in which the lines of force circle the conductor.

If we reversed the direction of current through the conductor, it would move the other way (or to the left in Fig. 14-10). The magnetic lines of force would circle the conductor in a clockwise direction; they would pile up to the right of the conductor and move it to the left.

§ 14-12 A SIMPLE MOTOR

If we formed a wire into a complete loop, as shown in Fig. 14-11, and connected the ends to a split ring, we could make the loop rotate. Current would be both coming and going in the loop, as shown by the arrows. Thus, one side of the loop would be pushed one way and the other side the other way. The current would enter the loop through one part, or seg-

ment, of the split ring and would leave through the other.

We could give the rotating loop a stronger rotary push if we strengthened the magnetic field between the magnetic poles. This can be done by adding coils to each magnetic pole. When current flows through these coils, they add to the magnetic field and thus produce a much stronger push on the rotating loop. Figure 14-12 shows one method of connecting the field coils to the loop. The same current flows through both, as shown by the arrows. Since the field coils and rotating loop are connected in series, the motor is called a **series motor.** Many starting motors are series motors.

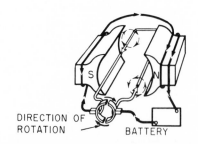

Fig. 14-12. A schematic drawing of a starting motor. Heavy arrows show the direction of current flow, and light circular arrows indicate the direction of the magnetic field around the conductors. Compare this with Fig. 14-11.

§ 14-13 STARTING-MOTOR OPERATION

The starting motor is operated by electro-magnetism. Earlier in the chapter, we said that electricity flowing through wires and through coils of wire, or windings, produces electro-magnetism. We also mentioned two important facts about magnets that you should know when you study starting motors:

- Like magnetic poles repel each other.
- Unlike magnetic poles attract each other.

Now, let's put all this together and see how the starting motor works. Look at Fig. 14-5. It doesn't look like much of a motor, does it? It really isn't, but it will show you how motors work. First, notice the three pieces of copper wire and the shapes they have been bent into: two supports and one swinging loop. Note how the two supports have been fastened to a piece of wood with thumbtacks. Notice also how the swinging loop has been hung from the two supports between the two poles of the horseshoe magnet.

If the dry-cell battery is momentarily connected to the two supports, the swinging loop swings out from under the poles of the magnet. Here's the reason: When current flows through the loop, the magnetic field caused by the cur-rent fights the magnetic field from the horse-shoe magnet. The result is that the horseshoe magnetic field tries to "throw out" the loop.

Now, in the electric motor, there are a lot of loops. And the magnetic field they move in is very strong. So there is a very strong push on the loops.

§ 14-14 STARTING-MOTOR CONSTRUCTION

In an actual starting motor, there are many rotating loops instead of one, with the loops connected to the many segments of the com-mutator. The rotating assembly, mounted on a shaft, is called an **armature** (Fig. 14-13). In the assembly, the armature is held in two bearings at the ends of the starting motor (Fig. 14-14). Brushes (made of conducting material such as copper and carbon) rest on the commutator and feed current to the armature. The current flows through the armature and through the field windings when the circuit to the battery is closed. This causes the armature to rotate, as already mentioned. There is a gear on the end of the armature shaft that meshes with a large ring gear on the engine flywheel during cranking. Thus, the crankshaft is rotated, and the engine is cranked and started. A disassem-bled view of a starting motor using an over-

Fig. 14-13. The two major parts of a starting motor: the armature and the field-frame assembly. (*Delco-Remy Division of General Motors Corporation*)

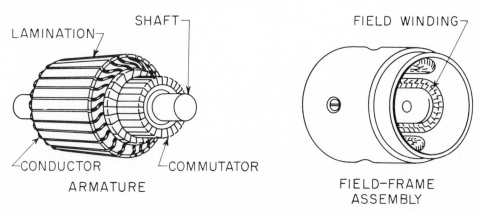

LAMINATION SHAFT FIELD WINDING

CONDUCTOR COMMUTATOR

ARMATURE

FIELD-FRAME ASSEMBLY

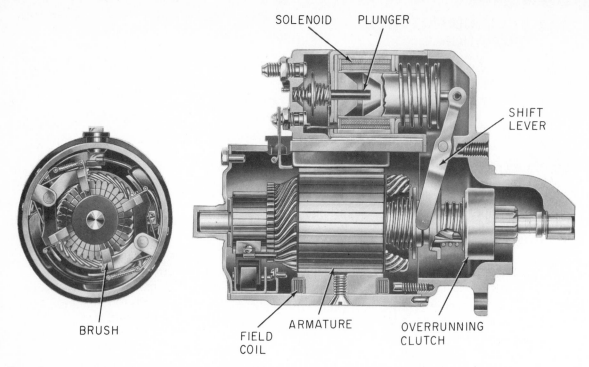

SOLENOID   PLUNGER

SHIFT
LEVER

BRUSH

FIELD
COIL

ARMATURE

OVERRUNNING
CLUTCH

Fig. 14-14. A sectional view of an enclosed shift-lever starting motor. (*Delco-Remy Division of General Motors Corporation*)

running clutch and solenoid is shown in Fig. 14-15.

§ 14-15 STARTING-MOTOR WIRING CIRCUIT

Figure 14-16 shows, in simplified form, the wiring circuit of a starting motor. The current flows into the motor terminal and through the two field windings. From there, it flows through one brush and then through the armature conductors to the other brush. This brush is connected to ground, which is the return circuit to the battery. Ground is explained in Chapter 13. The brushes make sliding contact with the armature commutator. The commutator is made up of separate pieces, or segments, each connected to the end of one of the loops, or conductors. To avoid making a complicated picture, we have not shown the armature conductors in Fig. 14-16.

When current flows through the starting motor, a very strong magnetic field is produced by the field windings. Another powerful magnetic field is produced by the armature conductors. The field-winding magnetic field tries to push the armature magnetic field out of the way. The result is a powerful rotary push on the armature, which causes the armature to revolve. When the armature is connected to the engine flywheel through the meshing gears, it spins the engine crankshaft so that the engine starts.

§ 14-16 DRIVE ASSEMBLY

The drive assembly contains a small pinion that, in operation, meshes with teeth on the flywheel ring gear. The starting motor mounts on the flywheel housing so that the starting-motor pinion can move into mesh with the flywheel and turn it for cranking.

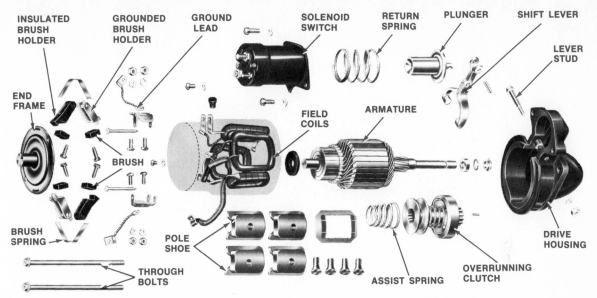

Fig. 14-15. A disassembled view of a starting motor using an overrunning clutch and solenoid. (*Delco-Remy Division of General Motors Corporation*)

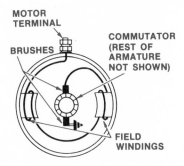

Fig. 14-16. A simplified wiring circuit of a starting motor.

There are about 15 times as many teeth in the ring gear as in the pinion. Thus, the pinion must turn about 15 times to turn the ring gear once. This large gear reduction makes it possible to use a comparatively small motor. The motor can turn at high speed, where it develops considerable power while turning the engine crankshaft at low speeds. Thus, the motor armature may revolve at 3,000 rpm (revolutions per minute) while turning the crank-

shaft at 200 rpm. This is ample cranking speed for starting.

After the engine starts, it may increase in speed to 3,000 rpm or more. If the starting-motor drive pinion remained in mesh with the flywheel ring gear, it would be spun at 45,000 rpm because of the 15:1 (15 to 1) gear ratio. This means that the armature would also be spun at this terrific speed. Centrifugal force would cause the conductors and commutator segments to be thrown out of the armature, ruining it. To prevent such damage, automatic meshing and demeshing devices are used. For passenger cars, there are two types: the Bendix and the overrunning clutch.

§ 14-17 BENDIX DRIVE

The Bendix drive mounts on the end of the armature shaft, as shown in Fig. 14-17. Figure 14-18 is a view of the drive with the pinion partly cut away so that its mounting arrangement can be seen. The drive pinion is mounted loosely on a sleeve that has screw threads matching internal threads in the pinion. When

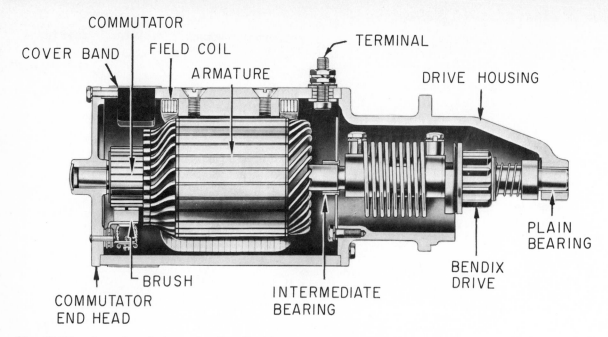

Fig. 14-17. A sectional view of a Bendix-drive-type starting motor. (*Delco-Remy Division of General Motors Corporation*)

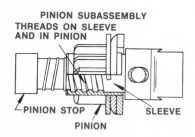

Fig. 14-18. A Bendix drive with the pinion partly cut away to show the internal and external threads of the pinion and sleeve. (*Delco-Remy Division of General Motors Corporation*)

the starting motor is at rest, the drive pinion is not meshed with the flywheel teeth. It is in the position shown in Fig. 14-17. When the starting motor is connected to the battery for cranking, the armature begins to rotate (Fig. 14-19). This causes the sleeve to rotate also, since the sleeve is fastened to the armature shaft through the heavy spiral drive spring. Inertia prevents the pinion from picking up speed instantly. As a result, the sleeve turns within the pinion. This forces the pinion to

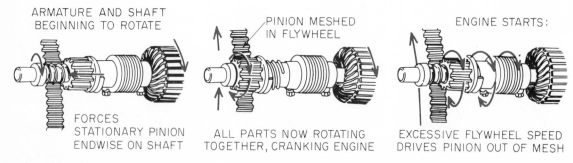

Fig. 14-19. The operation of a Bendix drive. (*Delco-Remy Division of General Motors Corporation*)

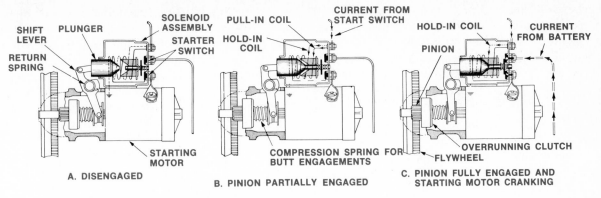

STARTING MOTOR

A. DISENGAGED

COMPRESSION SPRING FOR BUTT ENGAGEMENTS

B. PINION PARTIALLY ENGAGED

OVERRUNNING CLUTCH

FLYWHEEL

C. PINION FULLY ENGAGED AND STARTING MOTOR CRANKING

Fig. 14-20. The operation of the solenoid and overrunning clutch as the pinion meshes with the flywheel teeth. (*Delco-Remy Division of General Motors Corporation*)

move endwise along the sleeve and into mesh with the flywheel teeth. As the pinion reaches the pinion stop, its endwise movement stops. Now, the pinion must turn with the armature, causing the engine to be cranked. The spiral spring takes up the shock of meshing.

After the engine begins to run and increase in speed, the flywheel rotates the drive pinion faster than the armature is turning. This causes the pinion to be spun back and out of mesh from the flywheel. That is, the pinion turns on the sleeve, and the screw threads back the pinion out of mesh from the flywheel.

§ 14-18 OVERRUNNING CLUTCH

The overrunning clutch is operated by a lever that moves the pinion into and out of mesh with the flywheel (Fig. 14-20). Figure 14-14 is a sec-

tional view of a starting motor using an overrunning clutch with a solenoid. The solenoid is a small electromagnet with a moveable plunger, or core. When the solenoid is connected to the battery, the magnetism produced by its coil pulls the core in. The core is attached by a link to the lever. The core movement thus causes the lever to shift the pinion along the armature shaft and into mesh with the flywheel teeth. Figure 14-20 shows this action. As the solenoid core completes its travel, it closes heavy contacts that connect the starting motor to the battery. The armature now turns, causing the pinion to turn the flywheel so that the engine is cranked.

The clutch consists of the outer shell, with four hardened steel rollers fitted into four notches, plus the pinion and collar assembly (Fig. 14-21). The notches are not concentric

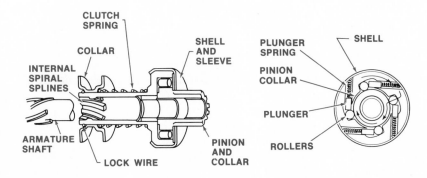

Fig. 14-21. The cutaway and end-sectional views of an overrunning clutch. (*Delco-Remy Division of General Motors Corporation*)

but are smaller in the ends opposite the plunger springs. When the armature and shell begin to rotate, the pinion is momentarily stationary. This causes the rollers to rotate into the smaller sections of the notches, where they jam tight. The pinion must now rotate with the armature, starting the engine. After the engine starts, it spins the pinion faster than the armature is turning. The rollers are rotated into the larger sections of the notches, where they are free. This allows the pinion to spin independently of, or **overrun,** the remainder of the clutch and the armature. Then, when the solenoid is disconnected from the battery, a spring pulls the lever back so that the pinion is demeshed from the flywheel. At the same time, the heavy contacts in the solenoid are separated so that the starting motor is disconnected from the battery.

§ 14-19 SOLENOID WINDINGS

Notice that there are two separate windings in the solenoid: a pull-in winding and a hold-in winding (Fig. 14-20). The pull-in winding has a few turns of a heavy wire. The hold-in winding has many turns of a relatively fine wire. The pull-in winding takes a high current through it and produces a strong magnetic field. A strong field is needed to pull the plunger in and shift the pinion into mesh. After the shift has been completed, a much smaller amount of magnetism is required to hold the plunger in. Notice that the pull-in winding is connected to the two heavy contacts. When the copper disk hits the two contacts to connect the starting motor to the battery, the copper disk shorts out the pull-in winding. The hold-in winding, however, continues to do its job. It is connected from one contact to ground.

Thus, the purpose of the pull-in winding is to provide enough magnetism to shift the pinion into mesh. Then, the pull-in winding is killed to reduce the load on the battery. The hold-in winding continues to hold the plunger in. This arrangement allows the battery to do its best on the starting job without having to waste current when it is no longer needed.

Review questions

1. How many cells make up a 12-volt battery?
2. What is the voltage of each cell in a battery?
3. Battery electrolyte is made up of what two substances?
4. Does battery voltage go up or down while the battery is discharging? Why?
5. What principle causes a starting motor to operate?
6. What is the name given to each end of a magnet?
7. Name two types of magnets.
8. What is determined by use of the left-hand rule?
9. Why is it important to demesh the motor drive pinion from the flywheel ring gear?
10. What are the two basic types of starting-motor drives?
11. How many windings are there in the solenoid? What are they?

Study problem

1. Carefully examine the simple electric motor shown in Fig. 14-5. Following the details shown in the illustration, build this simple electric motor. When you get the motor operating, take it to school and demonstrate the basic motor principles to your class.

15 | AUTOMOTIVE CHARGING SYSTEMS

§15-1 PURPOSE OF THE CHARGING SYSTEM

When you take current out of the battery, you have to put it back in. Otherwise, you will end up with a run-down battery. The generator, or alternator, does this job in the automobile. The battery is a storage place for electrical current, but its capacity is limited. It must always supply current to the starting motor when the engine is cranked. As part of every electrical circuit on the car, the battery must be ready to supply current for all these additional electrical loads. But when the engine is running fast enough, the generator, or alternator, takes over the job as the source of electrical current. At this time, the generator, or alternator, begins to put back into the battery the current taken out for starting. (This is how the battery is kept charged.) In fact, the generator, or alternator, takes over the job of handling the entire electrical load of everything that is turned on. The generator, or alternator, along with the regulator and wiring, is called the **charging system.** Here's what the charging system is all about.

§15-2 GENERATORS AND ALTERNATORS

The generator, or alternator, is a device that produces electric current when driven by an engine or other power source (Fig. 15-1). On the automobile, one of its jobs is to charge the battery. It also supplies current for operation

Fig. 15-1. Location of the alternator.

of electrical devices such as the ignition system, lights, radio, etc. Usually, the alternator is mounted to the side of the engine block. It is driven by the engine fan belt (which also drives the water pump). Both direct-current (dc) and alternating-current (ac) generators have been used on automobiles. Almost all modern cars use ac generators, called **alternators.** But we will describe both dc and ac generators.

§ 15-3 BASIC DC GENERATOR PRINCIPLES

When a current-carrying conductor is held in a magnetic field, it will move (motor principle, Fig. 14-9). The reverse of this is also true. That is, if a conductor is moved through a magnetic field, it will have a voltage induced in it. If a complete circuit is present, current will flow. We will explain later how this action takes place.

If the conductor is moved as shown in Fig. 15-2, a flow of current will be induced in it as shown by the arrow (and also by the dot in the center of the conductor). Figure 15-3 shows how the magnetic field is distorted by the conductor movement. Note that the magnetic lines of force pile up ahead of the conductor and tend to wrap around it in a clockwise direction (in the direction that clock hands turn). Use of the left-hand rule shows that the current would therefore flow toward you (as indicated by the dot in the conductor).

§ 15-4 MAKING ELECTRICITY

In Chapter 14, we explained that current flowing in a wire held in a magnetic field causes the wire to move. The reverse of this is also true. That is, if you move a wire in a magnetic field, you will cause, or **induce,** current to flow in the wire. This process is called **electromagnetic induction** because electricity is induced by magnetism. Let's see how this works.

If the conductor were bent into the shape of a loop and connected to the two halves of a split copper ring, we would have the elements of a generator (Fig. 15-4). Stationary contacts, called **brushes,** connect the rotating loop to the electrical load (a lamp or other electrical device). When the two halves of the loop move (that is, the loop rotates) as shown in Fig. 15-4, current flows as shown by the arrows.

§ 15-5 INCREASING THE AMOUNT OF CURRENT

The amount of current that is induced in the conductor is determined by the strength of the magnetic field and by the speed with which the conductor is moving. In order to strengthen the magnetic field, field windings are used in the generator. Part of the current from the loop

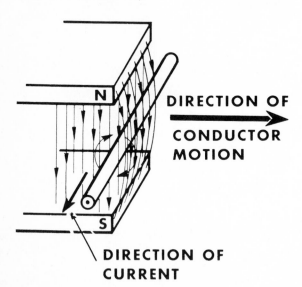

Fig. 15-2. A conductor moving through a magnetic field, as shown, will have a flow of current induced in it, as indicated.

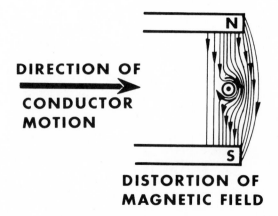

DISTORTION OF MAGNETIC FIELD

Fig. 15-3. Distortion of a magnetic field as a conductor is moved through it and current flows in the conductor.

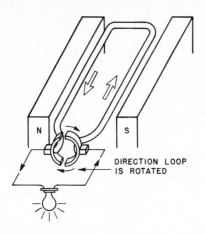

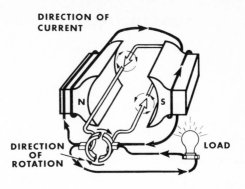

DIRECTION OF CURRENT

DIRECTION OF ROTATION

LOAD

Fig. 15-4. A simple generator showing the direction of induced current.

Fig. 15-5. A simplified schematic diagram of a generator. Heavy arrows show the direction of current flow, and light circular arrows show the direction of magnetic fields around the conductors.

flows through the field windings, increasing the strength of the magnetic field. Thus, the two sides of the loop are cutting through a stronger magnetic field, and more current is induced in the conductor. Generators having the field windings shunted across (or connected between) the main brushes, as shown in Fig. 15-5, are called **shunt** generators. Auto-

motive-type dc generators are usually shunt generators.

The generator must use many rotating loops in order to produce any appreciable amount of current. These loops, or conductors, are assembled into the armature (Fig. 15-6) and are connected to a segmented ring called a **commutator.**

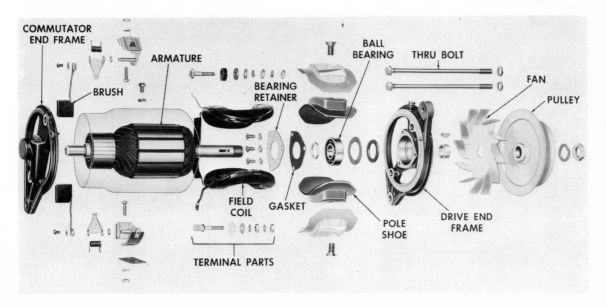

Fig. 15-6. A disassembled view of a passenger-car generator. (*Delco-Remy Division of General Motors Corporation*)

§ 15-6 COMMUTATION

Study Fig. 15-5 for a moment. Notice that as the loop rotates, the two sides cut the magnetic field first in one direction and then in the other. Thus, the current flows through the loop first one way and then the other. The current alternates.

The commutator is used to turn the alternating current into direct current. Notice, in Fig. 15-5, that the commutator, which has two segments in the simple example, also rotates. Note that the side of the loop that is moving down through the magnetic field is always connected through its segment of the commutator to the right-hand brush. So, every time that the current reverses in the loop, the commutator segments reverse positions. The current always flows from the same brush and always flows back through the other brush.

So, you see, the commutator and the brushes turn the ac of the armature loops into dc.

§ 15-7 GENERATOR CONSTRUCTION

The generator consists of the armature, which contains the moving conductors, and the field frame, which holds the field windings in posi-

tion. Two end frames support the armature in bearings and the brushes that are assembled into the commutator end of the generator. A drive pulley, mounted on the end of the armature shaft (Fig. 15-7), is rotated by a V belt. This same V belt may also drive the engine fan and water pump, in addition to the generator drive pulley.

§ 15-8 REGULATORS FOR DC GENERATORS

Generators have to be regulated. Otherwise, they would continue to increase their output until they burned up. Look at Fig. 15-5 again. Part of the current from the armature loop goes through the field coils. This increases the magnetic strength of the field coils. The stronger the magnetic field, the more current the armature produces. And the more current the armature produces, the stronger the magnetic field gets. Thus, in this situation the current would continue to increase until the generator burned up.

To prevent this, generators have regulators. The principle of the regulator is simple. The regulator puts a resistance into the field circuit when the current starts to go too high (Fig. 15-8). The resistance cuts down the amount of

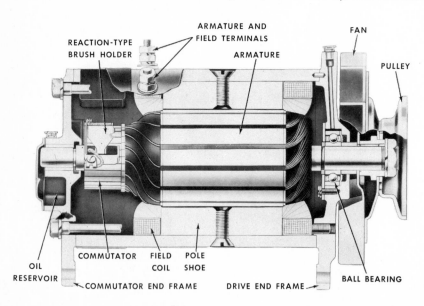

Fig. 15-7. A sectional view of a passenger-car generator. (*Delco-Remy Division of General Motors Corporation*)

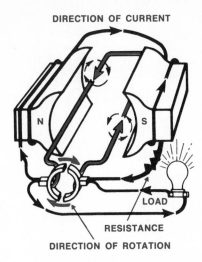

Fig. 15-8. Putting a resistance into the generator field circuit cuts down on the field current. This reduces the magnetic field and thus the generator output.

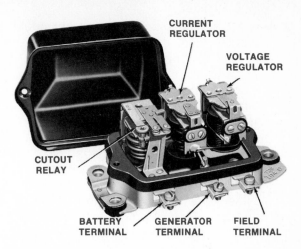

Fig. 15-9. A three-unit current-and-voltage regulator for a direct-current generator. (*Delco-Remy Division of General Motors Corporation*)

current flowing, thereby reducing the strength of the magnetic field. The resistance is used with a pair of contact points. When the points are closed, there is no resistance in the field circuit. When the points are open, the resistance goes into the field circuit. In operation, the points vibrate, or open and close, staying closed just the right amount of time to feed the field coils the amount of current they need.

§ 15-9 THREE-UNIT REGULATOR

The most common regulator used with dc generators has three units (Fig. 15–9). Let's review them quickly so that you will know what they are, in case you work on a dc regulator.

1 Cutout relay. The cutout relay is a unit that closes the circuit to the battery when the generator is running. This allows the generator to charge the battery. The cutout relay opens the circuit when the generator stops, preventing the battery from discharging back through the generator.

2 Current regulator. The current regulator uses a pair of contact points and a resistance.

When the generator current output goes too high, the generator load approaches the danger point, and so the contact points open. When the contact points open, the resistance goes into the generator field circuit and reduces the current. Actually, the points vibrate—open and close—hundreds of times a second. This leaves the resistance in the field circuit just the right amount of time to prevent too much generator current.

3 Voltage regulator. The voltage regulator works on voltage. Voltage, you remember, is electrical pressure. The higher the pressure, or voltage, the more current is pushed through electrical equipment.

When a battery is low, it will take a lot of current. But when a battery is fully charged, it will take only a very small current. The small current requires a high pressure, or voltage. Therefore, when the generator is working against a fully charged battery, it keeps pushing its voltage up and up as it tries to get current through the battery.

If this voltage were allowed to increase, the battery would be overcharged and thus ruined. At the same time, all the electrical equipment would have too much current pushed through it by the high voltage. Excess current could ruin the electrical equipment. For example, high voltage will burn out the headlights.

To prevent this, the voltage regulator has a pair of contact points and a resistance. When the voltage starts to go too high, the points open and the resistance goes into the generator field circuit. The magnetic field is weakened, and the generator voltage is held to a safe amount. Actually, the points vibrate, just as in the current regulator. The action of the points keeps the resistance in the field circuit just the right amount of time to prevent the generator voltage from going too high.

4 Combined action. Remember that the current regulator operates to prevent the generator from exceeding its rated output. For example, if the generator is rated at 40 amperes, the current regulator will prevent the output from going above this amount. Remember, also, that the voltage regulator operates to prevent too much voltage. When it operates, it cuts down the generator output to suit the requirements of the battery and the connected electrical load. For example, suppose that the battery is charged and that nothing is turned on but the ignition. In this case, the voltage regulator will cut output down to a few amperes. Remember this: Either the current regulator is working, holding output to a safe maximum, or the voltage regulator is working, holding the voltage down and thereby cutting output down. They do not both work at the same time.

§ 15-10 ALTERNATORS

As we said earlier, the alternator has come into widespread use in recent years to replace the dc generator in automobiles. In many ways, the alternator is similar to the dc generator. Both have a means of producing a strong magnetic field. Both have loops, or windings, that cut the magnetic lines of force in the magnetic field. However, the basic arrangement of the alternator is different, as you will see.

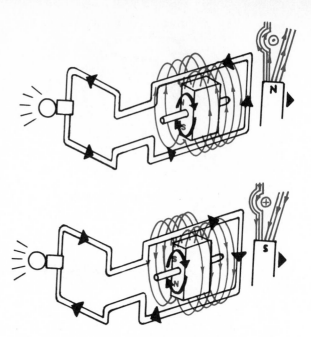

Fig. 15-10. A simplified alternator, consisting of a single stationary loop of wire and a rotating bar magnet. Distortion of moving lines of force around the legs of the loop (conductor) and the direction of current (electron) flow are shown at the right.

§ 15-11 BASIC ALTERNATOR PRINCIPLES

The dc generator rotates the conductors in a stationary magnetic field. The **alternator** rotates a magnetic field so that stationary conductors cut the moving magnetic lines of force.

In a simple one-loop alternator (Fig. 15-10), the rotating bar magnet furnishes the moving field. At the top, as the north pole of the bar magnet passes the upper leg of the loop, and as the south pole passes the lower leg of the loop, current is induced in the loop in the direction shown by the arrows. At the bottom, the magnet has rotated half a turn so that its south pole is passing the upper leg of the loop and its north pole, the lower leg. Now, magnetic lines of force are being cut by the two legs in the opposite direction, and current is induced in the loop in the opposite direction. Thus, as the magnet spins and the two poles

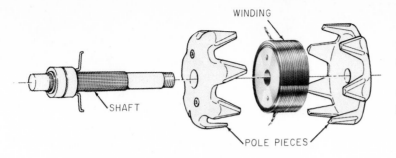

Fig. 15-11. The rotor of an alternator, in partly disassembled view. (*Delco-Remy Division of General Motors Corporation*)

alternately pass the two legs of the loop, current flows first in one direction and then in the other in the loop. In other words, alternating current flows.

§ 15-12 INCREASING ALTERNATOR OUTPUT

Two things will increase the strength of the current moving in the loop—increasing the strength of the magnetic field and increasing the speed with which the magnetic field moves past the two legs of the loop. A third method of increasing the current can be used—increasing the number of loops.

§ 15-13 ALTERNATOR ROTOR

In the actual alternator, both the strength of the magnetic field and the number of loops are increased. Instead of a simple bar magnet, the rotating part of the alternator (called the **rotor**) is made up of a pair of pole pieces assembled on a shaft over an electromagnetic winding (Fig. 15-11). More than two pole pieces are used in many alternators. The electromagnet is made up of many turns of wire. When current flows in the electromagnetic winding, a strong magnetic field is created so that the pointed ends of the two pole pieces become alternately north and south poles (Fig. 15-12). The winding is connected to the battery through a pair of insulated rings that rotate with the rotor shaft and a pair of stationary brushes that ride on the rings. The two ends of the winding are attached to the rings, and the brushes make continuous sliding (or slipping) contact with the slip rings (Fig. 15-12).

Fig. 15-12. The rotor of an alternator, showing brushes in place on slip rings. (*Delco-Remy Division of General Motors Corporation*)

§ 15-14 ALTERNATOR STATOR

Figure 15-13 shows the stationary loops of an alternator assembled into a frame. The assembly is called a **stator** because it is stationary—it does not revolve. The loops are interconnected so that the current produced in all loops adds together. Because this current is alternating, it must be treated in such a way as to convert it into direct current. Remember that the battery must be charged with direct current and that the starting motor is a direct-current unit. Also, the ignition system and other electrical units in the electrical system require direct current. Converting ac into dc is covered in the following sections.

Fig. 15-13. The stator of an alternator. (*Delco-Remy Division of General Motors Corporation*)

ternator and the generator shown in Fig. 15-7. The alternator is more compact, and the rotating part looks different.

In the assembled alternator, the rotor is positioned inside the stationary conductors. The subassembly that includes the stationary conductors is the stator (Fig. 15-13). When the rotor is rotated, the magnetic field it produces passes through the conductors in the stator. Current is induced in these conductors. The current is ac because north and south poles alternately pass the conductors. But the ac must be converted to dc. Diodes do this job. Now let us find out what diodes are all about.

§ 15-15 ALTERNATOR CONSTRUCTION

Figure 15-14 shows an end- and a side-sectional view of an alternator. You will notice immediately a number of differences between the al-

§ 15-16 DIODES

Diodes are one-way electrical valves (Fig. 15-15). A diode lets current through in one direction but not the other. For example, if you connect a battery in one direction to a light bulb and a diode, the light will come on. But if you reverse the connections to the battery terminals, the light will not come on, because no current will flow.

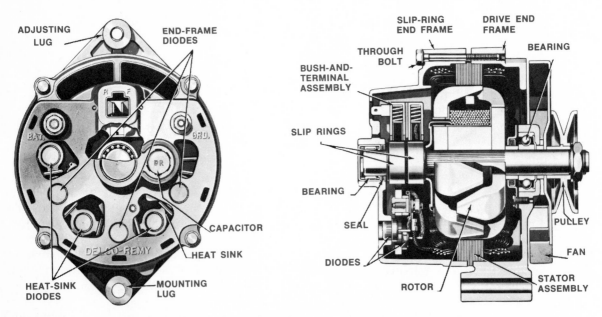

Fig. 15-14. End- and side-sectional views of an alternator with built-in diodes. The manufacturer calls this unit a Delcotron®. (*Delco-Remy Division of General Motors Corporation*)

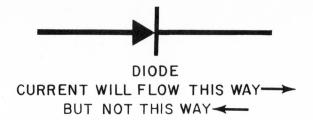

DIODE
CURRENT WILL FLOW THIS WAY→
BUT NOT THIS WAY←

Fig. 15-15. The symbol for a diode in a schematic wiring diagram.

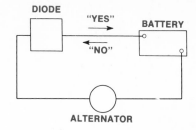

Fig. 15-16. Alternating current from an alternator can be changed to direct current by a diode. The diode allows current to flow in one direction only.

Now, let's go back to the alternator. It puts out alternating current. If we connected a simple alternator through a diode to a battery, we could get direct current and could charge the battery (Fig. 15-16). There are actually three sets of conductors in the alternator stator. Thus, there are three different alternating currents, one in each set of conductors. The alternator is known as a **three-phase unit,** which means that there are three sets of conductors in the stator.

The reason for having three phases in an alternator is the same as the reason for adding more cylinders to an automobile engine. Just as more cylinders in an engine provide more power strokes and a smoother flow of power, so do three phases in an alternator provide more current and a smoother flow of current.

Let's see how diodes can convert ac in a single phase—a single set of conductors—into

dc. Figure 15-17 shows how the diodes do this job. Notice that there are four diodes, numbered 1, 2, 3, and 4. Now, look at the left-hand picture in Fig. 15-17. Notice that current is coming up through the right-hand lead of the ac source. As the current moves up to the diode assembly, only diode 3 lets the current flow through. So the current moves up and flows to the negative (−) terminal of the battery. The current flows through the battery and back to the diodes. As the current returns to the diodes, only diode 1 lets the current flow through.

A moment later, when the current starts flowing in the opposite direction (it has alternated), the situation is as shown at the right in Fig. 15-17. Now, the current is flowing up the left-hand lead from the ac source. The

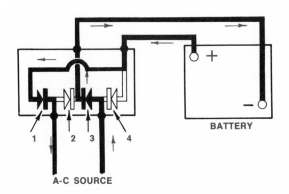

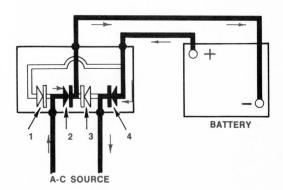

Fig. 15-17. Four diodes connected to an ac source to rectify the alternating current (change it to direct current) to charge the battery. "To rectify" means to change ac to dc.

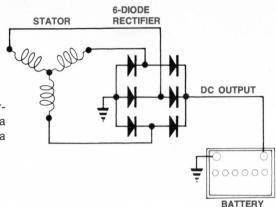

Fig. 15-18. The wiring circuit of an alternator with a six-diode rectifier and a Y-connected stator.

Fig. 15-19. (*Below*) An ac voltage regulator with the cover removed. (*Pontiac Motor Division of General Motors Corporation*)

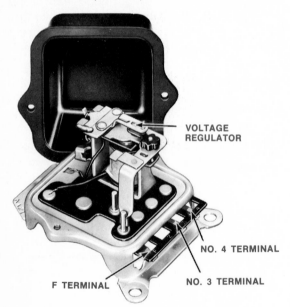

current cannot pass through diode 1, but it can move through diode 2 and from there to the battery. On its return from the battery, the current passes through diode 4 and then down to the ac source.

As you can see, the four diodes, working together as one-way valves, change the ac into dc, which can then be used to charge the battery.

An actual wiring circuit of an alternator with a six-diode rectifier is shown in Fig. 15-18. The stator has three circuits in it that overlap, in effect, to give a more even flow of current. Trace out the circuits, and see how the diodes act to change the ac from the stator to dc.

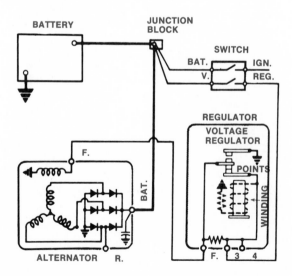

Fig. 15-20. (*Right*) The wiring circuit of an ac regulator and alternator. (*Pontiac Motor Division of General Motors Corporation*)

§15-17 ALTERNATOR REGULATORS

Many alternator regulators work in about the same way as the vibrating regulators for dc generators. Other alternator regulators are solid-state devices. They have no moving parts. Let's talk first about the vibrating type of alternator regulator. Figure 15-19 shows one of these, and Fig. 15-20 shows the wiring circuit for the regulator and the alternator.

The regulator has a winding that is connected through the ignition switch to the battery. When the ignition switch is turned on, current from the battery flows to the winding. Current can also flow through the lower contact points to the field winding, or rotor, in the alternator.

When the engine is running, the current flowing through the rotor winding produces a magnetic field. This field rotates as the rotor turns, and current is produced in the stator windings. The current is ac, which is changed to dc by the diodes. The current flows to the battery. As we noted earlier, if the battery is charged, it does not need much current. Thus, a charged battery will cause the alternator voltage to go up as the alternator tries to push current into the battery.

The way in which the regulator keeps the voltage from going too high is as follows: As the voltage increases, the magnetic field of the regulator winding also increases. The more voltage, the more current flows through the winding, and the more magnetism. When the maximum safe voltage is reached, the magnetism is strong enough to pull the armature down. The armature is a flat piece of steel with a pair of contacts mounted on it. When the armature is pulled down, the contacts open. Now, current can no longer flow directly to the alternator field winding. It has to flow through the resistance. The resistance cuts down the amount of current and the strength of the alternator magnetic field. As a result, the alternator cannot push as hard. The alternator's voltage drops, and the voltage at the regulator winding also drops.

Dropping the regulator-winding voltage weakens the magnetic field, and the magnetic field is no longer strong enough to hold the armature down. Spring tension pulls the armature up, the contact points come together, and current can flow directly to the field winding again. Voltage goes up, and the whole cycle is repeated. The points open and close several hundred times a second and thus limit the top voltage.

Note that there is an upper set of contacts in the regulator. The purpose of this second set is to control voltage during high-speed operation. The rotor is spinning so fast in the alternator that even with the resistance in the alternator field, the alternator voltage can still go too high. The upper set of contacts prevents this, because the higher voltage causes the armature to be pulled down farther so that the top set of contacts closes. When the top set of contacts closes, both ends of the field winding are connected to ground. Therefore, the field current is killed off completely. Thus, the magnetic field gets very weak so that the alternator can make only a weak push, and the voltage is kept from rising to a dangerous level.

§15-18 SOLID-STATE REGULATORS

The solid-state regulator has no moving parts. It does its job with transistors. You have heard of transistor radios, of course, and you know that many television sets use transistors. The transistors in the alternator regulator work in about the same way as they work in televisions and radios. However, in the alternator regulator, transistors do a different job. They prevent high voltage. They work to put resistance into or take resistance out of the alternator field. Figure 15-21 shows an alternator with a built-in solid-state regulator. Notice that this built-in, or integral, arrangement simplifies the system. No separately mounted regulator and extra wiring are required. Also, the solid-state regulator never needs adjustment.

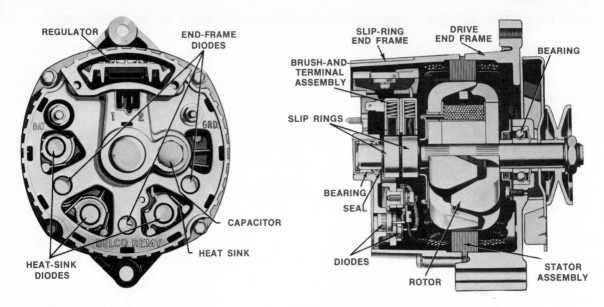

Fig. 15-21. End and sectional views of an alternator with integral (built-in) diodes and a solid-state voltage regulator. (*Delco-Remy Division of General Motors Corporation*)

Review questions

1. What are the jobs of the alternator, or generator, in an automobile?
2. What is the name of the generator device that is used to turn alternating current into direct current?
3. What device prevents the generator from producing too much current? How?
4. What is the name of the device that prevents excessive voltage from developing?
5. Do the current regulator and the voltage regulator work at the same time?

6. What is the name of the part in the alternator that has a winding and two pole pieces?
7. In the alternator, what are the two slip rings and brushes for?
8. What is the name of the stationary part of the alternator that contains the conductors?
9. What is the job of the diode?
10. What is the function of the transistors in a solid-state regulator?

Study problem

1. Obtain a good diode from an alternator. Using the information given in § 15-16 and in Fig. 15-16, determine in which direction the diode passes current and in which direction the diode blocks current flow. Demonstrate this to your class.

16 | AUTOMOTIVE IGNITION SYSTEMS

§ 16-1 PURPOSE OF THE IGNITION SYSTEM

The ignition system produces the sparks that ignite the compressed air-fuel mixture in the engine cylinders. The sparks have to be strong enough to ignite the mixture. Also, they have to arrive at the engine cylinders at exactly the right time. Let's look at the ignition system and find out how it does its job.

§ 16-2 COMPONENTS OF THE IGNITION SYSTEM

Figure 16-1 shows an ignition system. It consists of the battery, the ignition switch, the ignition coil, the ignition distributor, the spark plugs, and the wiring. The ignition coil has the job of producing high-voltage surges. The distributor has the job of getting these surges to the right spark plugs at the right time. The high-voltage surges produce the sparks at the spark plugs.

In Fig. 16-1, to make the picture easier to understand, only one of eight spark plugs is shown. The distributor cap has been removed from the distributor and placed below it. These parts have been rearranged so that you can see the two circuits in the ignition system. The two circuits are explained below.

§ 16-3 ACTION IN THE IGNITION PRIMARY CIRCUIT

There are two circuits in the ignition system: the primary circuit and the secondary circuit.

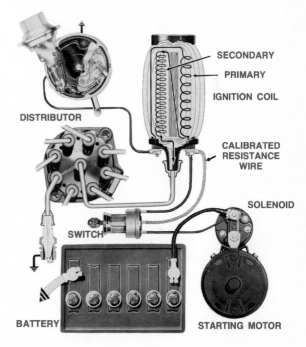

Fig. 16-1. A typical ignition system. It consists of the battery (source of power), ignition switch, ignition coil (shown schematically), distributor (shown in top view with cap removed and placed below it), spark plugs (one shown in sectional view), and wiring. The coil is shown schematically with magnetic lines of force indicated. (*Delco-Remy Division of General Motors Corporation*)

The primary circuit (Fig. 16-2) is composed of the following:

- The contact points in the ignition distributor
- The primary winding in the ignition coil
- The ignition switch
- The battery

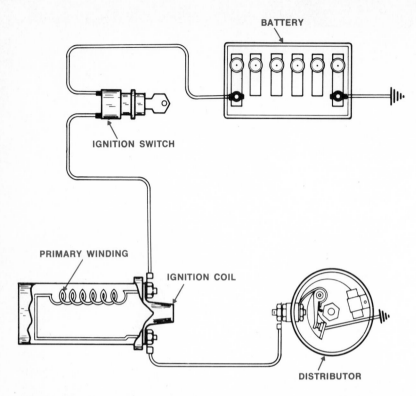

BATTERY

IGNITION SWITCH

PRIMARY WINDING

IGNITION COIL

DISTRIBUTOR

Fig. 16-2. A simplified diagram of the primary circuit of the ignition system.

The contact points consist of a stationary point that is connected to ground through the distributor and a movable point that is mounted on a contact arm (Fig. 16-3). The distributor has a shaft on which a cam is mounted. The cam has the same number of high points, or lobes, as there are cylinders in the engine.

When the engine is running, the distributor shaft turns, and the cam repeatedly opens and closes the contact points. Every time a cam lobe comes up under the rubbing block on the contact arm, it pushes the arm out. This separates the movable contact point from the sta-

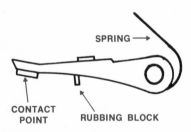

SPRING →

CONTACT
POINT

RUBBING BLOCK

Fig. 16-3. A distributor contact arm.

tionary contact point. Then, as the cam continues to move around, the cam lobe moves out from under the rubbing block. Now, the spring on the movable contact point pushes it back into contact with the stationary contact point.

The contact points are often called breaker points because they repeatedly break the circuit between the primary winding of the ignition coil and the battery. Likewise, the cam is often called the breaker cam.

Every time the contact points close, the primary winding of the ignition coil is connected to the battery (providing the ignition switch is turned on). This connection allows current to flow through the primary winding. A magnetic field builds up. (You remember that current flowing in a winding produces a magnetic field around the winding.) Then, when the contact points are separated, the current stops flowing, and the magnetic field collapses. It is this buildup and collapse of the magnetic field that produces the high-voltage sparks at the spark plugs, as we will see in a moment.

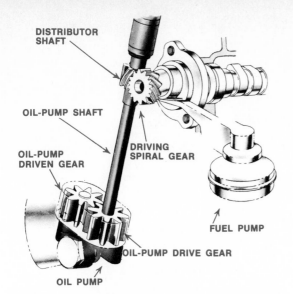

DISTRIBUTOR SHAFT

OIL-PUMP SHAFT

OIL-PUMP DRIVEN GEAR

DRIVING SPIRAL GEAR

FUEL PUMP

OIL-PUMP DRIVE GEAR

OIL PUMP

Fig. 16-4. The method of driving the distributor shaft. The spiral gear on the shaft meshes with the spiral gear on the camshaft. This turns both the distributor shaft and the oil pump shaft. There is an eccentric on the camshaft that operates the fuel pump.

§ 16-4 DRIVING THE DISTRIBUTOR

The distributor shaft is driven from the engine camshaft by a pair of spiral gears, as shown in Fig. 16-4. The driving gear is on the cam-

shaft, as shown. The driven gear is on the distributor shaft. This pair of gears also drives the oil pump, which you can see at the bottom of the picture. The camshaft turns at half the speed of the crankshaft. Because the camshaft gear is meshed with the distributor gear, the distributor shaft turns at half the speed of the crankshaft.

§ 16-5 ACTION IN THE IGNITION SECONDARY CIRCUIT

In Fig. 16-5, we add part of the secondary circuit. Shown here are:

- The secondary winding in the ignition coil
- The distributor cap with the wires, or cables, coming from the cap and heading toward the spark plugs

This part of the ignition system—the secondary circuit—produces the high-voltage sparks and distributes them to the spark plugs. Every time a piston is nearing TDC on the compression stroke, the secondary winding sends a high-voltage surge to the spark plug in that cylinder. A spark occurs at the spark plug, the

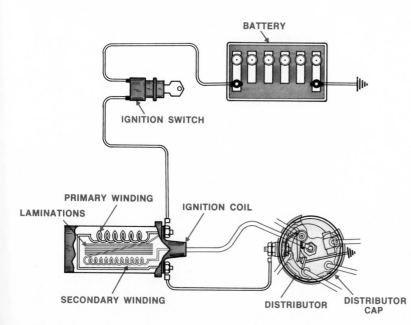

BATTERY

IGNITION SWITCH

PRIMARY WINDING

LAMINATIONS

IGNITION COIL

SECONDARY WINDING

DISTRIBUTOR

DISTRIBUTOR CAP

Fig. 16-5. Here, we have added part of the secondary circuit to the primary circuit.

compressed mixture fires, and the power stroke takes place.

§ 16-6 DISTRIBUTOR AND ROTOR ACTION

Let's take a closer look at the distributing action of the distributor. On top of the cam in the distributor is a rotor (Fig. 16-6). The rotor sits on the cam, as shown in Fig. 16-7. As the distributor shaft rotates, the cam and the rotor also rotate. The purpose of the rotor is to connect the center terminal of the distributor cap to the outside terminals of the distributor cap.

The terminals in the cap are insulated from one another and are held in place in the cap. You can see three of them cut away in Fig. 16-7. The center terminal of the cap has a carbon button on its lower end. This button rests on one end of the rotor blade. A small spring holds the carbon button and rotor blade in continuous contact. Therefore, the rotor blade is always connected to the secondary winding of the ignition coil. Whenever the coil secondary winding produces a high-voltage surge, the metal blade of the rotor sends the surge to the proper spark plug. In other words, the rotor has turned so that it points to the correct side terminal. That is, the rotor points to the side terminal that is connected to the

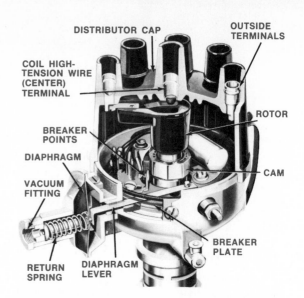

Fig. 16-7. A cutaway view of a distributor, showing how the rotor is mounted on top of the cam. This picture also shows the construction of the vacuum-advance mechanism. (*Ford Motor Company*)

spark plug in the cylinder which is ready to fire (Fig. 16-8).

Let's go over this again. First, the cam turns, and a lobe opens the contact points. This causes the magnetic field from the coil primary winding to collapse. The collapse produces a high-voltage surge in the secondary winding (we'll explain how later). The rotor is pointing to the side terminal that is connected to the spark plug which needs the high-voltage surge. The surge quickly comes out of the coil secondary winding. It goes through the high-tension cable to the center terminal of the distributor cap. The surge then jumps from the rotor to the side terminal and rushes to the spark plug. This causes a spark at the gap between the two electrodes in the spark plug. The compressed air-fuel mixture is ignited, and the power stroke results.

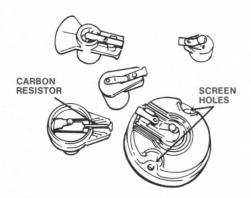

Fig. 16-6. Types of distributor rotors. The one at the lower left has a carbon resistor. The one at the lower right is attached to the advance mechanism by screws. (*Delco-Remy Division of General Motors Corporation*)

§ 16-7 SPARK PLUG

A spark plug (Fig. 16-9) has an insulator to hold the center electrode, a shell to support the insulator, and a ground electrode. When the

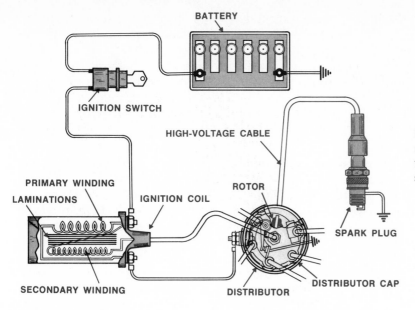

BATTERY

IGNITION SWITCH

HIGH-VOLTAGE CABLE

PRIMARY WINDING
LAMINATIONS

IGNITION COIL

ROTOR

SPARK PLUG

SECONDARY WINDING

DISTRIBUTOR

DISTRIBUTOR CAP

Fig. 16-8. Here, we add the rest of the secondary circuit—the high-voltage cable and spark plug.

high-voltage surge enters a spark plug, it flows down the center electrode and jumps from the lower end of the center electrode to the ground electrode just beneath it.

Some spark plugs have a built-in resistor, as shown in Fig. 16-9. This resistor limits interference on radio and television receivers. Every high-voltage surge that flows from the coil secondary through the cables to the spark plugs sends out a form of radio wave. The radio wave is actually a magnetic field that can make noise in the radio or blur the television picture. The resistor concentrates the high-voltage surge and therefore reduces the interference.

§ 16-8 SECONDARY WIRING

The secondary wiring needs heavy insulation to hold in the high-voltage surges. The secondary wires are sometimes called high-voltage cables. At one time, the high-voltage cables were made of copper wire covered with a heavy rubber or neoprene insulation. Later, the conducting core was made of linen impregnated with carbon. This core formed a high-resistance path that acted in the same

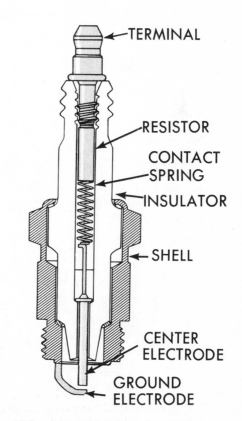

TERMINAL

RESISTOR

CONTACT
SPRING

INSULATOR

SHELL

CENTER
ELECTRODE

GROUND
ELECTRODE

Fig. 16-9. A sectional view of a resistor-type spark plug. (*Ford Motor Company*)

way as the resistor in the spark plug. About 1963, some car manufacturers started using fiber glass saturated with graphite as the conducting core. These cables do the same job as the linen-core cable, but are not damaged as easily.

§ 16-9 IGNITION-COIL ACTIONS

Let's take a closer look at the ignition coil and see how it produces the high-voltage surges. A coil partly cut away is shown in Fig. 16-10. The secondary winding is on the inside, and the primary winding is on the outside. The lamination, which consists of strips of iron, helps to concentrate the magnetic lines of force.

When the contact points in the distributor close, they connect the primary winding to the battery. Magnetic lines of force build up almost instantly around the primary winding. Then, when the magnetic field is built up, the contact points open. Now, with no current flowing through the windings, the magnetic field collapses rapidly. That is, the lines of force move through the secondary winding at high speed. This produces the high-voltage surge in the

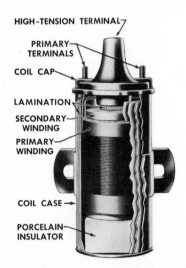

HIGH-TENSION TERMINAL
PRIMARY TERMINALS
COIL CAP
LAMINATION
SECONDARY WINDING
PRIMARY WINDING
COIL CASE →
PORCELAIN INSULATOR

Fig. 16-10. An ignition coil with the case cut away to show how the primary winding is wound around outside the secondary winding. (*Delco-Remy Division of General Motors Corporation*)

secondary winding. The collapse is aided by the condenser, which we will describe in a moment.

Now, let's review for a minute. In the alternator, which is explained in Chapter 15, magnetic lines of force move through the conductors in the stator to produce electric current. Well, the same thing happens in the ignition coil. The moving lines of force, as they collapse through the secondary winding, produce current in the secondary winding.

Every turn of wire in the secondary winding has electrical pressure, or voltage, produced by the moving lines of force. As you know, it is voltage that makes the current (electrons) flow. Since there are thousands of turns of wire in the secondary winding, a high electrical pressure of thousands of volts builds up.

The voltage builds up enough to send a high-voltage surge of current rushing out of the center, or high-voltage, terminal of the coil cap. The current flows through the high-voltage cable to the center terminal of the distributor cap. From there, the current passes through the rotor and the high-voltage cable to the spark plug that is ready to fire. The high-voltage surge leaps across the spark-plug gap to produce the spark that ignites the compressed mixture.

§ 16-10 CONDENSER

The condenser is also called the capacitor. It consists of two long strips of lead or aluminum foil, with insulating strips of paper between them, wrapped in a winding (Fig. 16-11). The winding is put into a metal can with a bracket so that the assembly is able to be mounted in the distributor, as shown in Fig. 16-12. The can is connected to one of the two metal strips. The other metal strip is connected to the lead. In the distributor, the can is attached to the plate on which the contact points are mounted. This plate is called the **breaker plate.**

The lead is connected to the movable contact arm. The condenser, therefore, is connected *across* the breaker points. One of the

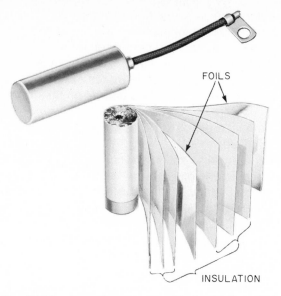

Fig. 16-11. A condenser assembled, with the winding partly unwound.

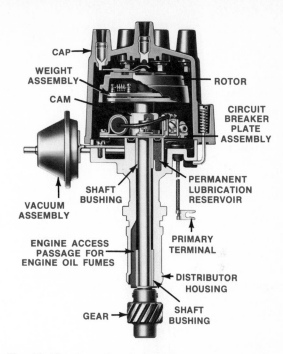

Fig. 16-12. A sectional view of an ignition distributor. (*Delco-Remy Division of General Motors Corporation*)

metal strips is connected to the movable contact point. The other metal strip is connected to the stationary (grounded) contact point.

Now, let's see what the condenser does. When the contact points begin to separate, the current stops flowing. However, it "wants" to continue to flow. In other words, current tries to keep flowing even as the distributor contacts start to separate. Without the condenser, it would jump across the separating contact points, making an arc. This arc would waste most of the high-voltage surge, and there might not be enough voltage left in the coil to produce the spark at the spark plug. Also, an arc across the points would soon burn the points so that they would no longer make good contact. The condenser prevents an arc from forming. It momentarily provides a place for the current to flow while the contact points separate. In doing this, the condenser brings the current flowing in the primary winding to a quick stop. This speeds up the magnetic-field collapse and increases the voltage in the secondary winding. With a higher voltage, the high-voltage surge is more intense, and the spark at the spark plug is stronger. A strong

spark assures good ignition of the compressed air-fuel mixture.

§ 16-11 DISTRIBUTOR CONSTRUCTION

Figures 16-12 and 16-13 show various views of ignition distributors. The distributor has a housing that protects and supports the other parts. Centered in the housing is the distributor shaft. The bottom of this shaft has a spiral gear through which the distributor shaft is driven. The upper end of the shaft supports a **centrifugal-advance** mechanism and the distributor cam on which the rotor is mounted. We will describe the centrifugal-advance mechanism in a moment. First, study the two illustrations of distributors. Note, in Fig. 16-13, that the rotor is a large, heavy molding of insulating material. The rotor is larger in this distributor to make sure that the high-voltage surge goes out to the spark plug instead of jumping around inside the distributor.

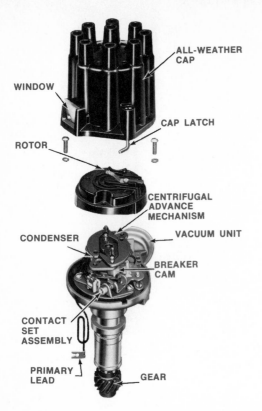

Fig. 16-13. A partly disassembled distributor. *(Delco-Remy Division of General Motors Corporation)*

ALL-WEATHER CAP

WINDOW

CAP LATCH

ROTOR

CENTRIFUGAL ADVANCE MECHANISM

CONDENSER

VACUUM UNIT

BREAKER CAM

CONTACT SET ASSEMBLY

PRIMARY LEAD

GEAR

§ 16-12 IGNITION TIMING

Ignition timing refers to the adjustment of the distributor so that the breaker points open at just the right moment. When the breaker points open at the right moment, the high-voltage surge arrives at the spark plug at just the right time. Ignition timing is adjusted by turning the distributor in its mounting, as we will learn later. This adjustment shifts the position of the breaker points around so that the cam lobe opens them at the proper moment.

§ 16-13 CENTRIFUGAL-ADVANCE MECHANISM

When the engine is idling or running at low speed, the high-voltage surge is *timed* to arrive

at the spark plug just before the piston reaches TDC on the compression stroke.

When the engine speeds up, things happen much faster in the cylinders. The pistons move up on the compression strokes, go past TDC, and start down in far less time when the engine is running at highway speed. However, the combustion process in the cylinder does not speed up.

Therefore, if the spark continued to appear at the spark plug at the same time at high speed as at low speed, this is what would happen: The spark would occur and combustion would start, but the piston would be past TDC and starting down again before the combustion really got going. The piston would move down so fast that it would get ahead of the pressure rise. This would mean a very weak push on the piston and a weak engine at high speed. Most of the power in the burning fuel would be wasted.

To prevent this, distributors have centrifugal-advance mechanisms that advance, or push ahead, the spark as the engine speed increases. You can see the advance mechanisms assembled in Figs. 16-12 and 16-13. The parts include a pair of advance weights, also called governor weights. These weights are held by pivots on the base; the base is attached to the distributor shaft. Both weights have advance springs, which hold them in when the engine is idling. Figure 16-14 shows the relationship. At the left, the engine cylinder is shown in sectional view. The engine is running at 1,000 rpm. The spark is occurring at 8 degrees before the piston reaches TDC on the compression stroke. This gives the compressed air-fuel mixture enough time to start burning and to develop high pressure.

At the right in Fig. 16-14, the positions of the two advance weights are shown with the engine operating at 1,000 rpm. There is no advance. This is the basic ignition timing. In other words, with no advance, the spark occurs at 8 degrees before TDC.

Now, look at Fig. 16-15. Here is the situation when the engine is running at 4,000 rpm (2,000 distributor rpm). The increased speed

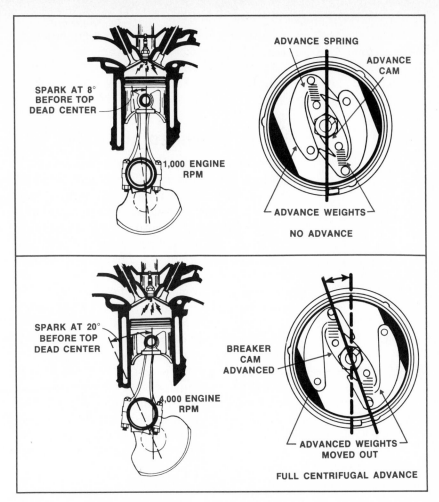

Fig. 16-14. The common position of the advance weights at low speed. There is no centrifugal advance action, so there is only the initial timing advance.

Fig. 16-15. The common position of the advance weights at 4,000 engine rpm. The weights have moved out against the spring tension to move the advance cam and breaker cam ahead. In the typical engine, the spark now occurs at 20 degrees before TDC, but these figures do not apply to all engines. Different engines have different advances.

has caused centrifugal force to push the advance weights out against the tension of the advance springs.

As the advance weights push out, their inner ends push against the advance cam, which is free to turn on the upper end of the distributor shaft. As the advance cam turns, it pushes the breaker cam forward in the direction of the rotation. Now, the breaker-cam lobes move under the breaker-arm rubbing block earlier. The points are opened earlier, and as a result, the spark is advanced. In Fig. 16-15, the spark has been pushed ahead, or advanced, 12 degrees for a total advance of 20 degrees (8 + 12 = 20) before TDC, when the engine is running at 4,000 engine rpm (2,000 distributor rpm). This arrangement gives the compressed

mixture more time to burn and build up pressure on the piston. The power in the fuel is therefore used more efficiently.

Note that the items shown in Figs. 16-14 and 16-15 are not for any particular engine. They are typical and are given to show how the advance mechanism works.

§ 16-14 VACUUM-ADVANCE MECHANISMS

There is another condition under which an advance should occur. This condition occurs when the engine is operating at part throttle. At part throttle, less air-fuel mixture gets into the cylinders. Thus, there is less air-fuel mix-

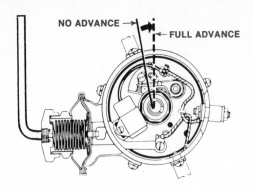

Fig. 16-16. The top view of a distributor with cap and rotor removed. The vacuum-advance mechanism is shown in sectional view. The solid line, arrow, and dashed line show the no-advance and full-advance positions of the breaker plate.

ture in the combustion chambers. With less fuel, the mixture takes longer to burn after it is ignited.

So we have the same problem here as we had before. If the mixture burns more slowly, the piston will be past TDC and moving down before the mixture has a chance to burn and produce high pressure. As a result, much of the power in the fuel will be lost. The vacuum-advance mechanism is designed to prevent this loss.

You can see vacuum-advance mechanisms in Figs. 16-12 and 16-13. Figure 16-7 shows a vacuum-advance mechanism partly cut away. A sectional view of a vacuum-advance mechanism is shown in Fig. 16-16. The mechanism contains a flexible diaphragm that is spring loaded. The center of the diaphragm is connected by a linkage to the breaker plate, on which the breaker points are mounted. The breaker plate is supported on a bearing so that it can rotate a few degrees one way or the other. Study Figs. 16-7 and 16-16 to get the construction of the vacuum-advance unit clear in your mind. Now, look at Figs. 16-17 and 16-18, which show how the vacuum advance works.

Note that with a wide-open throttle, there is very little vacuum in the intake manifold. Therefore, there will be no vacuum advance.

Vacuum advance occurs when there is a vacuum in the intake manifold and less air-fuel mixture is getting into the cylinders. In other words, vacuum advance occurs only during part throttle.

§ 16-15 COMBINED ADVANCE

In a distributor equipped with both centrifugal advance and vacuum advance, the two advances will combine to give the total advance needed for any operating condition. Centrifugal advance is based on engine speed. Vacuum advance is based on intake-manifold vacuum. Centrifugal advance always occurs as the engine speed increases. But vacuum advance is determined by the position of the throttle and the vacuum in the intake manifold.

A curve showing a typical combined advance is shown in Fig. 16-19. The vacuum advance is shown as being added to the centrifugal advance. In the example shown by a black line, the centrifugal advance is 16 degrees. Added to this is 10 degrees of vacuum advance that is possible if the throttle is only partly open at the engine speed indicated. The result is a total of 26 degrees possible advance.

§ 16-16 CONTROLLING THE VACUUM ADVANCE

In many engines today, there are emission-control devices that, in effect, turn the vacuum advance on or off to reduce emissions from the engine. We talk about this in Chapter 20, the chapter in which we describe emission-control devices used on modern automobiles.

§ 16-17 ELECTRONIC IGNITION SYSTEM

The electronic ignition system does not use breaker points. Instead, it uses a magnetic pickup device in the distributor and an electronic amplifying device. These two devices work together to start and stop the flow of current in the ignition coil primary winding.

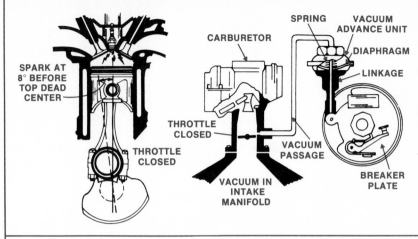

Fig. 16-17. When the throttle is closed and the engine is running at low speed, there is no vacuum advance or centrifugal advance.

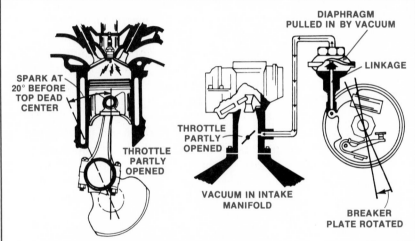

Fig. 16-18. When the throttle is partly opened so that vacuum is applied to the vacuum-advance unit, the breaker plate is rotated, or moved ahead. The cam closes and opens the points earlier to produce a vacuum advance.

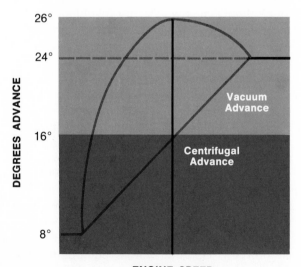

Fig. 16-19. Centrifugal- and vacuum-advance curves for one engine.

There are various designs of this system, and we will discuss two of them: the design used in cars made by the Chrysler Corporation and the design used in cars made by General Motors.

§ 16-18 CHRYSLER ELECTRONIC IGNITION SYSTEM

Chrysler uses an electronic ignition system on all its cars made in the United States since 1973. In this system, contact points are not

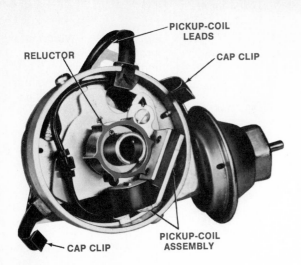

RELUCTOR · PICKUP-COIL LEADS · CAP CLIP · CAP CLIP · PICKUP-COIL ASSEMBLY

Fig. 16-20. The top view of the Chrysler electronic ignition distributor. The cap and rotor have been removed so that you can see the reluctor and the pickup coil. (*Chrysler Corporation*)

used. Instead, the distributor has a metal rotor with a series of tips on it. This rotor, called the **reluctor,** is shown in Fig. 16-20. The reluctor takes the place of the cam in the type of distributor we discussed previously. Notice that the reluctor shown in Fig. 16-20 has six tips. It is for a six-cylinder engine. Notice also that the distributor has a permanent magnet and a pickup coil.

The principle of operation is simple. Every time a tip of the reluctor passes the pickup coil, it carries a magnetic field from the magnet through the coil. The reluctor provides a path for the magnetic lines of force. This magnetic field, sweeping through the pickup coil, produces a pulse of electric current. The current is very small, but it is enough to trigger the control unit into action.

The control unit uses electronic devices—diodes and transistors—to control the flow of current to the ignition coil. When the pulse of current from the pickup coil arrives at the control unit, the control unit shuts off the flow of current to the ignition coil. This is the same job performed by the contact points in the other kind of distributor.

As soon as the tip of the reluctor passes the pickup coil, the pulse of current from the pickup coil ends. This allows the control unit

to close the circuit from the battery to the ignition coil. Now, primary current flows again, and a magnetic field builds up once more in the ignition coil. Then, the next tip of the reluctor passes the pickup coil, and the whole series of actions is repeated.

In this system, there are no contact points to adjust or wear out. Everything is automatic, and the only adjustment required on this system is the ignition timing, which we talk about later.

§ 16-19 GENERAL MOTORS HIGH ENERGY IGNITION (HEI)

General Motors calls its distributor a magnetic-pulse distributor. Figure 16-21 shows this HEI distributor. It looks a little different from the Chrysler unit, but it works the same way. The HEI distributor has a pole piece in the form of a ring with a series of teeth pointing inward. The pole piece has the same number of teeth as there are cylinders in the engine. Under the pole piece is a permanent magnet with a pickup coil. The timer core, made of iron, is placed on top of the distributor shaft in exactly the same way as the cam is placed in the contact-point system. The timer core has the same number of teeth as there are cylinders in the engine.

When the engine is running, the teeth on the timer core align with the teeth on the pole piece eight times with every revolution of the timer core (on an eight-cylinder distributor). Every time the teeth align, magnetic lines of force are carried through the pickup coil. This produces a pulse of current that flows to the ignition-pulse amplifier, which GM calls an electronic module, mounted in the distributor. There, the pulse electronically opens the circuit to the ignition coil primary. The magnetic field in the coil collapses, and a high-voltage surge is produced. This surge is carried by special silicone high-voltage leads, the distributor cap, and the rotor to the spark plug that is ready to fire.

For a couple of years prior to the introduction of HEI, General Motors equipped some

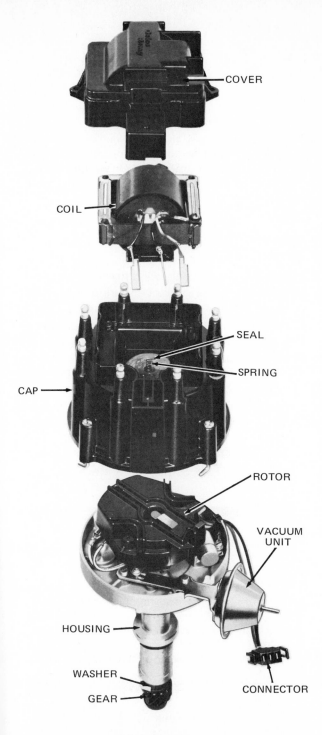

COVER

COIL

SEAL

SPRING

CAP

ROTOR

VACUUM UNIT

HOUSING

WASHER

GEAR

CONNECTOR

Fig. 16-21. Partly disassembled view of the General Motors high energy ignition distributor. (*Delco-Remy Division of General Motors Corporation*)

cars with a unit distributor. The unit distributor was similar to the HEI distributor. However, the unit distributor did not have the high secondary voltage of up to 35,000 volts that is available with the HEI system.

§ 16-20 IGNITION SWITCH

In late-model cars, the ignition switch is mounted on the steering column, as shown in Fig. 16-22. This arrangement locks the steering shaft at the same time that the ignition switch is turned off and the ignition key is removed. When this happens, a small gear on the end of the ignition switch rotates and releases a plunger. The plunger enters a notch in a disk on the steering shaft to lock the shaft. If a notch is not lined up with the plunger, the plunger rests on the disk. When the steering wheel, shaft, and disk are turned slightly, the plunger drops into a notch.

When the ignition key is inserted and the ignition switch is turned on, the plunger is withdrawn from the disk to unlock the steering shaft.

The ignition switch has an extra set of contacts that are used when the switch is turned past ON to START. The contacts connect the starting-motor solenoid to the battery so that the starting motor can operate. As soon as the engine is started and the switch is released, it returns to ON, and the starting motor is disconnected from the battery.

The alternator field circuit is connected to the battery through the ignition switch when it is turned to ON. When the ignition switch is turned to OFF, the alternator field circuit is disconnected so that the battery cannot run down through the field circuit.

Another job that the ignition switch does is to operate a buzzer if the key is left in the lock when the car door on the driver's side is open. This is a reminder to the driver to remove the key from the lock when leaving the car. This makes it harder for a thief to steal the car.

Such accessories as the radio and the car heater are also connected to the battery

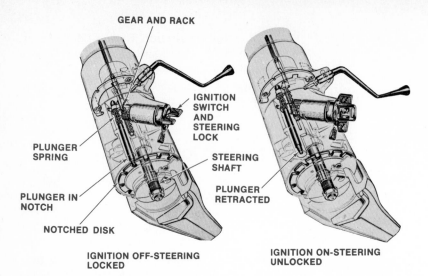

GEAR AND RACK

IGNITION
SWITCH
AND
STEERING
LOCK

PLUNGER
SPRING

STEERING
SHAFT

PLUNGER IN
NOTCH

PLUNGER
RETRACTED

NOTCHED DISK

IGNITION OFF-STEERING
LOCKED

IGNITION ON-STEERING
UNLOCKED

Fig. 16-22. A combination ignition switch and steering-wheel lock in phantom views, showing the two positions of the lock. (*General Motors Corporation*)

through the ignition switch. This arrangement prevents the driver from leaving these units running when he or she turns off the engine and leaves the car.

§ 16-21 IGNITION COIL RESISTOR

In many passenger cars with 12-volt ignition systems, there is a resistance wire in the primary circuit (see Fig. 16-1). This wire is shorted out by the ignition switch when the switch is turned to START. Now, full battery voltage is imposed on the primary winding of the ignition coil for good performance during cranking. With full battery voltage on the primary, the secondary voltage is higher. The sparks are stronger, and the engine starts more easily. After the engine is started, the ignition switch is turned to ON. Now the resistance goes into the ignition primary circuit. This protects the contact points from excess current. However, vehicles equipped with the General Motors high energy ignition system do not use a resistor in the ignition system. In these systems, full battery voltage is applied to the coil.

Review questions

1. What are the six major components of the ignition system?
2. How many windings are there in the ignition coil?
3. How is the distributor shaft driven?
4. Does the high-voltage surge from the coil secondary winding occur when the points close or when they open?
5. What is the job of the condenser?

6. What are the two types of advance mechanisms used on distributors?
7. How is the ignition timing adjusted?
8. To obtain a spark advance based on engine speed, is the breaker plate or the breaker cam moved ahead?
9. Does the electronic ignition system use breaker points?
10. What jobs are performed by the ignition switch?

Study problems

1. On a large sheet of paper, draw the complete automotive ignition system. Draw in every component and every wire used to connect components. When finished, check your work by comparing it with Fig. 16-8 in the book.

2. Obtain the components of the automotive ignition system. Using a large cardboard box, or a large flat board, mount the components and wire them just as the ignition system is wired on the car.

17 | AUTOMOTIVE FUEL SYSTEMS

§ 17-1 AUTOMOTIVE ENGINE FUELS

The most common fuel used in automobile engines is gasoline. However, some automobiles, trucks, and buses have diesel engines that require fuel oil. A limited number of cars are specially equipped to run on liquefied petroleum gas (LPG), which is liquid only when under pressure. In this chapter, we will talk about gasoline fuel and the systems that use gasoline.

§ 17-2 GASOLINE

Gasoline is a hydrocarbon (HC); that is, it is largely hydrogen and carbon. During combustion, the hydrogen unites with oxygen in the air to form water (H_2O). The carbon also unites with oxygen to form carbon dioxide (CO_2). In the engine, because not all the gasoline burns, some HC comes out the tail pipe of the automobile. Also, some of the carbon only partly burns, and so carbon monoxide (CO) is formed. HC and CO are atmospheric pollutants, and devices are used on modern cars to reduce them to a minimum (see Chapter 20).

§ 17-3 SOURCE OF GASOLINE

Gasoline is produced by a complicated refining process from crude oil, or petroleum. No one knows exactly how petroleum originated. It is generally considered to have come from animal and vegetable substances and to have formed and collected in underground pools or reservoirs over millions of years. When a well is drilled to one of these oil reservoirs, the underground pressure forces the petroleum up and out of the well. The petroleum is then put through the refining process. The resulting products include gasoline, fuel oil, lubricating oil, grease, and many other substances.

§ 17-4 GASOLINE VOLATILITY

Gasoline must evaporate easily. That is, it must have high volatility. This is necessary for easy starting, quick warm-up, and smooth acceleration. But it must not be too volatile or vapor lock could occur. Vapor lock is caused by fuel evaporating in the fuel pump or fuel line. The vapor prevents normal amounts of fuel from reaching the carburetor so that the engine is fuel-starved and stalls.

§ 17-5 ANTIKNOCK VALUE

Knocking, or detonation, occurs in an engine if the wrong fuel is used. The knocking is caused by the sudden explosion of the air-fuel mixture (Fig. 17-1). Some fuels knock more easily than others. The **octane rating** of a fuel measures its antiknock value. A high-octane gasoline will knock less than a low-octane gasoline. The ratings that you find posted on gasoline pumps at filling stations tell you how knock-free a gasoline will be. A high-compression engine requires a high-octane gasoline.

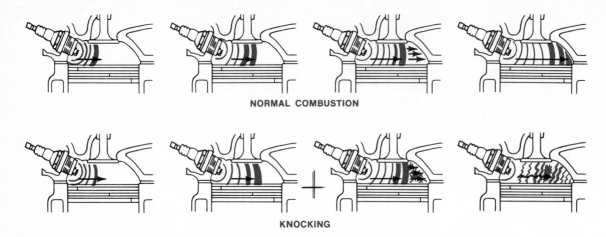

NORMAL COMBUSTION

KNOCKING

Fig. 17-1. Normal combustion without knocking is shown in the horizontal row on top. The fuel charge burns smoothly from beginning to end, providing an even, powerful thrust to the piston. Knocking is shown in the horizontal row on the bottom. The last part of the fuel explodes, or burns, almost instantaneously, to produce detonation, or knocking. (*General Motors Corporation*)

Note that tetraethyl lead, or ethyl, has been used in many gasolines to raise the octane rating. However, much of our gasoline is now being produced without lead for reasons explained in Chapter 20.

§ 17-6 OTHER ADDITIVES

Several additives are put into gasoline to improve its performance. These include anti-icers to prevent ice from forming in the carburetor, oxidation inhibitors to prevent formation of gum in the gasoline, antirust agents to protect engine parts from rusting, and other chemicals.

§ 17-7 PURPOSE OF THE FUEL SYSTEM

The fuel system has the job of supplying the engine with a mixture of air and gasoline vapor. The proportions of air and gasoline vapor must be changed as engine operating conditions change. When a cold engine is being cranked for starting, the mixture must be very rich. That is, it must have a high proportion of gasoline vapor. This proportion should be about 9 pounds of air to 1 pound of gasoline, or a ratio of 9 to 1 (9:1). When the engine is warmed up, the mixture should be leaner. That is, it should have a lower proportion of gasoline vapor. About 15 pounds of air per pound of gasoline (15:1) would be satisfactory for medium-speed driving with a warm engine. For accelerating and for high-speed operation, the mixture must again be enriched.

§ 17-8 THE FUEL SYSTEM

The fuel system is made up of:

1. The fuel tank, which stores the liquid gasoline.
2. The fuel filter, which filters out dirt particles from the gasoline.
3. The fuel pump, which delivers the gasoline from the tank to the carburetor.
4. The carburetor, which mixes the gasoline with air and delivers the combustible mixture to the engine.
5. The fuel lines between the tank and the fuel pump and between the fuel pump and the carburetor.

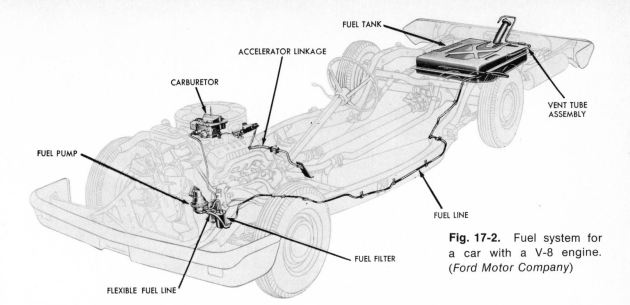

Fig. 17-2. Fuel system for a car with a V-8 engine. (*Ford Motor Company*)

A fuel system for an automobile with a V-8 engine is shown in Fig. 17-2.

§ 17-9 FUEL TANK

The fuel tank (Fig. 17-2) is normally located at the rear of the car. It is made of sheet metal and has two openings. Gasoline enters the tank through one opening and leaves the tank through the other opening. Cars manufactured since 1970 have an added system that prevents the escape of gasoline vapors from the fuel tank. It is called the vehicle vapor recovery (VVR) system. In this system, an extra connection to the gasoline tank sends any vapors to a storage tank or to the engine crankcase. The vapors cannot get out of the tank and pollute the air. You can read more about antipollution devices in Chapter 20.

The fuel tank has a filler cap on top that may be removed to add gasoline and an outlet near or at the bottom through which gasoline may flow to the carburetor. Often, the outlet is not placed at the bottom. Thus, dirt that has settled to the bottom of the tank will not be carried into the carburetor, where it could stop up openings and cause trouble. As an added safeguard, the outlet from the fuel tank is usually equipped with a screen-type filter that traps dirt and prevents it from leaving the tank.

§ 17-10 FUEL FILTER

Because some dirt does get into the fuel system, the fuel system has a fuel filter. The filter is made of a paper or gauze material that lets gasoline through but traps particles of dirt. Some filters, such as those shown in Fig. 17-3, are installed in the fuel line. Others are located in the carburetor, as shown in Fig. 17-4.

Fig. 17-3. In-line fuel filters. (*Ford Motor Company*)

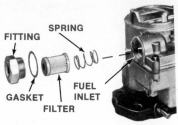

Fig. 17-4. The fuel filter located in the carburetor. (*Buick Motor Division of General Motors Corporation*)

§ 17-11 FUEL PUMP

The fuel pump (Fig. 17-5) is usually mounted on the side of the cylinder block. A rocker arm on the pump goes through an opening in the block and rests on an eccentric ring on the camshaft. In some engines, there is a pushrod between the eccentric ring and the end of the rocker arm. As the camshaft rotates, the eccentric ring causes the rocker arm to rock back and forth.

Inside the fuel pump (Fig. 17-5), there is a flexible diaphragm and two small valves. The center of the diaphragm is connected to the inner end of the rocker arm. A spring under the diaphragm maintains pressure on the diaphragm. As the rocker arm rocks, it pulls the diaphragm down and then releases it.

This diaphragm movement alternately produces vacuum and pressure in the chamber above it. When the diaphragm is pulled down by rocker-arm movement, a vacuum is created. Gasoline therefore is drawn in to fill this vacuum—that is, atmospheric pressure pushes it in. The gasoline opens the inlet valve on its way into the chamber above the diaphragm. Then, when the rocker arm rocks upward to release the diaphragm, the spring under the diaphragm pushes the diaphragm upward. The pressure now exerted on the gasoline in the chamber causes the inlet valve to be pushed closed and the outlet valve to be pushed open. Gasoline is therefore pushed out of the fuel pump and through the fuel line to the carburetor.

§ 17-12 VAPOR-RETURN LINE

Many cars have a vapor-return line. This line allows vapor that has formed in the fuel pump to return to the fuel tank. Figure 17-6 is a view from the top of a car frame, showing the location of the fuel tank, the fuel line, and the vapor-return line.

Why is the vapor-return line important? Vapor can form from the combination of vacuum

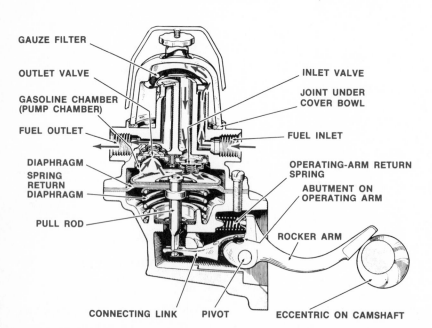

Fig. 17-5. A sectional view of a fuel pump. (*Hillman Motor Car Company, Limited*)

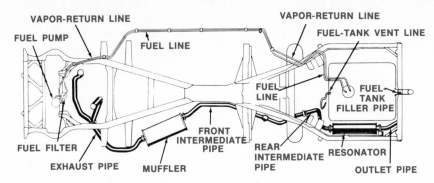

Fig. 17-6. The location of the vapor-return line. Only the frame and part of the fuel system are shown. (*Cadillac Motor Car Division of General Motors Corporation*)

and heat. Gasoline vaporizes more easily in a partial vacuum. The fuel pump, located on the engine, gets very hot. Thus, vapor can form. If vapor forms, the fuel pump can send the vapor back through the vapor-return line to the fuel tank. If vapor remains in the fuel pump, vapor lock can result. Vapor lock prevents normal pump action and keeps the carburetor from getting enough gasoline. When this happens, the engine stalls from lack of fuel.

Actually, the fuel pump keeps fuel circulating from the fuel tank, through the fuel pump, through the vapor-return line, and then back to the fuel tank. Since the fuel is cooler, it helps keep the fuel pump relatively cool so that vapor has less chance to form.

§ 17-13 VAPOR SEPARATOR

Some cars have a vapor separator located between the fuel pump and the carburetor, as shown in Fig. 17-7. The picture at the right shows the vapor separator in sectional view. Gasoline from the fuel pump enters the vapor separator through the inlet tube and exits through the outlet tube. If there are vapor bubbles, they rise to the top and enter the return tube. The return tube is connected to the vapor-return line so that the vapor goes back to the fuel tank.

§ 17-14 ELECTRIC FUEL PUMPS

Electric fuel pumps use electricity to draw fuel from the fuel tank and deliver it to the carburetor. There are several kinds of electric fuel pumps. One kind is installed inside the fuel

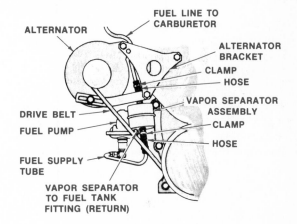

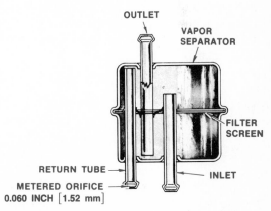

Fig. 17-7. At the top, the location of the fuel-vapor separator, in the fuel line between the fuel pump and the carburetor, on a V-8 engine. At the bottom, an enlarged sectional view of the fuel-vapor separator. (*Chrysler Corporation*)

tank. Figure 17-8 shows the mounting arrangement. The fuel pump is mounted on the same support that holds the fuel gage, which we talk

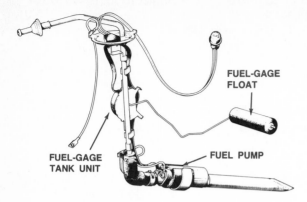

Fig. 17-8. The location of the electric fuel pump in the fuel tank. It is mounted on the same support as the fuel-gage tank unit. (*Buick Motor Division of General Motors Corporation*)

about later. This kind of electric fuel pump has a small electric motor that drives an impeller. The impeller has a series of blades that force the fuel out through the outlet pipe as the impeller spins. Figure 17-9 is a partial cutaway view of the fuel pump.

Compare the pictures of the different fuel pumps. You can see that the electric fuel pump works the same way as the fuel pump operated by the eccentric on the camshaft. The only difference is in the method used to produce the vacuum and the pressure in the pump chamber.

§ 17-15 AIR CLEANER

Air is apt to have many dust and dirt particles in it. If these particles entered the engine, they

could get on the cylinder walls, pistons, and rings and in engine bearings. Then, the moving surfaces would be scratched by the particles. This would cause the engine parts to wear very rapidly; the engine would soon fail.

To keep the dust particles out, the carburetor is equipped with an air cleaner.

Figure 17-10 is a cutaway view of an air cleaner. It is mounted on top of the carburetor. All air going through the carburetor and into the engine must first go through the air cleaner. The air cleaner has a filtering element made of fiber, special paper, or sponge-like polyurethane. The filtering element lets the air through but traps the dirt particles. After long use, the filtering element becomes filled with dirt and must be cleaned or replaced.

The air cleaner also muffles the noise of the air intake through the carburetor, intake manifold, and intake valves. Without the air cleaner, this noise would be quite noticeable. The air cleaner has a third function. It acts as a flame arrester in case the engine backfires through

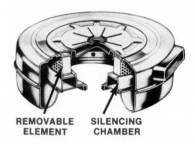

Fig. 17-10. A typical air cleaner, partly cut away to show the filter element. (*Ford Motor Company*)

Fig. 17-9. A cutaway view of the tank-mounted electric fuel pump. (*Buick Motor Division of General Motors Corporation*)

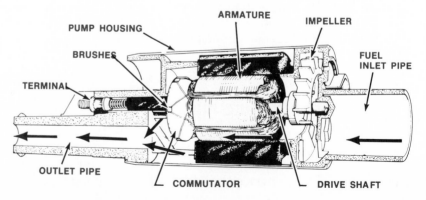

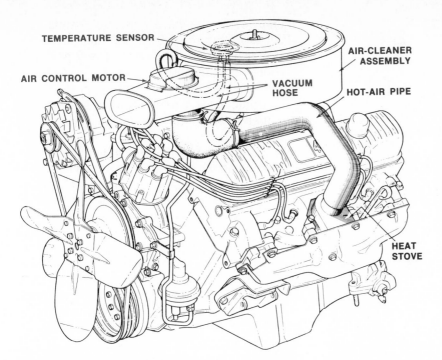

TEMPERATURE SENSOR

AIR CONTROL MOTOR

AIR-CLEANER ASSEMBLY

VACUUM HOSE

HOT-AIR PIPE

HEAT STOVE

Fig. 17-11. A heated air system installed on a V-8 engine. (*Buick Motor Division of General Motors Corporation*)

the carburetor. Backfiring may occur at certain times as a result of ignition of the air-fuel charge in the cylinder before the intake valve closes. When this happens, there is an explosive flashback through the intake manifold and carburetor. The air cleaner prevents this from erupting from the carburetor and possibly igniting gasoline fumes on the outside of the carburetor.

§ 17-16 THERMOSTATIC AIR CLEANER

Modern automobiles have thermostats in their air cleaners. Their purpose is to get hot air to the carburetor as quickly as possible when the engine is first started. For this reason, the system is often called a heated-air system. Figure 17-11 shows the system. The air inlet, or snorkel, of the air cleaner has a damper in it that is controlled by a thermostat. When the engine is cold, the thermostat keeps the damper closed. All air must now come from the heat stove around the exhaust manifold. This air becomes hot almost as soon as the engine starts, because the exhaust manifold heats up

quickly. As the engine warms up, the air in the engine compartment warms up too. The thermostat senses this and opens the damper. When the engine is warm, no additional heating of the air going into the carburetor is needed.

§ 17-17 CARBURETOR

As has already been explained, the carburetor has the job of mixing air and fuel in the proper proportions to form a combustible mixture. In the automotive engine, mixtures of different richness are required for various operating conditions. A rich mixture is required for starting, for accelerating, and for high speed. But for idling, a leaner mixture can be used. For medium-speed cruising, a still leaner mixture is desirable. The automotive carburetor must supply the proper mixture for all these conditions.

The carburetor is really a very simple gadget. It is basically an air horn, a fuel nozzle, and a throttle valve (Fig. 17-12). Let's take a look at these three parts again. The air horn is a round tube, or barrel. The venturi is a narrow

195

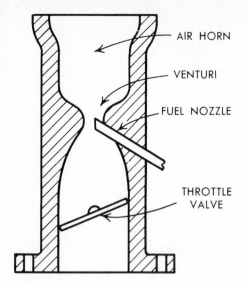

Fig. 17-12. Simple carburetor, consisting of air horn, fuel nozzle, and throttle valve.

section of the air horn. The fuel nozzle is a small tube through which fuel can flow from the float bowl, or reservoir, in the side of the carburetor. The throttle valve is a round disk mounted on a shaft. When the shaft is turned, the throttle valve is tilted. When it is tilted into the open position, air can flow through the air horn freely.

When the throttle valve is tilted into the closed position, as in Fig. 17-12, the throttle valve is closed. Little or no air can get through the air horn. The throttle valve, then, determines how much air gets to the carburetor.

The driver controls the position of the throttle valve by means of the right foot on the accelerator pedal. There is a linkage between this pedal and the throttle-valve shaft (Fig. 17-13). A spring tends to hold the throttle valve closed. But when the driver pushes down on

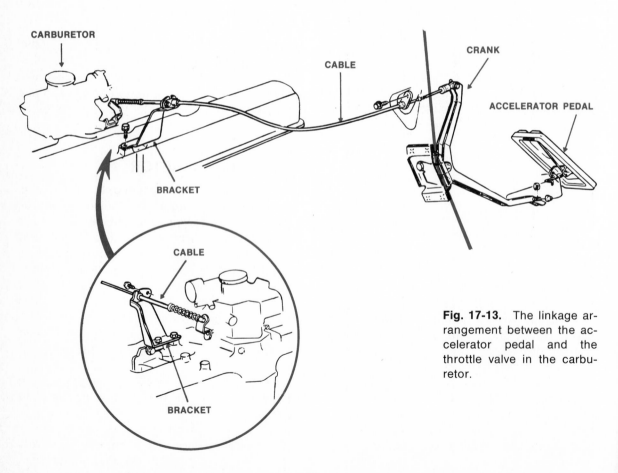

Fig. 17-13. The linkage arrangement between the accelerator pedal and the throttle valve in the carburetor.

the accelerator pedal, or "steps on the gas," the linkage causes the throttle shaft to turn so that the throttle valve is tilted toward the opened position. When the driver releases the pedal, the spring that tends to hold the throttle valve closed returns the valve to the closed position.

§ 17-18 CARBURETOR OPERATION

The carburetor has several systems to handle different operating requirements. These are the:

- Float system
- Idle system
- Main-metering system
- Power system
- Accelerator-pump system
- Choke system

Descriptions of these systems follow.

§ 17-19 FLOAT SYSTEM

The float system has the job of maintaining a constant level of fuel in the carburetor. If the level is too high, too much gasoline will flow from the fuel nozzles, and the mixture will be overrich. If the level is too low, not enough gasoline will feed from the nozzles, and the mixture will be too lean.

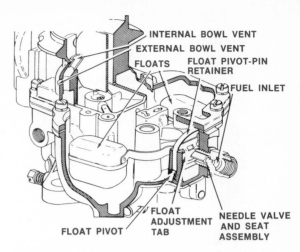

Fig. 17-14. A carburetor partly cut away to show the float system. (*American Motors Corporation*)

Figure 7-5 is a simplified drawing of a float system, and Fig. 17-14 shows a sectional view of an actual float bowl. As fuel flows into the bowl, the float rises, lifting the needle into its seat. When the proper height of fuel is reached, the needle is closed, stopping further fuel delivery. But as fuel is used, the float level drops and more fuel enters.

§ 17-20 IDLE SYSTEM

Figure 17-15 shows the idle system. The throttle is closed, and engine vacuum is pulling air

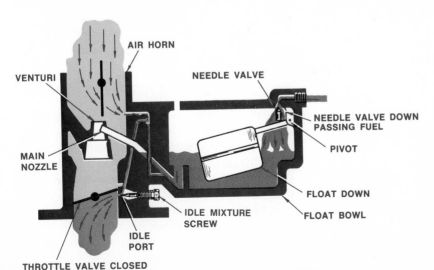

Fig. 17-15. Idle system in a carburetor. The throttle valve is closed so that only a small amount of air can get past it. All fuel is being fed past the idle mixture screw.

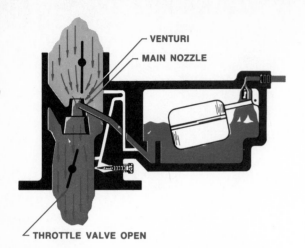

Fig. 17-16. The main-metering system in a carburetor. The throttle valve is open, and fuel is being fed through the high-speed, or main, nozzle.

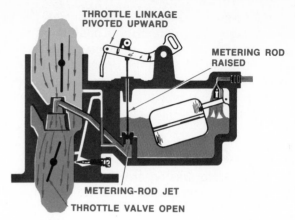

Fig. 17-17. A mechanically operated power system. When the throttle is opened, as shown, the metering rod is raised. This allows the smaller diameter of the rod to clear the jet so that additional fuel can flow.

through the passage, as shown. Fuel is also pulled through. It mixes with the air, and the mixture flows out through the opening around the idle mixture screw.

§ 17-21 MAIN-METERING SYSTEM

When the throttle is opened, more air flows through the air horn, and vacuum develops at the venturi. This causes fuel to feed from the main nozzle (Fig. 17-16). The fuel mixes with the air going through to produce the air-fuel mixture that the engine needs to operate at intermediate to high speeds.

§ 17-22 POWER SYSTEM

The power system comes into operation when the throttle is opened wide so that the engine can produce maximum power. There are two types, mechanical and vacuum. The mechanical system is shown in Fig. 17-17. Linkage to a metering rod raises the rod, allowing more fuel to flow through the opening, or jet, as it is called, for better performance at full throttle. Some carburetors use a vacuum diaphragm to control the metering rod. When the throttle is opened wide, vacuum is lost in the intake man-

ifold. This allows the vacuum device to lift the metering rod. Now, more fuel can flow for better performance.

§ 17-23 ACCELERATOR-PUMP SYSTEM

This system gives the engine an extra squirt of fuel when the throttle is pushed down. Figure 17-18 shows the system using a pump plunger. The pump is linked to the throttle linkage. When the throttle is opened, the linkage pushes the pump plunger down, forcing a jet of fuel out into the air passing through the carburetor air horn. This enriches the mixture momentarily for improved acceleration.

§ 17-24 CHOKE SYSTEM

Figure 17-19 shows the choke in action. With the choke closed, vacuum from the engine produces vacuum around the main nozzle. The main nozzle discharges fuel. The choke is necessary for starting an engine, especially a cold engine. More fuel is needed during starting. The choke provides this added fuel.

Chokes on most modern cars are automatic. They have a thermostatic spring that winds up

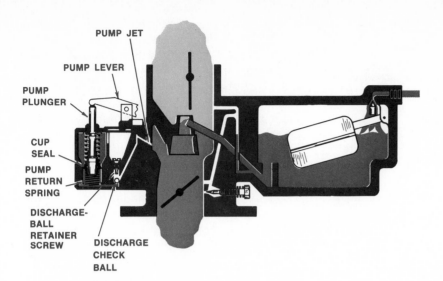

Fig. 17-18. The accelerator-pump system in a carburetor of the type using a pump plunger.

PUMP JET

PUMP LEVER

PUMP PLUNGER

CUP SEAL

PUMP RETURN SPRING

DISCHARGE-BALL RETAINER SCREW

DISCHARGE CHECK BALL

or unwinds with changing temperature. The thermostat is located in a well in the exhaust manifold in some engines, as shown in Fig. 17-20. As the engine warms up, the heat from the exhaust manifold causes the spring to unwind. This movement opens the choke valve.

Many late-model cars have electric automatic chokes. This type of choke includes an electric heating element (Fig. 17-21). The purpose of this heater is to assure faster choke opening. This helps reduce emissions from the engine. Emissions (HC and CO) are relatively high during the early stages of engine

warm-up. At low temperatures, the electric heater adds to the heat coming from the exhaust manifold. This reduces choke-opening time to as little as $1\frac{1}{2}$ minutes.

§ 17-25 TWO-BARREL AND FOUR-BARREL CARBURETORS

The carburetors we have described so far have a single barrel. In order to allow the engine to

CHOKE VALVE CLOSED

Fig. 17-19. With the choke valve closed, intake-manifold vacuum is introduced into the carburetor air horn, causing the main nozzle to discharge fuel.

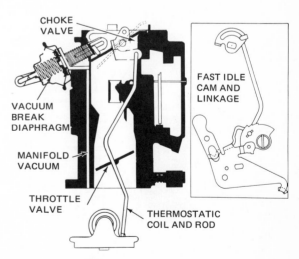

CHOKE VALVE

VACUUM BREAK DIAPHRAGM

MANIFOLD VACUUM

THROTTLE VALVE

FAST IDLE CAM AND LINKAGE

THERMOSTATIC COIL AND ROD

Fig. 17-20. The choke system with the thermostat located in a well in the exhaust manifold. Note the vacuum-break diaphragm. (*Chevrolet Motor Division of General Motors Corporation*)

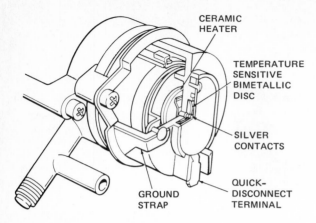

CERAMIC
HEATER

TEMPERATURE
SENSITIVE
BIMETALLIC
DISC

SILVER
CONTACTS

QUICK-
DISCONNECT
TERMINAL

GROUND
STRAP

Fig. 17-21. A cutaway view of an electric-assist choke. At low temperature, the ceramic heater turns on, adding heat to the choke so that it opens more quickly. (*Ford Motor Company*)

breathe more freely and to produce more power, carburetors with more than one barrel are often used. Thus, many carburetors have two barrels. Others have four barrels.

The **two-barrel carburetor** is essentially two single-barrel carburetors in a single assembly (Fig. 17-22). Each barrel handles the air-fuel requirements of half the engine cylinders. Figure 17-23 shows a top view of an intake manifold for a V-8 engine. The two-barrel carburetor mounts on top of the intake manifold so that each barrel aligns with one of the two openings in the manifold. One barrel takes care of cylinders 3, 4, 5, and 6. The other barrel takes care of cylinders 1, 2, 7, and 8. The arrows in Fig. 17-23 show the flow of air-fuel mixture to the cylinders. Each barrel has a complete set of fuel systems. Both throttle valves are fastened to a single throttle shaft so that both open together.

The **four-barrel carburetor** consists essentially of 2 two-barrel carburetors in a single assembly. One pair of barrels makes up the primary side, the other pair the secondary side (Fig. 17-24). Under most operating conditions, the primary side alone handles the air-fuel requirements of the engine. However, when the

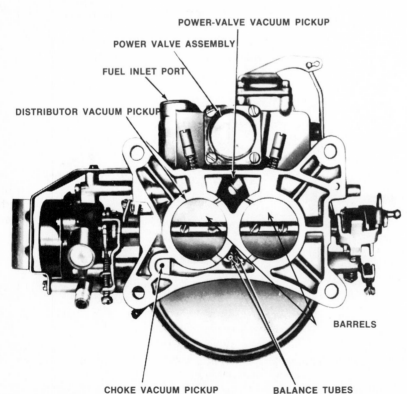

POWER-VALVE VACUUM PICKUP

POWER VALVE ASSEMBLY

FUEL INLET PORT

DISTRIBUTOR VACUUM PICKUP

Fig. 17-22. Bottom view of a dual carburetor. (*Ford Motor Company*)

BARRELS

CHOKE VACUUM PICKUP   BALANCE TUBES

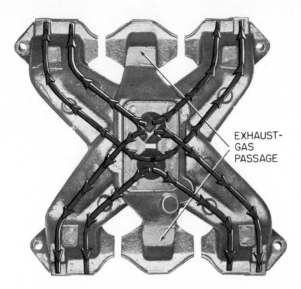

EXHAUST-GAS PASSAGE

Fig. 17-23. Top view of an intake manifold for a V-8 engine using a dual carburetor. Arrows show the paths the air-fuel mixture follows from each barrel of the carburetor to the cylinders.

throttle is moved toward the wide-open position for acceleration or for full-power operation, the secondary side comes into operation.

It supplies additional amounts of air-fuel mixture so that engine breathing and power output are improved.

§ 17-26 FUEL INJECTION

Engines with a fuel-injection system do not use carburetors. Instead, they use a pump to inject fuel into the cylinders or into the intake manifold. Figures 17-25 and 17-26 show these two arrangements. The diesel engine injects the fuel into the cylinders. The gasoline engine injects the fuel into the intake manifold. The injection takes place at the proper moment so that the fuel mixes with the air passing through the intake manifold. The mixture then enters the cylinders to be compressed, ignited, and burned in the usual manner.

The advantage of fuel injection, some experts say, is that the amount of fuel being fed to the engine can be more exactly controlled. Thus, better burning of the fuel is achieved. The results are better fuel economy and less hydrocarbon and carbon monoxide in the exhaust gases.

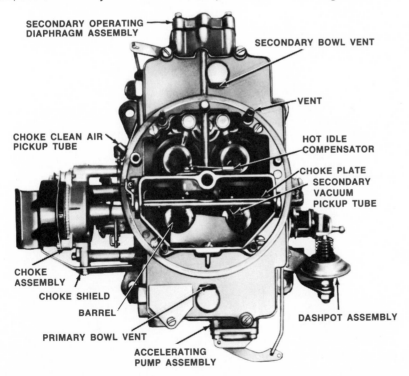

SECONDARY OPERATING DIAPHRAGM ASSEMBLY

SECONDARY BOWL VENT

VENT

CHOKE CLEAN AIR PICKUP TUBE

HOT IDLE COMPENSATOR

CHOKE PLATE SECONDARY VACUUM PICKUP TUBE

CHOKE ASSEMBLY

CHOKE SHIELD

BARREL

PRIMARY BOWL VENT

ACCELERATING PUMP ASSEMBLY

DASHPOT ASSEMBLY

Fig. 17-24. Top view of a four-barrel, or quad, carburetor. (*Ford Motor Company*)

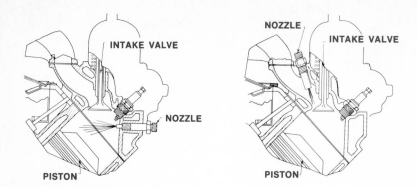

Fig. 17-25. (*Left*) A simplified view of the method of injecting fuel directly into the combustion chamber of an engine.

Fig. 17-26. (*Right*) A simplified view of the method of injecting fuel into the intake manifold just behind the intake valve.

§ 17-27 EXHAUST SYSTEM

After the air-fuel mixture has been burned in the engine cylinder, the exhaust valve opens, and the piston moves up on the exhaust stroke. The burned gases are forced out of the cylinder and into the exhaust manifold. From there they pass through the exhaust pipe, muffler, and tail pipe into the open air (Fig. 17-27).

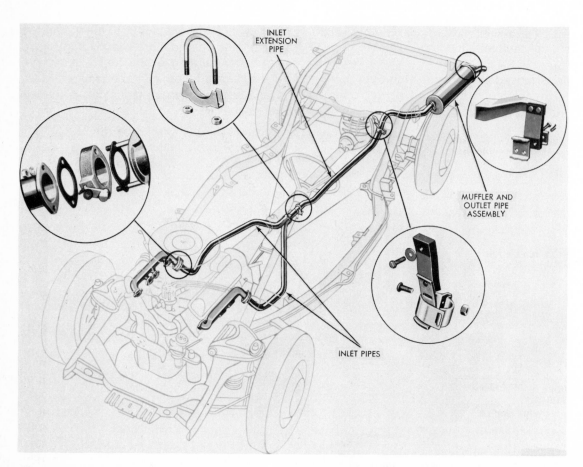

Fig. 17-27. Exhaust system for a V-8 engine. (*Ford Motor Company*)

Fig. 17-28. Exhaust muffler. Arrows show the path of the exhaust-gas flow. (*General Motors Corporation*)

The exhaust manifold is a metal casting with a series of passages connecting the exhaust ports of the cylinders with the exhaust pipe. In in-line engines, it is mounted on one side of the cylinder block or head, beneath the intake manifold. In V-8 engines, there are two exhaust manifolds, one on the outside of each bank of cylinders. On many cars, the two exhaust manifolds are connected to a common exhaust pipe, muffler, and tail pipe. On other cars, there are two separate exhaust systems, one for each bank; each exhaust system has its own muffler and tail pipe (Fig. 10-13). With two exhaust systems, there is less back pressure in the exhaust lines, and the engine develops a somewhat higher horsepower.

The muffler provides a series of passages and chambers through which the exhaust gases must pass before being discharged into the air (Fig. 17-28). These passages and chambers muffle the exhaust noise, thus quieting the engine.

§ 17-28 MANIFOLD HEAT CONTROL

The manifold heat control supplies heat to the intake manifold when the engine is cold. This helps vaporize the gasoline delivered by the carburetor to the intake manifold. As the engine warms up, the manifold heat control reduces the heat it supplies to the intake manifold; now, the intake manifold is warmed enough by the engine for adequate fuel vaporization.

The manifold heat control for an in-line engine contains a valve that is controlled by a thermostatic spring. Figure 17-29 shows the two valve positions. When the engine is cold, the spring has wound enough to turn the valve into the position shown. In this position, all the exhaust gases must pass through a jacket surrounding the intake manifold, as shown.

The hot exhaust gases heat the intake manifold to improve fuel vaporization. Then, when the engine warms up, the thermostatic spring unwinds. This causes the manifold-heat-control valve to rotate to the hot-engine position (right, Fig. 17-29). The exhaust gases then discharge directly into the exhaust pipe without circulating through the jacket. The hot exhaust gases therefore have no further heating effect on the intake manifold.

Note that many cars with thermostatic air cleaners (heated-air system) do not use manifold heat-control valves. The heated-air system supplies all the heat necessary to the air-fuel mixture when a cold engine is first started.

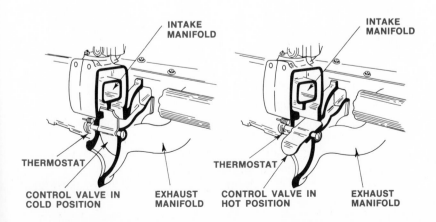

Fig. 17-29. Intake and exhaust manifolds of a six-cylinder in-line engine cut away so that the location and action of the manifold heat control can be seen. At the left, the heat-control valve is in the cold-engine, or HEAT ON, position, directing hot exhaust gases up and around the intake manifold. (*Ford Motor Company*)

INTAKE MANIFOLD

THERMOSTAT

CONTROL VALVE IN COLD POSITION

EXHAUST MANIFOLD

INTAKE MANIFOLD

THERMOSTAT

CONTROL VALVE IN HOT POSITION

EXHAUST MANIFOLD

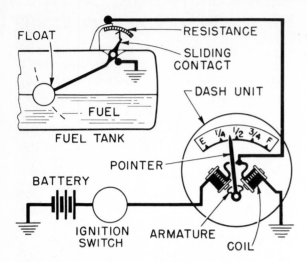

FLOAT — RESISTANCE — SLIDING CONTACT — FUEL — FUEL TANK — DASH UNIT — POINTER — BATTERY — IGNITION SWITCH — ARMATURE — COIL — E 1/4 1/2 3/4 F

Fig. 17-30. A schematic wiring circuit of a balancing-coil fuel-gage system.

§ 17-29 FUEL GAGE

The fuel gage includes two units, a measuring device in the fuel tank and an indicating device on the dash (Fig. 17-30). The measuring device, or tank unit, has a float that moves up and down as the fuel level changes. This causes a sliding contact to move back and forth along a resistance. The dash unit has a pair of coils that are connected in series to the battery through the ignition switch. When the ignition switch is turned on, current flows through the two coils, producing a magnetic pattern that acts on the armature to which the pointer is attached. The resistance of the tank unit is high when the tank is filled and the float is up. In this case, the current flowing through the left coil (on the E, or empty, side) also flows through the right coil (on the F, or full, side). This turns the armature so that the pointer points to full. But as the tank empties, the float drops and the resistance of the tank unit goes down. More and more of the current flowing through the E coil therefore flows to the tank unit instead of to the F coil. The E coil therefore becomes stronger than the F coil and pulls the armature around so that the pointer moves toward E, or empty.

§ 17-30 HYDROGEN: FUEL OF THE FUTURE?

In this chapter we have discussed the gasoline fuel system that is used almost universally on automobile engines. But as the oil stored in the earth is depleted, many questions arise as to which fuel will be the fuel of the future.

Today, we recognize that the ideal fuel must have several characteristics. It should be readily available in large quantities at relatively low cost. It should be nonpolluting and burn cleanly without smoke, odor, or any other air pollutant harmful to life or property. Also, the ideal fuel should be easily recyclable; that is, its combustion products should be able to create more fuel.

Sound impossible? Maybe so. But here are some facts to consider about the common gas hydrogen (H). According to some experts, hydrogen may be the fuel of the future. It has all the properties needed in the ideal fuel.

Hydrogen was discovered in 1766 by an English chemist, Henry Cavendish. Hydrogen is a colorless, tasteless, odorless, lighter-than-air gas. It makes up about 1 percent of the earth's crust. Most hydrogen can be found in water in conjunction with oxygen (H_2O). Hydrogen is extremely flammable. It burns in air or oxygen with a hot, colorless flame. When burned, hydrogen produces water vapor.

There are several ways in which hydrogen can be obtained. One is by heating certain elements and acids, such as zinc or carbon and hydrochloric acid. Another, better known, method of obtaining hydrogen is through electrolysis, that is, by running an electric current through water.

In the early 1800s, hydrogen was considered as a possible engine fuel, although engines as we know them today had not yet been developed. In the early 1900s, several engineers and inventors tried to use hydrogen in internal-combustion engines. But in these early applications, the engines knocked and backfired too much to operate satisfactorily.

In recent years, several hydrogen-fueled

Fig. 17-31. A hydrogen-fueled car. (*Billings Energy Research Corporation*)

cars have been built, and experiments are continuing (Fig. 17-31). Research into using hydrogen as a possible fuel is divided between two types of systems. One type of system adds hydrogen to the fuel that the engine is using. The second type of system enables an engine to operate solely on hydrogen as the fuel.

For its use as a fuel for an automobile engine, hydrogen must be handled and controlled in much the same way as liquefied petroleum gas (LPG). The same type of carburetor used for LPG can be used for hydrogen. Although there are a number of problems in mass-producing a hydrogen-fueled car, the advantages of hydrogen as a fuel are very great. One obvious advantage is the great abundance of hydrogen.

Hydrogen fuel, once used, vaporizes and rises to form clouds and eventually reappears as rain water. Hence, hydrogen is recyclable. It is nonpolluting, gives off no unpleasant odor, and does not create slicks on the road should

it leak out or spill. One great advantage is that hydrogen can be used as a fuel for any type of engine that is in use today. That is, hydrogen can be used in piston engines, Wankel engines, steam engines, jet engines, and rockets. With hydrogen as a fuel, some experts say that no expensive and prolonged search for an alternative automobile engine is needed. We should continue to use the types of engines we now have and change only the fuel system.

According to these experts, any country that has a large source of water can be completely self-sufficient in fuel. And that fuel is nonpolluting and recyclable! Of course, nuclear generating plants would be required to supply the huge amounts of electricity needed to obtain such a quantity of hydrogen from water.

As the earth's oil supplies continue to dwindle, and as we struggle with the problems of air pollution, hydrogen must be considered as a possible fuel for the future.

Review questions

1. If the last part of the compressed air-fuel mixture explodes before the spark reaches it, what is the result?
2. What is the purpose of the fuel pump?
3. In the thermostatic air-cleaner system, where does the air come from when the engine is cold?
4. What are the three basic parts of the carburetor?
5. Is a 9:1 air-fuel mixture rich or lean?
6. Name the six systems in the carburetor.
7. What operates the needle valve in the float bowl?
8. What is the purpose of adding barrels to a carburetor?
9. What are the two places that fuel is injected into in an engine?
10. What determines the position of the manifold heat-control valve?

Study problems

1. On an automobile, locate each part of the fuel system. Trace the flow of gasoline from the fuel tank filler neck, to the fuel pump, the carburetor, and through the engine to the tailpipe. Note the locations and types of fuel filters used, and see if the vehicle is equipped with a vapor-return line or a vapor separator.
2. Obtain an old fuel pump and disassemble it. Note carefully the flow of fuel through the pump, and how the pump creates pressure.
3. Disassemble as many different types of carburetors as you can. Trace the flow of fuel through the carburetor. Identify the six circuits to the carburetor. Determine how the basic carburetor adjustments are made.

18 ENGINE COOLING SYSTEMS

§ 18-1 PURPOSE OF THE COOLING SYSTEM

The burning fuel in the engine produces a lot of heat. Chapter 19 explains that some of this heat is removed by the lubricating oil. Some of the heat leaves the engine in the hot exhaust gases. But most of the heat is carried away by the engine cooling system. It has the job of keeping the engine from getting too hot, or overheating. Let's find out how the engine cooling system does its work.

The cooling system must cool the engine properly. If the cooling system cools the engine too much, gasoline will be wasted, and the engine will not have normal power. If the cooling system does not cool the engine enough, the engine will overheat. Overheating can burn the film of lubricating oil off the cylinder walls. This would damage the cylinder walls, pistons, and piston rings. The result could be a ruined engine!

An overheated engine also loses power. Let's take a look at engine heat. Combustion temperatures in the cylinders can reach 4500°F [2482°C]. That's a high enough temperature to melt the engine. Part of the heat goes into the metal parts around the combustion chamber: the cylinder heads, the cylinder block, and the pistons. Therefore, the engine must get rid of much of this heat before these parts melt.

Not all the heat is carried away, however. Some heat is left so that the engine stays just hot enough to run efficiently.

§ 18-2 TYPES OF COOLING SYSTEMS

There are two kinds of cooling systems: air cooling and liquid, or water, cooling. Small engines used in lawn mowers and similar equipment and some automotive engines are air cooled (Fig. 18-1). In air-cooled engines with more than one cylinder, the cylinders are usually separated so that air can flow around them. The cylinders and cylinder heads are also equipped with metal fins to help radiate the heat away from the engine.

§ 18-3 WATER-COOLED ENGINES

Almost all automotive engines have liquid cooling (Fig. 18-2). In these, water mixed with antifreeze circulates in spaces around the cylinder walls and through passages in the cylinder head. Antifreeze compounds are added to the water to prevent freezing and to improve cooling during hot weather. The mixture is called the **coolant.** As the coolant circulates, it picks up heat from the engine and carries this heat to a radiator. The radiator then cools the coolant by discharging the heat into the air that is passing through the radiator.

§ 18-4 WATER PUMP

Figure 18-3 shows a cooling system on a V-8 engine. A water pump, driven by the engine fan belt, keeps the coolant in continuous circulation while the engine is running. Figure 10-1

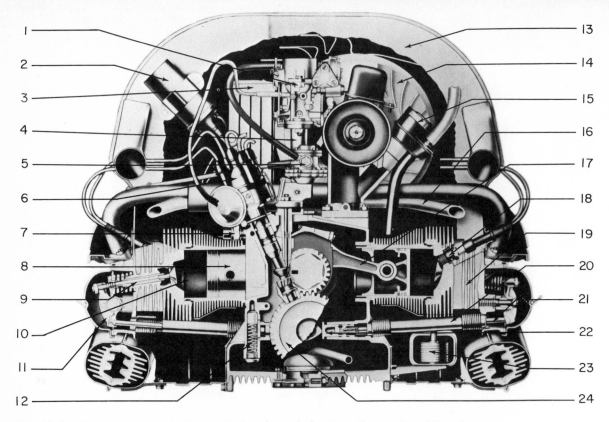

Fig. 18-1. A cutaway view of a four-cylinder air-cooled automotive engine. (1) carburetor, (2) ignition coil, (3) oil cooler, (4) intake manifold, (5) fuel pump, (6) ignition distributor, (7) oil-pressure switch, (8) piston, (9) valve, (10) cylinder, (11) rocker arm, (12) oil-pressure relief valve, (13) fan housing, (14) fan, (15) oil filter and breather, (16) preheating pipe, (17) connecting rod, (18) spark plug, (19) cylinder head, (20) valve pushrod tubes, (21) pushrod, (22) cam follower, (23) thermostat, (24) camshaft. (*Volkswagen*)

shows where the water pump is mounted on the cylinder block. The coolant enters the engine from the bottom of the radiator, passing up through the spaces in the block and head. These spaces are called the **water jackets.** Figures 18-4 and 18-5 show disassembled and sectional views of a water pump. There is an impeller on a shaft. The impeller has a series of blades. As it revolves, it **impels** (throws) the coolant outward by centrifugal force; the result of this is to cause the coolant to flow through the pump.

The pump is driven by the fan belt, a V-shaped belt that also drives the alternator. The belt is driven by a pulley on the front end of the crankshaft. The engine fan is mounted

on the water-pump pulley so that both the fan and the water pump turn together.

§18-5 ENGINE FAN

The engine fan helps pull air through the radiator. You can see the location and shape of the engine fan in several of the pictures in this chapter. For example, look at Figs. 18-3 and 18-5. The fan is located at the back of the radiator. It has four or more curved blades. The fan is driven by the belt from the pulley on the engine crankshaft. When the curved blades rotate, they "scoop up" air and push it back toward the engine. This action pulls air through the radiator.

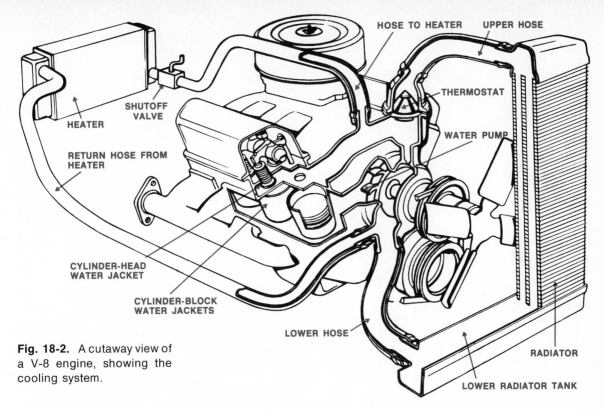

Fig. 18-2. A cutaway view of a V-8 engine, showing the cooling system.

HOSE TO HEATER

UPPER HOSE

SHUTOFF VALVE

HEATER

THERMOSTAT

WATER PUMP

RETURN HOSE FROM HEATER

CYLINDER-HEAD WATER JACKET

CYLINDER-BLOCK WATER JACKETS

LOWER HOSE

RADIATOR

LOWER RADIATOR TANK

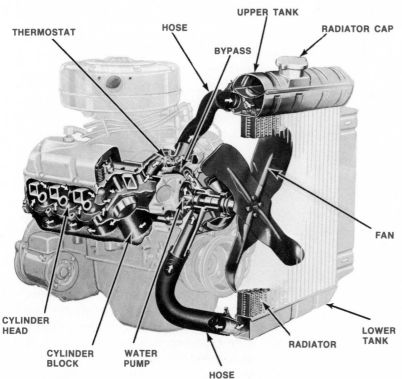

THERMOSTAT

HOSE

UPPER TANK

BYPASS

RADIATOR CAP

CYLINDER HEAD

CYLINDER BLOCK

WATER PUMP

HOSE

WATER PUMP

FAN

RADIATOR

LOWER TANK

Fig. 18-3. A cutaway view of a V-8 engine, showing the cooling system. Arrows show the direction of water flow through the engine water jackets. (*Ford Motor Company*)

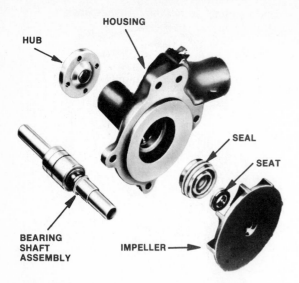

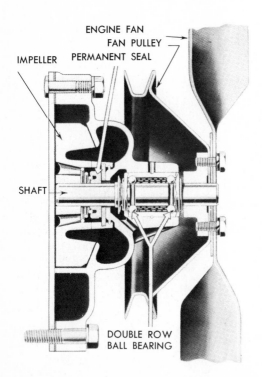

Fig. 18-4. A disassembled view of a water pump. (*Pontiac Motor Division of General Motors Corporation*)

Fig. 18-5. A sectional view of a water pump, showing the manner of supporting the shaft on a double-row ball bearing and the method of mounting the fan and pulley on the shaft.

§ 18-6 VARIABLE-SPEED FAN DRIVE

Many engines use a variable-speed fan drive. This drive increases fan speed when the engine gets hot and reduces speed when the engine cools off. This reduces the amount of power the fan uses up and also reduces fan noise.

§ 18-7 RADIATOR

Figures 18-6 and 18-7 show what radiators look like. The radiator has one series of passages for the coolant and another for air to flow through. The air removes heat from the coolant. In some radiators the coolant flows from top to bottom, as in Fig. 18-7. These radiators are called downflow. In the other type, the coolant flows from one side to the other. These are called crossflow radiators.

§ 18-8 THERMOSTAT

There is a thermostat in the passage between the cylinder head and the upper radiator tank (Fig. 18-3). The thermostat closes off this passage when the engine is cold. This prevents coolant flow to the radiator. The engine retains its heat and gets hot quickly. This is desirable because a cold engine wears fast and works poorly. The thermostat then opens as the engine reaches operating temperature to permit coolant circulation. Now, the cooling system prevents engine overheating. Figure 18-8 shows various types of thermostats.

§ 18-9 RADIATOR PRESSURE CAP

Most engines today have pressurized cooling systems. Water boils at about 212°F [100°C] at sea level. If the pressure is increased, the water will not boil until a higher temperature is reached. Without pressurizing, the cooling system must be designed to prevent the coolant from reaching 212°F [100°C]. But if the system is pressurized, the coolant temperature can be raised to about 260°F [126.7°C] with-

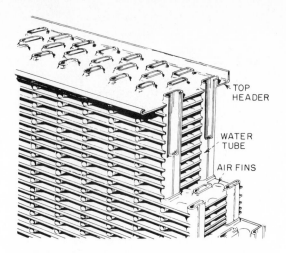

Fig. 18-6. Construction of tube-and-fin radiator core.

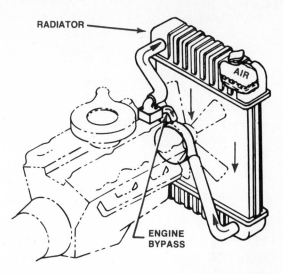

Fig. 18-7. A cooling system using a downflow radiator. (*Harrison Radiator Division of General Motors Corporation*)

out boiling. This higher temperature allows the cooling system to operate more efficiently. Since the coolant enters the radiator at a higher temperature, the temperature difference between the coolant and the surrounding air is greater. This causes a greater heat transfer.

The pressure cap contains two spring-loaded valves (Fig. 18-9). One, called the **pressure valve,** opens if the pressure gets too high

to allow the excessive pressure to escape. The other, the **vacuum valve,** operates when the engine cools off. When this happens, a partial vacuum could form in the cooling system due to condensation of steam. The vacuum valve prevents this by opening to admit air from the

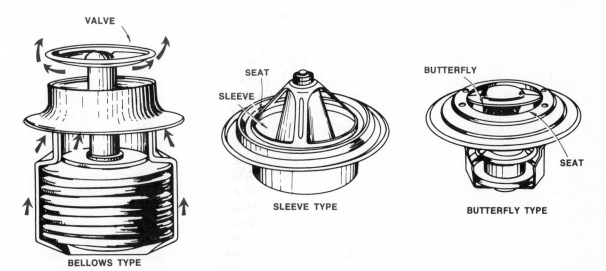

Fig. 18-8. Various thermostats for the engine cooling system. The bellows type, no longer widely used, is shown open, with arrows indicating the water flow past the valve. (*Chrysler Corporation*)

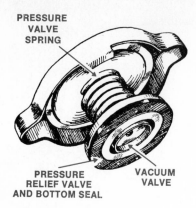

Fig. 18-9. A radiator pressure cap.

outside. If a high vacuum formed, it might cause the radiator to partly collapse, pushed in by atmospheric pressure.

§ 18-10 EXPANSION TANK

Many cooling systems have an expansion tank (Fig. 18-10). The expansion tank is partly filled with coolant and is connected to the radiator. Expansion of the coolant in the engine as the engine heats up sends part of the coolant into the expansion tank. Then, as the engine reaches operating temperature, the radiator-cap valve closes. Now, the cooling system is sealed as explained above. Pressure goes up, and the system operates more efficiently. When the engine cools off, the coolant in the engine contracts. This produces a vacuum that causes coolant to flow from the expansion tank back into the engine. The system keeps the engine water jackets and radiator filled at all times for maximum efficient operation.

§ 18-11 ANTIFREEZE

Antifreeze is added to the water in the cooling system to prevent freezing of the coolant and damage to the engine and radiator. The lower the expected temperature, the more antifreeze must be added. A mixture of half water and half antifreeze (ethylene glycol) will not freeze above $-34°F$ [$-36.7°C$]. The antifreeze solu-

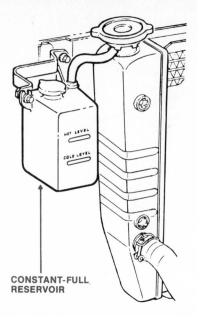

Fig. 18-10. An expansion tank, also called a constant-full reservoir, for a cooling system.

tion will not boil until a temperature higher than $212°F$ [$100°C$] (boiling point of water) is reached. The system is therefore more efficient in hot weather, too. The antifreeze also adds corrosion protection, since chemicals in the antifreeze fight corrosion.

§ 18-12 TEMPERATURE INDICATOR

The engine-temperature indicator gives the driver a continuous indication of the engine temperature. If the temperature goes too high, the driver knows that something is wrong and can stop the engine before some serious trouble develops that might ruin the engine.

The wiring circuit for the temperature indicator (Fig. 18-11) is very similar to that for the other indicators already discussed, except that the coils in the dash unit are connected differently. The engine unit decreases in resistance as temperature goes up. Thus, when the temperature is high, the right-hand coil (on the high-temperature side of the dial) will pass more current and the armature will be pulled around so that the pointer moves to the right to indicate the higher engine temperature.

Many late-model cars use a temperature indicator light mounted in the dash instead of a dial. The light system includes a light (usually red) that comes on when the engine temperature goes too high.

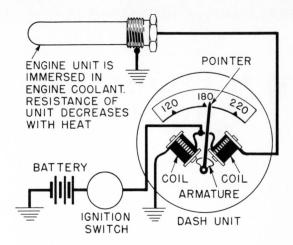

ENGINE UNIT IS IMMERSED IN ENGINE COOLANT. RESISTANCE OF UNIT DECREASES WITH HEAT

POINTER

120 180 220

BATTERY

COIL COIL

ARMATURE

IGNITION SWITCH

DASH UNIT

Fig. 18-11. A circuit diagram of the electric-resistance temperature-indicator system.

Review questions

1. What are two kinds of engine cooling systems?
2. What are the two series of passages through the radiator?
3. What is the name of the device that shuts off the circulation of the coolant to the radiator when the engine is cold?
4. What is the purpose of pressurizing the cooling system?
5. How many valves are there in the radiator pressure cap? What are their names?

Study problem

1. Obtain a thermostat and immerse it in a pan of water on a stove. Insert a cooking thermometer in the water, and then heat the water until it boils. Notice the temperature at which the thermostat opens and the temperature at which it closes.

19 | ENGINE LUBRICATING SYSTEMS

§ 19-1 PURPOSE OF THE LUBRICATING SYSTEM

The engine has many moving metal parts. If these metal parts rub against one another, they will wear out before very long. The purpose of the lubricating system is to keep metal from rubbing against metal. In this chapter, we explain how lubricating systems flood the moving parts with lubricating oil. You are going to learn how the oil works and how it gets to the moving parts in the engine. But first, let's find out more about oil.

§ 19-2 ENGINE OIL

Oil is the liquid used in the lubricating system. Oil was formed underground millions of years ago in various locations around the world. Oilworkers drill deep holes into these underground pools of oil and force the oil out of the ground. This oil—called crude oil—must be refined before it can be used. In the refining process, gasoline, kerosene, lubricating oil, and many other products are made.

Not all oil is the same. There are several grades of oil and several ratings. Oil made for automobiles contains a number of **additives** (chemical compounds that are *added* to the oil) that improve the performance of the oil. Let's take a look at these ratings and additives.

§ 19-3 OIL VISCOSITY AND SERVICE RATINGS

Oil viscosity is rated in two ways by the Society of Automotive Engineers (SAE). It is rated for (1) winter driving and (2) summer driving. Winter-grade oils come in three grades: SAE5W, SAE10W, and SAE20W. The "W" stands for winter grade. For other than winter grade, the grades are SAE20, SAE30, SAE40, and SAE50. The higher the number, the higher the viscosity (the thicker the oil). All these grades are called single-viscosity oils.

Many oils have multiple-viscosity ratings. For example, the SAE10W-30 oil has the same viscosity as SAE10W when it is cold and the same viscosity as SAE30 when it is hot.

Car manufacturers specify the viscosity of the oil to be used in their engines. Study Fig. 19-1. As you can see, the higher the outside temperature, the higher the viscosity rating specified.

Service ratings indicate the type of service for which the oil is best suited. For gasoline engines, the service ratings are SA, SB, SC, SD, and SE. Here, in brief, is what these service ratings stand for:

SA—Acceptable for engines operated under the mildest conditions
SB—Acceptable for minimum-duty engines operated under mild conditions

Multiple-viscosity oils	
Outside temperature consistently	Use SAE viscosity
Below +32°F [0°C]	5W30*
−10 to +90°F [−23.3 to +32.2°C]	10W-30
−10 to above +90°F [−23.3 to above +32.2°C]	10W-40
Above +10°F [−12.2°C]	20W-40

*For sustained high-speed driving, use 10W-30 or 10W-40.

Fig. 19-1. Recommended oil viscosities. (*Ford Motor Company*)

SC—Meets requirements of gasoline engines in 1964–1967 model passenger cars and trucks
SD—Meets requirements of gasoline engines in 1968–1970 model passenger cars and some trucks
SE—Meets requirements of gasoline engines in 1972 and later cars and certain 1971 model passenger cars and trucks

§ 19-4 OIL ADDITIVES

Oil alone will not do the job required in engines. Therefore, certain chemicals, called additives, are put into the oil during refining. Not all oils have all the following, but each of the additives mentioned is used in one engine oil or another.

1 Viscosity improver. The viscosity improver is what makes an oil a multiple-viscosity oil. It combats the tendency of the oil to get thick when cold and thin when hot.

2 Pour-point depressants. These help keep the oil thin enough to pour at low temperatures.

3 Inhibitors. These combat corrosion, rust, foaming, and oxidation. Oxidation of the oil can take place when it gets hot and mixes with the air in the crankcase. Likewise, as the oil is stirred by the revolving crankshaft, it could foam. The foam inhibitor added to the oil prevents this.

4 Detergent-dispersants. A detergent is similar to soap. The detergent in oil loosens particles of dirt, carbon, and gum from engine parts and carries them to the crankcase. There, the larger particles settle and are drained out when the oil is changed. Smaller particles are trapped by the oil filter. The dispersant action prevents the particles from forming clots that could clog oil passages.

5 Extreme-pressure compounds. These help the oil resist the high pressures that occur between the parts of the modern high-powered engine.

§ 19-5 ENGINE LUBRICATING SYSTEMS

All parts in the engine that move or slide over each other must be coated with oil so that friction and wear will be kept low. We have already explained what friction is (in **§** 6-10, "Friction") and how the engine oil spreads and acts between moving surfaces. (Fig. 19-2 shows the moving parts in a V-8 engine that must be lubricated.)

Small four-cycle engines use several methods to provide lubrication for the moving parts. The simplest means of lubrication is to splash the oil about so that all parts are drenched. The splashing is accomplished by attaching a dipper to the connecting-rod cap (Fig. 19-3). The oil in the crankcase is splashed every time the dipper reaches into it on the downstroke of the piston. Splash may also be accomplished by the use of an oil slinger that is rotated by the camshaft gear (Fig. 19-4).

A cam-operated barrel-type pump is used in certain small engines to provide a pressure-type lubrication system (Figs. 19-5, 19-6, and 19-7). A small gear-type pump operated by the camshaft gear is sometimes used on small vertical-shaft engines (Fig. 19-8).

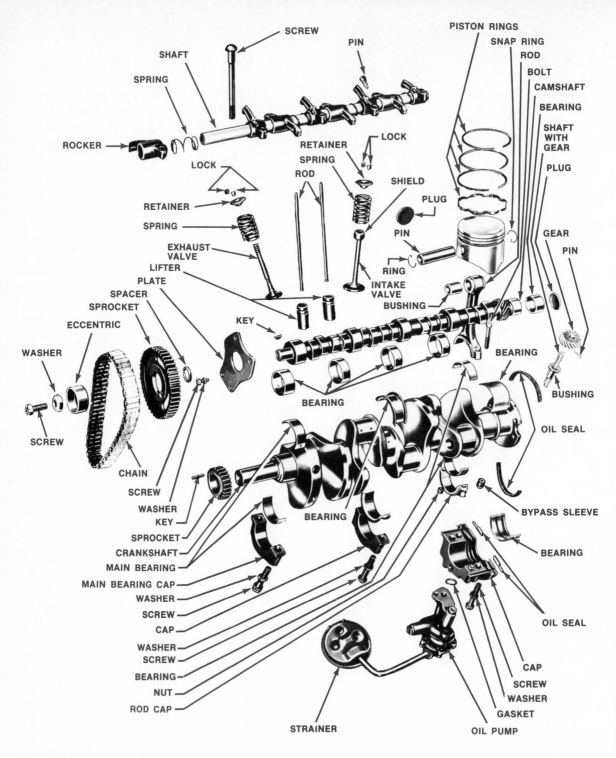

Fig. 19-2. In a V-8 engine, these are the moving parts that must be lubricated. (*Chrysler Corporation*)

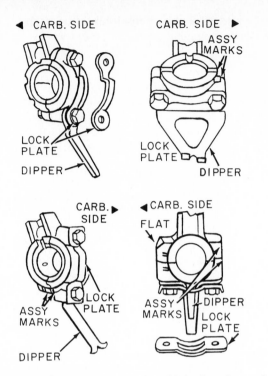

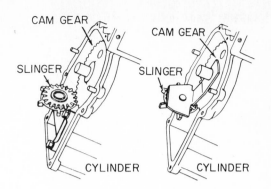

Fig. 19-4. Oil slinger is rotated by the camshaft gear.

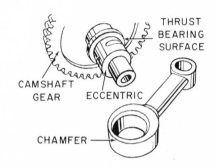

Fig. 19-6. Plunger-type oil pump is placed on camshaft eccentric. (*Lawson*)

Fig. 19-3. Different ways in which the dipper is mounted on the connecting-rod cap.

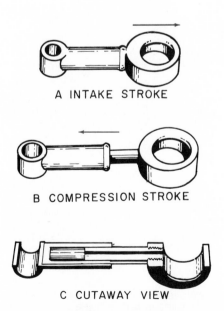

Fig. 19-5. Barrel-type lubrication pump used on the Lawson vertical-shaft engine.

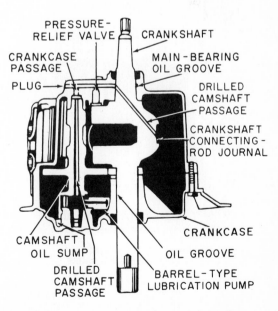

Fig. 19-7. Location of barrel-type lubrication pump and the oil passages in the engine. (*Lawson*)

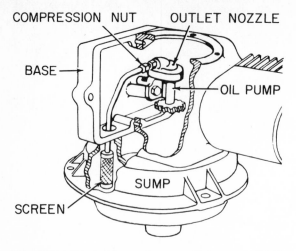

COMPRESSION NUT OUTLET NOZZLE

BASE

OIL PUMP

SUMP

SCREEN

Fig. 19-8. Gear-driven oil pump. (*Briggs and Stratton Corporation*)

Figure 19-9 shows the lubrication system on a small multicylinder air-cooled engine, which uses spray nozzles to direct oil to the moving parts.

The engine lubricating system forces or splashes oil on all moving surfaces and thus provides ample lubrication to keep friction low. But the oil does many other jobs. Here are the main jobs the engine oil does in the engine:

1. Lubricates to minimize wear.
2. Lubricates to minimize power loss from friction.
3. Acts as a cooling agent by removing heat from engine parts.
4. Absorbs shocks between bearings and other engine parts, acting as a cushion,

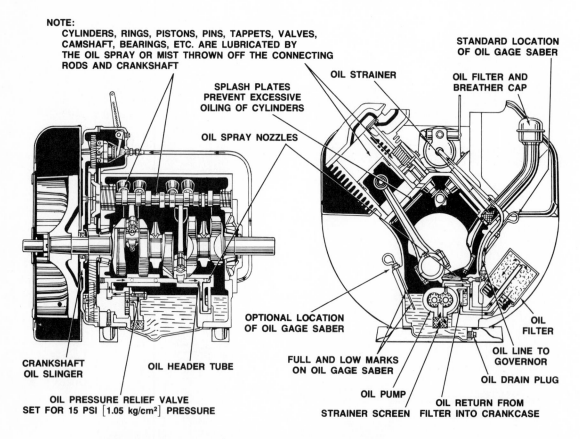

NOTE:
CYLINDERS, RINGS, PISTONS, PINS, TAPPETS, VALVES, CAMSHAFT, BEARINGS, ETC. ARE LUBRICATED BY THE OIL SPRAY OR MIST THROWN OFF THE CONNECTING RODS AND CRANKSHAFT

SPLASH PLATES PREVENT EXCESSIVE OILING OF CYLINDERS

OIL SPRAY NOZZLES

OIL STRAINER

STANDARD LOCATION OF OIL GAGE SABER

OIL FILTER AND BREATHER CAP

OIL FILTER

OIL LINE TO GOVERNOR

OIL DRAIN PLUG

OIL RETURN FROM FILTER INTO CRANKCASE

OIL PUMP

STRAINER SCREEN

FULL AND LOW MARKS ON OIL GAGE SABER

OPTIONAL LOCATION OF OIL GAGE SABER

OIL HEADER TUBE

CRANKSHAFT OIL SLINGER

OIL PRESSURE RELIEF VALVE SET FOR 15 PSI [1.05 kg/cm²] PRESSURE

WITH ENGINE AT OPERATING TEMPERATURE, OIL PRESSURE IN HEADER WILL BE APPROXIMATELY 5 PSI [0.35 kg/cm²]. AN OIL PRESSURE GAGE IS NOT REQUIRED.

Fig. 19-9. Multicylinder air-cooled-engine lubrication system. (*Wisconsin*)

and thereby reducing engine noise and extending engine life.

5. Forms a good seal between piston rings and cylinder walls. This reduces the amount of gas that can escape from the combustion chamber by passing between the piston and cylinder walls and entering the crankcase.

6. Helps keep engine parts clean by flushing away dirt and other foreign material. This dirt settles to the bottom of the oil pan, where it can be drained off when the oil is changed.

§ 19-6 AUTOMOTIVE-ENGINE LUBRICATING SYSTEMS

Modern automobile engines use pressure-feed oiling systems. That is, the oil flows under pressure to the moving engine parts. Figure 19-10 is a simplified drawing of the lubricating system for a V-8 engine. The oil pan at the bottom of the engine holds several quarts of oil. The oil pump sends oil from the oil pan through the oil filter and then through oil lines to the crankshaft bearings, camshaft bearings, valve lifters, pushrods, rocker arms, and valves. (These oil circuits are shown by lines in the picture.) Then, the oil drains off and flows back down into the oil pan.

§ 19-7 OIL PUMPS

The two common types of oil pumps are the gear type and the rotor type. The pump is driven by a pair of spiral gears from the camshaft, as shown in Fig. 19-11. The gears also drive the ignition distributor. An eccentric on the camshaft drives the fuel pump.

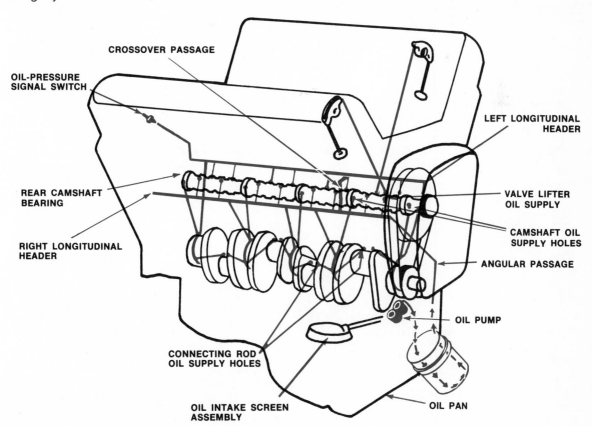

Fig. 19-10. A simplified drawing of the lubricating system for a V-8 engine.

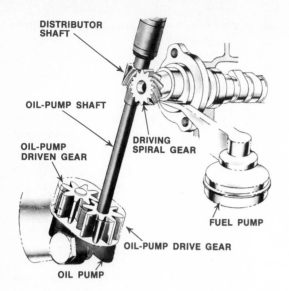

DISTRIBUTOR SHAFT

OIL-PUMP SHAFT

DRIVING SPIRAL GEAR

OIL-PUMP DRIVEN GEAR

FUEL PUMP

OIL-PUMP DRIVE GEAR

OIL PUMP

Fig. 19-11. Oil-pump, distributor, and fuel-pump drives. The oil pump is the gear type. A gear on the end of the camshaft drives the ignition distributor, and an extension of the distributor shaft drives the oil pump. The fuel pump is driven by an eccentric on the camshaft.

In the gear-type oil pump (Fig. 19-12), the drive gear drives the driven gear. The spaces between the gear teeth are filled with oil from the oil inlet of the pump. As the gear teeth mesh, the oil is forced out through the oil outlet.

In the rotor-type oil pump (Fig. 19-13), the inner rotor drives the outer rotor. The outer rotor is offset to one side so that there is space between the lobes of the two rotors on that side. Oil fills this space. Then, as the rotors rotate, the lobes move together, forcing the oil out. Figure 19-14 shows the lubricating system for a V-8 engine. The arrows show the flow of oil from the oil pump through the oil passages to the various moving engine parts.

§ 19-8 RELIEF VALVE

Since the oil pump is providing positive oil pressure, this pressure might go too high at high speed. To prevent this, there is a pressure relief valve in the oil line, often at the pump. It consists of a spring-loaded ball or plunger (Fig. 19-15). When the pressure goes too high, it forces the ball or plunger off its seat so that oil can flow past the valve and back into the crankcase.

§ 19-9 OIL FILTER

Many engines use an oil filter (Fig. 19-16). Part or all of the oil from the oil pump flows through this filter. Inside the filter, there is an element made of porous cellulose material, metal mesh, or similar material. As the oil passes through the element, solid particles of dust, carbon,

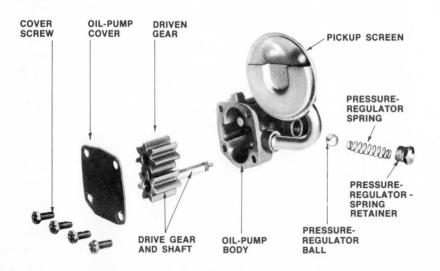

COVER SCREW

OIL-PUMP COVER

DRIVEN GEAR

PICKUP SCREEN

PRESSURE-REGULATOR SPRING

Fig. 19-12. A disassembled view of a gear-type oil pump. (*Pontiac Motor Division of General Motors Corporation*)

PRESSURE-REGULATOR-SPRING RETAINER

DRIVE GEAR AND SHAFT

OIL-PUMP BODY

PRESSURE-REGULATOR BALL

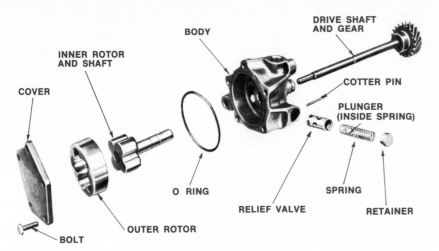

Fig. 19-13. A disassembled view of a rotor-type oil pump. (*Chrysler Corporation*)

Labels: COVER, INNER ROTOR AND SHAFT, BODY, DRIVE SHAFT AND GEAR, COTTER PIN, PLUNGER (INSIDE SPRING), SPRING, RETAINER, RELIEF VALVE, O RING, OUTER ROTOR, BOLT

and so on are trapped by the element and are filtered out. After a car has been driven many miles, the filter element tends to become clogged with impurities. It must then be replaced with a clean element.

§ 19-10 CRANKCASE-OIL DILUTION

As we mentioned in § 11-9 in our discussion of piston rings, some unburned air-fuel mixture and burned gases blow by the rings and enter the crankcase. In addition, water vapor is formed during the combustion process in the engine cylinders. Most of this water passes out through the exhaust system with the burned gases. However, when the engine is cold, some of the water condenses on the cylinder walls and works its way down past the piston rings into the oil pan. This is somewhat the same as water condensing on the window panes in your home on a cold day. If the air inside the house is moist enough, enough water will condense to trickle down the pane. In the engine, water that works its way down into the oil pan mixes with the oil. If the engine warms up enough, this water evaporates and does no harm. But in some kinds of driving, for instance short-trip city operation, the engine never gets really warm. In such a case, the oil and water are whipped into a thick sludge by the rotating crankshaft.

Gasoline also gets into the crankcase during cold-engine operation; this thins the oil and makes it less effective as a lubricant. The gasoline gets to the crankcase by condensing on the cylinder walls and working down past the piston rings.

§ 19-11 CRANKCASE VENTILATION

To aid in removing water, gasoline, and blow-by from the crankcase, engines have a crankcase ventilation system. In older model cars, the system had an air entrance (the oil filler pipe) through which air entered the crankcase. It passed through the crankcase, picking up water and gasoline vapors and discharging them into the atmosphere through a vent tube.

In recent years, the problem of smog has brought about a change in the crankcase ventilation system. Smog is fog-like mist that often hangs over large cities and is considered a health hazard. Authorities say that smog is partly due to the various gaseous compounds released by automobile engines. Therefore, engineers are trying to reduce these trouble-making compounds. They can be released from the engine through the exhaust system, which we will describe later, and through the crankcase ventilation system.

To eliminate atmospheric pollution from the crankcase ventilation system, the vent to the

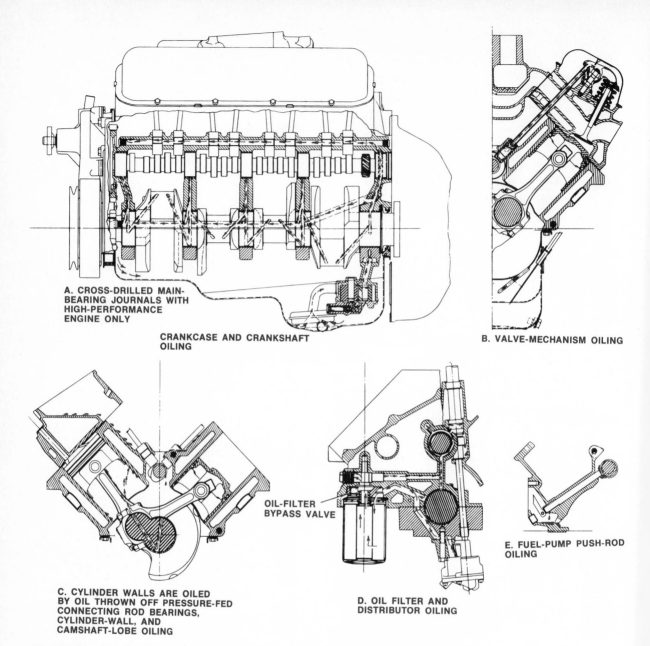

A. CROSS-DRILLED MAIN-
BEARING JOURNALS WITH
HIGH-PERFORMANCE
ENGINE ONLY

CRANKCASE AND CRANKSHAFT
OILING

B. VALVE-MECHANISM OILING

OIL-FILTER
BYPASS VALVE

E. FUEL-PUMP PUSH-ROD
OILING

C. CYLINDER WALLS ARE OILED
BY OIL THROWN OFF PRESSURE-FED
CONNECTING ROD BEARINGS,
CYLINDER-WALL, AND
CAMSHAFT-LOBE OILING

D. OIL FILTER AND
DISTRIBUTOR OILING

Fig. 19-14. The lubrication system of a V-8 engine. Arrows show the flow of oil to the moving parts in the engine. (*Chevrolet Motor Division of General Motors Corporation*)

atmosphere was closed. Modern engines have a closed system, called a **positive crankcase ventilating (PCV) system.** A typical system for a six-cylinder engine is shown in Fig. 19-17. The principle is simple. Filtered air from the carburetor air cleaner is drawn through the crankcase. In the crankcase, it picks up the water, fuel vapors, and blow-by. The air then flows back up to the intake manifold and enters the engine. There, unburned fuel is burned.

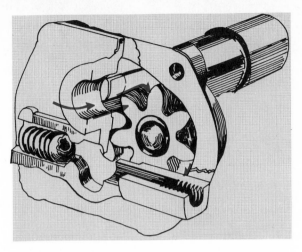

Fig. 19-15. Gear-type oil pump with built-in oil-pressure relief valve. Arrows indicate direction of oil through pump.

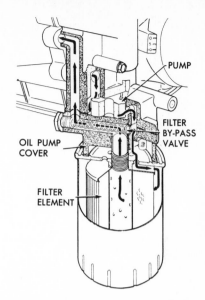

Fig. 19-16. A cutaway view of a full-flow oil filter with bypass valve. (*Buick Motor Division of General Motors Corporation*)

Too much air flowing through the intake manifold during idling would upset the air-fuel-mixture ratio and cause poor idling. To prevent this, the PCV system uses a regulator valve (Fig. 19-18). This valve is called a **positive crankcase ventilating (PCV) valve.** The valve allows only a little air to flow during idle. But as engine speed increases, the valve opens wider to permit more air to flow through.

§ 19-12 OIL-PRESSURE INDICATORS

Most multicylinder engines are equipped with some form of indicator to tell the driver when the engine oil pressure is too low for safe engine operation. If the pressure is too low, then not enough oil will get to moving parts. Without adequate lubrication, moving parts will rub hard against each other, overheat, and wear rapidly. They could even seize so that the parts would be broken and the engine ruined. Two types of indicators are used: a light that comes on to warn of low pressure or an indicating gage that has a needle which swings around to show the relative pressure.

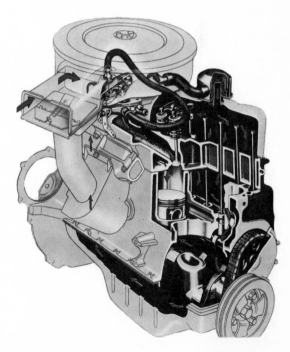

Fig. 19-17. A positive crankcase ventilating system on a six-cylinder engine.

In a V-8 engine the PCV valve is located in:
1. Rocker arm cover
2. Rear of engine
3. Carburetor base

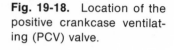

In a six-cylinder engine the PCV valve is located in:
1. Rocker arm cover
2. Base of carburetor
3. Hose

Fig. 19-18. Location of the positive crankcase ventilating (PCV) valve.

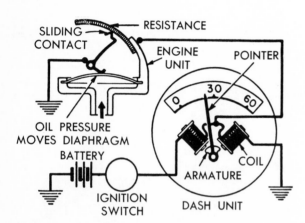

Fig. 19-19. Electric circuit of electric-resistance oil-pressure indicator.

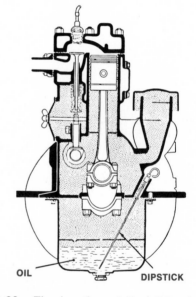

Fig. 19-20. The location of the oil-level stick, or dipstick, in the engine. (*Buick Motor Division of General Motors Corporation*)

The wiring circuit for one type of oil-pressure indicator (Fig. 19-19) is very similar to that for the fuel gage discussed in Chapter 18. Changing oil pressure changes the resistance of the engine unit, and this changes the magnetic pattern of the two coils in the dash unit. Thus, as oil pressure decreases, the resistance of the engine unit goes down, and more of the current flowing through the left-hand coil (the low-pressure side of the dial) flows to the engine unit rather than through the right-hand coil. Thus, the armature is swung to the left-hand coil, and this moves the pointer toward the low-pressure side of the dial.

Many cars use an oil-pressure indicator light. The light is off when the oil pressure is normal. But if the pressure drops too low, the light comes on.

§ 19-13 OIL-LEVEL INDICATORS

A dipstick is used to check the level of oil in the oil pan (Fig. 19-20). To use the dipstick, pull it out, wipe it off, and put it back in place. Then pull it out again and check the level of the oil shown on the dipstick.

In the positive crankcase ventilating (PCV) system used in modern engines, the dipstick has a seal to prevent crankcase gases from escaping through the dipstick tube.

Review questions

1. What prevents metal-to-metal contact of moving parts in the engine?
2. What are the compounds called that are added to the oil?
3. Does SAE10W or SAE20W oil have the higher viscosity?
4. What is a multiple-viscosity oil?
5. What are the five major oil additives?
6. Where does the oil pump get the oil that it sends to the engine parts?
7. How does the oil get to the various engine parts?
8. What are the two common types of oil pumps?
9. What is the job of the oil filter?
10. What is the name of the oil-level indicator in an engine?

Study problems

1. Ask at your local service station or garage for empty oil cans. Try to get one for each viscosity and rating of oil. Study every marking on the can and learn what each means.
2. Locate a disassembled engine. Trace the flow of oil from the oil pan, to the oil pump, through the engine oil passages, and back to the pan. Follow the flow of oil that is shown in Fig. 19-14.

20 | AUTOMOTIVE EMISSION CONTROLS

§ 20-1 REASON FOR AUTOMOTIVE EMISSION CONTROLS

In the automobile, gasoline is combined with air to produce the combustible mixture that makes the car run. Chapter 17 explains that complete combustion of the air-fuel mixture never occurs, and that gaseous chemicals are emitted from the tail pipe. These chemicals combine with chemicals from other sources and cause air pollution. You have all heard about smog. To help control this unhealthy condition, lawmakers in Washington, D.C., and in various states have passed laws that require car manufacturers to install pollution-control devices on their cars. These control devices cut down the amount of gaseous chemicals emitted that contribute to smog.

Automobiles are said to be responsible for about half of the air pollution in the United States. Smoke from power-plant and factory smokestacks, incinerators, and home-heating plants also contributes to air pollution. In this chapter, we will concentrate on automotive antipollution devices. They are called **automotive emission controls.**

§ 20-2 POLLUTION FROM AUTOMOBILES

The automobile gives off pollutants from four places, as shown in Fig. 20-1. Pollutants can come from the fuel tank, the carburetor, the crankcase, and the tail pipe. Pollutants from the fuel tank and carburetor consist of gasoline vapors. Pollutants from the crankcase consist of partly burned air-fuel mixture that has leaked past the pistons and rings into the crankcase. Pollutants from the tail pipe consist of partly burned air-fuel mixture that did not complete combustion in the engine cylinders. In addition, there is a pollutant, called nitrogen oxides, that forms in any high-temperature combustion process. To understand all these pollutants, let us first review the combustion process.

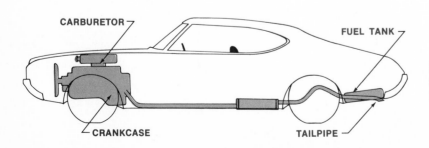

Fig. 20-1. Four possible sources of atmospheric pollution from the automobile.

§ 20-3 COMBUSTION

Gasoline is a hydrocarbon (HC), which is made up mostly of hydrogen and carbon. When a hydrocarbon burns completely, water (H_2O) and carbon dioxide (CO_2) are formed. Unfortunately, combustion is never complete; therefore, some HC remains and some CO (carbon monoxide) is formed. And, as you know, carbon monoxide is a poisonous gas.

There is another dangerous gas that the engine gives off—nitrogen oxides. Actually, there are several oxides of nitrogen, or nitrogen oxides. Nitrogen makes up about 80 percent of our atmosphere. Oxygen forms almost 20 percent. Usually, nitrogen is an **inert** gas; that is, it will not unite with any other element. But in the high combustion temperatures in the engine cylinders, some nitrogen will unite with oxygen to form nitrogen oxides. The chemical formula for these nitrogen oxides is NO_x. The "x" stands for different amounts of oxygen. NO_x unites with atmospheric moisture in the presence of sunlight to form an acid. This acid contributes to the eye-irritating, cough-producing effects of smog.

To sum up, there are three basic pollutants coming from the engine: unburned gasoline (HC), carbon monoxide (CO), and nitrogen oxides (NO_x). In addition, there is also loss of HC from the carburetor and fuel tanks. Let's find out how these pollutants are being controlled.

§ 20-4 CRANKCASE VENTILATION

Air must circulate through the crankcase when the engine is running. The reason for this is that water and liquid gasoline appear in the crankcase when the engine is cold. Also, there is some blow-by on the compression and power strokes. Blow-by is the name for the leakage of unburned gasoline vapor and burned gases past the pistons and the rings and down into the crankcase. Water appears as a product of combustion. Water is also carried into the engine as moisture in the air that enters the engine. When the engine is cold, this water condenses on the cold engine parts and runs down into the crankcase. Gasoline vapor also condenses on cold engine parts and runs down into the crankcase.

Unless the water, liquid gasoline, burned gases, and gasoline vapor are cleared from the crankcase, there will be trouble. Sludge and acids will form. Sludge is a gummy material that can clog oil lines and starve the engine lubricating system. This could mean a ruined engine. The acids corrode engine parts, and this, too, can ruin the engine.

In older engines, the crankcase was ventilated by an opening at the front of the engine and a vent tube at the back of the engine. The forward movement of the car and the rotation of the crankshaft moved air through the crankcase, as shown in Fig. 20-2. The air passing through removed the water and fuel vapors, discharging them into the atmosphere—and causing air pollution.

To prevent this atmospheric pollution, modern engines have a closed system, called a positive crankcase ventilating (PCV) system. A

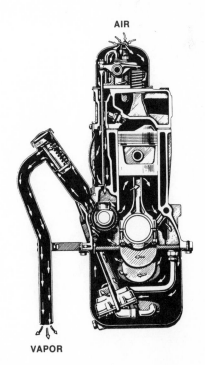

AIR

VAPOR

Fig. 20-2. An open crankcase ventilating system.

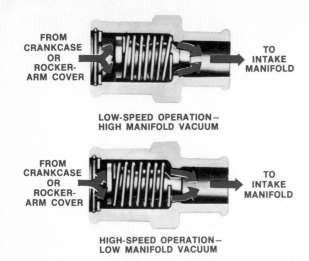

LOW-SPEED OPERATION—
HIGH MANIFOLD VACUUM

FROM CRANKCASE OR ROCKER-ARM COVER → TO INTAKE MANIFOLD

HIGH-SPEED OPERATION—
LOW MANIFOLD VACUUM

Fig. 20-3. Two operating positions of the PCV valve. (*Ford Motor Company*)

typical system is shown in Fig. 19-17. The principle is simple. Filtered air, usually from the carburetor air cleaner, is drawn through the crankcase. In the crankcase, the air picks up the water, fuel vapors, and blow-by. The air then flows back up to the intake manifold and enters the engine. There, unburned fuel is burned.

Too much air flowing through the intake manifold during the idle period would upset the air-fuel-mixture ratio and cause poor idling. To

prevent this, a regulator valve is used. This valve is the positive crankcase ventilation (PCV) valve. The PCV valve allows only a small amount of air to flow during idle. But as engine speed increases, the valve opens to allow more air to flow. Figure 20-3 shows the valve in the two positions.

§ 20-5 FUEL-VAPOR RECOVERY SYSTEMS

Modern automobiles have fuel-vapor recovery systems. These systems capture fuel vapor that might otherwise escape from the fuel tank or the carburetor when the engine is stopped and the car is sitting idle. Escaping fuel vapors add to atmospheric pollution. There are laws against pollution, and that is the reason for the vapor recovery system. One system is shown in Fig. 20-4. The fuel tank is sealed. Any gasoline vapor that tries to escape flows through the vapor recovery line to the charcoal canister. Gasoline vapor from the carburetor float bowl also flows to the charcoal canister. There, the charcoal particles capture the vapor.

When the engine is started the next time, air passes through the charcoal canister on its way to the carburetor. This air picks up the trapped gasoline vapor so that it is carried into the engine and burned.

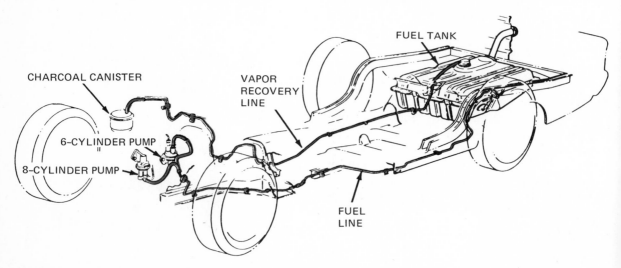

CHARCOAL CANISTER

6–CYLINDER PUMP

8–CYLINDER PUMP

VAPOR RECOVERY LINE

FUEL TANK

FUEL LINE

Fig. 20-4. Fuel-vapor recovery system. (*Ford Motor Company*)

§ 20-6 CLEANING UP THE EXHAUST GASES

As we mentioned, complete combustion never occurs in the engine cylinders. So, some HC and CO are contained in the exhaust gases. It is possible to increase the combustion temperature and thus burn more of the HC. But then, more NO_x is produced. Thus, simply increasing combustion temperatures is not the answer.

Let's see how the automotive engineers worked out this problem. They did several things. We will discuss each of these in detail later in this chapter.

1 Controlling the air-fuel mixture. First, the engineers changed the gasoline to make it burn better and to reduce pollutants in the exhaust gas. Second, the engineers used leaner carburetor settings and faster warm-up so that more of the fuel is burned during start-up.

2 Controlling combustion. Engineers have altered the combustion chambers of engines to improve combustion. By lowering compression ratios, they have also lowered the engine power but helped the emission problem. They have modified ignition timing to prevent excessive emissions during certain operating conditions. And they have circulated part of the exhaust gases through the engine to modify the combustion process.

3 Treating the exhaust gases. One procedure is to pump fresh air into the exhaust gases as they leave the engine. This supplies additional oxygen so that unburned HC and CO can burn. Another way to treat the exhaust gases is to use a **catalytic converter** in the exhaust line. Catalysts are chemicals that cause a chemical reaction, such as combustion, without actually becoming part of that reaction.

§ 20-7 CONTROLLING THE AIR-FUEL MIXTURE

Gasoline has been changed to make it burn better. One of the changes is the elimination of lead. Chapter 17 points out that octane can be increased by adding tetraethyl lead (ethyl) to gasoline. Thus, engines with higher compression ratios can be built. The reason that lead is now being removed is to let the catalysts work better to control pollutants. We'll get to catalysts in a moment.

Carburetors have been modified to deliver a leaner air-fuel mixture, particularly during idling. Modern carburetors have an idle limiter (Fig. 20-5), which is preset at the factory and prevents an excessively rich idle mixture.

Faster warm-up and quicker choke action are also important ways of reducing pollutants. During choking, a very rich mixture leaves the carburetor. Some of the excess gasoline has no chance to burn, and it exits in the exhaust gas as unburned HC. One way to reduce the amount of unburned gasoline in the exhaust is to preheat the mixture during warm-up. The thermostatic air cleaner does this job. Figure 17-11 shows the system. It includes a heat stove at the exhaust manifold, a hot-air pipe, and a thermostatic arrangement in the snorkel tube of the air cleaner. The **snorkel tube** is the tube through which air enters the air cleaner.

Figure 20-6 shows the parts of the air cleaner. It has a thermostatic spring, called a temperature-sensing spring in Fig. 20-6. When the entering air is cold, this spring holds the air-bleed valve closed. Now, intake-manifold vacuum can work on the vacuum chamber in the snorkel tube. The vacuum chamber, called the motor in Fig. 20-6, has a diaphragm, which

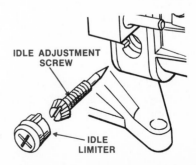

IDLE ADJUSTMENT SCREW

IDLE LIMITER

Fig. 20-5. The location of the idle limiter in one model of carburetor. (*Ford Motor Company*)

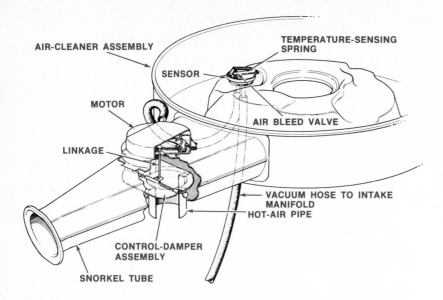

Fig. 20-6. An air cleaner with thermostatic control. (*Chevrolet Motor Division of General Motors Corporation*)

Labels in figure:
AIR-CLEANER ASSEMBLY
TEMPERATURE-SENSING SPRING
SENSOR
MOTOR
AIR BLEED VALVE
LINKAGE
VACUUM HOSE TO INTAKE MANIFOLD
HOT-AIR PIPE
CONTROL-DAMPER ASSEMBLY
SNORKEL TUBE

is raised by the vacuum. This movement of the diaphragm tilts a damper so that the snorkel tube is blocked off. Now, all air must come from the heat stove through the hot-air pipe.

As the engine warms up, the thermostat in the air cleaner begins to open the damper. Now air is taken from the engine compartment and not from the heat stove. When the engine is warm, no additional heat is needed to assure good fuel vaporization.

§ 20-8 CONTROLLING THE COMBUSTION PROCESS

Late-model engines have been changed somewhat to improve the combustion process. Until recently, the compression ratios of engines went up a little each year; until about 1969, the average was a little above 9.5:1. Now, however, the average is down around 8:1 or a little higher. We mentioned earlier in the chapter that NO_x is formed in the high-temperature combustion process. Reducing the compression ratio reduces top temperatures and, therefore, reduces the amount of NO_x that forms. Unfortunately, reducing the compression ratio also reduces engine performance and efficiency to some extent, but that is the price we must pay for clean air.

We mentioned earlier that lead (tetraethyl lead) has been added to gasoline to raise its octane. However, lead "poisons" the catalytic converter so that it stops doing its job. Thus, cars with a catalytic converter cannot use leaded gasoline. They must use lead-free gasoline. We will discuss catalytic converters in detail in **§** 20-9.

Another method of controlling the formation of NO_x in the engine is to recirculate some of the exhaust gas back through the engine. Figure 20-7 shows this system. It is called the exhaust gas recirculation (EGR) system. A part of the exhaust gas is picked up from the exhaust manifold and sent up through the carburetor and back through the engine. Usually, less than 10 percent of the exhaust gas is recirculated this way. The exhaust gas, mixing with the air-fuel mixture, lowers the combustion temperature enough to greatly reduce the amount of NO_x that forms.

Another way to mix some of the exhaust gases with the incoming air-fuel mixture is to leave part of the exhaust gases in the cylinders. The Chrysler NO_x control system (Fig. 20-8) does this by increasing the exhaust-valve and intake-valve overlap. The camshaft has been ground so that the cams provide this additional valve overlap.

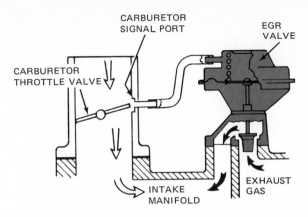

Fig. 20-7. Schematic view of an exhaust-gas recirculation system. (*Chevrolet Motor Division of General Motors Corporation*)

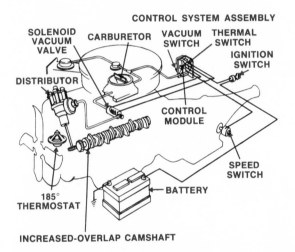

Fig. 20-8. The Chrysler Corporation NO$_x$ control system for cars with automatic transmissions. (*Chrysler Corporation*)

Figure 20-9 shows one General Motors version of the NO$_x$ control system. It is called the transmission-controlled spark (TCS) system. It uses a TCS solenoid that is connected electrically to a switch in the transmission. The switch is open only when the transmission is in high gear. When the transmission is in a lower gear, the switch is closed, connecting the TCS solenoid to the battery. The solenoid therefore lifts its plunger. This shuts off the

vacuum line from the carburetor to the vacuum-advance unit on the distributor.

The result is no vacuum advance. Then, when the transmission shifts into high gear, the transmission switch opens, and the solenoid is disconnected. Now, the vacuum line from the carburetor to the vacuum unit on the distributor opens so that vacuum advance results.

Vacuum advance is desirable when the engine starts as well as when the engine overheats. With vacuum advance, cold-engine operation is improved. Also, vacuum advance helps to prevent engine overheating, especially during long idling periods.

There are other vacuum-advance controls to prevent or to allow vacuum advance under different operating conditions. When you work in the shop and use the manufacturers' shop manuals, you will find out about these controls and how to service them.

§ 20-9 TREATING THE EXHAUST GAS

As we mentioned, not all the gasoline burns in the engine. Because of this, there is some HC and CO in the exhaust gases. One way to help burn these gases is to inject air into the exhaust gas as it enters the exhaust manifold. An air-injection system is shown in Fig. 20-10. It includes an air pump that forces air into the exhaust manifold through a series of injection tubes. The tubes are located just opposite the exhaust valves. This additional air supplies additional oxygen that helps burn up the HC and CO still in the exhaust gas.

A second method of treating exhaust gas is to use catalytic converters. These convert the gaseous pollutants into harmless gases. A catalyst is a material that causes a chemical change without entering into the chemical reaction. In effect, the catalyst "stands by" and "encourages" two chemicals to react, forming harmless gases.

A late-model system, engineered by General Motors, is shown in Fig. 20-11. The converter is filled with pellets of metal. They are coated with a thin layer of platinum or a similar cata-

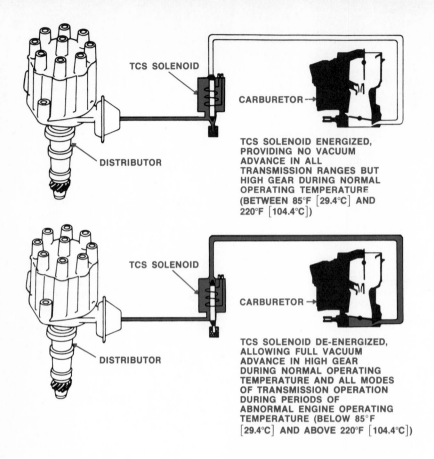

TCS SOLENOID

CARBURETOR

TCS SOLENOID ENERGIZED, PROVIDING NO VACUUM ADVANCE IN ALL TRANSMISSION RANGES BUT HIGH GEAR DURING NORMAL OPERATING TEMPERATURE (BETWEEN 85°F [29.4°C] AND 220°F [104.4°C])

DISTRIBUTOR

TCS SOLENOID

CARBURETOR

DISTRIBUTOR

TCS SOLENOID DE-ENERGIZED, ALLOWING FULL VACUUM ADVANCE IN HIGH GEAR DURING NORMAL OPERATING TEMPERATURE AND ALL MODES OF TRANSMISSION OPERATION DURING PERIODS OF ABNORMAL ENGINE OPERATING TEMPERATURE (BELOW 85°F [29.4°C] AND ABOVE 220°F [104.4°C])

Fig. 20-9. The General Motors NO_x control system, called by General Motors the transmission-controlled spark (TCS) system. (*Pontiac Motor Division of General Motors Corporation*)

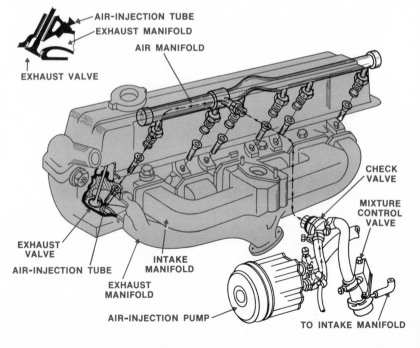

AIR-INJECTION TUBE
EXHAUST MANIFOLD
AIR MANIFOLD

EXHAUST VALVE

CHECK VALVE

MIXTURE CONTROL VALVE

Fig. 20-10. Cylinder head with manifolds and the parts of the Thermactor system detached. The cylinder head has been cut away at the front to show how the air-injection tube fits into the head.

EXHAUST VALVE

AIR-INJECTION TUBE

INTAKE MANIFOLD

EXHAUST MANIFOLD

AIR-INJECTION PUMP

TO INTAKE MANIFOLD

232

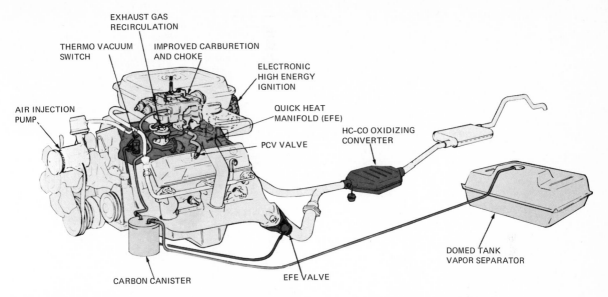

Fig. 20-11. One 1975 emission-control system using an under-the-floor catalytic converter. Note that the system uses air injection and other emission-control features described previously. (*General Motors Corporation*)

lytic metal. The pellets form a matrix (surrounding material) through which the exhaust gas must pass. Figure 20-12 shows one design of catalytic converter.

One of the problems with the catalysts is that the lead in leaded gasoline soon coats the catalysts. When this happens, the catalysts are no longer effective. Thus, if an engine runs on leaded gasoline, the catalysts stop doing their job in a very short time. This is one of the basic reasons that gasoline producers are getting the lead out of gasoline.

Removing the lead from gasoline can have a bad effect on the engine valves. The lead acts as a lubricant for the exhaust-valve faces and seats. A thin coat of lead covers the valve

Fig. 20-12. Cutaway view of the catalytic converter shown in Fig. 20-11. (*General Motors Corporation*)

faces and seats and helps to reduce wear. To prevent troubles, manufacturers are hardening the valve seats and coating the valve faces with special metallic compounds. This overcomes the problems brought on by the loss of lead in the gasoline.

§ 20-10 PREVENTING RUN-ON

Some of the emission-control devices have a tendency to allow the engine to keep running after the ignition is turned off. This is sometimes called dieseling, because the engine acts a little like a diesel engine and runs without any spark from the ignition system. The high idle-speed setting leaves the throttle partly open. This allows enough air-fuel mixture to enter the cylinders to continue combustion. The combustion is caused by hot spots in the combustion chamber, and the engine continues to run.

In one method used to stop the engine once the ignition has been turned off, there is a solenoid-operated throttle stop on the carburetor, as shown in Fig. 20-13. When the ignition is on, the solenoid is connected to the battery, and it keeps the idle speed up around 750 rpm.

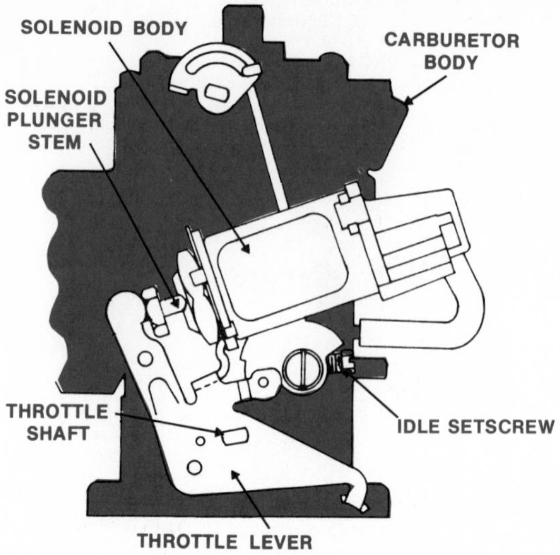

Fig. 20-13. A throttle-stop solenoid mounted on a carburetor.

This means that the throttle is partly open. However, when the ignition switch is turned off, the solenoid is disconnected. This allows the throttle to close fully, and thus practically close off the flow of air-fuel mixture to the engine. The engine therefore stops.

Review questions

1. What are the four places in the automobile from which pollutants can be given off?
2. What are the three pollutants from automobiles?
3. What is the name of the valve in the crankcase ventilation system?
4. What two types of vapor recovery systems are used in automobiles today to control automotive emissions?
5. What are the three basic things that can be done to clean up the exhaust gas?
6. When you start a cold engine equipped with a thermostatic air cleaner, where does the air that goes to the air cleaner come from?
7. How does the EGR system work?
8. How does the Chrysler NO_x control system mix some of the exhaust gases with the incoming fresh air-fuel mixture?
9. In operation, what does the air-injection system do?
10. What is the name of one type of catalyst that is used to control auto emissions?

Study problem

1. Inspect various vehicles, trying to locate and identify each emission-control system. Then, name every part of each system, and describe what each part does. On a sheet of paper, list the year, make, and model of vehicle. Then list each emission-control system that you found on that vehicle.

21 OTHER AUTOMOTIVE POWER SOURCES

§ 21-1 POSSIBLE ALTERNATIVE POWER SOURCES

Ever since the internal-combustion engine got its start more than a century ago, there have been challengers to its supremacy. Engineers have proposed and sometimes developed power sources that in some respect could rival the gasoline-burning internal-combustion engine. But so far, none of the rivals has been able to beat out the conventional piston engine in the market place. In this chapter, we will discuss several other power sources. These include the steam engine, the Stirling engine, electric cars, hybrid power systems, and the kinetic energy wheel (flywheel propulsion). Separate chapters in the book cover Wankel and gas-turbine rotary engines, and diesel engines.

§ 21-2 STEAM ENGINES

In 1769, a French Army engineer, Nicholas Cugnot, built the first steam-powered road vehicle at the Paris Arsenal. A drawing of Cugnot's steam carriage is shown in Fig. 1-11. Designed as a tractor to tow artillery, it had a top speed of about 2 miles per hour (mph) [3.2 kilometers per hour (km/h)]. When the vehicle overturned in 1771, thoughts of a steam road vehicle were temporarily abandoned. But Cugnot's setback didn't stop the spread of the idea of using machines instead of people and animals to do heavy work.

As the twentieth century began, the steam engine became a strong competitor of the internal-combustion engine for use in automo-

Fig. 21-1. Charles Randolph's steam carriage of 1872. This vehicle, built in Glasgow, weighed $4\frac{1}{2}$ tons [4,050 kilograms] and had a top speed of 6 miles [9.66 kilometers] per hour. It carried eight passengers, the driver, and the engineer. (*A History of Techology, vol. 5, Oxford*)

biles. An example of an early steam carriage is shown in Fig. 21-1. But development of the electric starting motor caused the internal-combustion engine to win over both steam and electric power. In 1932, Doble built the last of the production steam cars.

§ 21-3 ADVANTAGES OF STEAM ENGINES

An important advantage today for the steam engine is that, as an external-combustion engine, it will burn almost any hydrocarbon fuel. As we discussed in Chapter 1, in a steam engine, the fuel does not explode periodically in the cylinders, as in the conventional gasoline piston engine. Burning takes place continuously in a burner outside the cylinders with consistent flame, combustion, and emission characteristics. This makes emission control much easier with the steam engine than with the conventional engine.

§ 21-4 DISADVANTAGES OF STEAM ENGINES

Automotive piston-type steam engines also have many disadvantages for passenger car use. The first steam cars used the wasteful open steam system. In the open steam system, the steam pushed the piston from one end of the cylinder to the other, transmitting power to the crankshaft. Then the exhaust valve opened, and the steam was exhausted into the outside air. Because steam is vaporized water, this meant that the car lost its water quickly and had to have water added at frequent intervals. This feature created yet another problem common to the steam locomotive (Fig. 3-1). If the fresh water had a high mineral content, it left a scaly deposit as it evaporated, and the boiler tubes soon clogged shut.

Scaling was greatly reduced by using a closed steam system. In this system, after the used steam came out of the exhaust valve, it would be condensed back into water in a condenser or heat exchanger instead of being exhausted into the outside air. The water was saved for reuse within the engine by using this method, and excessive scaling was reduced.

Another problem with almost all steam engines is that it takes time to "get up steam." To start the engine, light the boiler. That doesn't take long. But then there is a wait—now usually less than 1 minute—until sufficient steam pressure has built up to move the car.

§ 21-5 CURRENT DEVELOPMENTS IN STEAM

Two widely publicized men who have worked on perfecting the automotive steam engine are industrialist William P. Lear and Wallace L.

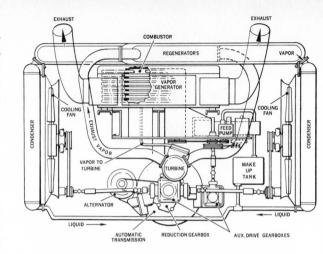

Fig. 21-2. A schematic of the engine compartment of a steam bus. (*General Motors Corporation*)

Minto, a Sarasota, Florida, inventor. Lear's company, Lear Motors of Reno, Nevada, built one of three steam buses tested in California in 1972. A schematic of the engine compartment is shown in Fig. 21-2.

Minto's engine, shown in Fig. 21-3, uses freon as the working fluid instead of water. Freon is the refrigerant used in most air-conditioning systems. The evaporated freon, at high pressure, drives a rotary type of engine. This engine has meshing gears, somewhat like an oil pump. The high-pressure freon gas rotates the gears, and this rotary motion is carried through the transmission and propeller shaft to the differential and wheels.

Figure 21-4 shows a 160-horsepower [120-kW] steam engine built by General Motors and installed in a Pontiac passenger car. General Motors reported fuel consumption to be 3.8 miles per gallon (mpg) in city driving. This problem was compounded by the fact that the steam engine weighs 450 pounds [204.12 kilograms] more than the engine it replaced, and it produces less than half the horsepower.

§ 21-6 STEAM-ENGINE EMISSIONS

After Congressional hearings in 1968, public interest in steam cars was aroused once again. To determine if a modern steam engine could

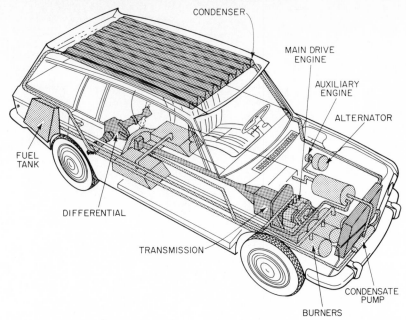

CONDENSER

MAIN DRIVE ENGINE

AUXILIARY ENGINE

ALTERNATOR

FUEL TANK

DIFFERENTIAL

TRANSMISSION

CONDENSATE PUMP

BURNERS

Fig. 21-3. The Minto automobile.

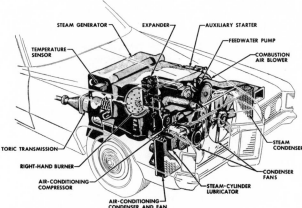

STEAM GENERATOR EXPANDER AUXILIARY STARTER

TEMPERATURE SENSOR

FEEDWATER PUMP

COMBUSTION AIR BLOWER

TORIC TRANSMISSION

RIGHT-HAND BURNER

AIR-CONDITIONING COMPRESSOR

AIR-CONDITIONING CONDENSER AND FAN

STEAM-CYLINDER LUBRICATOR

CONDENSER FANS

STEAM CONDENSER

Fig. 21-4. A steam engine. (*General Motors Corporation*)

serve as an alternate power plant for the piston engine, contracts were given to three different companies to build new steam engines for buses. After acceptance testing, the buses were placed in limited service on routes in Los Angeles, Oakland, and San Francisco, California. Tests indicated that the bus emission levels beat the 1975 California exhaust emission standards for the diesel engines that they replaced. But there were other problems.

General Motors designed, built, and tested two steam-powered passenger cars. One of these is shown in Fig. 21-4. In their vehicles,

NO_x levels exceeded federal passenger-car engine standards, the engines used almost three times as much fuel as conventional engines, production costs were much higher, maintenance requirements were also high, and vehicle performance was so poor as to be unacceptable to the average driver.

§ 21-7 FUTURE OF THE STEAM CAR

Why today's interest in steam power for automobiles? A continuous combustion process—like that of the steam engine—has the potential to burn almost any fuel completely, thereby producing fewer exhaust emissions. If steam engines replaced the internal-combustion engine in millions of cars, proponents claim that cleaner air would result. Most tests indicate that HC and CO levels are very low in the steam engine. But experienced automotive engineers have learned one thing only too well: Success in designing an ideal engine—steam or any other kind—and getting it into mass production to replace the conventional piston engine are two entirely different and highly speculative undertakings.

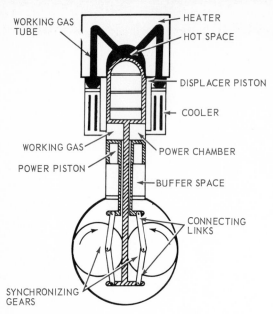

Fig. 21-5. Schematic drawing of Stirling engine.

Labels: WORKING GAS TUBE, HEATER, HOT SPACE, DISPLACER PISTON, COOLER, WORKING GAS, POWER CHAMBER, POWER PISTON, BUFFER SPACE, CONNECTING LINKS, SYNCHRONIZING GEARS

Fig. 21-6. Front ¾ view of mock-up of Philips Stirling engine for a Ford Torino. Preheater and burner are at front. Exhaust pipes show at bottom, with exhaust-gas recirculation tube extending upward to intake air blower. A combined water pump and power steering pump, belt-driven, is located near the front, and the alternator is shown at the rear corner. (*Ford Motor Company*)

§ 21-8 STIRLING ENGINE

Society's continuing quest for the perfect engine includes the consideration that no one be injured in any way by the engine's operation. Low noise, low smoke, low odor, and low exhaust emissions are necessary standards for such an engine. Only one known engine comes close to meeting these standards, and that is the Stirling engine.

§ 21-9 HISTORY OF THE STIRLING ENGINE

This engine, shown in Fig. 21-5, was patented in 1816 by Robert Stirling, a minister of the Church of Scotland. Like the steam engine, it is a hot-air engine. It converts heat energy into mechanical energy by using the closed-cycle external continuous combustion process.

Since 1938, the N. V. Philips Company of the Netherlands has been working on the engine in attempts to refine and simplify its design. But little was heard about it until 1960, when General Motors and Philips announced a joint development agreement, a project that was terminated in 1970. Then, in 1972, Ford Motor Company announced a licensing and develop-

ment agreement with Philips for the Stirling engine (Fig. 21-6).

While no car manufacturer has a Stirling engine in production, the engine has been built in small sizes and power outputs for experimental and development work. Let us take a look at how this engine operates.

§ 21-10 STIRLING-ENGINE OPERATION

The pressure of gas in a container goes up when the gas is heated and goes down when the gas is cooled. Since the Stirling engine is a closed-cycle engine, a small amount of working gas is sealed inside (Fig. 21-5). This gas is alternately heated and cooled. When heated, its increasing pressure pushes a power piston down. When cooled, its lowered pressure, in effect, helps pull the power piston up.

§ 21-11 STIRLING-ENGINE POTENTIAL

Experimental engines have operated at 3,000 rpm with an efficiency of 30 percent. This is a little better than the efficiency of most conventional piston engines. Stirling engines have less pumping losses—those compression- and exhaust-stroke power losses—than the conventional piston engines. However, with two synchronized pistons for each cylinder, transmission of power from the power piston to the output shaft requires use of a two-gear rhombic drive (shown in Fig. 21-5) or a swash-plate drive system.

The Stirling engine requires a source of heat. This heat source can be the burning of almost any form of fossil fuel, or it can be nuclear energy. The use of heat from the sun (solar heat) has even been proposed.

§ 21-12 STIRLING-ENGINE EMISSIONS

Exhaust emissions of HC and CO are so low that almost any standard can be met. However, the high peak temperatures when petroleum fuel is used causes high NO_x emission levels. As with the piston engine, exhaust gas recirculation lowers the NO_x formation, but it also lowers the engine thermal efficiency.

The Stirling engine is an operating reality. It shows a potential for low emissions and low noise levels. With further development in mechanical simplification and size, weight, and cost reduction, the Stirling engine may someday be mass-produced.

§ 21-13 ELECTRIC CARS

Years ago, there were many electric cars in use throughout the world. An early electric cab is shown in Fig. 21-7. They were powered by the same type of conventional lead-acid storage battery used in the electrical system of every car today. But instead of using just one battery per car, each old-time electric vehicle needed a dozen or more batteries. These old-time electric cars were very simple in design. As Fig. 21-7 shows, most had no gears, simple brakes, and tiller steering. They gave off no smog-producing pollutants and were almost silent in operation. Their acceleration and

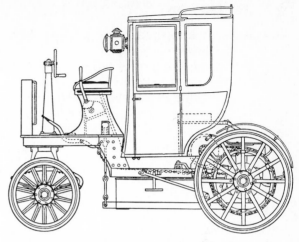

Fig. 21-7. Bersey electric cab of 1897. A number of these cabs operated for several years in London. They had a range of up to 50 miles [80.50 kilometers] before requiring battery recharging. (*A History of Technology, vol. 5, Oxford*)

speed were very low. They had a range of 40 to 50 miles [64 to 80 kilometers] of "around-town" driving before the car had to be garaged for a long battery recharging. These old electric cars were ideal for the purpose for which they were designed—to provide limited, low-speed transportation in town.

§ 21-14 ELECTRIC VEHICLE USAGE

For many years, the electric car has been a museum piece, its lack of range and performance making its use unacceptable in today's world. But concern with the air-pollution problem and the search for possible solutions is causing many individuals, states, and the federal government, as well as automobile manufacturers, to take a fresh look at electric cars. Records show that in England, an estimated 50,000 electric vans are in daily use for short-mileage delivery service around towns and cities. This is exactly the type of service performed by millions of air-polluting internal-combustion-engined vehicles here in the United States.

§ 21-15 THE BATTERY PROBLEM

The problem that plagues the electric car designer today is the same problem that the designer fought at the turn of the century. The familiar lead-acid battery is heavy. After enough of these batteries have been loaded on a vehicle to provide acceptable speed and range, little power is available to move a payload of either packages or people.

Many different kinds of batteries are being investigated by scientists and engineers. But no other type of battery has proved as reliable or as cheap to manufacture as the lead-acid battery used in our present cars.

§ 21-16 FUEL CELLS

The fuel cell has been thought to have some possibilities for automotive applications (Fig. 21-8). In the fuel cell, fuel is burned in such a way that the energy is directly converted into electricity. But the cost is high, the crash hazards great, and the problems of refueling a fuel cell are complicated. In Fig. 21-8, note that the fuel cell and batteries take up almost all the space in the van, even using part of the roof. As a result, the van, when fuel-cell powered, becomes a two-passenger vehicle.

§ 21-17 GM ELECTROVAIR

General Motors built an electric car called the Electrovair, which uses a silver-zinc battery. This car, shown in Fig. 21-9, had fair acceleration but an operating range between charges of only 40 to 80 miles [64 to 129 kilometers]. Because of the amount of silver and zinc required per car, the cost of these batteries is prohibitive.

§ 21-18 FUTURE OF ELECTRIC CARS

The electric car still awaits a major breakthrough in battery design and construction. Until that happens, no mass production of a modern electric car is likely. Present estimates indicate that an electric car built today would cost much more than the current production-model piston-engined vehicle. In addition, an electric car is essentially a one-purpose car. The family owning an electric car today still needs a second car for longer distance and high-speed trips.

Another problem would arise if many people began driving electric cars. Recharging the batteries in thousands of electric cars would require the electric companies to build new and larger generating plants. Probably these new plants would burn fossil fuel—coal or oil. Ac-

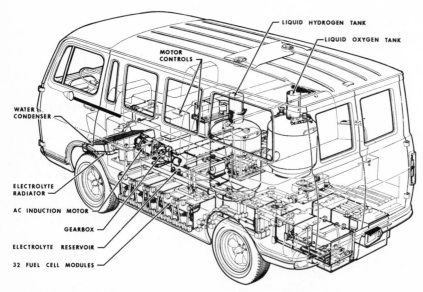

Fig. 21-8. Phantom view of Chevrolet Electro-Van, powered by a fuel cell, showing arrangement of major components. (*General Motors Corporation*)

LIQUID HYDROGEN TANK
LIQUID OXYGEN TANK
MOTOR CONTROLS
WATER CONDENSER
ELECTROLYTE RADIATOR
AC INDUCTION MOTOR
GEARBOX
ELECTROLYTE RESERVOIR
32 FUEL CELL MODULES

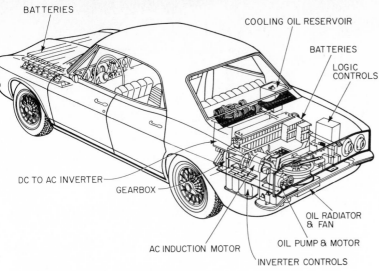

BATTERIES

COOLING OIL RESERVOIR

BATTERIES

LOGIC CONTROLS

DC TO AC INVERTER

GEARBOX

AC INDUCTION MOTOR

INVERTER CONTROLS

OIL PUMP & MOTOR

OIL RADIATOR & FAN

Fig. 21-9. The General Motors prototype electric car using silver-zinc storage batteries. (*General Motors Corporation*)

cording to some experts, burning greater amounts of these fuels would add more pollutants to the air than if we continued to drive our conventional gasoline-engine cars.

§ 21-19 HYBRID CARS

There has been a considerable amount of research on hybrid cars—usually cars that are partly electric and partly internal- or external-combustion. These are interesting from the engineering standpoint. But at present, no practical combination shows up well competitively with the internal-combustion engine.

§ 21-20 VARIOUS HYBRID CAR DESIGNS

Many interesting ideas have been advanced as design concepts for hybrid cars (Fig. 21-10). One frequently used design is to add a small combustion engine to an electric car. The purpose of the small engine is to power a built-in battery charger. With the engine running all the time, or at least when needed, the batteries will remain charged longer. This will improve both vehicle speed and range. In some proposed designs, the charging engine can be mechanically connected to the drive wheels for high-speed freeway driving or hill climbing.

Figure 21-10 shows a hybrid gasoline-electric vehicle built by General Motors. The

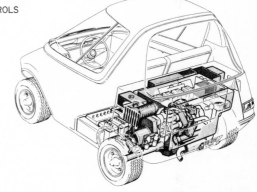

Fig. 21-10. A hybrid gasoline-electric vehicle. (*General Motors Corporation*)

power system consists of a small gasoline engine coupled with a dc series electric motor through an electromagnetic clutch. Figure 21-11 shows another hybrid car, combining a Stirling engine with an electric motor. Also built by General Motors, this car is called the Stir-Lec II. In this vehicle, an 8-horsepower [5.97-kW], single-cylinder Stirling engine runs at constant speed. It drives an alternator to charge a lead-acid battery pack of 14 batteries connected in series. A 20-horsepower [14.92-kW] dc motor drives the vehicle.

Another system frequently found in hybrid car designs is a regenerative braking and charging system. With this system, the deceleration and braking energy of the vehicle could be used to power the battery charger. Regenerative braking would enable the batteries to

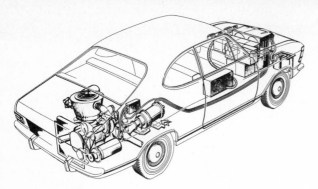

Fig. 21-11. A hybrid car combining a Stirling engine with an electric motor. (*General Motors Corporation*)

be charged as a result of the vehicle coasting or going downhill.

§ 21-21 ADVANTAGES AND DISADVANTAGES OF HYBRID CARS

Analysis of the hybrid vehicle shows that it may combine the best advantages of several power systems. But it also usually combines serious disadvantages of those same systems. On-board battery packs and recharging systems have the same shortcomings discussed previously. Hybrid car performance, speed, and range characteristics are similar to those discussed for the electric car.

One advantage in most hybrid car designs is that controlling emissions from the small constant-speed-and-load engine is simpler and cheaper than from the conventional-size piston engine. Often, this auxiliary power source is proposed to be a fuel cell or a steam or Stirling engine, as we discussed. A basic disadvantage of the hybrid car is that there are two separate power systems to contend with instead of one. This increases the complexity, cost, manufacturing, and service problems tremendously. Emission controls must be applied to both power systems. Therefore, total hardware costs may exceed the price of controlling emissions from the typical automobile engine. As a result, no production of a hybrid car is anticipated in the immediate future by any major automotive manufacturer.

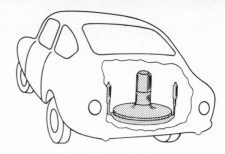

Fig. 21-12. The driving flywheel could be located at the back of the car.

§ 21-22 FLYWHEEL PROPULSION

This concept for operating an automobile has been studied and, in addition, has had tryouts in Europe. The idea is as simple as a child's windup toy. A large flywheel weighing several hundred pounds is mounted in a vehicle, somewhat as shown in Fig. 21-12. It is "charged" by electrically spinning the flywheel at high speed. The energy thus stored in the flywheel operates a generator to produce electric current that then powers electric motors at the car wheels.

§ 21-23 THE FLYWHEEL BUS

In 1953, the 70-passenger Oerlikon bus was introduced in Europe (Fig. 21-13). It was driven by a 3,300-pound [1,496.9-kilogram] flywheel and was considered a technical success. However, it had to pull up to a utility pole about every half mile [0.8 kilometer]. There electrical connections were made to "recharge" the flywheel, that is, to speed up the flywheel again. Since each recharge took about 2 minutes, passengers were hardly pleased about the bus. After use in Switzerland, Germany, and Leopoldville, the last Oerlikon bus was removed from service in 1969.

In 1972, Lockheed Missiles & Space Company announced a plan to build a flywheel-powered (kinetic energy) trolley coach for use in San Francisco. (A trolley coach is a type of electric bus that has rubber tires and gets its power from overhead streetcar wires.) Figure 21-14 shows a cutaway view of the proposed installation.

§ 21-24 FUTURE OF FLYWHEEL-POWERED VEHICLES

With relatively heavy flywheels spinning at supersonic speeds, and relatively light vehicles, the theoretical predicted performance for buses and cars might be achieved. But other problems await solutions. For example, in areas where streetcar or trolley coach wiring does not exist, a system of recharging stations would have to be built. And some people are concerned about the hazard that exists in a collision when a flywheel weighing several hundred pounds is spinning at high speed in the back of the vehicle. Also, charging stations require extra electrical generating capacity. This means building more generating plants that will probably burn fossil fuel. The problems in controlling electrical generating plant emissions were covered earlier. In short, the use of flywheel propulsion may create as many problems as it solves.

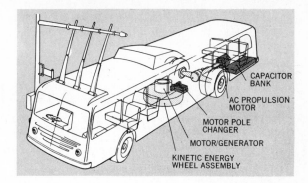

Fig. 21-13. The Oerlikon bus. (*Lockheed Missiles and Space Company, Inc.*)

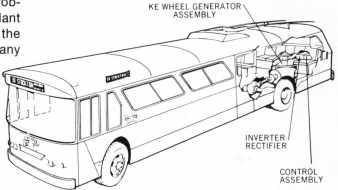

Fig. 21-14. Electric coach with kinetic energy propulsion kit. (*Lockheed Missiles and Space Company, Inc.*)

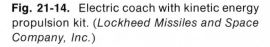

Review questions

1. Name five different engines or power sources that could be used to replace the conventional automobile engine.
2. Is the steam engine an internal- or external-combustion engine? Why?
3. What is the main advantage of using a steam engine in an automobile?
4. What pollutant is very difficult to control when petroleum is burned?
5. What is the engine called that is named for a minister?
6. Which engine could theoretically use the sun as the source of heat?
7. What is the main disadvantage of the electric car today?
8. What would be the effect on air pollution if large numbers of people began using an electric car?
9. What is the danger when a fuel cell is used in a car?
10. What is a hybrid car?

Study problem

1. Take a trip to a museum where there are some old cars. Make a list of the cars and the type of engine or power plant used in each. Try to decide for yourself why the conventional piston engine became the standard automotive engine.

22 | ROTARY ENGINES

§ 22-1 ROTARY-ENGINE TYPES

All the engines we have discussed so far are reciprocating engines. That is, they have pistons that move up and down (reciprocate) in cylinders. In this section, we are going to look at another type of internal-combustion engine— the type in which rotors spin. These are **rotary engines,** and there are two general kinds, the gas turbine and the Wankel.

The gas turbine burns fuel in a combustion chamber at a steady rate and shoots the resulting high-pressure gas against the curved blades of a turbine or rotor wheel (Fig. 22-1). This spins the turbine rotor at high speed. The rotary motion is carried to the car wheels by gears and shafts. Note that this is a steady process and that speed is controlled by the amount of fuel that is burned in the combustion chamber.

The Wankel engine is often called a rotary-combustion, or RC, engine because the combustion chambers rotate. For each rotor in the Wankel engine, there are three combustion chambers, and all are in constant rotation. We will explain this in more detail later.

§ 22-2 GAS TURBINE

Figure 22-2 is a cutaway view of a gas-turbine engine. Figure 22-3 is a simplified drawing of the turbine engine. The turbine has two separate shafts. The shaft to the left (in Fig. 22-3) carries the compressor impeller and the gasifier rotor. As the shaft spins, it causes the impeller to force air into the turbine. This air, under fairly high pressure, flows through the regenerator, which we will describe in a moment. From here, it flows into the combustor, or combustion chamber. At the same time, fuel

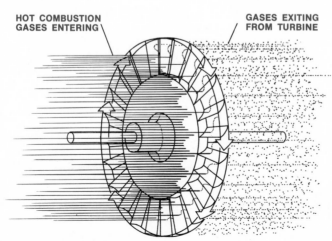

HOT COMBUSTION GASES ENTERING

GASES EXITING FROM TURBINE

Fig. 22-1. The pressure of the gas hitting the blades of the turbine rotor forces the rotor to spin.

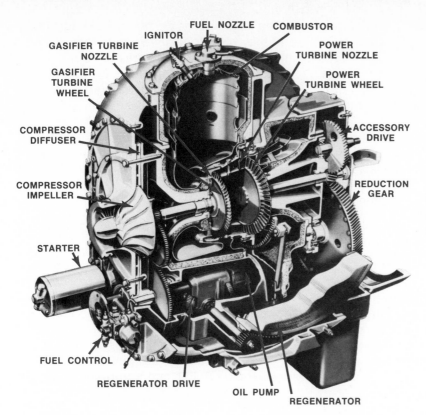

Fig. 22-2. A cutaway view of a turbine engine. (*Ford Motor Company*)

Labels on Fig. 22-2:
GASIFIER TURBINE NOZZLE
IGNITOR
FUEL NOZZLE
COMBUSTOR
POWER TURBINE NOZZLE
GASIFIER TURBINE WHEEL
POWER TURBINE WHEEL
COMPRESSOR DIFFUSER
ACCESSORY DRIVE
COMPRESSOR IMPELLER
REDUCTION GEAR
STARTER
FUEL CONTROL
REGENERATOR DRIVE
OIL PUMP
REGENERATOR

Fig. 22-3. A schematic view, showing the location of components in a turbine engine. (*Ford Motor Company*)

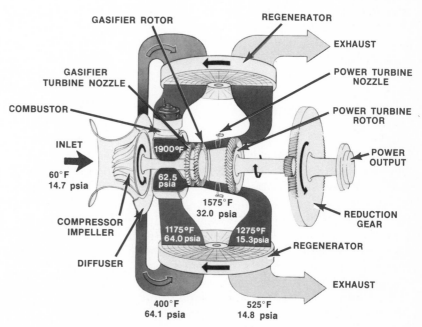

Labels on Fig. 22-3:
GASIFIER ROTOR
REGENERATOR
EXHAUST
GASIFIER TURBINE NOZZLE
POWER TURBINE NOZZLE
COMBUSTOR
POWER TURBINE ROTOR
INLET
60°F 14.7 psia
1900°F
62.5 psia
POWER OUTPUT
1575°F 32.0 psia
REDUCTION GEAR
COMPRESSOR IMPELLER
1175°F 64.0 psia
1275°F 15.3psia
REGENERATOR
DIFFUSER
EXHAUST
400°F 64.1 psia
525°F 14.8 psia

is sprayed into the combustion chamber and ignited. The fuel burns and produces high pressure. This high-pressure gas flows through the gasifier rotor and causes it to spin. This rotor is on the other end of the shaft from the compressor impeller. This is how the impeller is forced to spin.

As the hot high-pressure gas leaves the gasifier rotor, it flows into the power-turbine rotor. This rotor is spun by the high-pressure gas. The rotary motion is carried by the turbine-rotor shaft to reduction gears and then to the vehicle wheels.

§ 22-3 REGENERATOR

The regenerator makes the turbine more efficient. The gases are still hot when they leave the power-turbine rotor. Then they pass through the regenerator wheels. The regenerator wheels are made of porous ceramic material, and they revolve very slowly (15 rpm at high turbine speed). The hot gases heat the ceramic material as they pass through. Then the hot parts of the wheels move around to where the air is flowing in from the compressor impeller. This incoming air has to pass through the ceramic material, and it picks up heat. In other words, the air is preheated before it goes into the combustor. The result is that a good part of the heat that would otherwise be wasted is saved. This means that the turbine runs more efficiently.

§ 22-4 GAS-TURBINE APPLICATIONS

Gas turbines are at their most efficient when they can run at constant speed, without speedups and slowdowns. This means that they are not at their best in passenger cars. They do better in intercity trucks that run at a more or less constant speed for many miles. They are at their best in stationary power plants, where they run at a constant speed.

The automotive manufacturers have experimented with turbines in passenger cars, but there are certain drawbacks. The turbines spin at very high speeds (36,000 rpm, for instance), and they have to take very high temperatures and pressures. This requires special, and expensive, materials. In addition, the turbine provides only slow acceleration. It takes time for the rotor to increase in speed from a few hundred rpm to many thousands of rpm. This means relatively slow acceleration. Nevertheless, experimental work continues, and we may see turbines in passenger cars some day.

§ 22-5 WANKEL ENGINE

The Wankel engine was developed by a German engineer, Felix Wankel, in 1957. Since then, many automotive manufacturers have begun experimenting with the engine, and some have started using this engine instead of the piston engine. For example, the Toyo Kogyo Company, a Japanese firm, has produced hundreds of thousands of Wankel engines for its Mazda cars. Several other firms in the United States and Europe are also producing Wankel engines.

§ 22-6 CONSTRUCTION OF THE WANKEL ENGINE

Figure 22-4 is a cutaway view of a two-rotor Wankel engine, with an attached transmission. Since both rotors work the same way, let us concentrate on what happens with a single rotor. A rotor is shown in Fig. 22-5. Note that it has three lobes, or apexes. Figure 22-6 shows how the rotor fits into the rotor housing. Note that the housing is shaped like a fat figure "8." The rotor rotates off center (eccentrically) in the housing (Fig. 22-7). The three lobes are in constant contact with the inside face of the rotor housing.

The sides of the rotor slide on flat side housings. Figure 22-8 shows one of the side housings. The sides of the rotor are sealed against the flat surfaces of the side housings. The stack-up looks as shown in Fig. 22-9. As you can see, the rotor, with its seals, makes three separate chambers. These are the combustion chambers, and when the rotor rotates, the

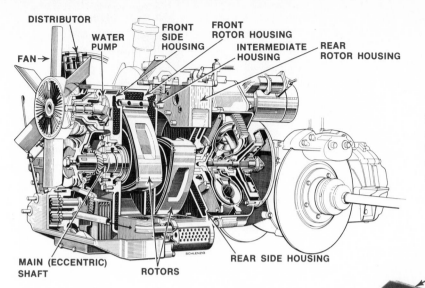

DISTRIBUTOR

FAN →

WATER
PUMP

FRONT
SIDE
HOUSING

FRONT
ROTOR HOUSING

INTERMEDIATE
HOUSING

REAR
ROTOR HOUSING

MAIN (ECCENTRIC)
SHAFT

ROTORS

REAR SIDE HOUSING

Fig. 22-4. A cutaway view of a two-rotor Wankel engine with attached torque converter and transmission. (*NSU of Germany*)

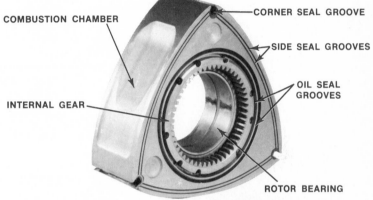

COMBUSTION CHAMBER

INTERNAL GEAR

APEX SEAL GROOVE

CORNER SEAL GROOVE

SIDE SEAL GROOVES

OIL SEAL GROOVES

ROTOR BEARING

Fig. 22-5. Rotor for Mazda Wankel engine. (*Toyo Kogyo Company, Limited*)

Fig. 22-6. (*Above*) This is how the rotor fits into the rotor housing. (*Toyo Kogyo Company, Limited*)

Fig. 22-7. The rotor rotates eccentrically so that the three apexes are always in sliding contact with the inner face of the rotor housing. (*Toyo Kogyo Company, Limited*)

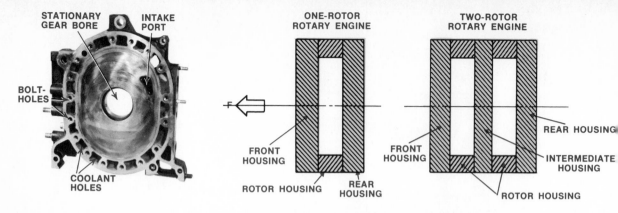

Fig. 22-8. (*Left*) Front side housing. Note the large holes around the outer edge for coolant to flow through and the intake port to the right of the stationary gear bore.

Fig. 22-9. (*Right*) How the side and rotor housings stack up in a one-rotor and a two-rotor Wankel engine. (*Toyo Kogyo Company, Limited*)

chambers also rotate. This is the reason the Wankel engine is known as a rotary-combustion (RC) engine.

§ 22-7 OPERATION OF THE WANKEL ENGINE

Figure 22-10 shows the principle of Wankel-engine operation. There are four stages in the action cycle. Start at A (upper center in Fig. 22-10). Here, the rotor has moved around so that one of the rotor lobes has cleared the intake port, as shown at 1. As the rotor continues to rotate (clockwise in Fig. 22-6), the space between the rotor and housing increases, as shown at 2 in B (Fig. 22-10). This produces a vacuum, which causes the air-fuel mixture to enter, as shown by the small arrow under 2. This is the same action as in the piston engine when the piston moves down on the intake stroke. As the rotor moves farther around, the space between it and the housing continues to increase. See 3 in C and 4 in D. When the rotor reaches the point shown in D, the trailing lobe passes the intake port. Now, the mixture is sealed between the two lobes of the rotor, as shown at 5 in E.

Now, let us see what happens as the rotor continues to turn. Look at F. Here, the mixture

(6 in F) is starting to be compressed. The compression continues through 7 in A. At 8 in B, the mixture is nearing maximum compression. This is the same as the piston approaching TDC on the compression stroke.

Next, combustion takes place. At 9 in C, the spark plugs fire, and the compressed mixture is ignited. Now, the hot gases push against the rotor and force it to turn, as shown at 10 in D. The hot gases continue to expand, as shown at 11 in E and 12 in F. This is the same as the power stroke in piston engines.

At 13 in A, the leading lobe of the rotor is clearing the exhaust port in the housing. Now, the burned gases begin to exhaust from the space between the rotor lobes. The exhaust continues through 14 in B, 15 in C, 16 in D, 17 in E, and 18 in F. By that time, the leading lobe is clearing the intake port again, as shown at 1 in A. Now, the whole sequence of events takes place again.

We have looked at the actions taking place between one pair of rotor lobes. But there are three lobes and three chambers between the lobes. Therefore, there are three sets of actions going on at the same time in the engine. In other words, there are three power thrusts for every rotor revolution. With a two-rotor engine, there are six power thrusts for every revolution of the two rotors.

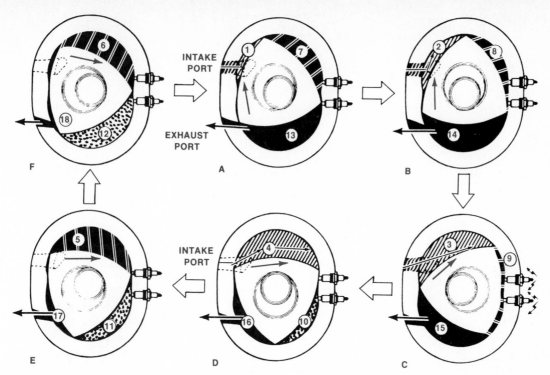

Fig. 22-10. Principle of Wankel-engine operation. By following the actions from A through F, and from numbers 1 through 18, you can follow the complete actions taking place between two apexes of the rotor. (*Toyo Kogyo Company, Limited*)

PRINCIPLE OF THE ROTARY ENGINE

1-4	INTAKE
5-9	COMPRESSION
10-12	POWER
13-18	EXHAUST

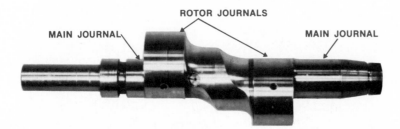

Fig. 22-11. Eccentric shaft. (*Toyo Kogyo Company, Limited*)

§ 22-8 GETTING THE POWER TO THE CRANKSHAFT (ECCENTRIC SHAFT)

The eccentric shaft (Fig. 22-11) in the Wankel engine serves the same purpose as the crankshaft in the piston engine. It is rotated by the power thrusts on the rotor. Figure 22-12 shows how the pressure on the rotor due to combustion makes the eccentric shaft turn. The thrust is off-center from the center line of the eccentric shaft. Thus, most of the push is below the

center line of the shaft. This means that the shaft turns. The bearing in the rotor turns on the eccentric rotor journal. This is the same as the crankpin journal of the crankshaft turning in the connecting-rod bearing.

To keep the rotor rotating in the proper manner, there is a stationary gear that meshes with the internal gear of the rotor. You can see the internal gear of the rotor in Fig. 22-5. Figure 22-13 shows the stationary gear. Figure 22-14 shows how the rotor gear and stationary gear react with each other as the rotor rotates.

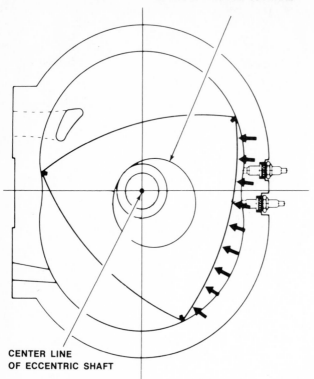

CENTER LINE
OF ECCENTRIC SHAFT

Fig. 22-12. How the combustion pressure, acting off center on the eccentric of the eccentric shaft, forces the shaft to rotate. (*Toyo Kogyo Company, Limited*)

Fig. 22-13. Stationary gear that is mounted in the side housing. (*Toyo Kogyo Company, Limited*)

Fig. 22-14. How the rotor rotates eccentrically around the stationary gear, following an orbit that keeps all three apex seals in sliding contact with the rotor housing. (*Toyo Kogyo Company,* Limited)

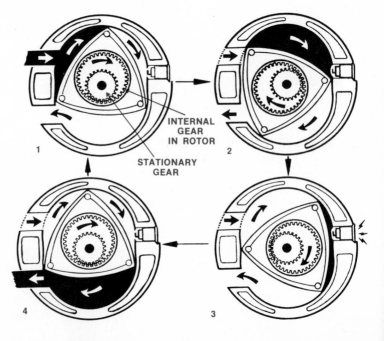

INTERNAL GEAR IN ROTOR

STATIONARY GEAR

1

2

3

4

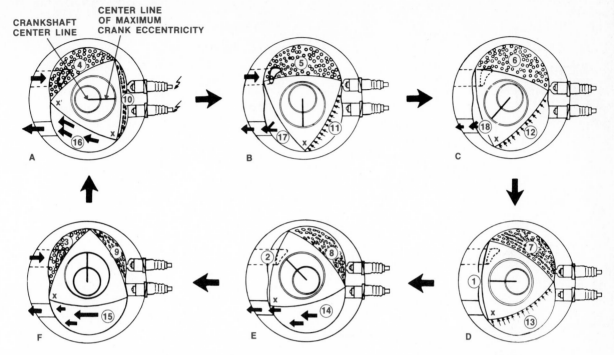

Fig. 22-15. Sequence of actions in the engine, showing how the rotor, as it moves from A to F and back to A again, causes the eccentric shaft to rotate one full revolution, or 360 degrees, even though the rotor has turned only 120 degrees. (*Toyo Kogyo Company, Limited*)

§ 22-9 THE THREE-TO-ONE RATIO

The eccentric shaft rotates three times while the rotor rotates once. It is hard for you to picture this in your mind, and so we have included an illustration (Fig. 22-15) to make it easier. Start at A in the upper left corner. The lobe marked "X" is at the lower right in this part of the picture. The center line of maximum crank (journal) eccentricity is pointing to the right. Note that in B the rotor has moved only 30 degrees. That is, rotor lobe X has advanced only 30 degrees. But, at the same time, the center line and the eccentric shaft have advanced 90 degrees. Now, notice how the center line and the eccentric shaft continue to move ahead of the X-marked lobe as it moves through C, D, E, and F. In F, lobe X has moved only 90 degrees from its position in A. But the center line and the eccentric shaft have turned

270 degrees. When the rotor lobe moves 30 degrees more, or to the position X′ in A, it will have turned 120 degrees, or one-third of a complete revolution. Meantime, however, the eccentric shaft will have made a complete revolution of 360 degrees. By the time the rotor makes one complete revolution, the eccentric shaft will have turned three times. That is, there is a 3 to 1 (3:1) ratio.

§ 22-10 IGNITION SYSTEM

Some Wankel engines have two spark plugs per rotor. With a two-rotor engine, there could be two semiseparate ignition systems, as shown in Fig. 22-16. One distributor and coil fire the leading plugs. The other distributor and coil fire the trailing plugs. The purpose of the two-plug-per-rotor system is to assure improved ignition. However, many Wankel engines use only a single plug per rotor.

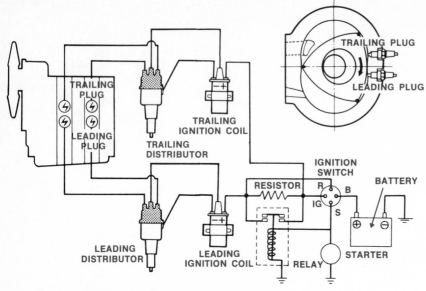

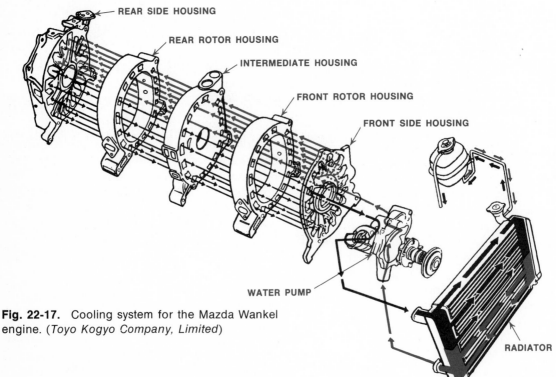

Fig. 22-16. Ignition system for a two-rotor Wankel engine. Note that there are really two ignition systems, one that fires the trailing plugs, and one that fires the leading plugs. (*Toyo Kogyo Company, Limited*)

Fig. 22-17. Cooling system for the Mazda Wankel engine. (*Toyo Kogyo Company, Limited*)

§ 22-11 FUEL SYSTEM

The fuel system for the Wankel engine is much like those used in piston engines. Remember, however, that the Wankel engine has no valves. The rotor serves as the valve, uncovering and then covering up the intake and outlet or exhaust ports.

§ 22-12 COOLING SYSTEM

Figure 22-17 shows the cooling system for the Wankel engine. It is much like the cooling system for the piston engine, except for the water jackets in the housings. The housings are shown separated in Fig. 22-17 so that the flow of coolant, shown by the arrows, can be seen. The water pump sends coolant from the bottom of the radiator through the upper set of water jackets in the housings. In the rear side housing, the coolant reverses directions and flows back through the lower set of water jackets on its way back to the radiator.

Review questions

1. What is the basic principle on which the turbine operates?
2. Are the compressor impeller and gasifier rotor in the left or right shaft of the turbine engine shown in Fig. 22-2?
3. What is the basic purpose of the regenerator?
4. How many lobes, or apexes, does the Wankel rotor have?
5. How many chambers are formed by the rotor and the rotor housing?

6. The Wankel engine is known as an RC engine. What does RC stand for?
7. How many sets of actions go on at the same time in the Wankel engine?
8. How many power thrusts are there for each complete rotor revolution in a Wankel engine?
9. How many times does the eccentric shaft turn as the rotor makes one complete revolution?
10. What is the benefit of having two plugs per rotor?

Study problem

1. If you do not know a friend or neighbor who has a Mazda, visit the local Mazda dealer. Look under the hood and name the parts that you know. Then, identify each of the housings and the leading and trailing spark plugs.

23 | DIESEL ENGINES

§ 23-1 THE DIESEL CYCLE

Until now, we have been talking about gasoline engines. All gasoline engines have this in common: The fuel is mixed with air before the air goes into the cylinder. Then, the mixture is compressed and is ignited by an electric spark at the spark plug.

In the diesel engine, air alone enters the cylinder. The air is compressed and becomes very hot. This is called heat of compression. Then, as the piston nears TDC on the compression stroke, the fuel is injected, or sprayed, into the compressed air. The air is so hot that it ignites the fuel.

As we said, when air is compressed, it gets hot. The more it is compressed, the hotter it gets. In the diesel engine, the compression ratio may be as high as 21:1. This means that the air is compressed to only $\frac{1}{21}$ of its original volume. This is like compressing a quart into less than one-fourth of a cup (less than 2 ounces). When the air is compressed this much, it gets very hot—up to 1000°F [538°C]. Water boils at 212°F [100°C], so 1000°F [538°C] is really hot! This temperature is high enough to ignite the fuel as it is sprayed into the compressed air.

The fuel used in diesel engines is normally a light oil. It is special in several ways. It ignites easily when sprayed into high-temperature air, and it burns cleanly, leaving little residue such as carbon.

Diesel engines are two-cycle and four-cycle engines, just like gasoline engines. The four-cycle diesel engine requires the usual four piston strokes—intake, compression, power, and exhaust. The essential difference between the gasoline engine and the diesel engine is the way the fuel is put into the cylinder and the way it is ignited.

§ 23-2 HISTORY OF THE DIESEL ENGINE

Doctor Rudolf Diesel patented and built his first compression ignition engine in 1892. When he tried to start the first engine, it actually blew up. But by 1897, his engine was running and was pronounced a scientific success. Figure 23-1 shows an early diesel engine built in 1903. An engine of this type was in operation until 1955.

Doctor Diesel gained fame and made a fortune from his diesel-engine patents in the years following 1892. But in the fall of 1913, he disappeared one night from a steamship traveling from Antwerp to London. While his disappearance remains a mystery to this day, his engine survived to become the backbone of the heavy transportation and construction industries. Diesel engines have powered everything from airplanes to submarines. The diesel has even appeared several times at Indianapolis; a Cummins-powered car started the race in the pole position in 1952. Today, diesel-powered passenger cars are imported and sold in the United States. A typical four-cylinder engine for use in passenger cars is shown in Fig. 23-2. Other foreign companies

Fig. 23-1. An early diesel engine. (*Smithsonian Institution*)

build diesel-powered automobiles that are not imported here.

§ 23-3 COMPARISON OF DIESEL AND GASOLINE ENGINES

In the gasoline engine, the fuel and air are mixed outside the engine in the carburetor. The mixture then goes through the intake manifold and past the intake valve to enter the cylinder, where it is compressed and ignited by a spark plug.

In the diesel engine, only air is drawn into the cylinder. The fuel is sprayed into the air in the cylinder by a special fuel system *after* the air is compressed. The diesel engine does not require an ignition system with spark plugs, because the air becomes so hot when it is

compressed that the heat itself ignites the fuel. The fuel used in diesel engines is oil having certain special characteristics that will be described later.

§ 23-4 TWO-CYCLE DIESEL ENGINE

There are several types of two-cycle diesel engines. In one type, there are intake and exhaust ports very similar to those in the two-cycle gasoline engine described in Chapter 6. When the piston moves down on the power stroke, it clears these ports. Now, air under pressure from a pump, or blower, flows into the cylinder through the intake port. This air forces the burned gases out through the exhaust port and thus **scavenges** (clears) the cylinder (Fig. 23-3). Now, when the piston moves back up, it seals off the two ports and traps the charge of fresh air ahead of it. This air becomes greatly compressed and, therefore, very hot.

When air is compressed to about $\frac{1}{15}$ of its original volume, the air pressure goes up to about 500 psi [35.15 kg/cm²]. Air quickly compressed to this pressure will increase in temperature to as much as 1000°F [538°C]. Then, when fuel oil is sprayed into the compressed air, it will ignite and burn.

The pressure created by this combustion forces the piston down, and the piston moves down past the intake and exhaust ports. Now, a fresh charge of air flows in, and the burned gases flow out. This sequence of events is repeated as long as the engine operates.

§ 23-5 TWO-CYCLE DIESEL ENGINE WITH EXHAUST VALVES

The engine described above is not widely used because of the problem of getting a full charge of fresh air in and of getting the burned gases out. The two-cycle diesel engine must use a blower, or high-pressure air pump, to ensure fast entry of air and fast exhaust of the exhaust gases. One type of two-cycle diesel engine has an exhaust valve in the cylinder head. A

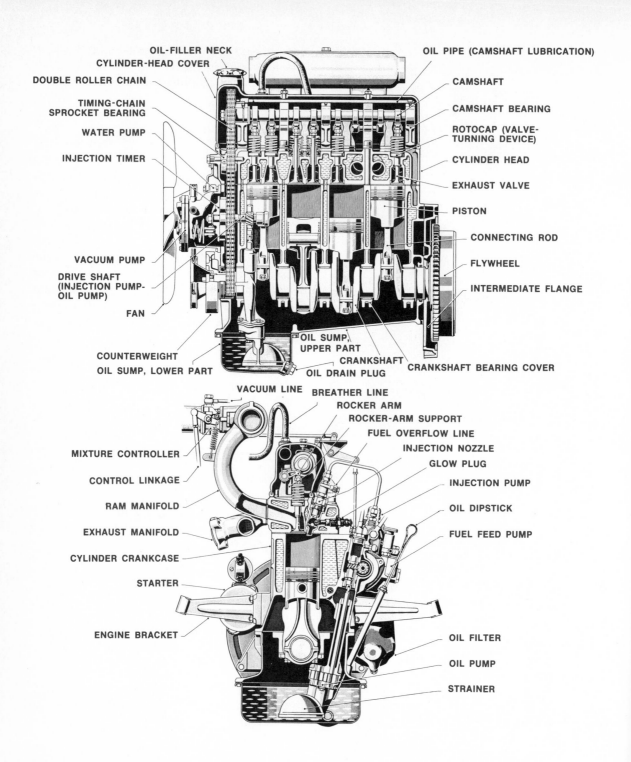

Fig. 23-2. Sectional views of a four-cylinder diesel engine for passenger cars. (*Mercedes-Benz*)

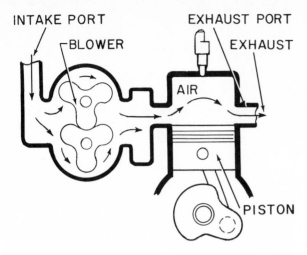

Fig. 23-3. Action in a two-cycle diesel engine with intake and exhaust ports during intake of a fresh charge of air.

cutaway view of this engine is shown in Fig. 23-4. The arrows in Fig. 23-5 show the direction of air flow from the blower to the exhaust manifold.

As shown in Fig. 23-6, air under pressure is delivered to the cylinder through a series of intake ports in the cylinder wall. This action is shown in Fig. 23-6A. As we said, the air is delivered by an air pump, or blower. As the air pours in, the exhaust valve opens, and the fresh air sweeps the burned gases out of the cylinder past the open exhaust valve. Next, as the piston moves up, it closes off the intake ports (Fig. 23-6B). Meanwhile, the exhaust valve has closed. The air is compressed, the fuel is sprayed in and ignited (Fig. 23-6C), and the power stroke (Fig. 23-6D) and the exhaust stroke (Fig. 23-6E) follow. The fuel system is described in **§** 23-7.

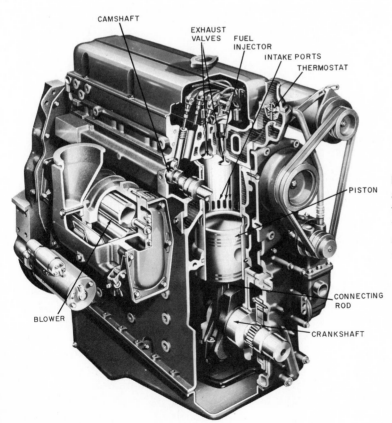

Fig. 23-4. A cutaway view of a two-cycle diesel engine that uses exhaust valves in the cylinder head. (*Detroit Diesel Allison Division of General Motors Corporation*)

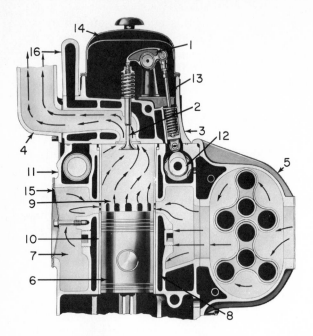

Fig. 23-5. (*Left*) Two-stroke-cycle diesel engine with an exhaust valve in the top of the cylinder. (1) exhaust-valve rocker, (2) exhaust valve, (3) cylinder head, (4) exhaust manifold, (5) blower, (6) piston, (7) air box, (8) cooling-water passage, (9) port admitting air to cylinder, (10) cylinder liner, (11) cylinder block, (12) camshaft, (13) pushrod, (14) rocker cover, (15) hand-hole cover, (16) water manifold. (*Detroit Diesel Allison Division of General Motors Corporation*)

Fig. 23-6. (*Below*) Sequence of events in a two-stroke-cycle diesel engine. (*Detroit Diesel Allison Division of General Motors Corporation*)

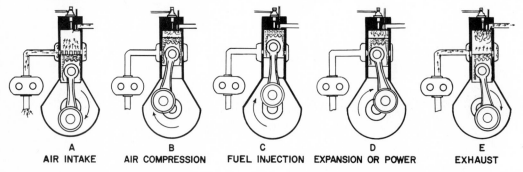

A	B	C	D	E
AIR INTAKE	AIR COMPRESSION	FUEL INJECTION	EXPANSION OR POWER	EXHAUST

§ 23-6 FOUR-CYCLE DIESEL ENGINES

Four-cycle diesel engines have an intake valve and an exhaust valve in each cylinder. In one type, there are two intake valves and two exhaust valves in each cylinder. The valves work the same way as the valves in the four-cycle gasoline engine. Figure 23-7 shows the actions that take place during the four piston strokes. When the piston is moving down on the intake stroke, the intake valve is open, and fresh air is being drawn into the cylinder. As the piston starts back up on the compression stroke, the intake valve closes, and the air in the cylinder is trapped and compressed.

At the end of the compression stroke, the air has been compressed to about $\frac{1}{15}$ of its original volume, and its temperature has gone up to as much as 1000°F [538°C]. Now, the fuel system delivers a spray of fuel oil to the cylinder. The fuel oil is ignited by the high temperature. It burns, and the further increase of pressure caused by this combustion drives the piston down on the power stroke. Finally, the piston starts back up on the exhaust stroke, and the exhaust valve opens so that the burned gases are pushed out of the cylinder. The four strokes continue in the same order as long as the engine operates.

Figure 23-8 shows in cutaway view one cylinder of a four-cycle diesel engine using four

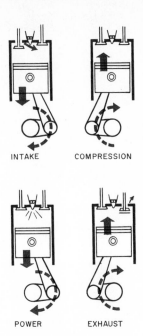

INTAKE   COMPRESSION

POWER   EXHAUST

Fig. 23-7. Actions in a four-cycle diesel engine during the four piston strokes.

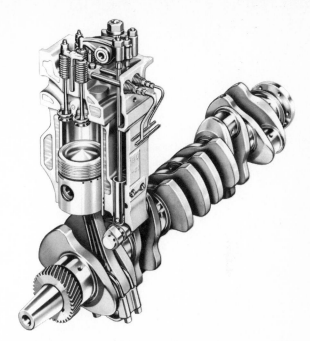

Fig. 23-8. A cutaway view of one cylinder of a four-cycle diesel engine using two intake and two exhaust valves in each cylinder. (*Cummins Engine Company, Incorporated*)

valves per cylinder, two intake valves and two exhaust valves. Figure 23-9 shows two cylinders of the head taken from a similar engine. Notice that there are four valves for each cylinder. Figure 23-10 shows a sectional view, from the end, of a V-8 four-cycle diesel engine that uses four valves per cylinder. The right-hand cylinder in this illustration has been cut away to show a sectional view of the fuel nozzle and its operating mechanism. Fuel systems are described in § 23-7. Figure 23-11 shows a V-8 diesel engine partly cut away to show the four valves per cylinder. Note that the fuel-injection nozzle is centered within the cluster of valves.

§ 23-7 DIESEL-ENGINE FUEL SYSTEM

Remember, the compressed air in a diesel engine is under high pressure. Thus, the fuel must be at a still higher pressure in order for it to be injected into the compressed air. The fuel system must also vary the amount of fuel injected so as to meet the operating requirements. When the engine demands are low, as,

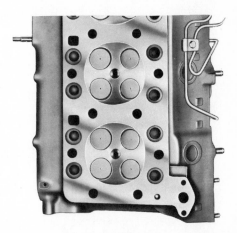

Fig. 23-9. View from inside the cylinder head, showing the locations of the four valves in the cylinder head. Two of these are intake valves and two are exhaust valves. (*Caterpillar Tractor Company*)

for instance, when cruising at medium speed along the highway, a relatively small amount of fuel is injected before each power stroke.

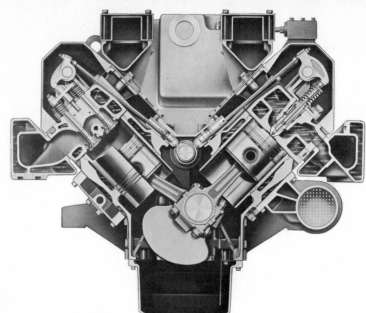

Fig. 23-10. A sectional view of a V-type four-cycle diesel engine. (*Cummins Engine Company, Incorporated*)

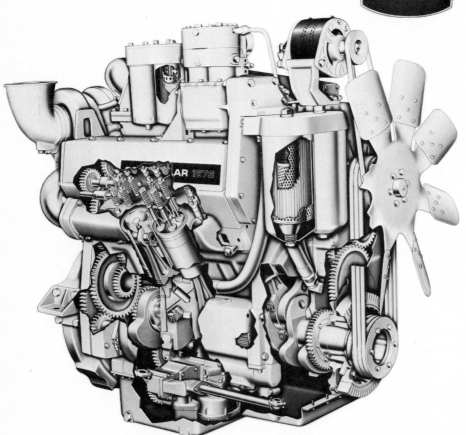

Fig. 23-11. A partial cutaway view of a V-8 diesel engine using overhead camshafts. (*Caterpillar Tractor Company*)

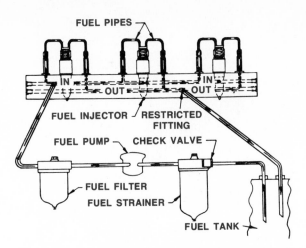

Fig. 23-12. Typical fuel system for a diesel engine. (*Detroit Diesel Allison Division of General Motors Corporation*)

But if the load requirements are high, as, for instance, when climbing a hill, the fuel system increases the amount of fuel injected.

One type of diesel-fuel system is shown in Fig. 23-12 in schematic view. The fuel pump delivers fuel oil at a comparatively low pressure to the injectors. There is an injector in each cylinder. Figure 23-13 shows the injector and operating arrangement at one cylinder. The injector has a plunger that is operated and forced downward by a rocker arm. The rocker arm is actuated by a pushrod and cam on a camshaft. This arrangement is much like the valve train used in overhead-valve engines. When the plunger is forced down, oil is forced from the spray tip at high pressure. It sprays into the compressed air in the cylinder and ignites from the heat of the air. Figure 23-14 is a cutaway view of a fuel injector.

Power output is varied by altering the length of the plunger stroke. For full power, the plunger is forced to take a full effective stroke. If only a small amount of power is required, the effective stroke is reduced. This action results from the movement of a rack that is linked to the accelerator pedal in the driver's compartment. The rack moves back and forth as the driver moves the accelerator pedal down or up.

As the rack moves, it changes the effective length of the plunger.

Now, let's see how this works. Refer to Fig. 23-15, which shows how the fuel injector meters fuel for the engine. When the engine is turned off, the plunger is turned by the rack so that no injection occurs (Fig. 23-15A). In Fig. 23-15B, the cutaway part, called the helix, of the plunger is turned so that the plunger has a very short effective stroke. This is typical of an idling engine. At half load (Fig. 23-15C), the effective stroke has increased in length, which causes more fuel to be injected. With the accelerator pedal to the floor, the effective stroke is at its maximum length, as shown in Fig. 23-15D.

A variety of fuel-injection systems are used on diesel engines. Not all work exactly like the system just described. However, all fuel-injection systems have one common characteristic: To get higher power output from an engine, more fuel must be injected.

§ 23-8 DIESEL-ENGINE FUEL

Diesel engines use fuel oil, a comparatively light oil with a rather low viscosity and certain special characteristics. The fuel oil used in a diesel engine must have a relatively low viscosity (Chapter 19) so that it will flow easily through the pumping and injection system that supplies the fuel to the engine cylinders. It must also be of relatively low viscosity so that it will atomize thoroughly and easily as it is injected into the cylinder. If it is too viscous, it will not break up into fine enough particles. This means that it will not burn rapidly enough, and engine performance will be poor. On the other hand, it must be of sufficiently high viscosity to lubricate the moving parts in the fuel system satisfactorily and to help seal the moving parts and prevent leakage.

§ 23-9 CETANE NUMBER

Diesel fuel oil must have the correct cetane number for good combustion. The cetane number might be compared, in a way, to the octane number of gasoline. **Cetane number**

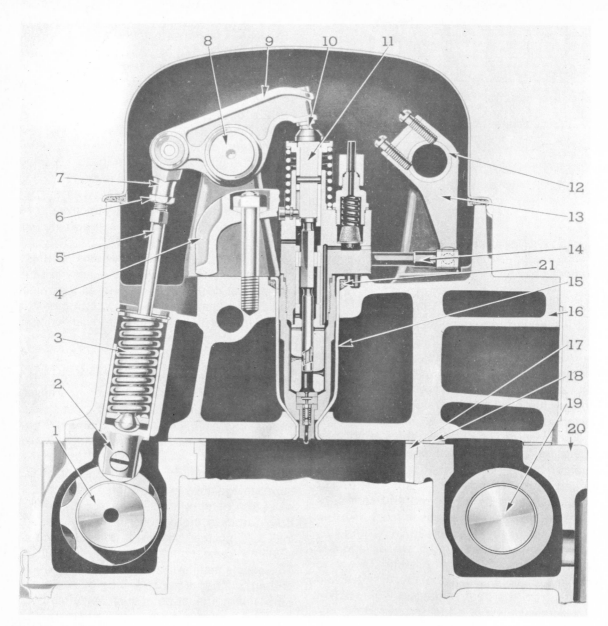

Fig. 23-13. Mounting of fuel injector above the engine cylinder. (1) camshaft, (2) cam follower, (3) following spring, (4) injector clamp, (5) pushrod, (6) lock, (7) clevis, (8) rocker-arm shaft, (9) injector rocker arm, (10) ball stud and seat, (11) injector assembly, (12) control tube, (13) rack-control lever, (14) injector control rack, (15) copper tube, (16) cylinder head, (17) cylinder liner, (18) cylinder-head gasket, (19) balancer shaft, (20) cylinder block, (21) copper-tube sealing ring. (*Detroit Diesel Allison Division of General Motors Corporation*)

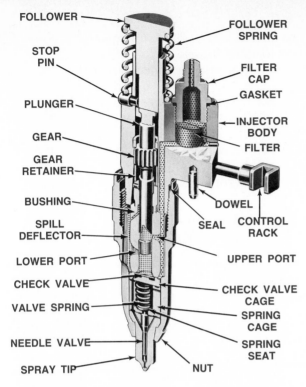

FOLLOWER

STOP
PIN

PLUNGER

GEAR

GEAR
RETAINER

BUSHING

SPILL
DEFLECTOR

LOWER PORT

CHECK VALVE

VALVE SPRING

NEEDLE VALVE

SPRAY TIP

FOLLOWER
SPRING

FILTER
CAP

GASKET

INJECTOR
BODY

FILTER

DOWEL

SEAL

CONTROL
RACK

UPPER PORT

CHECK VALVE
CAGE

SPRING
CAGE

SPRING
SEAT

NUT

Fig. 23-14. A cutaway view of a fuel injector. (*Detroit Diesel Allison Division of General Motors Corporation*)

refers to the ignition quality, or ease of ignition, of the fuel. The lower the cetane number, the higher the temperature required to ignite the fuel. Or, to say it the other way around, the higher the cetane number, the lower the auto-ignition point (or temperature required to ignite the fuel). And the higher the cetane number of diesel fuel, the less the tendency of the fuel to knock in the engine. To understand how cetane number and knocking are related, let us see what causes knocking in a diesel engine.

At the end of the compression stroke, the fuel system injects a spray of oil into the compressed air. The oil is not delivered all at once. The oil spray starts, continues for a fraction of a second, and then stops. If the oil does not start to burn almost instantaneously, oil will continue to accumulate. Then, when the oil does ignite, there will be a considerable amount of oil present that will ignite and burn almost at the same instant. This will cause a sudden pressure rise and knock. At the same time, ignition will not be complete, and smoke will appear in the exhaust gas.

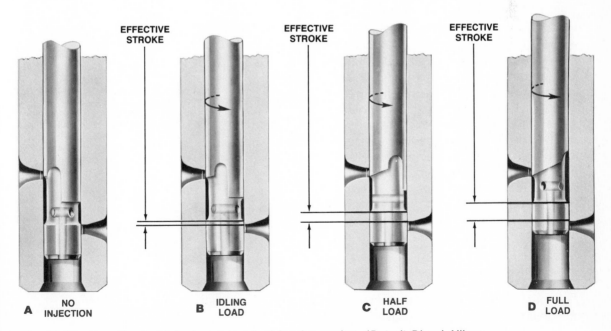

EFFECTIVE
STROKE

EFFECTIVE
STROKE

EFFECTIVE
STROKE

A NO
   INJECTION

B IDLING
   LOAD

C HALF
   LOAD

D FULL
   LOAD

Fig. 23-15. How the fuel injector meters fuel for the engine. (*Detroit Diesel Allison Division of General Motors Corporation*)

If the cetane number of the fuel is high (ignition temperature low), the sprayed oil will ignite as soon as injection begins. In this case, there will be no accumulation of unburned fuel to ignite. Ignition continues evenly as the spray continues, and an even combustion-pressure rise results. But if the cetane number of the fuel is too low, there will be an ignition delay. It takes longer for the low-cetane fuel to ignite. This then results in the sudden ignition of accumulated fuel and a consequent combustion knock. Since the fuel may not have sufficient time to burn after it has started, not all of it may be burned. Some will exit from the engine as smoke.

Note that fuel of excessively high viscosity will also smoke. A heavy, or viscous, fuel will not atomize properly. The oil particles will be too large to burn completely, and full combustion will not take place.

§ 23-10 CETANE-NUMBER REQUIREMENTS

The cetane number of diesel fuel must be high enough to prevent knock, as noted above. With low water-jacket temperatures, low atmospheric temperature, low compression pressures, and light-load operation, a higher cetane fuel is required. All these conditions tend to reduce compression temperature. The fuel must, therefore, have a sufficiently high cetane number (or sufficiently low ignition point) to ignite satisfactorily at these low temperatures. High-speed diesel engines require high-cetane fuels. At high speed, there is less time for the fuel to ignite. It must ignite promptly without ignition delay to prevent knocking and smoking. For starting, the lower the atmospheric temperature, the higher the cetane requirements.

§ 23-11 FUEL-OIL PURITY

The oil must have as little sulfur as possible, since sulfur tends to form sulfur acids. These acids will corrode engine and fuel-system parts. Furthermore, the oil must be clean. Even

small amounts of dirt or foreign matter are liable to cause trouble in the fuel system. The fuel system has passages and nozzles of very small size. Small particles can clog them and prevent normal fuel-system and engine operation. Also, dirt particles can scratch injector parts and cause serious damage. Thus, suppliers of diesel fuel oil are very careful to hold sulfur to a minimum and to use great care in handling the oil to prevent it from becoming contaminated.

§ 23-12 DIESEL-ENGINE APPLICATIONS

Diesel engines are used in trucks, buses, tractors, construction equipment, railroad locomotives, ships, and to drive the electric generators in many electric-power stations. A versatile 430-horsepower [320.78-kW], turbo-supercharged V-8 diesel engine is shown in Fig. 23-16.

The diesel engine has certain advantages over the gasoline engine in many of these applications. For one thing, it is a very efficient engine. It gets more power out of the fuel it

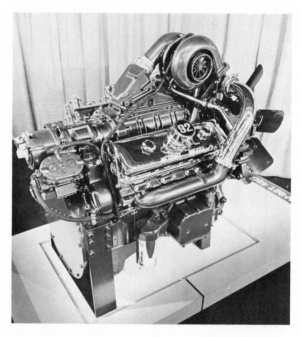

Fig. 23-16. A V-8 diesel engine. (*Detroit Diesel Allison Division of General Motors Corporation*)

burns than does the gasoline engine. This is especially true during part-throttle operation under light loads. With this condition, the diesel-engine fuel system is injecting only a small amount of fuel into the cylinders. But in the gasoline engine, a rich mixture is required, which reduces the efficiency of the engine.

Diesel fuel is also safer than gasoline. Diesel fuel, being a form of oil, will not ignite easily. On the other hand, gasoline vaporizes readily to mix with air and form an explosive mixture. For this reason, the diesel engine is often preferred in such applications as motorboats, where the engine is in a closed compartment and where gasoline vapors could gather and become a potential menace.

The diesel engine, however, is considerably heavier and more expensive than the gasoline engine. One reason for this is that the diesel engine must be made heavier and stronger to stand up under the higher compression and combustion pressures that are reached in the engine cylinders. Notice the size of the crankshaft being checked in Fig. 23-17, for use in the engine shown in Fig. 23-16. This also means that the cost of manufacturing the diesel engines is greater. Another condition that increases the cost of diesel engines is that, often, not very many of any one model of a diesel engine are made. In the automotive field, in contrast, an elaborate assembly line with many expensive automatic tools may be set up to run only one model of engine. This line may produce hundreds of thousands of one model of engine in a year. With this high volume, great manufacturing efficiency can be reached using very costly manufacturing equipment.

On the other hand, because the diesel engine is not generally made in such large volume, the manufacturer cannot afford to install highly expensive automatic equipment.

Another reason why diesel engines are more expensive is that they require a rather costly fuel-injection system to pump the fuel, at high pressure, into the engine cylinders. Also, because they require higher cranking power to get them started, they must have more elaborate and expensive starting systems than gas-

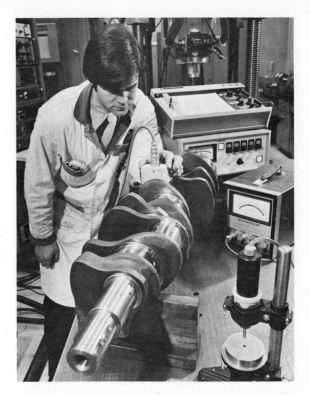

Fig. 23-17. Crankshaft for diesel engine shown in Fig. 23-16. (*Detroit Diesel Allison Division of General Motors Corporation*)

oline engines. The reason that higher cranking power is required is that the air in the diesel-engine cylinders must be much more highly compressed than the air-fuel mixture in the gasoline engine. The gasoline engine may have a compression ratio of 9:1, whereas the diesel engine will have a compression ratio of 15:1 or higher.

§ 23-13 DIESEL-ENGINE EMISSIONS

The diesel engine does not use a throttle valve to control the amount of intake air entering the engine. Therefore, plenty of air is always available in the cylinder to burn all the fuel. This reduces the emissions of carbon monoxide from the diesel engine to a very low level. Fewer unburned hydrocarbons appear in the diesel exhaust gas than in the exhaust of a gasoline engine. But the diesel engine has a problem with the creation of fairly high amounts of NO_x. This is because of the high combustion temperature and pressure. However, exhaust emissions from diesel engines do vary widely with differences in combustion chamber design and the fuel-injection system used.

Some people object to three other problems with some diesel engines: smoke, odor, and noise. The smoke and odor are usually considered nuisances with questionable ill effects on human health. They are not serious contributors to the formation of smog. Engine designers are working on designing and building quieter running diesel engines.

§ 23-14 WHY DIESEL ENGINES ARE USED

You might think that, because diesel engines are heavier, more complicated, and more expensive than gasoline engines, that everyone would prefer to use gasoline engines. However, diesel engines are preferred for railroad locomotives, many trucks, buses, tractors, and construction equipment for several reasons. For one thing, diesel engines are much cheaper to operate than railroad steam engines. The steam engine requires more frequent service and is thus out of action a larger part of the time. Diesel engines use less fuel, do not require frequent stops to take on coal and water, and run more smoothly with less smoke.

Truck and bus operators often prefer diesel engines, despite their higher initial cost. This is because they operate on fuel oil that is cheaper than gasoline, and because they are more efficient and deliver more mileage per dollar of fuel cost. Also, diesel engines have greater pulling power at low and intermediate speeds. In contrast, as a gasoline engine slows down, it loses most of its power.

Review questions

1. How do the operating principles of the diesel engine differ from the operating principles of the gasoline engine?
2. What causes the air in the combustion chamber of a diesel engine to reach a very high temperature just before fuel injection?
3. Why doesn't a diesel engine need a carburetor?
4. Why do many two-cycle diesel engines have exhaust valves?
5. How is the power output varied in a diesel engine?
6. Why must the fuel oil used in diesel engines have low viscosity?
7. What is cetane number?
8. Name several advantages diesel engines have over gasoline engines.
9. Which engine is heavier and more expensive, the diesel engine or the gasoline engine? Why?
10. What are the exhaust-emission problems with the diesel engine?

Study problem

1. Find a repair shop or trucking company that services diesel engines. Examine several diesel-powered trucks and tractors. Note whether the engine is two-stroke or four-stroke. Determine the type of fuel-injection system used. Locate the blower, supercharger, or turbocharger, if used. Identify the fuel pumps and the fuel filters.

24 | MOTORCYCLE ENGINES

§ 24-1 TYPES OF MOTORCYCLE ENGINES

About 5 million Americans own motorcycles and vehicles powered by motorcycle engines. Although many owners repair their own motorcycles, most rely on skilled motorcycle mechanics. There are about 46 different makes of motorcycles sold in this country. These range in size from children's minibikes (Fig. 24-1) to large multicylinder touring machines (Figs. 24-2 and 24-14). These big motorcycles often have engines larger than those used in some small cars. Each manufacturer usually markets several different sizes and types of motorcycle. As we will discuss, there are more differences among motorcycle engines than there are among automobile engines. But there are still some classifications of motorcycle engines that can be made.

Almost all motorcycle engines are air-cooled. They use gasoline for fuel and have simple carburetors. Many motorcycles today have electric starting motors, in addition to the conventional kick starter. Some still use a magneto ignition similar to that used on small gasoline engines. But most use an automobile-type battery ignition system.

Motorcycle engines may be either two-cycle or four-cycle piston engines, having from one to four cylinders. A few motorcycles are built using an air-cooled Wankel engine. The Wankel engine is covered in Chapter 22.

Fig. 24-1. A minibike. (*Kawasaki Heavy Industries, Ltd.*)

§ 24-2 TWO-CYCLE MOTORCYCLE ENGINES

Some small two-cycle motorcycle engines have only one cylinder, with a displacement of 75 cu cm or less (Fig. 24-1). Other models may have one, two, or three cylinders. The two-cycle engine used in motorcycles does not use intake and exhaust valves. Instead, the air-fuel mixture travels through the crankcase into the combustion chamber through ports that are

Fig. 24-2. A multicylinder motorcycle. (*Kawasaki Heavy Industries, Ltd.*)

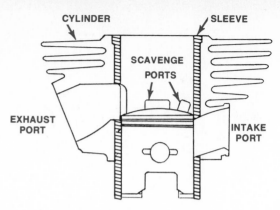

Fig. 24-3. Ports being opened and closed by the piston. (*Kawasaki Heavy Industries, Ltd.*)

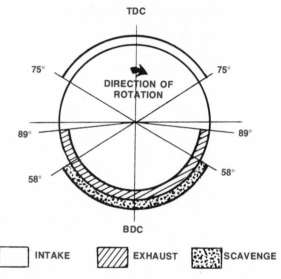

Fig. 24-4. When the piston opens and closes the ports, they will always open and close the same number of degrees before and after the piston reaches TDC or BDC. (*Kawasaki Heavy Industries, Ltd.*)

opened and closed by the piston, as shown in Fig. 6-11. The fact that the two-cycle engine has few moving parts has made it very popular for use in motorcycles. Early model two-cycle engines were considered to be less powerful than a four-cycle engine of comparable size. However, developments during the last few years have made the two-cycle engine more powerful than earlier models. This increased power is largely the result of changes in the design of the ports and of changes in port timing.

In a conventional type of two-cycle engine, the piston opens and closes the ports in the engine cylinder (Fig. 24-3). This means that the intake, transfer or scavenge, and exhaust ports must always open and close the same number of degrees before and after the piston reaches TDC or BDC (Fig. 24-4). This limits the power and speed of the engine. However, now there are devices that can be used to vary the intake-port timing. One such device is the rotary disk valve.

In an engine using the rotary disk valve (called a rotary valve engine), the disk valve rotates with the crankshaft (Fig. 24-5). As the disk rotates, it opens and closes the inlet port into the crankcase. The disk valve can be designed so that the intake port opens and closes at different times before and after TDC. An increase in the length of time that air-fuel mixture can enter the engine will result in increased engine power.

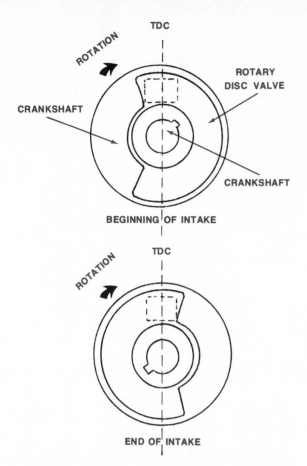

Fig. 24-5. A rotary disk valve. (*Kawasaki Heavy Industries, Ltd.*)

§ 24-3 MULTICYLINDER TWO-CYCLE ENGINES

Multicylinder two-cycle engines can be thought of as two or more separate engines that transmit power to the same crankshaft. The two-cycle motorcycle engine is always of the crankcase compression type. That is, the pressure caused in the crankcase by the downward travel of the piston slightly pressurizes the crankcase. Then, this pressure is used to force the air-fuel mixture through the transfer or scavenge port into the combustion chamber. Therefore, each cylinder of a multicylinder two-cycle engine must have its own airtight section of the crankcase.

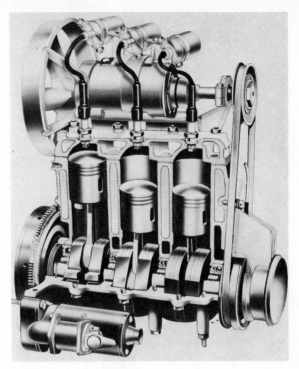

Fig. 24-6. A sectional view of a three-cylinder, two-cycle engine. (*Daimler-Benz*)

Figure 24-6 shows a sectional view of a three-cylinder, two-cycle engine for automobiles. In this engine, the crankpins are 120° apart. As you can see by examining Fig. 24-6, the crankcase is separated into three compartments. This allows the engine to utilize the crankcase pressure as the piston moves down to transfer the air-fuel mixture from the crankcase to the cylinder. Multicylinder two-cycle engines used in motorcycles have the same type of separated crankcase as shown in Fig. 24-6.

§ 24-4 TWO-CYCLE-ENGINE LUBRICATION

As we discussed in Chapter 6, in a two-cycle engine, the air-fuel mixture passes through the crankcase on the way to the combustion chamber. Because of this, any oil in the crankcase would be carried into the combustion

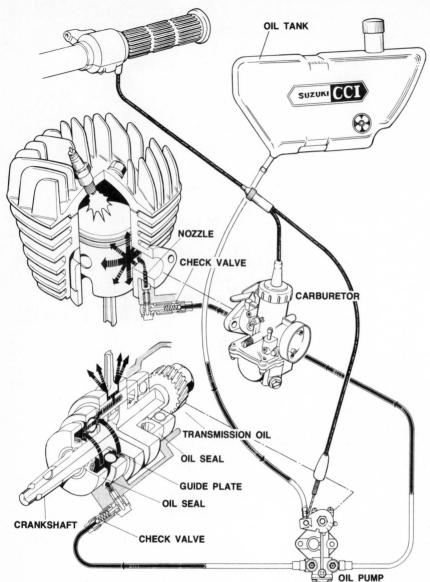

OIL TANK

SUZUKI CCI

NOZZLE

CHECK VALVE

CARBURETOR

TRANSMISSION OIL

OIL SEAL

GUIDE PLATE

OIL SEAL

CRANKSHAFT

CHECK VALVE

OIL PUMP

Fig. 24-7. Oil-injection system for a single-cylinder, two-cycle motorcycle engine. (*Suzuki Motor Co., Ltd.*)

chamber and burned. To provide the needed lubrication between the piston and the cylinder wall, a small amount of oil is mixed with the gasoline. This system is very simple, and the engine using it does not require an oil pump. However, engine failure can result if oil is not added to the gasoline in the correct amount.

The lubrication system in a two-cycle engine is sometimes called a "total loss" system. Remember that in the two-cycle engine the air-fuel mixture travels through the crankcase. This prevents the engine from using a lubrication system in which oil circulates through the engine and is reused. As we mentioned earlier, two-cycle engines use up their lubricating oil as the engine runs. In other words, in a two-cycle engine, the lubricating oil is a total loss.

In an engine that lubricates by mixing oil with gasoline in the fuel tank, the air-fuel and oil mixture is drawn into the crankcase as a vapor. The oil is contained in the vapor in the form of very small drops, almost a mist.

When the port to the combustion chamber opens, the air-fuel mixture leaves the crankcase. However, some of the oil mist remains behind in the crankcase. This is because the oil is heavier than the air-fuel mixture. Some of the oil mist settles on the parts of the engine, providing the needed lubrication. The remaining oil vapor in the crankcase is pulled into the combustion chamber and is burned.

Because lubricating oil is consumed as the two-cycle engine runs, it always is lubricated by clean new oil. Once the lubricating oil has settled inside the crankcase, and then is drawn into the combustion chamber and burned, new oil must be delivered to replace the lost oil. This new oil comes from the fresh oil vapor contained in the next incoming charge of air-fuel mixture. Special "two-cycle" oil is available that mixes quickly with gasoline in the fuel tank to simplify the mixing problem. Too much oil, or large drops that have not mixed completely with the gasoline, can block the carburetor jet or form carbon that will plug the cylinder ports.

To prevent these problems today, instead of the system just described, many motorcycles use a separate oil system to inject the needed amount of oil into the air-fuel mixture. A typical oil-injection system for a single-cylinder, two-cycle motorcycle engine is shown in Fig. 24-7. This system requires the use of an engine-driven oil pump whose output can be varied. The advantage of this system is that the rider is not required to do any mixing. Gasoline is put into the gasoline tank, and the correct oil level is maintained in the separate oil tank (Fig. 24-7). The oil-injection system does the rest by metering the correct amount of oil to the moving parts of the engine.

In two-cycle engines having an oil-injection system, any amount of oil can be supplied to any moving engine part. This can be seen by studying Fig. 24-7. Notice that a single nozzle injects oil into the intake port, while an oil passage delivers more oil to the main bearings, crankpin, and connecting-rod bearing. Note also that the problem of lubricating oil plugging the carburetor jet is eliminated, since only fuel passes through the jet.

Oil-injection systems usually use engine speed and throttle position to control the output of the oil pump. An oil that burns with a low ash content is recommended for use in two-cycle-engine oil-injection systems.

§ 24-5 FOUR-CYCLE MOTORCYCLE ENGINES

Four-cycle motorcycle engines are made in many varieties of single, twin, three, and four cylinders. Also, many different types of valve trains are used. This means that there is a wide choice of engine types and sizes for use in motorcycles.

The four-cycle motorcycle engine operates the same as the small four-cycle engines that we discussed in Chapter 9. For motorcycle use, most of these engines have overhead valves (Fig. 24-8), and many have overhead camshafts to operate the valves (Fig. 24-9). In general, motorcycle engines, both the two-cycle and the four-cycle models, operate at

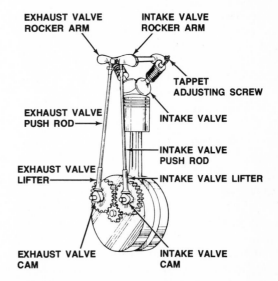

Fig. 24-8. Overhead-valve train for a four-cycle motorcycle engine. (*American Honda Motor Company, Inc.*)

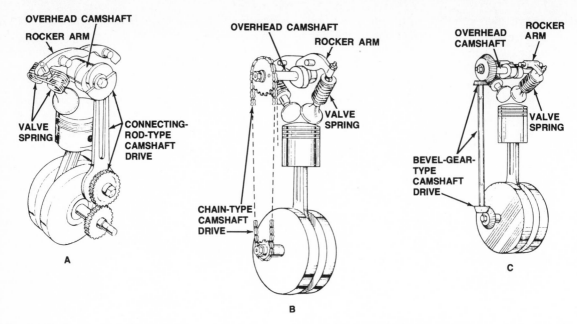

Fig. 24-9. Types of overhead-camshaft drives used to operate the overhead valves. (*American Honda Motor Company, Inc.*)

much higher rpm than the typical automobile engine. Because motorcycle engines are designed for high-speed operation, lugging will usually damage these engines more than high rpm. Many motorcycle engines can turn 7,000 rpm continuously without damage. Other models can safely turn 10,000 rpm, and some even higher. Figure 24-10 is a curve showing the relationship between engine speed and horsepower for one model of motorcycle engine. Compare the engine speed at which maximum horsepower occurs to the horsepower curve of a typical automobile engine shown in Fig. 12-7.

In general, most four-cycle motorcycle engines use the dry-sump type of lubrication system. That is, extra lubricating oil for the engine is not carried in the sump or oil pan forming the bottom of the crankcase. Instead, the oil pump scavenges, or picks up, any oil that collects in the crankcase. This oil is pumped from the crankcase to a separate oil tank for storage. As a result, the crankcase is kept dry of any excess oil. One advantage to the dry-sump

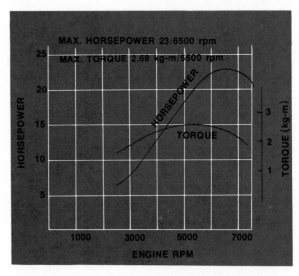

Fig. 24-10. Relationship between engine speed and horsepower. (*Suzuki Motor Co., Ltd.*)

system is that it enables the engine to operate in any position, even upside down.

Oil pumps are used in all four-cycle motorcycle engines. These pumps may be the famil-

272

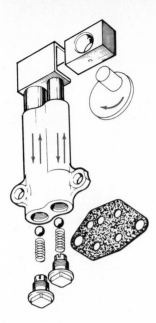

Fig. 24-11. A plunger-type oil pump. (*Triumph Norton International*)

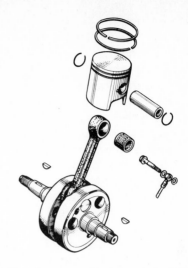

Fig. 24-12. An assembled crankshaft for a one-cylinder motorcycle engine. (*Suzuki Motor Co., Ltd.*)

iar gear type (Fig. 19-12) or rotor type (Fig. 19-13) used in automobile engines. However, many motorcycle engines use a plunger-type oil pump (Fig. 24-11). The plunger-type pump shown in Fig. 24-11 is driven by an extension on the intake camshaft. This pump has two plungers of different diameters to pump the oil. The larger diameter plunger (left plunger in Fig. 24-11) is used to scavenge the crankcase. That is, the large plunger removes oil from the bottom of the crankcase and pumps that oil into the oil tank. The job of the smaller plunger (right plunger in Fig. 24-11) is to pressurize the oil and deliver it into the engine oil passages. Oil gets to the small plunger by gravity-feed from the oil tank, since the oil tank is mounted above the level of the oil pump.

§ 24-6 MOTORCYCLE-ENGINE CRANKSHAFTS

Many motorcycle-engine crankshafts are built-up assemblies formed by positioning and fastening several separate parts to the crankpin. Figure 24-12 shows an assembled crank-

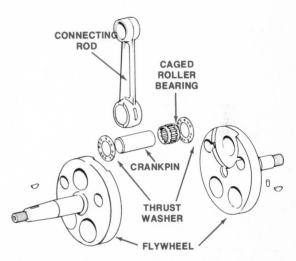

Fig. 24-13. Disassembled view of the parts that form the crankshaft of a motorcycle engine. (*Kawasaki Heavy Industries, Ltd.*)

shaft for a one-cylinder motorcycle engine. This crankshaft differs from the single-piece crankshaft widely used in other small engines. It also differs from the crankshaft used in automotive engines (Fig. 10-14). As you can see in Fig. 24-13, instead of using a two-piece connecting rod, the built-up crankshaft uses a single-piece connecting rod. For a connecting-rod bearing, in place of the split sleeve bearing typical of automotive usage (Fig. 6-14), the motorcycle engine uses a caged roller bearing.

Some motorcycle engines have a flywheel mounted on one end of the crankshaft and turning outside the crankcase, as in automobile engines. However, many motorcycle engines use the type of split flywheel shown in Figs. 24-12 and 24-13. This flywheel is split into two parts. Each of the two flywheels forms an integral part of the crankshaft inside the crankcase. In some ways, these split flywheels resemble the counterweights on the automobile-engine crankshaft. Figure 24-13 shows in detail the parts that are assembled together to form the crankshaft of a motorcycle engine. Great care must be exercised in assembling a crankshaft of this type. If the flywheels are out of alignment on the crankpin, engine vibration and early bearing failure will occur.

To assemble a crankshaft of the type shown in Fig. 24-13, the crankpin is placed in position through the bearing. The bearing is placed in the big end of the connecting rod, with a thrust washer on each side. Then the two flywheels are accurately positioned. The assembly is pressed together, maintaining the correct position and alignment of the flywheels, until the correct side clearance between the connecting rod and the flywheels is obtained.

§ 24-7 MOTORCYCLE-ENGINE APPLICATIONS

A motorcycle is defined as any motor vehicle, other than a tractor, designed to operate on no more than three wheels in contact with the ground. It must weigh less than 1,500 pounds [680 kilograms] curb weight. Curb weight is the total weight of a vehicle, including a full load of fuel, oil, and water, but without any passengers or cargo.

The first motorcycles were little more than steam-powered bicycles. As early as 1869, one of these machines was operated in the United States, and in 1880, a steam-powered tricycle was built here. But the first motorcycle powered by an internal-combustion engine was built in

Fig. 24-14. A powerful motorcycle engine in operation on the highway. (*Kawasaki Heavy Industries, Ltd.*)

Germany by the automotive engineer Gottlieb Daimler. In 1885, he became the first person to mount an internal-combustion engine on a bicycle-like machine.

Today, motorcycle engines are, of course, used to power the motorcycles that we see on the highways (Fig. 24-14). In addition, millions of other vehicles are also powered by the highly developed and reliable motorcycle engine. It is estimated that there are more than 5 million off-road recreation vehicles in use in the United States. The motorcycle has been a popular recreation vehicle for many years. However, the motorcycle designed for off-road use did not really develop until about 1960. About that time, manufacturers began redesigning the motorcycle into a much smaller machine with a shorter wheelbase and heavily treaded tires. From these early models evolved the highly versatile and adaptable minibike (Fig. 24-1), trailbike, and dirt bike (Fig. 24-15).

Fig. 24-15. A dirt bike. (*Kawasaki Heavy Industries, Ltd.*)

The all-terrain vehicle (ATV) is the newest adaptation of a wheeled recreation vehicle (Fig. 24-16). Generally, it has six or eight wheels and is designed to travel over very rough terrain, marshes, and wet ground. Some of these vehicles can even be propelled through calm water.

In addition to recreation, there are other important uses of off-road recreation vehicles Snowmobiles are extremely important to the rancher in winter, providing a vehicle by which to reach cattle for care and feeding. Rural mail carriers and line workers also find the snowmobile very useful in carrying out their tasks. We will discuss snowmobiles in § 24-8. Off-road motorcycles, and other two-wheeled vehicles, are equally important to the law-enforcement officer and to other workers who place a premium on flexibility, economy, and diversity in their transportation needs. There is one

Fig. 24-16. An all-terrain vehicle (ATV). (*American Honda Motor Company, Inc.*)

Fig. 24-17. A snowmobile. (*Yamaha International Corporation*)

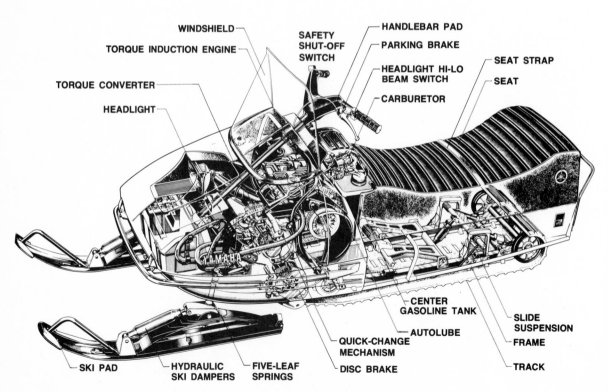

WINDSHIELD

TORQUE INDUCTION ENGINE

TORQUE CONVERTER

HEADLIGHT

SAFETY SHUT-OFF SWITCH

HANDLEBAR PAD

PARKING BRAKE

HEADLIGHT HI-LO BEAM SWITCH

CARBURETOR

SEAT STRAP

SEAT

CENTER GASOLINE TANK

AUTOLUBE

SLIDE SUSPENSION

FRAME

TRACK

QUICK-CHANGE MECHANISM

DISC BRAKE

SKI PAD

HYDRAULIC SKI DAMPERS

FIVE-LEAF SPRINGS

Fig. 24-18. A schematic view of a snowmobile. (*Yamaha International Corporation*)

characteristic that all these off-road vehicles just discussed have in common. They are all powered by motorcycle-type engines.

§ 24-8 SNOWMOBILES

The most popular off-road recreation vehicle, and one of the newest attractions in the outdoor recreation field, is the snowmobile (Fig. 24-17). A snowmobile is a self-propelled vehicle intended for off-road travel primarily on snow, having a curb weight of not more than 1,000 pounds [454 kilograms]. It is driven by tracks that are in contact with the snow.

Although the first snowmobile was built early in this century, its popularity as a recreation vehicle began little more than 10 years ago. Early snowmobile designs ranged from Model A Fords with tracks over the wheels to a motor-driven toboggan. The motorized-toboggan concept was the most practical design. From it has evolved the highly sophisticated snowmobile of today. Figure 24-18 shows a schematic view of a popular model of snowmobile.

Review questions

1. How does the air-fuel mixture get into the combustion chamber in a two-cycle motorcycle engine?
2. What is the purpose of using a rotary disk valve in a two-cycle engine?
3. What is a total loss oil system?
4. What is an oil-injection system? What type of engines use it?
5. How does a motorcycle engine differ from an automobile engine?

Study problems

1. If you, or anyone in your family, has a motorcycle, make some comparisons. Park the motorcycle beside an automobile that has the hood up. Then, using pencil and paper, list the differences between the two engines. When that list is complete, make another list, this time of the similarities between the two engines. By comparing the lists, you will notice that there are certain parts that all piston engines have in common.
2. Try to locate a two-cycle motorcycle engine and a four-cycle motorcycle engine. Note carefully the differences between the two. Notice how the air-fuel mixture gets into and out of each cylinder. If possible, disassemble each engine to learn how these engines differ.

25 | SMALL MARINE ENGINES

§ 25-1 TYPES OF MARINE ENGINES

Ocean-going cargo or passenger ships use huge steam or diesel engines to turn their propellers. But on lakes, rivers, and at the seashore, there are many smaller boats that are used for fishing, water skiing, and pleasure boating. The engines used in these smaller boats are the ones discussed in this chapter.

All these powered boats use internal-combustion engines, and the majority of these engines are of the two- or four-cycle type using gasoline as fuel. Some motorboats use diesel engines, however, even if these engines are of heavier construction. Because diesel fuels are less explosive than gasoline, there is less danger of fumes from the fuel being ignited in the engine compartment. However, if the engine compartment is properly ventilated, and the engine and fuel system are in good condition, the danger of fire or explosion from gasoline is practically nonexistent. The internal-combustion engines used in motorboats can be classified in another way, as **outboard, inboard,** or **stern-mounted.** The **outboard** engine is mounted on the outside, or outboard, of the boat hull (Fig. 25-1). It is mounted on the **transom** (stern board) of the boat (Fig. 25-2). This is a relatively simple arrangement, and most boats can be fitted with an outboard engine. All that is required is to make sure the transom can support the engine and that the engine is properly mounted. For a boat that has no transom (a canoe, for example), a mounting bracket can be used.

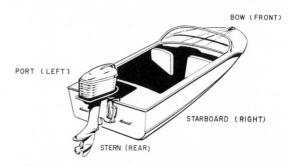

Fig. 25-1. Location of outboard engine on a boat.

The outboard engine has a pair of clamps that are used to mount the engine. Figure 25-1 shows a motorboat with an outboard engine mounted in place on the transom. (Note that this illustration also identifies the bow, stern, and port and starboard sides of the boat.) One advantage of the outboard engine is that it can be taken off the boat at any time and serviced, put away for safekeeping, or stored for the winter.

For higher boat speeds, and to pull a heavy load—several water skiers, for example—two outboard engines may be mounted on the transom (Fig. 25-3). With this arrangement, the two are linked so that both turn together for steering. Outboard engines are described in detail in § 25-3 through 25-6.

The **inboard** engine is mounted inside the hull, usually toward the **stern** (rear) (see Fig. 25-4). The engine drives a propeller that is mounted on the end of a propeller shaft which passes through an opening in the bottom of the hull. In most installations, the engine is

enclosed in a separate compartment, often below deck. Inboard engines are permanently mounted in place, just as automobile engines are permanently mounted in automobiles. Minor engine services can be performed in the boat, but for major service the engine is removed.

Most gasoline-burning inboard engines are four-cycle engines. These four-cycle marine

Fig. 25-2. (*Left*) Boat with outboard engine mounted on the transom.

Fig. 25-3. (*Below*) Boat using two outboard engines mounted on the transom. (*Chrysler Outboard Corporation*)

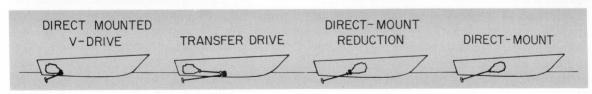

DIRECT MOUNTED V-DRIVE   TRANSFER DRIVE   DIRECT-MOUNT REDUCTION   DIRECT-MOUNT

Fig. 25-4. Inboard mounting of engine in a boat.

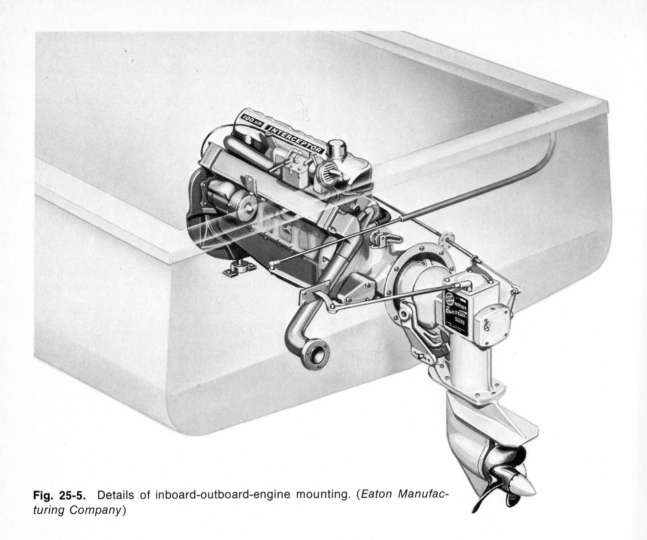

Fig. 25-5. Details of inboard-outboard-engine mounting. (*Eaton Manufacturing Company*)

engines are much alike, but there are some important differences. Inboard engines are described in **§** 25-16.

The third type of small marine engine is **stern-mounted** (the inboard-outboard). It is, in a way, a combination of the two other arrangements for marine engines. With the inboard-outboard arrangement, the engine is mounted inside the hull and drives through a special opening in the transom (Fig. 25-5). There is a gearing arrangement that permits the propeller and drive assembly to be pivoted, as with the outboard engine, for steering. This engine is described in **§** 25-18.

§ 25-2 WHY THE NAME "MOTORBOATS"?

Have you ever wondered about the fact that the boats are called *motor*boats, even though they are powered, not by motors, but by engines? You may have wondered about the same thing with automobiles. Automobiles are sometimes called *motor*cars even though they are driven by engines. Actually, motors are electrical devices that produce power as electicity flows through them. A long time ago, however, the word "motor" was often used instead of "engine," and thus we have **motor-**

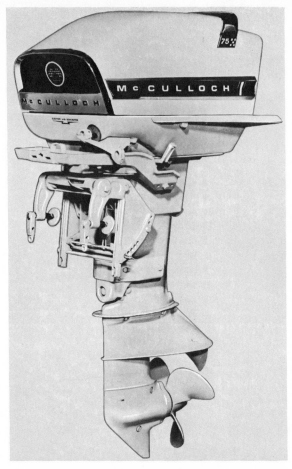

Fig. 25-6. An outboard engine. (*Scott Division of McCulloch Corporation*)

boats, motorcars, and **motoring** (for driving in your car). Strictly speaking, these should be **engine boats, engine cars,** and **engining.** However, the word "motor" has been used for so long in referring to engine-powered boats and cars that it will probably never be changed.

§ 25-3 POWER FOR MOTORBOATS

Figures 25-1 and 25-2 show the mounting arrangement of the outboard engine. Figure 25-6 shows a typical outboard engine. Outboard engines range from small one-cylinder units producing only a few horsepower to six-cylin-

der units capable of producing over 100 horsepower. Most outboard engines have two or four cylinders positioned horizontally and operating a crankshaft that is vertical. This permits the crankshaft to be coupled to the drive shaft that connects the engine and the gears in the bottom of the lower unit. This gearing provides a right-angle drive so that the horizontally placed propeller shaft is rotated when the vertical crankshaft turns.

A few of the outboards are four-cycle units, but most are two-cycle. Most are water-cooled, but several are air-cooled. Figure 25-7 shows a cutaway view of a three-cylinder, two-cycle outboard engine. Figure 25-8 is a cutaway view of a V-4, two-cycle engine. Figure 25-9 shows a four-cylinder, in-line, four-cycle outboard engine. The V-4, two-cycle engine and the four-cylinder, in-line, four-cycle engine are described in following sections. Now, we will discuss the lower unit, including the propeller and the gearing arrangement that connects the propeller shaft with the drive shaft. A description of the water pump is also included, because it is housed in the lower unit.

§ 25-4 LOWER UNIT

Outboard manufacturers divide their engines into two basic parts: the **powerhead,** which includes the engine, and the **lower unit,** which includes the gearing, the mounting case, the drive shaft, the water pump, the propeller shaft, and the propeller.

§ 25-5 THE PROPELLER

The propeller has curved blades—usually three of them—that "bite" into the water, thrusting it rearward. The boat moves forward as a reaction against this rearward thrust. Figure 25-10 shows a typical propeller for a motorboat. For efficient engine operation and good boating, the appropriate propeller for the particular application must be used. The two basic measurements of a propeller are **diameter** and **pitch** (Figs. 25-11 and 25-12). **Diameter** is the diame-

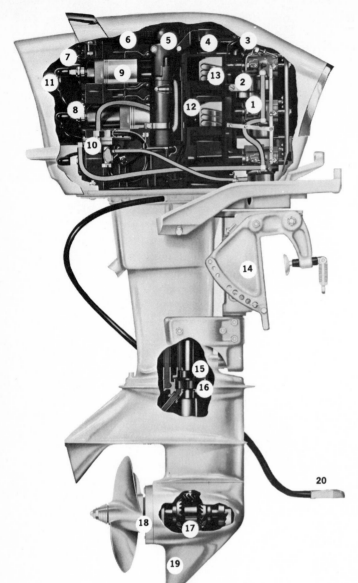

1. Fixed-Jet Carburetors
2. Choke Solenoid
3. Vacuum-actuated Anti-flood Device
4. Distributor Timing Belt
5. Positive Engagement 12 volt Starter
6. Silicon Diode Rectifier (10 amp Alternator)
7. Thermostat Control
8. Surface Gap Spark Plugs (UJ17V)
9. Permanent-mold, Aluminum-alloy Pistons
10. Automotive-type Fuel Pump
11. Acoustical Silencing
12. Balanced Crankshaft
13. Spring-steel Reed Valves
14. Remote-controlled ShalloWater® Drive
15. Bail-A-Matic® Pump
16. Silicon-bronze Water Pump
17. Quiet Spiral Gears
    (Forward-Neutral-Reverse)
18. Sacrificial Metal Plug
19. 2° Offset Aqua-Blade Lower Unit
20. Bail-A-Matic® Pickup

Fig. 25-7. A cutaway view of a three-cylinder, two-cycle outboard engine. (*Scott Division of McCulloch Corporation*)

ter of the circle made by the tips of the blades as the propeller rotates. **Pitch** is the amount of twist that the blades have. A propeller with high pitch has the blades twisted considerably. The more the pitch, the larger the bite the blades will take, and the farther the propeller will go through the water with each revolution.

The propeller can be mounted directly on the propeller shaft by means of a nut and cotter pin. It is necessary in such units to include a pin made of soft metal—the shear pin. The shear pin has two purposes: It serves to hold the propeller to the shaft in positive drive, and it protects the propeller and engine from damage if the propeller strikes an underwater object. In such cases, the pin is sheared off and the engine is protected from the shock of the impact.

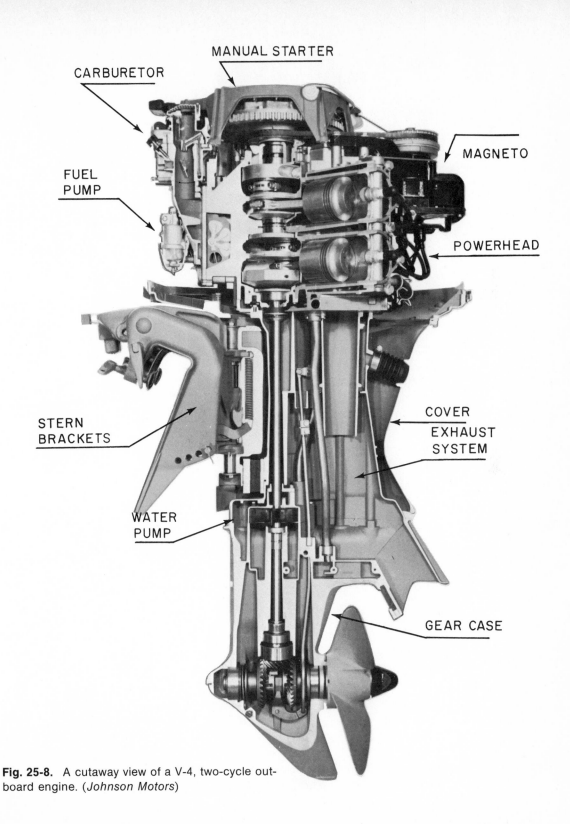

CARBURETOR

MANUAL STARTER

MAGNETO

FUEL
PUMP

POWERHEAD

STERN
BRACKETS

COVER
EXHAUST
SYSTEM

WATER
PUMP

GEAR CASE

Fig. 25-8. A cutaway view of a V-4, two-cycle outboard engine. (*Johnson Motors*)

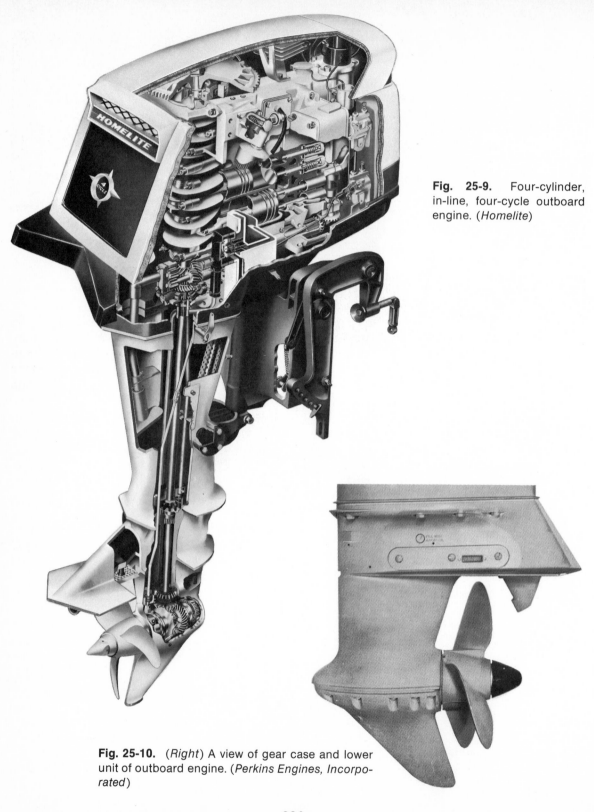

Fig. 25-9. Four-cylinder, in-line, four-cycle outboard engine. (*Homelite*)

Fig. 25-10. (*Right*) A view of gear case and lower unit of outboard engine. (*Perkins Engines, Incorporated*)

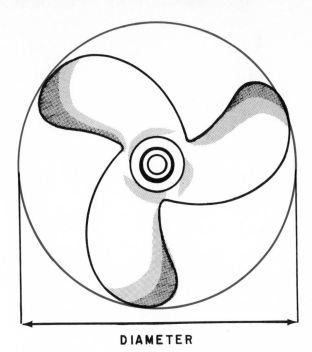

Fig. 25-11. Diameter of a propeller. (*Johnson Motors*)

On many engines, a propeller clutch has been added to give the propeller and the engine extra protection. There are numerous such designs, some with shear pins and some without. All are designed to disengage the propeller.

§ 25-6 OUTBOARD DRIVES

There are four major types of gear drives in the outboard unit: direct drive, gear-shift drive, automatic transmission, and neutral clutch. In the direct-drive units, the drive shaft is connected to the propeller shaft (usually at right angles) through bevel gears. It is possible to swing the whole unit, including the powerhead, 180° to reverse the motion of the boat. In gear-shift units, forward, neutral, and reverse gears are provided, usually through a toothed clutch system, which includes two bevel gears, a clutch dog, and the drive pinion. In most of these units, the two bevel gears and the clutch dog are mounted on the drive shaft. Such a unit is described in **§ 25-7.**

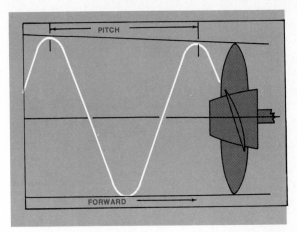

Fig. 25-12. Pitch of a propellor.

In the automatic-transmission units, the gears are shifted by twisting the steering handle. Such transmissions usually contain a neutral clutch spring and a forward-and-reverse spring. In neutral, the neutral clutch spring unwinds and frees the lower end of the drive shaft from the upper, or engine, end. In forward gear, the spring for the neutral clutch tightens around the upper drive shaft, and the whole unit turns with the drive shaft. In reverse gear, the forward-and-reverse spring uncoils, and the idler gears, lower shaft, and gear revolve in the opposite direction.

The neutral clutch units allow the engine to be started in neutral. No forward or reverse gears are included. The drive shaft is divided into two parts, an upper and a lower. They are joined by a spring. In forward, the spring joins the two tightly, and they revolve as a unit with the drive shaft. In neutral, however, the shift latch (operated manually) causes the spring to release the upper bushing. This means that the lower end of the drive shaft and the propeller shaft are disconnected from the engine.

§ 25-7 GEAR-SHIFT DRIVES

Figure 25-13 is a sectional view of a gearing and shifting arrangement, and Fig. 25-14 is a

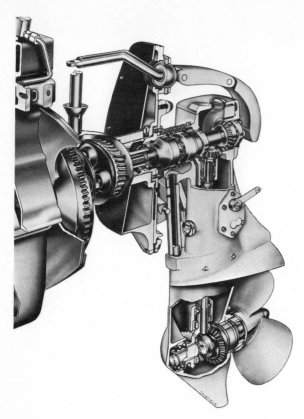

Fig. 25-13. A sectional view of a gearing and shifting assembly. (*Perkins Engines, Limited*)

disassembled view of a similar lower unit. The shifting arrangement is made up of a shifter or clutch dog and two bevel gears and permits shifting into forward, neutral, or reverse. The bevel gears are meshed with the drive pinion that is attached to the lower end of the drive shaft. When the engine is running, the drive pinion is turning, and both bevel gears are also turning. Note that one of the bevel gears turns in one direction, and the other bevel gear turns in the opposite direction.

§ 25-8 THE WATER PUMP

The water pump, which supplies water for cooling the powerhead in water-cooled engines, is sometimes located in the gear head just ahead of the propeller, and sometimes on the drive shaft above the gears (Fig. 25-8).

There are several major types: eccentric rotary, centrifugal, and vane. In some units, the intake for the water pump is located just below the exhaust outlet. Most outboard water pumps are driven by the drive shaft, in contrast to inboard-engine water pumps, which are usually belt or gear driven from the crankshaft.

Outboard water pumps usually consist of the housing, the impellers, and the inlet and outlet. In some pumps, the impeller is made of a rubber-like material that has a series of flexible blades (Fig. 25-15). At low and intermediate speeds, the impeller blades flex to follow the outer housing and to create sufficient pressure to keep water circulating. At high speeds, however, the water pressure begins to build up, and the blades flex still more to avoid excessive pressures that could damage engine parts and also use up too much power. On some installations, the water pump also serves as a bilge pump. The bilge is the part of the boat or ship between the bottom and the sides. Water, called **bilge water,** collects here and must be pumped out. The bilge pump serves this purpose. On many engines, proper operation of the water pump is indicated by a stream of water that discharges from the powerhead. Other engines indicate pump failure or overheating by use of a warning buzzer or a warning light.

§ 25-9 THE V-4 OUTBOARD ENGINE

The engine shown in Fig. 25-16 is a cutaway view of a four-cylinder, two-cycle unit. The four cylinders are arranged in two banks set at an angle to form a **V.** The cylinders are horizontal, and the pistons work to the vertical crankshaft. The engine is water-cooled and has a magneto ignition system. Figure 25-18 is a sectional view of the engine through two of the cylinders, looking down from the top. Notice that this engine uses reed valves (see **§** 6-8). Figures 25-17 and 25-19 show the actions that take place in one of the cylinders.

In Fig. 25-17, the piston is moving up on the compression stroke. Air-fuel mixture is passing

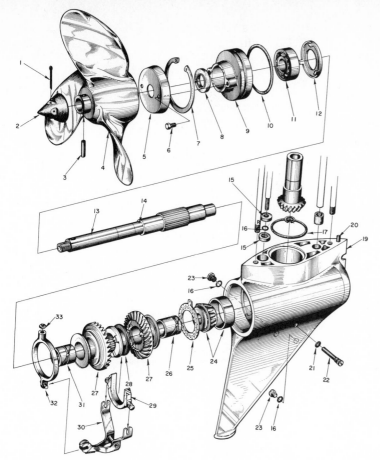

1. cotter pin
2. nut—propeller
3. pin—propeller drive
4. propeller—accessory
5. guard—weed
6. screw
7. ring—retaining
8. seal—propeller shaft
9. housing—bearing
10. seal—O ring, bearing housing
11. bearing—ball, rear shaft
12. ring and tab
13. propeller shaft assembly
14. pin
15. housing—O ring
16. O ring
17. O ring—gear housing
18. drive pinion
19. gear housing assembly
20. pin
21. O ring—shift fork screw
22. screw—shift fork
23. screw—vent
24. roller bearing assembly
25. bearing—thrust
26. bushing—gear, front
27. gear—propeller shaft
28. dog—clutch
29. yoke—shift
30. shift fork assembly
31. bushing—gear (rear)
32. connector—lower shift rod
33. nut

Fig. 25-14. A disassembled view of gear housing. (*Homelite*)

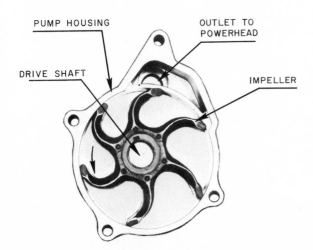

Fig. 25-15. Outboard water pump.

one of the reed-valve assemblies and entering the crankcase. It is pulled in by the vacuum created as the piston moves upward.

Figure 25-19 shows the actions as the piston moves down on the power stroke. The downward-moving piston increases the pressure in the crankcase. This pressure closes the reed valve. As the piston moves on down past the intake and exhaust ports, the burned gases begin to pour out through the exhaust port. At the same time, the air-fuel mixture, under pressure in the crankcase, begins to pour into the cylinder through the intake port. Then, after the piston has passed BDC (bottom dead center) and is moving upward again, it moves past the intake and exhaust ports, sealing them off and trapping the air-fuel mixture above it. The

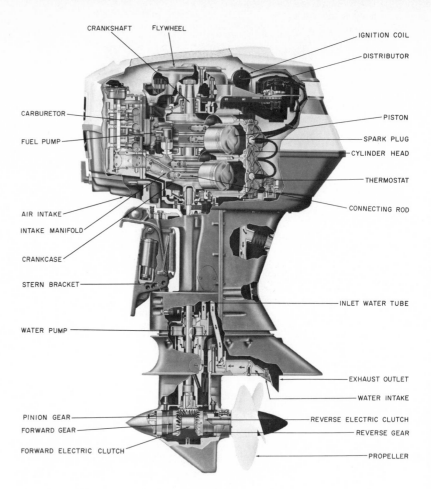

CRANKSHAFT FLYWHEEL

IGNITION COIL

DISTRIBUTOR

CARBURETOR

FUEL PUMP

PISTON

SPARK PLUG

CYLINDER HEAD

THERMOSTAT

AIR INTAKE

INTAKE MANIFOLD

CRANKCASE

CONNECTING ROD

STERN BRACKET

INLET WATER TUBE

WATER PUMP

EXHAUST OUTLET

WATER INTAKE

PINION GEAR

FORWARD GEAR

FORWARD ELECTRIC CLUTCH

REVERSE ELECTRIC CLUTCH

REVERSE GEAR

PROPELLER

Fig. 25-16. A cutaway view from the side of a V-4, two-cycle outboard engine. (*Evinrude Motors Division of Outboard Marine Corporation*)

mixture is compressed and ignited, and the power stroke starts. At the same time, air-fuel mixture is again pouring into the crankcase, as shown in Fig. 25-17. These actions continue as long as the engine runs. Every other piston stroke is a power stroke.

§ 25-10 V-4 ENGINE CONSTRUCTION

The crankcase is made up of four compartments, each sealed from the others. This is

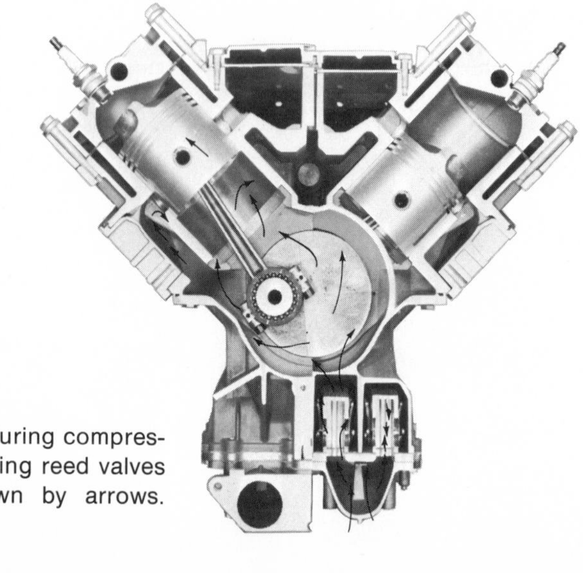

Fig. 25-17. Action in left cylinder during compression stroke. Air-fuel mixture is passing reed valves and entering crankcase, as shown by arrows. (*Johnson Motors*)

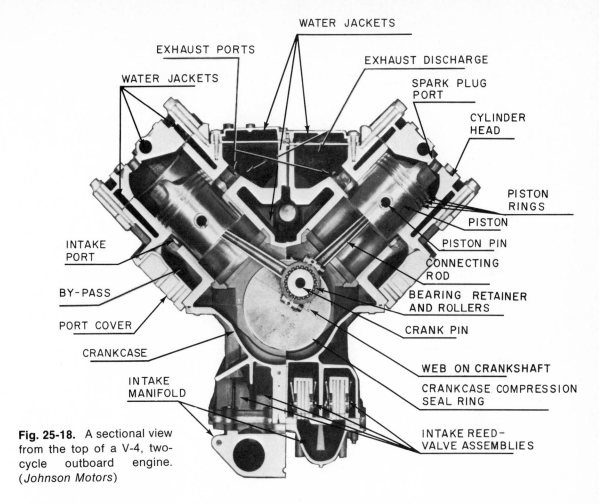

WATER JACKETS

EXHAUST PORTS

EXHAUST DISCHARGE

WATER JACKETS

SPARK PLUG PORT

CYLINDER HEAD

PISTON RINGS

PISTON

PISTON PIN

CONNECTING ROD

INTAKE PORT

BY-PASS

PORT COVER

CRANKCASE

BEARING RETAINER AND ROLLERS

CRANK PIN

WEB ON CRANKSHAFT

CRANKCASE COMPRESSION SEAL RING

INTAKE MANIFOLD

INTAKE REED-VALVE ASSEMBLIES

Fig. 25-18. A sectional view from the top of a V-4, two-cycle outboard engine. (*Johnson Motors*)

necessary because each of the four cylinders must have its own crankcase compartment in which air-fuel mixture is compressed. The sealing is accomplished by means of circular webs on the crankshaft and by the center main bearing. Figure 25-20 shows the crankshaft. Each of the two circular webs has a crankcase compression ring. In the engine, these rings are compressed between half-round sections in the crankcase and in the engine block. Thus,

Fig. 25-19. Action in left cylinder as piston nears BDC on power stroke. Pressure has closed reed valves and forced air-fuel mixture in crankcase to move through bypass and intake port and enter the cylinder. Burned gases exit through exhaust port. (*Johnson Motors*)

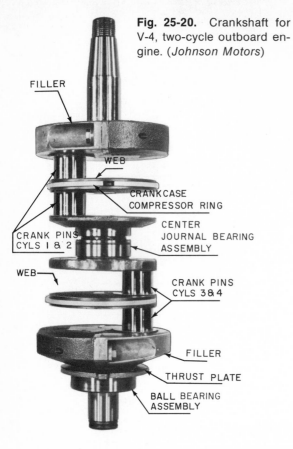

Fig. 25-20. Crankshaft for V-4, two-cycle outboard engine. (*Johnson Motors*)

FILLER

WEB

CRANKCASE
COMPRESSOR RING

CRANK PINS
CYLS 1 & 2

CENTER
JOURNAL BEARING
ASSEMBLY

WEB

CRANK PINS
CYLS 3 & 4

FILLER

THRUST PLATE

BALL BEARING
ASSEMBLY

has a roller bearing at the crankpin and a needle bearing at the piston pin. Figure 25-22 is a disassembled view of the cylinder block and heads with related parts. When the parts shown in Fig. 25-21 are added to the parts shown in Fig. 25-22, the engine is complete except for the fuel system and the ignition system. These are described later. Study the parts shown in these two illustrations as well as the cutaway views. Note how they go together to make the complete engine.

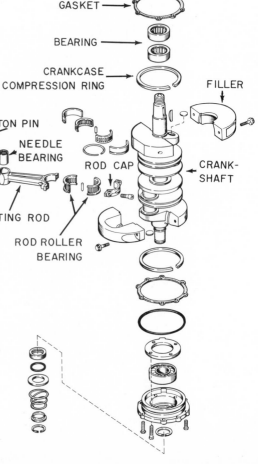

END COVER

SEALING RING

GASKET

BEARING

CRANKCASE
COMPRESSION RING

FILLER

CRANK-
SHAFT

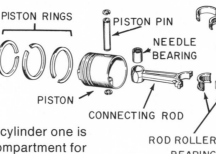

Fig. 25-21. (*Right*) Crankcase and related parts for V-4, two-cycle outboard engine. (*Johnson Motors*)

PISTON RINGS

PISTON PIN

NEEDLE
BEARING

ROD CAP

PISTON

CONNECTING ROD

ROD ROLLER
BEARING

the crankcase compartment for cylinder one is sealed off from the crankcase compartment for cylinder two. Likewise, the crankcase compartments for cylinders three and four are sealed off from each other. Notice that the crankcase compartments for cylinders two and three are sealed off from each other by the center main bearing.

The crankshaft is supported by roller bearings at the top and center and by a ball bearing at the bottom. Figure 25-21 shows the crankshaft and one piston and connecting rod with related parts. Notice that the connecting rod

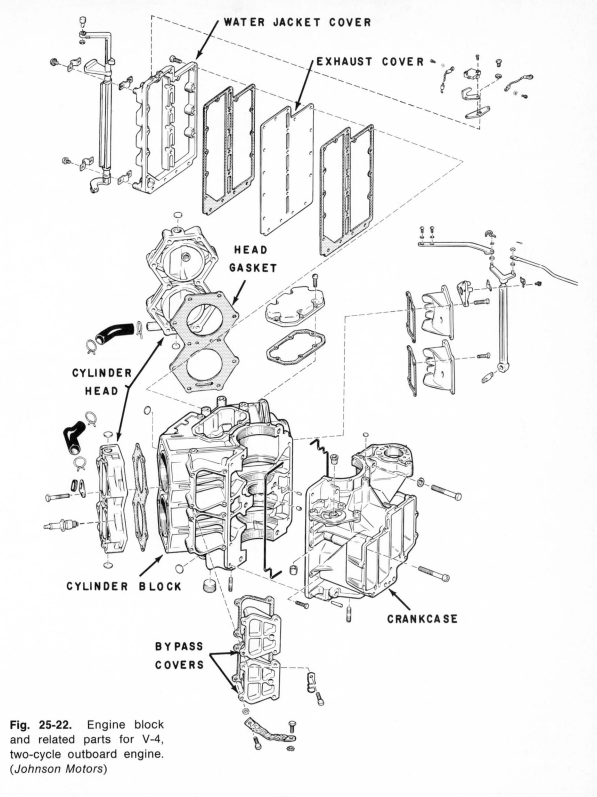

WATER JACKET COVER

EXHAUST COVER

HEAD GASKET

CYLINDER HEAD

CYLINDER BLOCK

CRANKCASE

BYPASS COVERS

Fig. 25-22. Engine block and related parts for V-4, two-cycle outboard engine. (*Johnson Motors*)

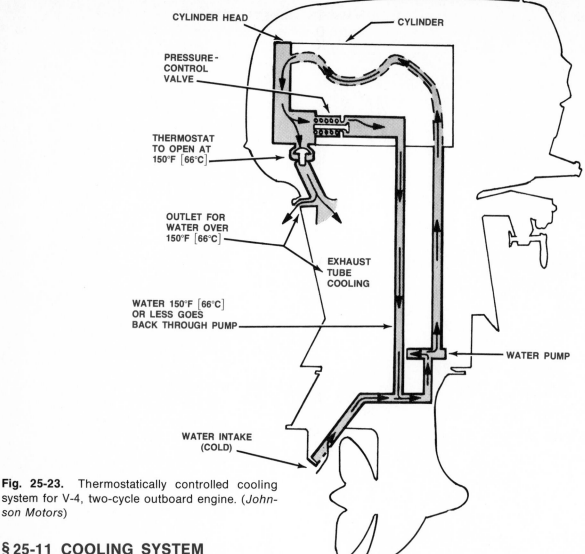

Fig. 25-23. Thermostatically controlled cooling system for V-4, two-cycle outboard engine. (*Johnson Motors*)

Labels in figure:
CYLINDER HEAD
CYLINDER
PRESSURE-CONTROL VALVE
THERMOSTAT TO OPEN AT 150°F [66°C]
OUTLET FOR WATER OVER 150°F [66°C]
EXHAUST TUBE COOLING
WATER 150°F [66°C] OR LESS GOES BACK THROUGH PUMP
WATER PUMP
WATER INTAKE (COLD)

§ 25-11 COOLING SYSTEM

The engine is water-cooled by a pump that draws cool water from the lower unit, as shown in Fig. 25-23. The cool water flows from the pump upward through the water jackets surrounding the cylinders and into the cylinder head. From there, it passes down and goes through either the pressure-control valve or the thermostatic valve. If the water has heated only a little, the thermostatic valve remains closed, and the water goes through the pressure-control valve and back through the pump. In this way, the engine quickly heats up to an efficient operating temperature.

When the water coming out of the cylinder head has increased to 150°F [66°C], the thermostat opens to allow the hot water to discharge through the outlet or past the exhaust tube. The water flowing past the exhaust tube prevents overheating of the tube.

Figure 25-24 is a sectional view of the thermostat in the closed or cold-engine position. When the engine is cold, the thermal element is not expanded. But when the engine heats up, the thermal element expands and raises

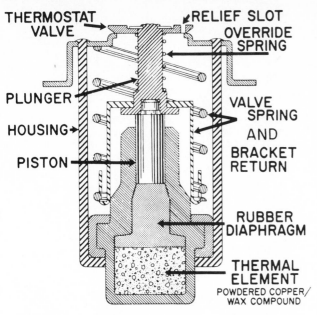

THERMOSTAT VALVE

RELIEF SLOT

OVERRIDE SPRING

PLUNGER

HOUSING →

PISTON →

VALVE SPRING AND BRACKET RETURN

RUBBER DIAPHRAGM

THERMAL ELEMENT
POWDERED COPPER/ WAX COMPOUND

Fig. 25-24. Sectional view of thermostat. (*Johnson Motors*)

the piston. This lifts the thermostat valve off its seat so that water can flow past the valve and through the outlet. The thermostat takes the position at which it lets just enough hot water out so that the exiting water is around 150°F [66°C]. This keeps the engine operating at an efficient hot temperature.

§ 25-12 V-4 OUTBOARD-ENGINE FUEL SYSTEM

The fuel system consists of a fuel tank, fuel pump, and carburetor, together with connecting fuel lines. There is no camshaft on the two-cycle engine, and thus another means must be provided to operate the fuel pump. (An eccentric on the camshaft of four-cycle engines drives the fuel pump.) On the two-cycle engine, therefore, the varying pressures in the crankcase are used to operate the fuel pump.

Figures 25-25 and 25-26 are schematic views of the fuel pump at work. In Fig. 25-25, the piston is moving up in the cylinder, and a partial vacuum is created in the crankcase. This

vacuum lifts the rubber diaphragm in the upper part of the fuel pump, creating a partial vacuum below the diaphragm. This partial vacuum raises the inlet valve and draws fuel from the tank into the chamber under the diaphragm.

Then, on the downward stroke of the piston, the air-fuel mixture in the crankcase is put under pressure. This pressure forces the rubber diaphragm down in the fuel pump, as shown in Fig. 25-26. The pressure thus put on the fuel under the diaphragm causes the inlet valve to be pushed closed and the discharge, or outlet, valve to be opened. Now, fuel flows from the chamber under the diaphragm, past the discharge valve to the carburetor. These actions continue as long as the engine runs.

On many outboards, a fuel-pump primer is used to make sure that fuel is forced into the pump and through it to the carburetor for starting. One type of primer consists of a rubber bulb in the fuel line from the tank to the fuel pump. Before a start is attempted, the bulb is squeezed so as to force gasoline to flow from the fuel tank to the pump.

When fueling an outboard engine, use only grades of gasoline approved for the engine by its manufacturer. In general, unleaded automotive gasoline should not be used in outboard engines. Use of regular or low-lead fuel is recommended by most manufacturers. Do not use premium fuel unless it is specifically listed as acceptable for the engine.

§ 25-13 CARBURETOR

The carburetor operates on the same principles as the carburetors used in automotive engines, described in Chapter 17. Figure 25-27 is a schematic sectional view of the carburetor. It has a float bowl, a low-speed system, and a high-speed system. This carburetor does not have a power system or an accelerator pump as do automotive carburetors. The reason for this is that the outboard engine is not required to provide quick acceleration and great variations in speed. As a rule, the throttle is set to operate at one speed, and the engine runs at this speed for a considerable time.

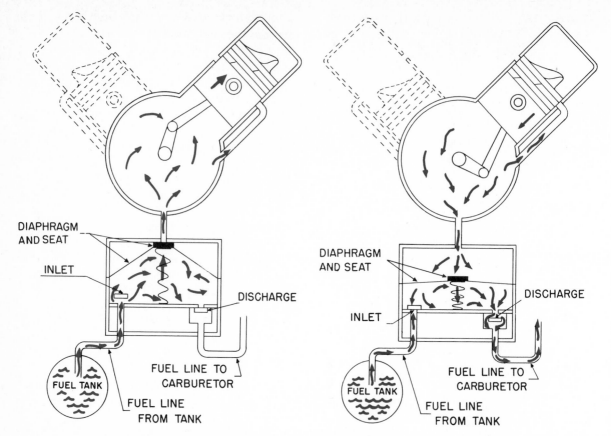

Fig. 25-25. A schematic sectional view of fuel pump, showing action during compression stroke of piston. (*Johnson Motors*)

Fig. 25-26. A schematic sectional view of fuel pump, showing action during power stroke of piston. (*Johnson Motors*)

The float bowl has a float that operates a needle valve. When the carburetor withdraws gasoline from the float bowl, the float drops. This allows the needle valve to move down to admit fuel from the pump (Fig. 25-28).

At low speeds, the throttle valve is opened only a little. The relatively high vacuum under the throttle valve causes fuel to feed from the low-speed orifices (Fig. 25-28).

At higher speeds, the throttle valve is opened more. The high-speed system begins to feed fuel into the air passing through the high-speed venturi (Fig. 25-29). When air flows through a venturi, a vacuum is created in the venturi. Review Chapter 17, which describes carburetor fundamentals.

The carburetor has a choke (Fig. 25-27) that can be closed to cause the carburetor to deliver a rich air-fuel mixture for starting. On many engines, the choke is manually operated. Some engines, however, use an automatic choke similar to the automatic chokes used on carburetors for automotive engines.

§ 25-14 V-4 OUTBOARD-ENGINE IGNITION SYSTEM

The ignition system for the engine is of the magneto type. The mounting of the magneto on the engine is shown in Fig. 25-8. A schematic drawing of the ignition system is shown in Fig. 25-30.

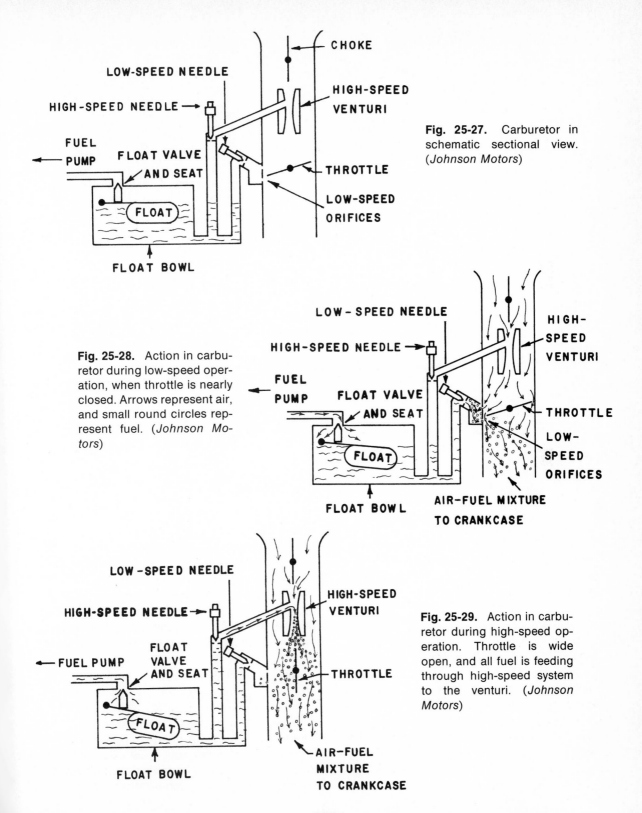

CHOKE

LOW-SPEED NEEDLE

HIGH-SPEED NEEDLE →

HIGH-SPEED VENTURI

FUEL PUMP

FLOAT VALVE AND SEAT

THROTTLE

LOW-SPEED ORIFICES

FLOAT

FLOAT BOWL

Fig. 25-27. Carburetor in schematic sectional view. (*Johnson Motors*)

Fig. 25-28. Action in carburetor during low-speed operation, when throttle is nearly closed. Arrows represent air, and small round circles represent fuel. (*Johnson Motors*)

LOW-SPEED NEEDLE

HIGH-SPEED NEEDLE →

FUEL PUMP

FLOAT VALVE AND SEAT

FLOAT

FLOAT BOWL

HIGH-SPEED VENTURI

THROTTLE

LOW-SPEED ORIFICES

AIR-FUEL MIXTURE TO CRANKCASE

LOW-SPEED NEEDLE

HIGH-SPEED NEEDLE →

FUEL PUMP

FLOAT VALVE AND SEAT

FLOAT

FLOAT BOWL

HIGH-SPEED VENTURI

THROTTLE

AIR-FUEL MIXTURE TO CRANKCASE

Fig. 25-29. Action in carburetor during high-speed operation. Throttle is wide open, and all fuel is feeding through high-speed system to the venturi. (*Johnson Motors*)

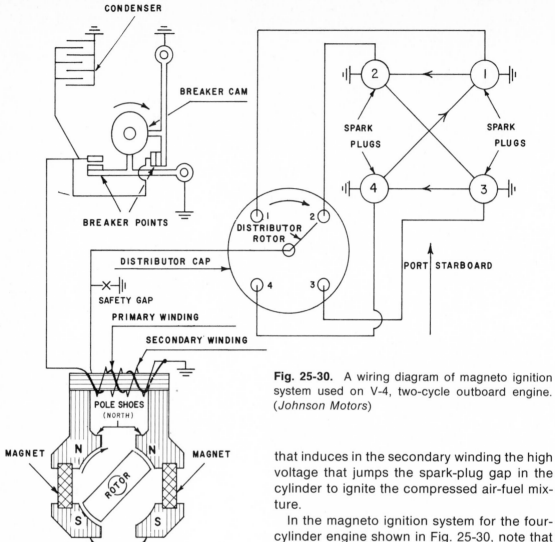

Fig. 25-30. A wiring diagram of magneto ignition system used on V-4, two-cycle outboard engine. (*Johnson Motors*)

that induces in the secondary winding the high voltage that jumps the spark-plug gap in the cylinder to ignite the compressed air-fuel mixture.

In the magneto ignition system for the four-cylinder engine shown in Fig. 25-30, note that a somewhat different arrangement of the magnets and windings is used. The rotor, which is of soft iron, spins inside magnetic pole shoes. As it does so, it passes magnetic lines of force first in one direction and then in the other (see Fig. 25-31). At the left, as the rotor moves from position 1 (broken lines) to position 2, the magnetic lines of force in the pole shoes change direction, or change from the direction shown by the broken arrows to the direction shown by the solid arrows. This change in direction of the magnetic field causes a rapid movement of the magnetic field through the primary winding assembled between the pole

Chapter 8 describes the operation of a magneto. It explains how magnets are whirled past a coil of wire to induce voltage and a flow of current in the coil. This coil, called the **primary winding,** is connected through a set of contact points so that, when the points separate, or open, the primary winding circuit is opened. When this happens, the current flowing through the primary winding stops, and the magnetic field produced by the current collapses. It is this collapse of the magnetic field

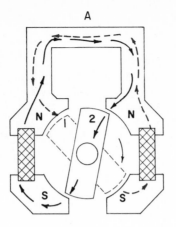
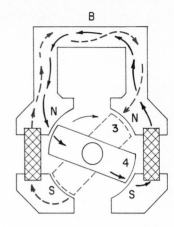

Fig. 25-31. Action of rotor in magneto as it rotates. (*Johnson Motors*)

shoes (Fig. 25-30). A moment later, as the rotor continues to spin, it reaches the position shown on the right in Fig. 25-31. Here, as it moves from position 3 to position 4, the magnetic field again reverses direction. Thus, there is a continuous buildup and collapse of the magnetic field as the rotor spins.

The primary circuit is periodically interrupted by two sets of breaker points, working alternately. Whenever the primary circuit is opened, the magnetic field produced by current flowing in the primary winding collapses. This induces the high voltage in the secondary winding that produces a high-voltage spark at the spark-plug gap in the engine cylinder. The high-voltage surge from the secondary winding is sent to the spark plug in the cylinder that is ready to fire. The distributor rotor rotates with the breaker cam and connects, in proper firing order, the secondary winding of the magneto to the spark plugs in the engine. Thus, every time a piston reaches TDC (top dead center), the rotor is in position to connect the spark plug in that cylinder with the secondary winding. At the same time, the breaker cam causes the breaker points to open so that the high-voltage surge is produced in the secondary winding. Thus, the plug fires and ignites the compressed air-fuel mixture.

Two sets of breaker points are used because of the high speed with which the points must work. Remember that, in the two-cycle engine, every cylinder spark plug fires on every revolu-

tion. Therefore, in the four-cylinder, two-cycle engine, 16,000 spark impulses are required every minute when the engine is running at 4,000 rpm. The breaker points are arranged to operate alternately and thus share the load of breaking the primary-winding circuit.

Available on many outboard engines is an electronic, or capacitor discharge, ignition system. These systems have helped to eliminate frequent spark-plug fouling in two-cycle outboard engines.

§ 25-15 STARTING MOTOR

Some of the smaller outboard engines use a manual starter that consists of a pull cord and ratchet arrangement which can be pulled to turn the engine over and get it started. This is similar to the arrangement used on many power mowers. As engine size and horsepower increase, it becomes increasingly difficult to turn the engine over manually. Thus, most of the larger engines have electric starting motors. In addition, electric starting motors are available for many smaller engines. They operate in the same way as the starting motors used on automotive engines (see Chapter 14).

Most outboards have an electric interlock that prevents cranking at wide throttle opening. Some of the interlocks will allow cranking only in neutral. This safety device prevents jack-rabbit takeoffs, which can be dangerous in a boat.

§ 25-16 INBOARD ENGINES

Inboard engines are mounted inside the marine hull. Such an engine drives the propeller by means of a shaft that passes through an opening in the bottom of the hull (Fig. 25-4). A great variety of internal-combustion engines have been used for inboard mounting in boats. They vary from small one-cylinder engines producing only a few horsepower to V-8 engines producing several hundred horsepower. Some of these engines are gasoline fueled, others are diesel engines. A few are air-cooled, but most are water-cooled. Figures 25-32 through 25-34 show several types of inboard engines. Most of the gasoline-burning inboards are four-cycle. Diesel-engine inboards may be either two- or four-cycle. For a discussion of diesel engines, refer to Chapter 23.

The four-cycle, gasoline-burning marine engines are generally similar to automotive engines in construction and operation. There are some important differences, however. For instance, water-cooled marine engines using a closed cooling system have, instead of a radiator, a heat exchanger (Fig. 25-35). The hot

Fig. 25-32. One-cylinder, four-cycle, inboard engine with starter-generator and reversing gear. (*Palmer Engine Company*)

Fig. 25-33. Two-cylinder, four-cycle, inboard engine. (*Wisconsin Motor Corporation*)

Fig. 25-34. V-8, eight-cylinder, four-cycle, inboard engine. This engine is rated at 300 horsepower [223.80 kW] at 3,600 revolutions per minute. (*Palmer Engine Company*)

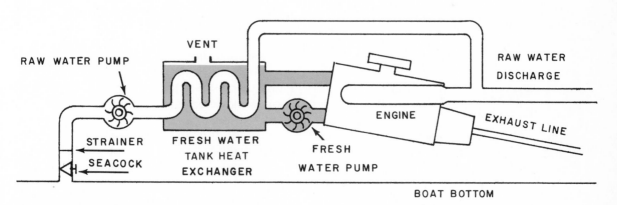

Fig. 25-35. A schematic of a two-pump closed cooling system using a fresh water tank heat exchanger.

water from the engine moves through the heat exchanger in one set of passages. Another set of passages circulates cool water from the river, lake, or ocean. This cool water picks up heat from the hot engine water and then is discharged overboard. A system such as this is desirable because the outside water may be salty or dirty and could clog the engine water passages or cause engine overheating and damage. Other engines have keel coolers (Fig. 25-36), while some use fresh lake or river water directly.

There are two other very basic differences in marine and automotive cooling systems. In marine engines, it is necessary to cool the engine lubricating oil and also to cool the ex-

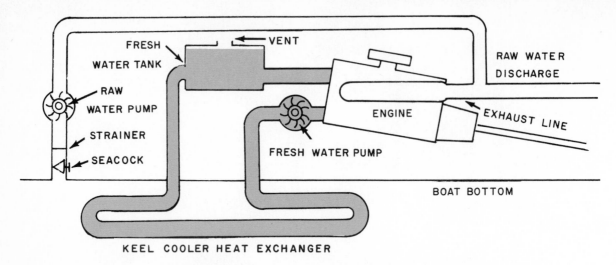

Fig. 25-36. A schematic of a closed system using a keel cooler as the heat exchanger. Note the water system used to cool and quiet the exhaust.

haust manifold. In the automobile, the passage of air around the engine, especially the airflow past the crankcase, serves in large measure to cool the lubricating oil. The inboard marine engine, on the other hand, operates in an enclosed area. There is no chance for air to cool the oil. It is thus necessary to use a water–heat-exchanging system to cool the oil.

§ 25-17 INBOARD STARTING SYSTEMS

Most inboard starting systems are more like automotive starting systems than they are like outboard systems. There are a number of such inboard systems available, and they vary considerably in their complexity. Diesel engines often use hydropneumatic starters. Some diesels, particularly the larger ones, employ a small gasoline-engine starter that is itself hand-cranked.

Some of the larger inboard gasoline engines use an inertia starter like the ones used on airplanes. This starter may be either hand-cranked or electric-powered. And many inboard gasoline engines, of course, use automotive-type electric starters. Some diesel units also employ this starting system. Marine elec-

tric starting systems are virtually the same as those used in automobiles. Automotive starting units are discussed in Chapter 14.

§ 25-18 INBOARD-OUTBOARD ENGINES

The stern-mounted, or inboard-outboard, engine combines some of the features of both the inboard and the outboard engines. The engine itself is mounted inside the hull, just forward of the transom. Figure 25-37 shows how the engine and the drive are mounted and interconnected through the transom. Note that this engine has four ignition coils, one for each cylinder. This is a special arrangement used on some engines that, in effect, provides a separate ignition system for each cylinder. There is a separate set of breaker points, as well as a coil and a condenser, for each cylinder.

Figure 25-38 is a side view of a typical mounting with a four-cylinder engine. To mount the engine and drive, two holes must be cut in the transom, a rectangular one to accommodate the tiller (or steering) linkage and a round one to accommodate the sealing boot (Fig. 25-38). Figure 25-39 shows how the assembly looks installed on the transom.

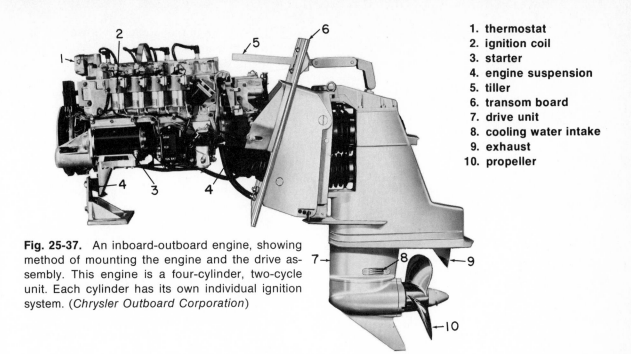

1. **thermostat**
2. **ignition coil**
3. **starter**
4. **engine suspension**
5. **tiller**
6. **transom board**
7. **drive unit**
8. **cooling water intake**
9. **exhaust**
10. **propeller**

Fig. 25-37. An inboard-outboard engine, showing method of mounting the engine and the drive assembly. This engine is a four-cylinder, two-cycle unit. Each cylinder has its own individual ignition system. (*Chrysler Outboard Corporation*)

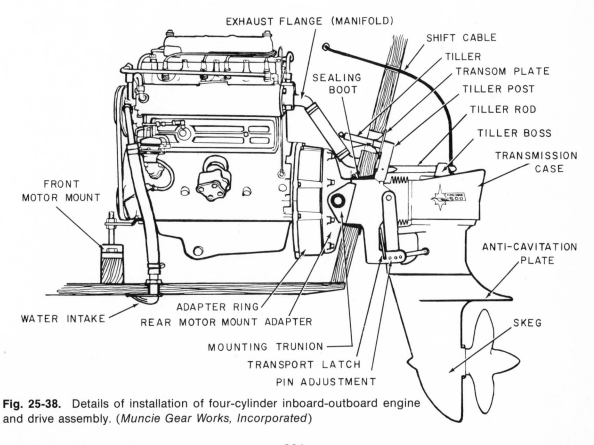

EXHAUST FLANGE (MANIFOLD)

SHIFT CABLE

TILLER

TRANSOM PLATE

TILLER POST

TILLER ROD

TILLER BOSS

TRANSMISSION CASE

SEALING BOOT

FRONT MOTOR MOUNT

ANTI-CAVITATION PLATE

WATER INTAKE

ADAPTER RING

REAR MOTOR MOUNT ADAPTER

MOUNTING TRUNION

TRANSPORT LATCH

PIN ADJUSTMENT

SKEG

Fig. 25-38. Details of installation of four-cylinder inboard-outboard engine and drive assembly. (*Muncie Gear Works, Incorporated*)

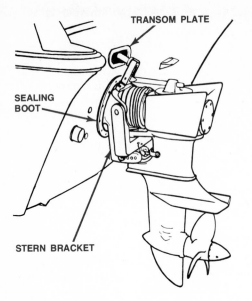

Fig. 25-39. Details of installation of drive unit for inboard-outboard engine on transom board. (*Muncie Gear Works, Incorporated*)

Almost any engine of sufficient power can be used with the stern-mounted drive. Usually, the size of the boat with which this type of drive is used is large enough to require an engine of at least four cylinders. V-4, four-in-line; V-6, six-in-line; and V-8 engines are all in use with this type of drive. These engines are very similar in construction, operation, and maintenance to automotive engines and the inboard marine engines.

Some of these engines are water-cooled from water drawn in through the drive itself (Fig. 25-40). On others, the water is drawn in from an intake at the bottom of the hull.

§ 25-19 THE DRIVE ASSEMBLY

Figure 25-41 is a cutaway view of one drive assembly, showing the drive gears and the universal joint. Universal joints are described in detail in Chapter 28. They are used in all

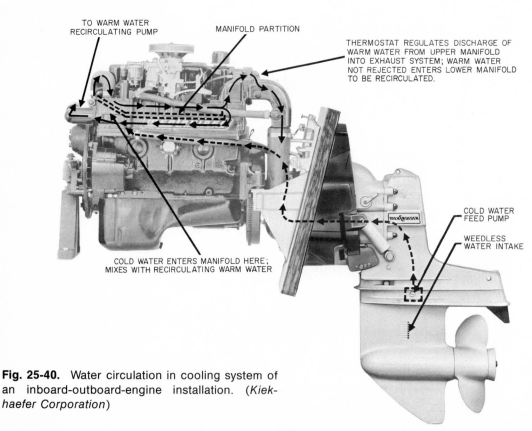

Fig. 25-40. Water circulation in cooling system of an inboard-outboard-engine installation. (*Kiekhaefer Corporation*)

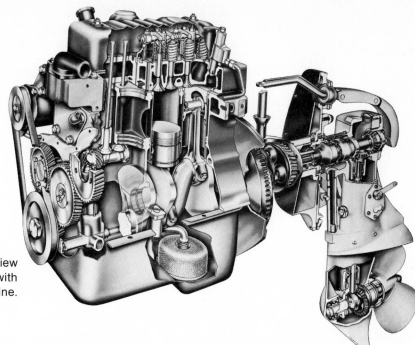

Fig. 25-41. A cutaway view of drive assembly used with inboard-outboard engine. (*Perkins*)

automotive drive lines, their purpose being to allow the angle of drive to change as the car axle moves up and down with spring flexing.

In the stern drive, the universal joint allows the drive assembly to be pivoted to the right or left by movement of the tiller. Even though the angle of drive is changed in this way between the drive shaft from the engine and the driven shaft to the rear bevel gear, power can still flow through.

As the rear bevel gear is turned, it turns a second bevel gear mounted on the vertical drive shaft. This shaft, in turn, drives two bevel gears mounted on the propeller shaft. There is a shifter, or clutch dog, that can be moved forward or aft to lock one or the other of the two bevel gears to the propeller shaft. This action then causes the propeller shaft to spin either forward or backward. The arrangement is very similar to that used on outboard engines.

Some models have an electric shifter arrangement operated with a switch by the boat pilot. This switch, turned one way or the other, closes a circuit to a magnetic coil in the drive, which then makes the shift electrically.

Review questions

1. Describe the differences between the inboard, the outboard, and the inboard-outboard marine engines.
2. What components make up the lower unit of the outboard engine?
3. Why is it necessary to seal each cylinder and crank-throw compartment in a multicylinder, two-cycle outboard engine?
4. Explain why carburetors on outboard engines do not have a power or an accelerator-pump system as do carburetors on automotive engines.
5. Study the schematic drawing of the ignition system in Fig. 25-30. How does it differ from the ignition system on an automotive engine?
6. Review § 25-14 and briefly explain, in your own words, the operation of a magneto.

7. What are the basic differences in marine and automotive cooling systems?

8. What are the advantages of having a stern-mounted inboard-outboard?

Study problems

1. Compare the application of power in an outboard-engine propeller with that in a propeller-driven aircraft.
2. Select a specific outboard engine; diagram and briefly explain the operation of its cooling system.

3. Select a specific outboard engine; diagram and briefly explain the operation of its fuel system.

26 | AIRCRAFT ENGINES

§ 26-1 HISTORY OF POWERED FLIGHT

One of the dreams of truly air-minded people had been to attach an engine to a glider so that it could soar upward in powered flight. Sir Hiram Maxim in England, Clement Ader in France, and Dr. Samuel P. Langley in the United States all tried—the first two with steam-powered airplanes, and Langley with a plane powered by a gasoline engine. Of the three, Langley probably came closest to success, but in 1903 his two attempts at flight failed.

In the same year that saw Langley's failure, the Wright brothers, Wilbur and Orville, of Dayton, Ohio, succeeded. On December 17, 1903, at Kitty Hawk, North Carolina, their awkward, cumbersome craft (Fig. 26-1), powered by a gasoline engine of their own making, lifted from the ground with Orville at the controls, and flew 120 feet [36.58 meters]. At that moment, their years of hard work were crowned with success, and the hopes and aspirations of others who had worked and planned, years before, were realized.

During the next few years, airplane development went on at a furious pace. In 1909, Louis Bleriot flew across the English Channel, a distance of 31 miles [49.89 kilometers], at an average speed of 50 miles per hour [80.47 km/h]. In 1913, in Russia, Sikorsky's *Aerobus* was put into operation. This was an enormous plane for its time. It weighed about 3,000 pounds [1,360.8 kilograms] and had four

THE WRIGHTS' FIRST SUCCESSFUL PLANE

Fig. 26-1. The Wright brothers' plane flew 120 feet [36.58 meters].

engines and a cabin with accommodations for 16 passengers. But, this was a small plane, indeed, when compared with our huge modern planes weighing more than 700,000 pounds [317,510 kilograms], measuring more than 230 feet [70.10 meters] in length, and carrying several hundred passengers at speeds of several hundred miles per hour (Fig. 26-2).

§ 26-2 WHAT MOVES THE PLANE FORWARD

An airplane in flight obviously moves through air. But before it can begin to fly, it must be pulled forward through the air. The air through which the plane rushes then acts on the airplane's wings, causing the plane to lift, or to become airborne.

It is clear, then, that the first step in understanding how an airplane flies is to understand how the plane is pulled forward through the

Fig. 26-2. A Boeing 747. (*Boeing Airplane Company*)

air. Two different devices are used to do this. One is a propeller that is spun by an internal-combustion engine (Fig. 26-3). The other is a jet, or reaction, engine, which we will look at later in this chapter. The propeller, or "prop," as it is called, was once used on all planes. Today, the larger, commercial-type planes almost all have reaction, or jet, engines.

The aircraft internal-combustion engines that spin airplane propellers work on the same principle as automobile engines. But they are almost always air-cooled and are made as light as possible. Also, they are usually flat pancake engines (Fig. 26-3), or **radial** engines (Fig. 26-4), with each cylinder located along a radius of a circle. The propeller is mounted on the engine shaft (Fig. 26-4) so that the propeller

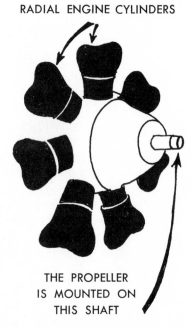

RADIAL ENGINE CYLINDERS

THE PROPELLER IS MOUNTED ON THIS SHAFT

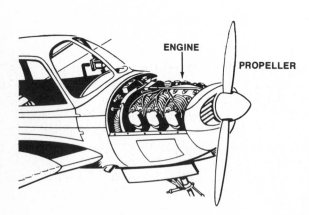

ENGINE

PROPELLER

Fig. 26-3. The engine spins the propeller.

Fig. 26-4. A radial engine.

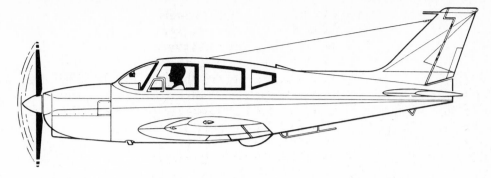

Fig. 26-5. The plane is pulled through the air by the turning propeller. (*Federal Aviation Agency*)

spins when the engine runs. As the propeller spins, it acts on the air in such a way that it is pulled rapidly through the air. Because the engine and the airplane are attached to the propeller, the whole plane (Fig. 26-5) is pulled through the air by the propeller when it turns. Let us see just how the propeller does its part in making the plane fly.

The propeller may have two, three, or four blades (Fig. 26-6). Each blade may be considered a sort of wing, like an airplane wing. Let us look at propeller-blade action from the scientific point of view. The molecules of air just behind the propeller blades are momentarily pushed close together. This produces a high-pressure area behind the blades so that the molecules bombard the backs of the blades harder than they do the front. The resulting pressure on the backs of the blades pushes them forward, and so the propeller drags the plane through the air.

§ 26-3 WHAT HOLDS THE PLANE UP?

The plane is held in the air by a lifting effect on the wings as the wings are moved forward through the air (Fig. 26-7). The wings are slightly curved on top, heavier at the front and trailing off to a thin tip. As the wing moves forward, it pushes the air out of the way. The air passing over the wing has to follow a curved path, creating a low-pressure area, or lift, on the top of the wing. Since this curved

path is longer, the air molecules, in effect, have to "spread out" a little. This means that there are fewer air molecules directly above the wing than there are directly below the wing. The air pressure on any surface is determined by the number of air molecules at that surface. Because there are more molecules below the

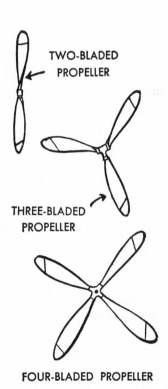

Fig. 26-6. A propeller may have two, three, or four blades.

307

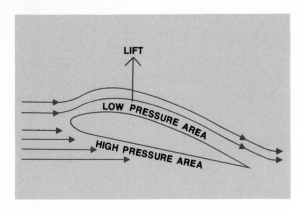

Fig. 26-7. Difference in pressure between upper and lower wing surfaces produces lift. (*Federal Aviation Agency*)

wing surface, the air pressure below the wing is higher than that above the wing. In other words, there is an upward push on the wings when they are moved through the air, and this provides the lift that holds the plane in the air.

§ 26-4 FORCES ACTING ON THE AIRPLANE

The airplane in straight-and-level unaccelerated flight is acted upon by four forces (Fig. 26-8). These forces are lift (the upward-acting force), weight (gravity, the downward-acting force), thrust (the forward-acting force), and drag (the backward-acting, or retarding, force of wind and air resistance). As you can see in

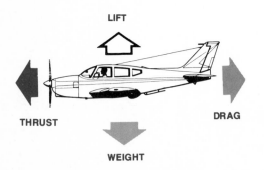

Fig. 26-8. Relationship of forces in flight. (*Federal Aviation Agency*)

Fig. 26-8, lift opposes weight, and thrust opposes drag. These four forces are acting on the airplane at all times during flight.

Drag and weight are the names for natural forces that act on any object lifted from the earth and moved through the air. Thrust and lift are artificially created forces used to overcome the forces of nature (drag and weight) to enable an airplane to fly. The engine-propeller combination is designed to produce lift to overcome weight, or gravity.

In the straight-and-level unaccelerated flight of an airplane, lift equals weight, and thrust equals drag (Fig. 26-8). If lift and weight are not equal, the airplane will begin to climb or descend. If thrust and drag are not equal, while maintaining straight-and-level flight, the airplane will go faster or slower until the two forces become balanced.

§ 26-5 THE HELICOPTER

The helicopter was suggested by Leonardo da Vinci some 400 years ago (Fig. 26-9). It is a rotating-wing aircraft that spins a three- or four-bladed "propeller" above the plane to provide the necessary lift. Control is achieved by shifting the blades slightly so that they provide more lift on one side than the other. For example, with a forward tilt, the helicopter will be pulled forward. If, however, the blades are not tilted, the helicopter can be made to hover motionless in the air.

§ 26-6 AIRCRAFT POWER PLANTS

Aircraft power plants fall into three broad categories: reciprocating engines, jet engines, and rocket engines. **Reciprocating aircraft engines** are used primarily on small craft designed to fly at relatively slow speeds and low altitudes (Fig. 26-10). The reciprocating engine, an internal-combustion engine like an automotive engine, is a very efficient power plant for such an application. Reciprocating aircraft engines reach maximum high efficiency at low speeds, and they also provide maximum lift at such speeds.

Fig. 26-9. A helicopter. (*Sikorsky Aircraft Division of United Aircraft Corporation*)

Jet engines, which are also internal-combustion engines, are widely used today for aircraft flying at subsonic, transonic, and supersonic speeds. They are most efficient at high speeds. (Subsonic means less than the speed of sound, which is about 700 miles per hour [1,126.5 km/h] at sea level. Transonic means traveling at about the speed of sound, and supersonic means traveling faster than the speed of sound.) Jet engines are limited to flight within the atmosphere, since they take in the surrounding air to provide oxygen for combustion. The **turbojet engine** (Fig. 26-11) provides its power (called **thrust**) by expelling jets of hot gases from an exhaust nozzle. However, a jet engine that, instead of providing all its thrust in this manner, uses some thrust to turn a propeller shaft is called a **turboprop** (Fig. 26-12). Jet engines that do not use turbines at all are much simpler engines. There are two basic types of such engines: the **ramjet** and the **pulsejet.** We will discuss both in detail later in this chapter.

Rocket engines operate on the same principles as jet engines. There is one difference, though, that is basic: Rocket engines are not dependent on the oxygen in the surrounding air, as are jet engines. Rocket engines carry their own oxygen with them in the form of an oxidizer. They are independent of the air and may therefore operate at altitudes above the earth's atmosphere. Space vehicles are all powered by rocket engines.

Fig. 26-10. A small aircraft. (*Cessna Aircraft Company*)

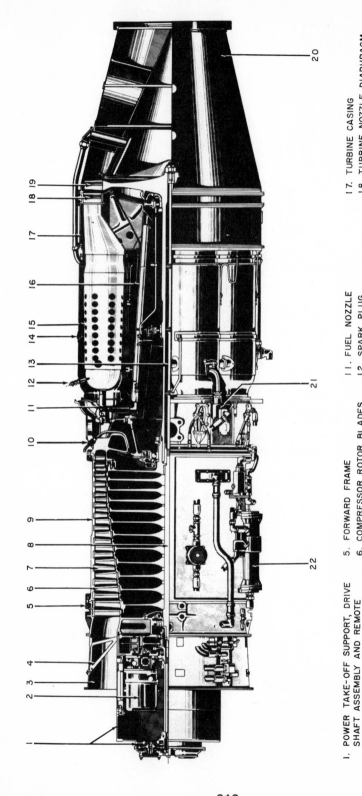

1. POWER TAKE-OFF SUPPORT, DRIVE SHAFT ASSEMBLY AND REMOTE GEAR BOX.
2. AUXILIARY COMPONENTS
3. INLET AIR GUIDE
4. INLET AIR SCREEN
5. FORWARD FRAME
6. COMPRESSOR ROTOR BLADES
7. COMPRESSOR STATOR BLADES
8. COMPRESSOR ROTOR
9. COMPRESSOR CASING
10. MID FRAME
11. FUEL NOZZLE
12. SPARK PLUG
13. TURBINE SHAFT BOLT
14. WATER INJECTION MANIFOLD
15. COMBUSTION CHAMBERS
16. AFT FRAME
17. TURBINE CASING
18. TURBINE NOZZLE DIAPHRAGM
19. TURBINE WHEEL
20. EXHAUST CONE
21. SCAVENGE PUMP
22. OIL COOLER

Fig. 26-11. A cutaway view of General Electric's J 47 turbojet engine. (*General Electric*)

Fig. 26-12. The TPE 331 is a fixed-shaft turboprop engine. It has a system for setting forward and reverse propeller-blade angles, a one-piece centrifugal compressor, and a negative torque sensing system (to reduce prop drag). This engine is designed to provide positive direct-shaft governing to control windmilling. (*Airesearch Manufacturing Company*)

Fig. 26-13. The rocket engine of the X-15 A-2 uses liquid propellant. One of its two propellant tanks contains liquid oxygen, and the other contains anhydrous ammonia. This engine delivers 57,000 pounds [25,855 kilograms] of thrust and enables the aircraft to attain a speed of Mach 8 at 100,000 feet [30,480 meters]. (*NASA and North American Aviation, Inc.*)

There are two types of rockets: those that use solid propellants and those that use liquid propellants (Fig. 26-13). (The oxidizer and the fuel together are called the **propellant.**) **Solid-propellant rockets** have so far proved unsuitable for satisfactory aircraft propulsion (except in the form of boosters), because a satisfactory method for stopping and then restarting them has not been developed. **Liquid-propellant rocket engines** have been used to power aircraft. For example, the X-15 (Fig. 26-13) and the M2-F2 (Fig. 26-14) have rocket engines that can provide enough thrust to drive an aircraft at hypersonic speeds. Rocket engines are most commonly used to place payloads in orbit. Much work is going into the development of rocket aircraft capable of hypersonic speeds, because such aircraft can be used both in space and in the atmosphere. An astronaut returning to earth in a rocket-powered re-entry vehicle could return to the atmosphere and land on a runway. Figure 26-14 shows such a re-entry vehicle.

Aircraft intended for use both in and above the earth's atmosphere sometimes have both jet and rocket engines. Rockets are used above the atmosphere not only to provide forward thrust but also to control the plane. The control surfaces customarily used in the atmosphere are useless in space. Rocket engines are discussed in detail in Chapter 27.

§ 26-7 AIRCRAFT PROPELLERS

Powered aircraft must have forward motion. This forward motion is caused by thrust. Thrust, in turn, is supplied in reciprocating-engine airplanes by the propeller. The job of the airplane propeller is to change engine **torque** (turning force) to **thrust.** Propellers

Fig. 26-14. This M2-F2 lifting-body research vehicle was successfully dropped from the wing of a B-52 bomber at 45,000 feet [13,716 meters] and glided to a high-speed conventional landing. Such powerless flights are made to demonstrate how future lifting-body space vehicles would behave in the earth's atmosphere during the critical period between re-entry and landing. (*Northrop Norair*)

may be either the **tractor** or the **pusher** type. The tractor propeller is mounted on the forward side of the wing and "pulls" the airplane forward. The pusher propeller is mounted behind the wing (and the engine) and "pushes" the plane through the air.

The propeller blades are, in effect, small wings that rotate. The tips of the propellers rotate faster than the elements near the hub.

The angle of the blade is altered along the length of the blade to give each section of blade the same angle of attack. This gives the blade its twisted appearance (Fig. 26-15). The thickness of the blade is also tapered toward the tip. Both of these devices help to offset the great speeds at the tip end of the blade. It is desirable to keep the blade tip moving at subsonic speeds (speeds below the speed of

Fig. 26-15. Checking oil in the engine of a light plane. Note the twisted appearance of the propeller.

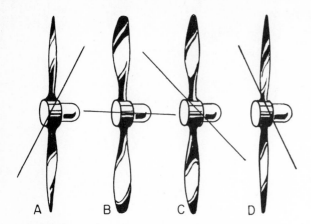

Fig. 26-16. The pitch of the propeller is altered for different flight conditions: (A) pitch angle for reversing the airflow to brake the aircraft; (B) pitch angle for feathering to stop windmilling; (C) high pitch angle for cruising; and (D) low pitch angle for take-off.

sound). If the blade speeds exceed the speed of sound, drag is increased greatly and lift is correspondingly decreased. It is this factor that limits the speeds of propeller aircraft. (Poor lift/drag ratios at supersonic speeds are also characteristic of wings.)

Many airplanes, especially the larger ones, have variable-pitch propellers. **Pitch** is the amount of advance of the blades per revolution—how big a ''bite'' of air the blades take as they rotate. It is desirable to vary the pitch for different operating conditions. For example, during takeoff, when maximum lift is required, the propellers are set at a fairly small, or **fine,** pitch. This gives good thrust without overloading the engine with rotational drag. For high-speed flight, the pitch is increased. The greater pitch increases thrust greatly and increases rotational drag at the same time (thus keeping the engine from overspeeding). This pitch would have caused the blades to stall on take-off. When landing, the pitch may be reversed so that the blades will thrust backward and act as a brake. The propeller can be feathered, too, by mechanically increasing the pitch angle until the blade is turned to the direction in which the airplane is moving. Figure 26-16

shows various pitch settings for different operating conditions.

§ 26-8 AIRCRAFT PISTON ENGINES

The reciprocating power plant used in aircraft operates very much like an automotive engine. Pistons reciprocate (move up and down or back and forth) in the cylinders. The intake and exhaust valves operate in much the same way as automotive valves. The ignition system produces sparks in the cylinders to start the power strokes of the pistons. Aircraft piston engines are of the four-cycle type, and they operate with the usual four strokes: intake, compression, power, and exhaust.

Aircraft piston engines, however, are somewhat different in construction from most other piston engines. They are generally air-cooled, and every effort is made to make them as light as possible. On the other hand, they must be designed and built for maximum reliability in performance. Engine failure in an automobile, while it can be serious, is almost never fatal. But engine failure in an airplane is very serious.

One of the major differences between many aircraft piston engines and other piston engines is the cylinder arrangement. They may be placed in a single row (in-line) or in two rows (either a V or flat-opposed). Figure 26-17 shows typical in-line, V, and horizontally opposed cylinder arrangements.

Cylinders may also be arranged like the spokes on a wheel, with all the pistons working inward to a common crankpin on the crankshaft (Fig. 26-18). This type of engine is called a **radial engine** because the pistons are placed along the radii of a circle.

Figure 26-19 shows the pistons, connecting rods, and valve train for a nine-cylinder radial engine. The cylinders are arranged so that they do not all complete their power strokes simultaneously. This makes for smoother operation. Figure 26-20 shows the cylinder assembly and associated parts for one cylinder. Note that all the connecting rods except one are attached by pins to a master rod. The series of gears

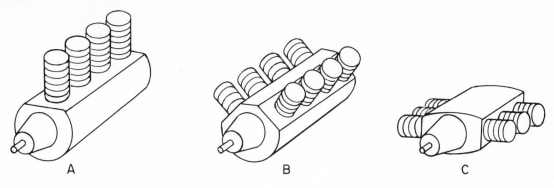

Fig. 26-17. Typical cylinder arrangements for (*A*) in-line, (*B*) V, and (*C*) horizontally opposed cylinder groupings.

shown in Fig. 26-19 is used to drive the valve train as well as the various accessories such as the fuel pump, oil pump, generator, magnetos, supercharger, and tachometer.

§ 26-9 VALVE TRAIN

The valve train is driven by gears that turn the cam gear. The cam gear has, as part of its assembly, two large cam rings. When the cam gear rotates, the cam rings also rotate. Lobes on the cam rings pass under the valve lifters, causing them to lift the pushrods and actuate the rocker arms. The rocker arms then rock and force the valves to move down, or open. Note that the pistons have depressions cut in their heads to provide room for the valves to open when the pistons are near or at TDC (top dead center).

One of the cam rings operates the intake valves; the other operates the exhaust valves. The cam rings operate much more slowly than the crankshaft is turning in the radial engine. Remember that each cam ring has nine valve lifters riding on it. Therefore, with one complete rotation of the intake-valve cam ring, each cam lobe will have operated all the intake valves. The actual speed of rotation of the cam rings is determined by the number of cylinders in the engine and the number of cam lobes. In some nine-cylinder radial engines, for example, the cam ring turns at only one-tenth the speed of the crankshaft.

§ 26-10 FUEL SYSTEM

Most reciprocating engines employ a carburetor for metering fuel and air, but some use a fuel-injection system. In the latter system, the airflow passes through the carburetor and the proper amount of fuel is forced by injection pumps into the cylinders under pressure.

The carburetors used on small-horsepower aircraft engines are of the float type and are quite similar to automotive carburetors. It was necessary, however, to provide a device to alter fuel flow at high altitudes, where the air density drops.

In fuel-injection systems, a pressure carburetor is used. (Actually, pressure carburetors are used on virtually all modern high-horsepower reciprocating aircraft engines, even those that do not use a fuel-injection system. The float carburetor cannot meet the greater fuel requirements of these powerful engines.)

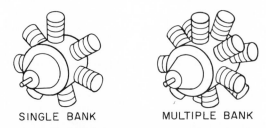

SINGLE BANK   MULTIPLE BANK

Fig. 26-18. Single- and multiple-bank radial cylinder arrangements.

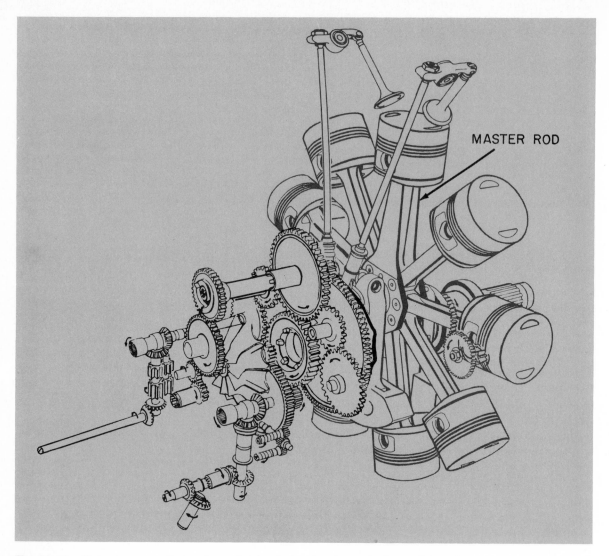

MASTER ROD

Fig. 26-19. Pistons, connecting rods, and valve train of a nine-cylinder radial engine.

Fuel is pumped into the carburetor under pressure and is metered. It is then sent to the injection pumps, which increase its pressure to force it into the cylinders.

§ 26-11 COOLING SYSTEMS

The majority of airplane engines in use today are air-cooled. Such aircraft also employ a circulating oil system. The circulating oil is used to diminish internal engine heat. Then the oil itself must be cooled. On some aircraft engines, the oil cooling is accomplished in a separate air-heat exchanger, and in others the passage of air over the sump is sufficient to cool the oil.

In air-cooled engines, the transfer of heat from the cylinders is made to the surrounding air, rather than to a liquid coolant. To accomplish this, an intricate and carefully planned series of hoods, baffles, and dams is attached

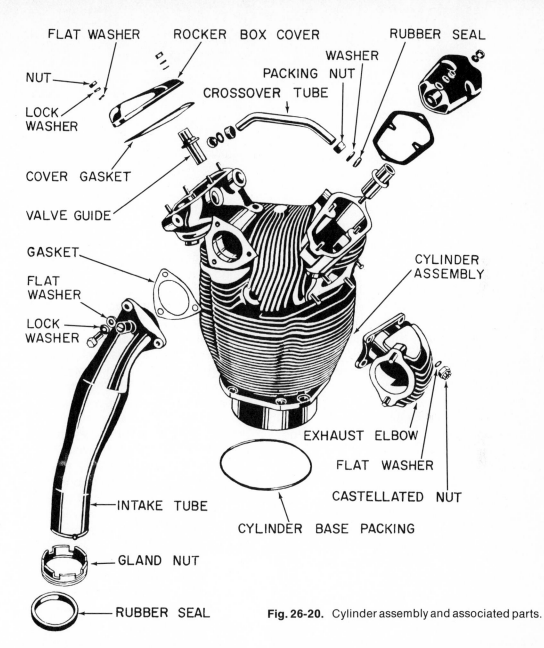

FLAT WASHER
ROCKER BOX COVER
RUBBER SEAL
WASHER
NUT
PACKING NUT
CROSSOVER TUBE
LOCK WASHER
COVER GASKET
VALVE GUIDE
CYLINDER ASSEMBLY
GASKET
FLAT WASHER
LOCK WASHER
EXHAUST ELBOW
FLAT WASHER
CASTELLATED NUT
INTAKE TUBE
CYLINDER BASE PACKING
GLAND NUT
RUBBER SEAL

Fig. 26-20. Cylinder assembly and associated parts.

to the surfaces of the cylinders. A large surface is needed for the air-heat exchange, because air is not as efficient as are liquids at heat absorption.

The air used for air cooling naturally loses some of its speed. This is called **cooling drag,** and manufacturers take great pains to minimize it. The air used for cooling can be regulated to some extent while the airplane is in

flight through the manipulation of flap openings at the air inlets and cowl flaps at the outlets.

§ 26-12 SUPERCHARGERS

The supercharger, used on many reciprocating aircraft engines over 200 horsepower [149.2 kW], increases the pressure on the air-

 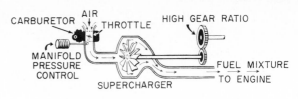

A B

Fig. 26-21. (*A*) Supercharger for use at sea level. (*B*) High-altitude supercharger.

fuel mixture going into the engine cylinders. There is a very important reason for using superchargers on powerful reciprocating engines, especially if they are to be used at high altitudes. Air pressure drops off as altitude increases. For example, at 40,000 feet [12,192 meters], an engine gets only about one-fourth as much air per intake stroke as it gets at sea level. This means that without supercharging, engine power will drop off rapidly as the plane climbs. Such a drastic power loss would greatly limit the altitude that such an airplane could reach.

It is of great importance for many aircraft to be able to operate at altitudes well above 40,000 feet [12,192 meters]. It was realized, long before World War II, that military planes that could not maneuver above enemy planes would prove to be easy targets. In commercial aviation, too, it is important for airplanes to have high-altitude capability.

The supercharger is, in effect, a rotary pump installed in the line between the carburetor and the engine (Fig. 26-21). The supercharger impeller, or rotor, is driven by the engine. As it spins, it compresses the air-fuel mixture going to the engine intake manifold. The pressure is thus boosted, and more of the air-fuel mixture gets to the engine. Engine power therefore rises.

The arrangement shown in Fig. 26-21A would be all right for low-altitude operation, but not for both low- and high-altitude flying. The reason for this is that the supercharger cannot be operated at varying speeds to suit the altitude. If it operates fast enough to boost intake-manifold pressure and engine power

satisfactorily for low altitudes, it will not do a good enough job at high altitudes. That is, it would not be turning fast enough for satisfactory boosting at high altitudes. On the other hand, if it turned fast enough to do a good job at high altitudes, it would turn too fast at low altitudes. It would overboost the engine, and the pressures in the engine would go too high. This could cause serious engine damage.

One solution to this problem is to drive the supercharger at a high enough speed to do a good job at high altitudes. In addition, a throttling device must be provided that will partly shut the throttle in the carburetor at low altitudes. Thus, at low altitudes the partly closed throttle keeps manifold pressures down to safe values. The device used for throttle control is a bellows that expands or contracts as manifold pressure changes. Whenever manifold pressure tends to go too high, the bellows expands enough to move the throttle toward the closed position. This then holds the pressure to a safe value.

Still more complicated superchargers have been used, including the arrangement shown in Fig. 26-22. In this arrangement, the air entering the carburetor is first passed through an auxiliary supercharger that boosts the pressure of the ingoing air. The amount of the boosting depends on the impeller speed, and this in turn depends on the action of the variable-speed hydraulic clutch. The clutch is controlled by altitude and power requirements. It boosts more at higher altitudes and when power requirements are high. Then, the air-fuel mixture is boosted in pressure as it passes through the engine-stage supercharger.

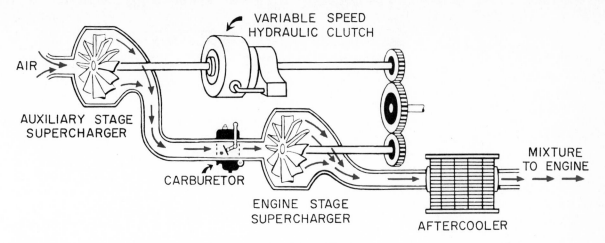

Fig. 26-22. Supercharger designed for use at various altitudes.

Finally, the mixture passes through an after-cooler that cools the mixture. (Compressing the air heats it.) To avoid overheating the air and thus causing detonation in the engine, the air must be cooled. This action permits the compression of still more air-fuel mixture into the engine cylinders so that further increases in engine power are made possible.

Many engines use a turbine driven by the exhaust gas to drive the supercharger. This arrangement is called a turbosupercharger, or turbocharger (Fig. 26-23). Some turbochargers used on aircraft engines have an integrated turbine, a compressor, and an absolute pressure controller. This assures the desired manifold pressure at all altitudes. A manifold-pressure relief valve prevents excessive overboost from the turbocharger.

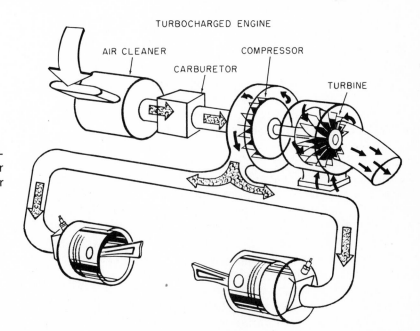

Fig. 26-23. Schematic layout of a turbosupercharger system on a flat six-cylinder engine.

§ 26-13 JET ENGINES

Jet propulsion, like rocket propulsion, depends upon reaction for the generation of thrust. As we will mention again later, Sir Isaac Newton's third law of motion states that for every action, there must be an equal and opposite reaction. In terms of the jet engine, this means that the action of the hot gases escaping rearward from the exhaust nozzle produces an equal and opposite reaction in the combustion gases. These combustion gases pressing in a forward direction on the interior of the combustion chamber produce the forward motion of the jet aircraft.

The amount of thrust developed by a jet engine is directly related to the mass-flow of air through the engine from the intake to the exhaust. The temperatures reached in the engine and the velocity of gases play an important part in the **efficiency** of the engine. However, the thrust developed depends upon the mass-flow.

There are two major types of jet powerplants: those that contain gas turbines and those that do not. Gas-turbine jets can further be divided into three groups: turbojets, turboprops, and turbofans. Jet engines that do not use turbines are either ramjets or pulsejets.

§ 26-14 RAMJETS

The **ramjet** is one of the simplest of all engines. In essence, it is simply a tube into which air is rammed, then compressed, burned (with fuel), and exhausted. Except for a fuel system that delivers fuel at high pressure, there are no moving parts. Figure 26-24 illustrates, in sectional view, the essential parts of a ramjet engine. The front end has a series of fuel nozzles located just inside the intake opening. The combustion section has a spark plug to ignite the air-fuel mixture.

When the ramjet is in operation, it is moving forward rapidly, and air is flowing through the front opening and into the combustion section (Fig. 26-25). As it flows past the fuel nozzles, the air picks up a charge of fuel sprayed from

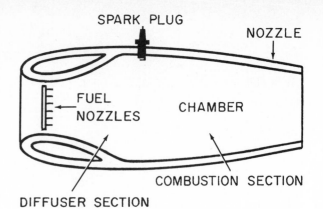

Fig. 26-24. Basic parts of a ramjet engine.

the nozzles. The combustible mixture then flows back into the combustion section, where it is ignited and burned. Such combustion produces very high pressures. The gases are then swept out of the exhaust nozzle.

The ramjet engine normally cannot take off from a stationary position. It will operate only after it has been accelerated to a relatively high speed so that the intake air (the ram air) has reached a high velocity and is being rammed into the intake. Once the ramjet has reached a high enough velocity, fuel is sprayed into the entering air and ignited. Then, the jet begins to function and furnishes forward thrust.

The ramjet now has only a limited use as an auxiliary power plant for high-speed aircraft. The ramjets are usually cut in after high speed is reached to provide additional forward thrust.

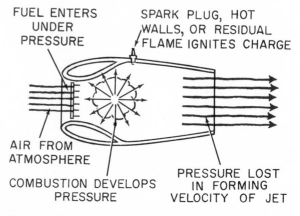

Fig. 26-25. Action in a ramjet engine.

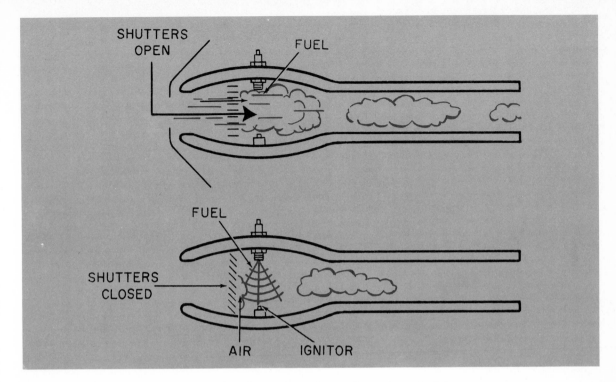

SHUTTERS OPEN

FUEL

FUEL

SHUTTERS CLOSED

AIR   IGNITOR

Fig. 26-26. A schematic view, showing the parts and operation of a pulsejet.

§ 26-15 PULSEJETS

The **pulsejet** engine, shown in Fig. 26-26, is very similar to the ramjet. In the pulsejet, however, the combustion process takes place intermittently, or in pulses. This is the reason for the name **pulsejet.** After the engine has been boosted to high speed, the pressure of the intake airflow opens a series of shutters (top, Fig. 26-26). The airflow now moves past the shutters into the combustion section. Fuel is sprayed into the air, and a spark plug ignites the air-fuel mixture. The sudden rise in pressure causes the intake shutters to close, and the intake airflow is cut off (bottom, Fig. 26-26). At the same time, the burned gases flow rearward, and the reaction to their motion on the closed shutter surfaces causes the pulsejet to move forward.

The power impulses, or pulses, follow one another rapidly, each adding its thrust to the engine. During World War II, the German V-1 ballistic missile used a pulsejet engine.

§ 26-16 AIRCRAFT GAS TURBINES

The gas turbines used for aircraft power plants are usually composed of three major sections. The first section is called the **compressor.** Air is brought in through the engine intake (at the forward end of the engine) and is compressed. There are two types of compressors: axial and centrifugal.

The **centrifugal compressor** (Fig. 26-27) is the simpler and less expensive of the two, but it is not capable of achieving the high compression ratios of the axial compressor. Basically, the centrifugal compressor consists of a bladed wheel and a diffuser. The air enters and is brought into contact with the vanes mounted on the compressor rotor (wheel). The spinning motion of the wheel forces the air outward by centrifugal force, and the air is then diffused by its passage over the vanes of the diffuser. Such diffusion raises the pressure of the air before it enters the combustion chamber. The diffuser accomplishes this by slowing the ve-

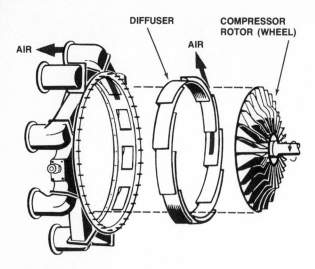

DIFFUSER | COMPRESSOR ROTOR (WHEEL) | AIR

Fig. 26-27. A centrifugal compressor.

locity of the air. (Remember that the pressure of the airflow increases to make up for the loss of velocity.) It is possible to install more than one stage in a centrifugal compressor, but multistaged centrifugal compressors become very unwieldy.

The **axial compressor** (Fig. 26-28) consists of a series of bladed rotors. The air on such bladed rotors behaves in much the same way as the airflow over a wing, and is also subject, therefore, to stall. Each blade on a single compressor is widened so that the air is diffused. After it passes the rotor, the airflow is directed over a stator (stationary bladed wheel). This

causes further air diffusion. Most axial compressors consist of a whole series of such rotors and stators.

In many engines, two series of rotors (each series is called a **spool**) are used. The first spool is the low-speed spool, and the second is the high-speed spool. The rotors of the two spools are mounted on two different shafts. When the aircraft (and the airflow) is at low speed, as it would be at takeoff, for example, the low-speed spool is in operation. Presently, when the aircraft speed is high enough, the high-speed spool is cut in, and the diffused airflow from the low-speed spool operates the second spool.

Both axial and centrifugal compressor rotors are subjected to extremely high temperatures, especially in high-speed operation. Gas-turbine blading for jet-engine applications is frequently manufactured of titanium (a light, strong, metallic element). In the future, lighter, thinner blades may be made. If new materials for blades can be developed, they will permit increased blade speed and efficiency, a critical consideration in jet-engine turbines.

§ 26-17 COMBUSTION CHAMBER AND TURBINE

The second section of the gas turbine is the combustion chamber (see combustion section in Fig. 26-29). There are several slightly different designs for combustion chambers, but the basic action of each, the ignition of the compressed air by burners, is the same. The gases are then exhausted. But in the case of all jets employing turbines, the gases must first pass through the bladed wheels of the turbine (Figs. 26-30 and 26-31), which change the energy of the gases to engine torque (turning power). The flow of the gases across the turbine rotors spins the rotors. Since the rotors are mounted on the turbine shaft, the shaft also turns. The engine torque produced is delivered to the compressor, where it serves to turn the compressor rotors. Finally, the gases are exhausted to the atmosphere through an exhaust nozzle.

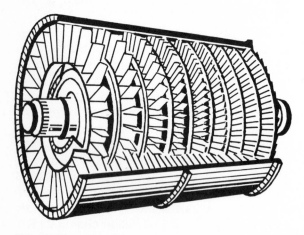

Fig. 26-28. An axial compressor.

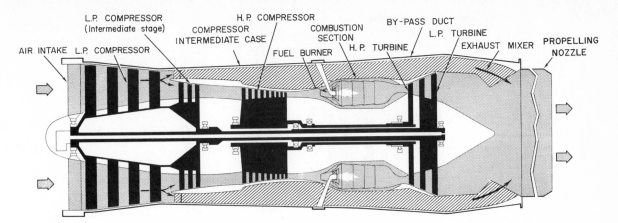

Fig. 26-29. A plan view of the Rolls-Royce Conway, showing position of combustion section in the engine. (*Rolls-Royce*)

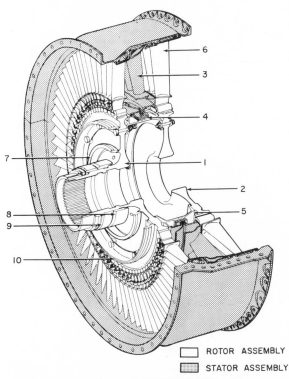

ROTOR ASSEMBLY
STATOR ASSEMBLY

1. FIRST STAGE WHEEL
2. SECOND-STAGE WHEEL
3. SECOND-STAGE NOZZLE
4. INTERSTAGE SEAL
5. TORQUE RING

6. TURBINE BLADE
7. TURBINE INNER LABYRINTH SEAL
8. SEAL RUNNER
9. NO. 3 BEARING INNER RACE
10. TURBINE OUTER LABYRINTH SEAL

Fig. 26-30. Turbine section for the CJ610. (*General Electric*)

§ 26-18 TURBOJETS

Turbojets (Fig. 26-31) are sometimes called **pure jets** because all their thrust comes from their exhaust gases. Turboprops (Fig. 26-34), on the other hand, get about 90 percent of their thrust from their propellers and the remaining 10 percent from their exhaust gases.

Mechanically, the turbojet is far lighter and more simple than the turboprop. Lubrication and cooling problems are much simpler in the turbojet than in either the turboprop or the reciprocating engine. Most turbojets use a heat exchanger to cool the oil. Fuel often is used as the coolant.

§ 26-19 TURBOJET FUEL SYSTEMS

The turbojet fuel system (Fig. 26-32) is a rather simple one. An engine-driven fuel pump brings fuel from the storage tanks, puts it under pressure, and delivers it to the injection nozzles in the combustion chamber. To ensure proper atomization, many engines have two manifolds, one for low-pressure and one for high-pressure fuel delivery. A governing valve shunts the fuel into the low-pressure manifold until fuel-pump pressures build up sufficiently to make possible the use of the high-pressure manifold. In other engines, valves within the fuel-injection nozzle itself alter the pressure to ensure atomization.

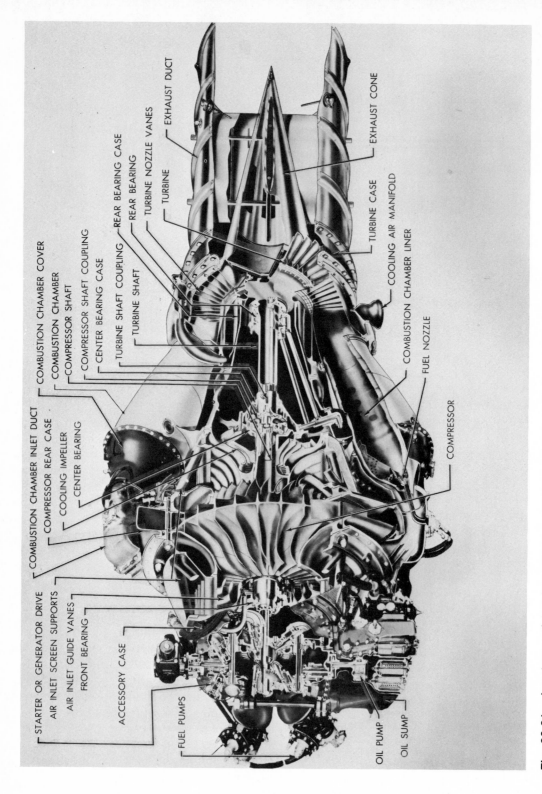

Fig. 26-31. A cutaway view of the Pratt and Whitney J42-P-4. (*Pratt and Whitney*)

STARTER OR GENERATOR DRIVE
AIR INLET SCREEN SUPPORTS
AIR INLET GUIDE VANES
FRONT BEARING

COMBUSTION CHAMBER INLET DUCT
COMPRESSOR REAR CASE
COOLING IMPELLER
CENTER BEARING

COMBUSTION CHAMBER COVER
COMBUSTION CHAMBER
COMPRESSOR SHAFT
COMPRESSOR SHAFT COUPLING
CENTER BEARING CASE
TURBINE SHAFT COUPLING
TURBINE SHAFT
REAR BEARING CASE
REAR BEARING
TURBINE NOZZLE VANES
TURBINE
EXHAUST DUCT

TURBINE CASE
EXHAUST CONE
COOLING AIR MANIFOLD
COMBUSTION CHAMBER LINER
FUEL NOZZLE
COMPRESSOR

ACCESSORY CASE
FUEL PUMPS
OIL PUMP
OIL SUMP

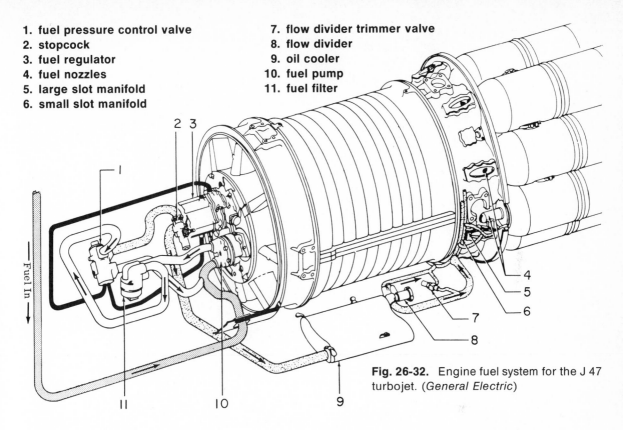

1. fuel pressure control valve
2. stopcock
3. fuel regulator
4. fuel nozzles
5. large slot manifold
6. small slot manifold
7. flow divider trimmer valve
8. flow divider
9. oil cooler
10. fuel pump
11. fuel filter

Fig. 26-32. Engine fuel system for the J 47 turbojet. (*General Electric*)

The flow of fuel to the combustion chamber is controlled differently in different engines. One method is to regulate fuel flow by measuring compressor airflow.

§ 26-20 AFTERBURNERS

One of the major modifications made in turbojets for supersonic flight is the addition of an afterburner. The afterburner is a special device fitted to many turbojet engines. In these applications, it is used for short periods of time to provide considerable additional thrust. In the case of supersonic turbojet aircraft, the turbine section of the powerplant is shortened, and the major engine area is composed of the afterburner. Figure 26-33 shows how an afterburner section has been added to the rear of a turbo-

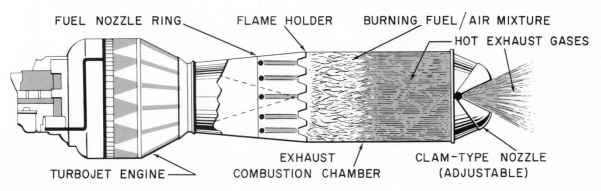

FUEL NOZZLE RING   FLAME HOLDER   BURNING FUEL / AIR MIXTURE
HOT EXHAUST GASES
TURBOJET ENGINE
EXHAUST COMBUSTION CHAMBER
CLAM-TYPE NOZZLE (ADJUSTABLE)

Fig. 26-33. One method of adding an afterburner to a turbojet engine.

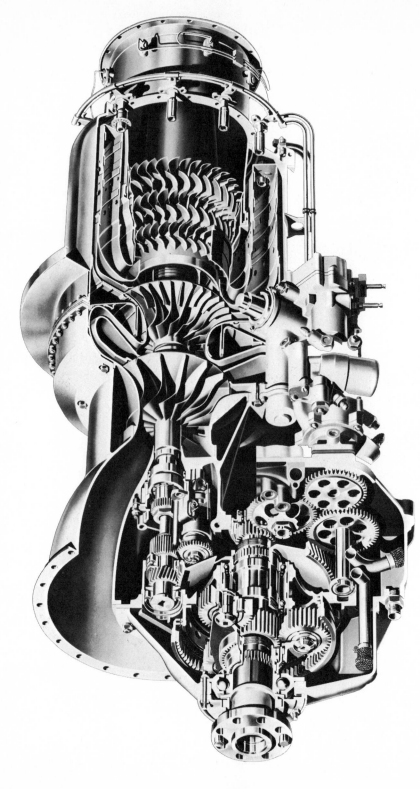

Fig. 26-34. A cutaway view of engine and reduction-gear assembly for a turboprop engine. (*Airesearch Manufacturing Company*)

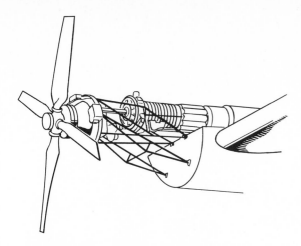

Fig. 26-35. One method of wing mounting for a turboprop engine.

jet engine. The afterburner acts much like the ramjet engine. That is, the burned gases from the combustion section of the turbojet engine are rammed into the afterburner, and fuel is sprayed into these gases. There is still much free oxygen left in these gases, even after combustion, and the fuel sprayed into the afterburner uses this free oxygen for combustion. The resulting thrust gives added power to the engine. The purpose of the flameholder in the afterburner is to stabilize the flame and prevent its flashing back into the combustion section of the engine.

§ 26-21 TURBOPROPS

The **turboprop engine** consists of a turbojet engine combined with a gear-reduction system that drives a propeller. Figure 26-34 is a cutaway view of such an engine. Figure 26-35 shows how the engine is mounted on the wing of the aircraft. Figure 26-36 is a cutaway view of the engine and reduction-gear assembly. The turbine rotor, as it spins, drives the compressor and, through reduction gears, the propeller. Most of the thrust comes from the propeller, although the turbojet exhaust provides some power. It is not possible to give the power output of a turboprop engine in pounds of thrust, since some of its thrust is provided by a propeller. The turboprop power output is

given in horsepower. The amount of thrust delivered by the exhaust jet is changed from pounds of thrust to horsepower by dividing it by 2.6.

Vastly increased speeds cause efficiency losses in turboprop engines, and so these engines are confined to subsonic applications, where they are more efficient. Some turboprops do not gain any thrust from jet exhaust. These power plants deliver all their thrust through the propeller.

It is necessary in the turboprop to provide more lubrication than that needed in the turbojet (Fig. 26-36). In the turbojet, the principal parts needing lubrication are the shaft bearings. In the turboprop, on the other hand, the gear-reduction system also needs lubrication.

§ 26-22 TURBOFANS

The **turbofan engine** (Fig. 26-37) is very similar to the conventional turbojet engine. Turbofans were developed to increase the thrust and efficiency of the conventional turbojet. In the turbofan, the compressor airflow is increased by placing a fan ahead of the compressor section and making provision for a secondary airflow. This secondary airflow does not pass through the combustion chamber or the turbine. It is directed toward the rear by ducts on the outside of the engine (Fig. 26-38). The airflow, being slightly compressed, supplies additional thrust as it is exhausted into the air stream.

§ 26-23 JET-ENGINE ACCESSORIES

There are several possible systems for powering the engine accessories on jet aircraft. One such system uses small turbines mounted in the wings of the aircraft to drive the engine accessories. This system drives its small turbines with compressed air bled from the jet-engine compressor. A second method is to operate these accessories by gear trains from the engine. Figure 26-39 shows the accessory section and components of a turbojet engine.

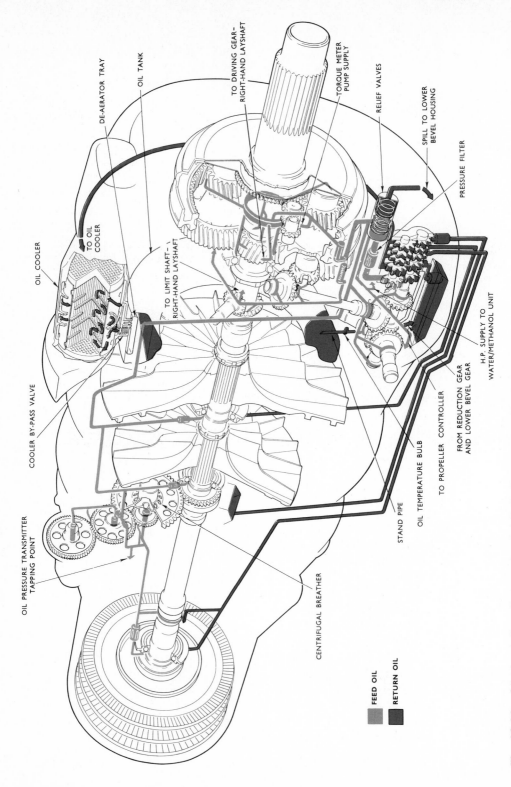

TO DRIVING GEAR–
RIGHT-HAND LAYSHAFT

TORQUE METER
PUMP SUPPLY

RELIEF VALVES

SPILL TO LOWER
BEVEL HOUSING

PRESSURE FILTER

DE-AERATOR TRAY

OIL TANK

TO OIL
COOLER

OIL COOLER

TO LIMIT SHAFT –
RIGHT-HAND LAYSHAFT

H.P. SUPPLY TO
WATER/METHANOL UNIT

FROM REDUCTION GEAR
AND LOWER BEVEL GEAR

TO PROPELLER CONTROLLER

OIL TEMPERATURE BULB

STAND PIPE

COOLER BY-PASS VALVE

CENTRIFUGAL BREATHER

OIL PRESSURE TRANSMITTER
TAPPING POINT

FEED OIL

RETURN OIL

Fig. 26-36. Oil-circulation diagram for the Rolls-Royce Dart. (*Rolls-Royce*)

Fig. 26-37. Pratt and Whitney JT-3D turbofan engine, partially cut away. (*Pratt and Whitney*)

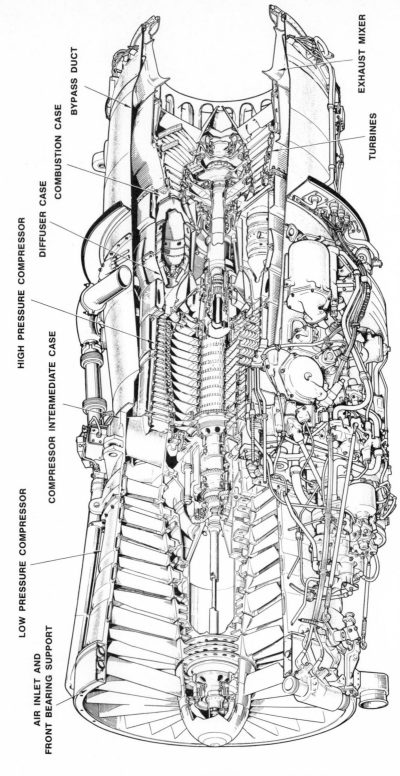

Fig. 26-38. Internal construction of the Rolls-Royce Spey turbofan, showing bypass ducts. (*Rolls-Royce*)

AIR INLET AND
FRONT BEARING SUPPORT

LOW PRESSURE COMPRESSOR

COMPRESSOR INTERMEDIATE CASE

HIGH PRESSURE COMPRESSOR

DIFFUSER CASE

COMBUSTION CASE

BYPASS DUCT

EXHAUST MIXER

TURBINES

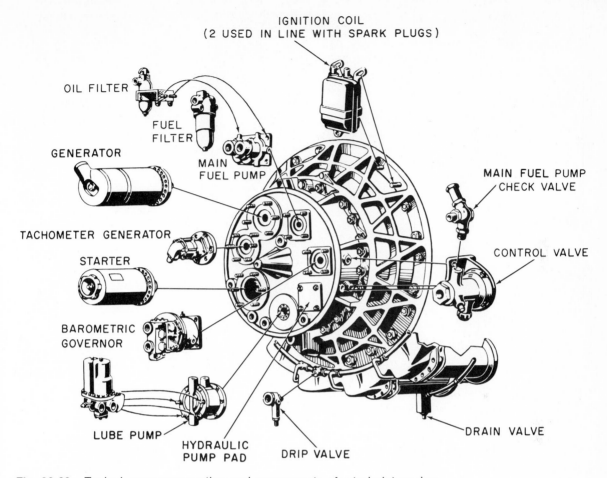

IGNITION COIL
(2 USED IN LINE WITH SPARK PLUGS)

OIL FILTER

FUEL FILTER

GENERATOR

MAIN FUEL PUMP

MAIN FUEL PUMP CHECK VALVE

TACHOMETER GENERATOR

CONTROL VALVE

STARTER

BAROMETRIC GOVERNOR

LUBE PUMP

HYDRAULIC PUMP PAD

DRIP VALVE

DRAIN VALVE

Fig. 26-39. Typical accessory section and components of a turbojet engine.

§ 26-24 SUPERSONIC TRANSPORT PLANES

New shapes for airplanes have been designed that enable a plane to go through the sonic barrier with little difficulty. Such planes are called **supersonic** planes because they fly above, or faster than, sonic barrier speed (travel faster than the speed of sound). Several countries have built supersonic fighter planes for their air forces (Fig. 26-40). And some countries are now producing supersonic transport planes for carrying passengers (Fig. 26-41). They are called SSTs (for Supersonic Transport). Such planes can approach speeds of 2,000 miles per hour [3,219 km/h].

There are certain objections to SSTs. One objection is what is known as the sonic boom. The sonic boom is a sharp sound, like an explosion, that can be quite startling and annoying to people. Windows are sometimes broken, and actual structural damage, such as cracks, may appear in buildings. The booms are caused by the plane thrusting through the air and pushing the air violently apart. This action sets up shock waves that follow behind the plane in the shape of a cone, with the plane at the apex, or point, of the cone. When the shock waves reach our ears, we hear the explosive sound we call a sonic boom. Although research continues into the cause and possible cure of the sonic boom from SSTs, there

Fig. 26-40. Convair's supersonic F-102A Delta Dagger has a maximum speed of over Mach 1 in level flight with a full load. The design of this aircraft conforms with the area rule. Note the pinched-in waist of the fuselage, which keeps the total cross section from exceeding acceptable drag limits. (*Convair Division of General Dynamics*)

is no easy solution in sight. It appears that the only solution at present is for the planes to fly below the sonic barrier until high altitudes are reached. The sonic boom from planes flying at high altitudes is much less noticeable.

However, high-altitude flying presents another problem about the SSTs that worries many scientists. At very high altitudes, the exhaust from their engines could have a very damaging effect on the upper atmosphere. At altitudes of 10 to 30 miles [16 to 48 kilometers] above the earth—the altitudes at which SSTs are to fly—there is considerable oxygen in the form of ozone. Ozone is made up of three atoms of oxygen linked together (the oxygen we breathe is made up of two atoms linked together). This unusual form of oxygen serves a very vital function so far as life on earth is con-

cerned. The ozone in the upper atmosphere protects the earth from most of the ultraviolet light coming from the sun. Without this protection, almost all the ultraviolet light would get through to the earth. Some scientists predict that the high-flying SSTs would give out enough nitric oxide in their exhausts to largely destroy the protective ozone. The ultraviolet light would then hit the earth, and, say some scientists, human beings could not go outdoors without heavy protection of skin and eyes. The ultraviolet light would burn the skin and blind the eyes. Other animals might also suffer and die from overexposure to the deadly concentrations of ultraviolet light. Also, continue some scientists, the searing ultraviolet light could kill all plant life on earth except underwater. This would mean the end of our civilization.

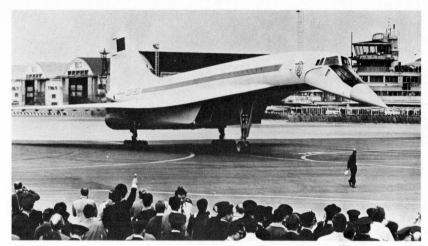

Fig. 26-41. A TU-144 Soviet supersonic passenger plane. (*Information department of Union of Soviet Socialist Republics*)

The question here is whether anyone has the right to risk this terrible possible fate for all people just to make a few hundred planes that could carry travelers across continents and oceans at three times the speed of sound. One eminent scientist, a leading authority in the field, has calculated that 500 SSTs, operating an average of 7 hours a day, could reduce the ozone content of the atmosphere by half in less than a year. And the result of this could be the end of life as we know it on this earth.

Much of what we have just said is speculation based on studies made by some prominent scientists. This potential danger may or may not exist. But until more is really known about the effects of SSTs in the upper atmosphere, some scientists are saying that SSTs should not be allowed to fly at these high altitudes.

Review questions

1. Explain the forces that act upon an airplane in flight. Explain how lift is developed.
2. Several types of power plants are used to propel aircraft. Name the three basic types of power plants.
3. How does the reciprocating engine used in an airplane differ from those used in automobiles?
4. The supercharger used on aircraft engines increases the pressure on the air-fuel mixture going into the engine cylinders. Why is this an important accessory on high-altitude aircraft?
5. What is the purpose of the aftercooler?
6. Explain the scientific principle that is the basis for jet-engine operation.
7. How do axial and centrifugal compressors work? What are the advantages of these compressors?
8. Describe the operation of the conventional turbojet engine.
9. What is the purpose of the afterburner on a jet engine?
10. Describe the turboprop engine; the turbofan engine.

Study problems

1. Explain how Newton's third law of motion applies to jet-engine power.
2. Sketch and label a simple cross-sectional view of a turbojet with an axial compressor and afterburner.
3. Design and draw a set of plans for a ramjet engine.
4. Construct a model pulsejet engine.

27 | ROCKETS AND SPACE TRAVEL

§ 27-1 HISTORY OF ROCKETS

The first record we have of rockets having been used is found in the writings of ancient Chinese historians. It would seem from these writings that the Chinese invented rockets more than 700 years ago, or around 1200 A.D. The ancient writings show that the Chinese were at war in 1232 and that they were using "arrows of flying fire," or what we now call rockets (Fig. 27-1).

Later, rockets were used in warfare in Europe, but they were inaccurate, clumsy, and unpredictable. These early rockets were packed with gunpowder (which the Chinese are credited with inventing). Even though the rocket was improved through the Middle Ages, its use as a weapon of war remained minor. Instead, guns were developed that proved to be more accurate and reliable.

However, the rocket continued to be used for fireworks displays and in some military operations. The British used rockets during the War of 1812 with the United States, when the British were attacking Fort McHenry.

In 1838, an Englishman, John Dennet, obtained a patent for a rocket that would carry a lifeline to a disabled ship from another ship or from shore. Since then, many lives have been saved by this device.

§ 27-2 K. E. TSIOLKOVSKY

It was a Russian teacher, Konstantin E. Tsiolkovsky, born in 1857, who laid the groundwork for space travel and became its first real working advocate. He did not just dream about going out into space, he actually worked out some of the mathematical laws on which the

Fig. 27-1. A skyrocket.

design of space rockets and space vehicles are based. In 1903, he published a paper outlining his work. For the rest of his life he continued to write on the subject of space exploration. Amazingly enough, even while the first airplanes were struggling to fly, he was writing about artificial satellites, the use of plants in space vehicles to provide oxygen for people to breathe, and putting colonies of people on the moon and the planets. He foresaw the problem of weightlessness in satellites and spacecraft and what this might mean to people. Many people ridiculed or ignored Tsiolkovsky, but he lived to see his prophetic work acclaimed. Before he died in 1935, Konstantin E. Tsiolkovsky had become a national hero in Russia.

§ 27-3 ROBERT GODDARD

Robert H. Goddard, an American physics professor born in 1882, was another man who became interested in rockets and space flight. Even before World War I, he was working on the mathematics of space flight. In 1919, he published a pamphlet, "A Method of Reaching Extreme Altitudes," which analyzed the problems of sending test instruments to high altitudes. But Goddard did far more than this. He began to experiment with rockets. He experimented with different fuels and various methods of constructing rockets. He very early became convinced that the best propellants were liquid. He had many failures, however, and many people ridiculed his efforts.

But, in 1926, he had his first real success; he launched a liquid-propellant rocket that flew a distance of 184 feet [56.08 meters]. This was a few feet better than the first flight of the Wright brothers' airplane. Even with this initial success, and many others, most people could see no practical value in his work. He had difficulty getting funds to continue his work. Nevertheless, he persisted. His successes were really not very spectacular. They measured only thousands of feet of flight. At present, our vast rockets lift space vehicles weighing tons into space. Yet Goddard laid the

foundations for our modern space technology. As a modern aeroscientist, Jerome Hunsaker, has said, "Every liquid-fueled rocket that flies is a Goddard rocket." Goddard received 214 patents covering nearly all aspects of liquid-fueled rockets. He was, indeed, a great pioneer of rocketry.

§ 27-4 HERMAN OBERTH

Oberth, a Romanian teacher, published a small pamphlet in 1923 dealing with the use of rockets to travel in interplanetary space. This pamphlet extended the analysis of rocketry beyond that of Goddard and Tsiolkovsky. His work inspired many people to take up experimental work on rockets. Rocket societies were formed in many countries. Their members were mostly students and young engineers who, with limited resources, started to design and build rockets.

§ 27-5 GERMAN V-2

Many Germans were actively working on the development of rockets during the 1930s. By 1939, when World War II started, preliminary work had begun on the V-2 rocket (Fig. 27-2). A leader in this work was Wernher von Braun, who later was brought to the United States, where he has contributed significantly to the

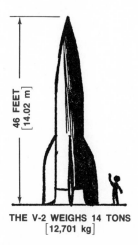

THE V-2 WEIGHS 14 TONS [12,701 kg]

Fig. 27-2. A V-2 rocket.

development of rocketry in this country. By late 1942, the first successful V-2 flight took place, but it was not until 2 years later that the V-2 began to be used to carry explosive warheads to England.

Still later, after the war was over, the United States captured and brought to America many of the German scientists who worked on the V-2 as well as many completed V-2 rockets and components. For a number of years, research with these various rockets was carried out. This research in turn led to further development of much more powerful and larger rockets.

§ 27-6 SATELLITES

One aim of the work on more powerful rockets was to produce rockets that would carry atomic bombs across the ocean. The nation that could make rockets to do this would have a tremendous advantage over its enemies. Also, there was the desire to make rockets so powerful that they would lift satellites into orbit around the earth and send space vehicles with people aboard to the moon and the planets.

The major race for rocket power was between the United States and Russia. At first, the United States had the lead, but through a combination of events, work in this country lagged while the Russians went full-speed ahead. For a number of years, the United States did very little research on rockets, and most of this work was with the captured V-2 rockets. Then, in late 1952, the nuclear scientists in the United States demonstrated the first hydrogen bomb. This new weapon, far more powerful and deadly than anything that had come before, changed the whole complexion of the rocket program. For now it was plain that the hydrogen bomb, a relatively small device, could be carried across the ocean by a rocket only slightly larger than the V-2.

The United States jumped back into the rocket business, and research and development of rockets was started again on a large scale. Within two years, the *Atlas* program began. This led to the development of the *Atlas* rocket that put the first American astronauts in orbit in 1962.

Meantime, however, the Russians had gone steadily ahead with their program. They won the race to put the first satellite in orbit when, on October 4, 1957, *Sputnik I* was launched. *Sputnik I* was followed in December with *Sputnik II*, which weighed half a ton and carried the first live creature, a dog, into space. These achievements astounded the world and were a spur to American scientists and engineers to develop larger and more powerful rockets.

Then, in March, 1958, the United States launched its first satellite, a 3-pound [1.36-kilogram] test sphere. There followed in quick succession many other satellites of varying sizes and shapes, with various objectives.

Today there are many artificial satellites circling the earth (Fig. 27-3). People have walked on the moon and returned to earth safely. Furthermore, unoccupied space vehicles have taken close looks at Mercury, Venus, Mars, and other bodies in space. No one can reasonably say now what the future of space exploration will bring, but space flight itself is now a reality.

§ 27-7 OVERCOMING GRAVITY

We would be lost without gravity. Everything would fly off the earth, including air, water, automobiles, people, and other unattached objects. In fact, the earth itself would fly apart. Without gravity, there would be no earth, no sun, no universe. So, it is fortunate that we have that attractive force between material objects that we call gravity.

On the other hand, people who dream of shaking the dust of earth from their feet are perhaps a little annoyed with gravity. For it is gravity that they have to overcome if they are to get away from the earth. After you throw a ball upward, gravity pulls it back to earth. The harder you throw the ball, the higher it will go. How fast would you have to throw it in order to escape gravity entirely, that is, in order for the ball to keep on going and never return to earth? You would have to throw it at a speed of 25,200 miles per hour [40,556 km/h]!

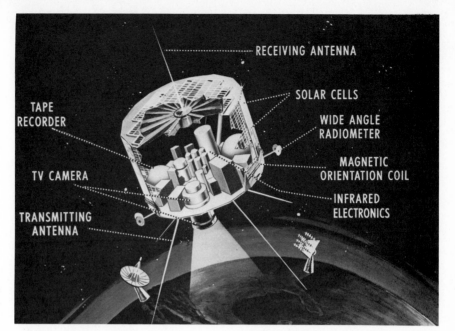

Fig. 27-3. A cutaway view of the *Tiros* weather satellite. (*NASA*)

Even the great pitcher Bob Feller could never fire a baseball that fast. His maximum was about 100 miles per hour [160.93 km/h]. And a high-powered rifle can fire a bullet at only about 1,800 miles per hour [2,896.8 km/h]. So, you can see that we have a tremendous problem facing us when we want to send anything upward fast enough for it to escape from the gravitational pull of the earth. If an object can be given a speed of 25,200 miles per hour [40,556 km/h], it will escape. This speed is called **escape velocity** (Fig. 27-4).

Of course, there are many times when scientists do not want to send the "payload" (the satellite or the capsule with passengers) out into space. Instead, they merely want to push it up into orbit around the earth (Fig. 27-5). This requires a considerably lower speed of around 18,000 miles per hour [28,968 km/h]. Many satellites have been put into orbit for various experiments and studies of the earth, our atmosphere, and space. Some of these have carried people. The first person in space was the Russian Yuri Gagarin, who went up on April 12, 1961, and circled the earth once. The first American was Alan Shepard, who made a suborbital flight on May 5 of the same year.

25,200 MILES PER HOUR [40,556 km/h] ESCAPE VELOCITY

Fig. 27-4. An object given escape velocity will escape the earth's gravitational pull.

18,000 MILES PER HOUR [28,968 km/h] ORBITAL VELOCITY

Fig. 27-5. An object given orbital velocity will circle the earth.

"Suborbital" means that the spacecraft did not go into orbit but returned to earth, as planned, in this case after a 15-minute flight. More recently, much longer flights in much larger spacecraft have been achieved, some with three people aboard. Some of these flights took spacecraft to the moon, where astronauts landed for exploration. More on this later.

There are now hundreds of unoccupied satellites circling the earth. Many of these are still active, but many are "dead." That is, their batteries or other equipment have worn out so that the satellites can no longer do the jobs for which they were sent aloft. Now, let us look at the rockets that made all these orbital and moon flights possible.

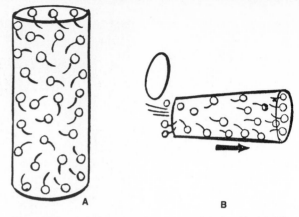

Fig. 27-6. (A) A closed tin can filled with highly compressed air would have no tendency to move. (B) When one end of the cylinder is removed, however, the cylinder would tend to move.

§ 27-8 REACTION ENGINES

About 300 years ago, Sir Isaac Newton, the famous English scientist, said, "For every *action,* there is an equal and opposite *reaction.*" Yet no one could be more surprised than he would have been to see rockets and jet planes zooming across the sky at speeds of many miles per minute, in demonstration of this basic physical law. As we discussed in Chapter 26, the jet engine used in such airplanes is a **reaction** engine. Reaction pushes the plane through the air, just as reaction sends skyrockets and space vehicles into the heavens.

§ 27-9 WHAT IS REACTION?

If we had a long cylinder that was shaped, for example, like a closed tin can and filled with highly compressed air, the molecules of air would bombard all sides and both ends of the cylinder (Fig. 27-6A). The cylinder would not have any tendency to move, because the same number of molecules would be bombarding each of its ends.

However, if we suddenly removed one end of the cylinder, the cylinder would tend to move (Fig. 27-6B). The movement would be in the direction of the closed end. The reason is that the molecules of air would no longer have anything to bump against in the open end. But

they would, for a moment, still bombard the closed end. This bombardment would tend to move the cylinder. In other words, the *action* of the molecules bombarding the cylinder end produces the *reaction* of cylinder movement.

Actually, the air would escape so quickly that the movement would be slight. But if it were possible to keep up a high pressure inside the cylinder, it would continue to be pushed along by the bombardment of molecules on the closed end. This is what takes place in the reaction engine. Fuel is burned to maintain the high pressure. A sustained push is the result.

You can make a sort of reaction engine with an ordinary sausage-shaped rubber balloon. Blow up the balloon, lay it on the palm of your hand, and then let go of the mouth (Fig. 27-7).

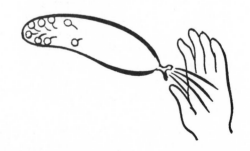

Fig. 27-7. Molecules of compressed air in a balloon push the balloon forward.

The balloon will take off and move around quite briskly for a moment or two, until all the air is released. Remember, the balloon does not move because the air is rushing out of the mouth. The balloon moves because the molecules of compressed air in the balloon are bombarding the closed front end of the balloon. They push the balloon forward.

§ 27-10 TWO TYPES OF REACTION ENGINES

There are two types of reaction engines. The basic difference between them lies in whether or not they carry a supply of oxygen along with the other fuel. As you will recall from earlier chapters, the burning of any fuel requires oxygen. The first type of reaction engine, known as the **chemical-fuel,** or **rocket,** reaction engine, carries along its own supply of oxygen. The second type of reaction engine, known as the **airstream** or **jet** engine, depends upon the surrounding air to supply necessary oxygen. These we discussed in Chapter 26.

§ 27-11 CHEMICAL-FUEL REACTION ENGINES

The skyrocket is probably the best-known example of the chemical-fuel reaction engine. It carries along its own oxygen, locked up in molecules of some chemical compound, such as gunpowder. Gunpowder is composed of sulfur, carbon, and potassium nitrate. This last compound has the chemical formula KNO_3 (one atom of potassium, one atom of nitrogen, and three atoms of oxygen to each molecule). When gunpowder is ignited, the oxygen and the carbon combine to form carbon dioxide (CO_2). Also, nitrogen is freed by the breakup of the potassium nitrate molecules.

Both the carbon dioxide and the nitrogen are gases, and their molecules move swiftly about, bombarding the confined space in the rocket cylinder. Since the back end of the cylinder is open, the molecules can fly out of that end freely (Fig. 27-8). Therefore, the *action* of the molecules bombarding the closed front end of

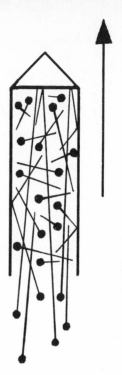

Fig. 27-8. Action of molecules in the skyrocket cylinder.

the cylinder produces the *reaction* that pushes, or **thrusts,** the rocket forward.

To make this clear, let us think of one swift-moving molecule that happens to be rushing forward toward the closed front end of the rocket cylinder (Fig. 27-8). It strikes the front end and bounces off in the opposite direction. This molecule thus gives the cylinder a tiny forward push. If the rear end of the cylinder were closed, the molecule might then strike the rear end and give the cylinder a tiny backward push. These two pushes would more or less balance, and the cylinder would have no tendency to move. But since the rear end is open, the molecule simply flies out of the rocket cylinder after it has given the cylinder a forward push. Think of billions upon billions of molecules doing this. Now, you can see how these billions of molecules that rush forward and bombard the front end of the cylinder give the rocket a powerful forward push.

In the next section we will begin to look more closely at rockets. We will find out how they are used to explore space and to increase our knowledge of the moon, Mars, Venus, and the other planets as well as outer space.

§ 27-12 ROCKET BOOSTERS

The rockets that speed payloads into orbit around the earth or out into space are called **booster rockets,** or boosters. They vary greatly in design, size, and performance, according to the mission they must perform. A booster designed to send a spacecraft with several people aboard to the moon would have to be much bigger than one that would put a basketball-sized satellite into orbit.

We have already explained, in previous sections, that the rocket is a reaction engine and that it works by the forward thrust of burning fuel in its combustion chambers. There are two types of rocket engines, liquid fuel and solid fuel.

In the liquid-fuel booster, there are two tanks, one for fuel and the other for the oxidizer (Fig. 27-9). The booster has to carry its own oxygen with it because it passes beyond the atmosphere and cannot, like internal-combustion engines, take its oxygen from the air. There are also pumps and valves that control the flow of the fuel and oxidizer into the rocket-engine combustion chambers.

Notice that the booster is mostly fuel tanks (Fig. 27-9). An enormous amount of fuel must be burned to put a relatively small amount of weight into orbit, or out into space. The actual payload that is orbited or sent into space is only a small percentage of the total weight of the booster and payload at launch. For instance, the total weight of the *Apollo 15* at launch was around 6,400,000 pounds [2,903,000 kilograms (kg)], loaded with fuel and ready to go (Fig. 27-10). Nearly 6,000,000 pounds [2,721,600 kilograms] of the total weight was fuel! The other 400,000 pounds [181,440 kilograms] included the empty shells of the boosters (or stages, as they are called), and the 96,000 pounds [43,545 kilograms] of the spacecraft itself. The entire *Apollo,* standing on the launching pad and ready to go, is

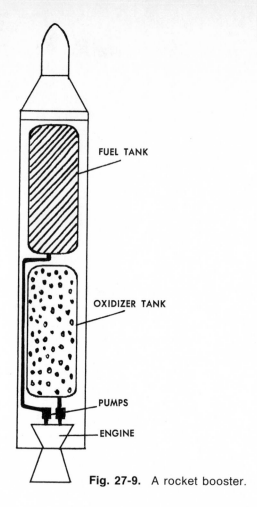

FUEL TANK

OXIDIZER TANK

PUMPS

ENGINE

Fig. 27-9. A rocket booster.

363 feet [110.64 meters] high—more than the length of a football field. Note that we said "boosters" and "stages." The booster to put the *Apollo* spacecraft into space is divided into three parts, or **stages,** as we will explain later.

You can understand that it requires enormous power to lift the 6,400,000-pound [2,903,000-kilogram] *Apollo* upward against the pull of gravity and to accelerate it to an ever higher speed. When the driver of an automobile "steps on the gas" to increase car speed, the engine must burn more gasoline to produce the acceleration. In the same way, it requires tremendous quantities of fuel to accelerate the launch vehicle with its spacecraft. The more weight that is accelerated, the more fuel that is required.

Even though the boosters are mostly fuel when ready for launching, the shells still weigh

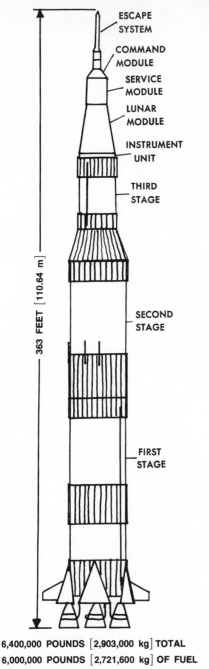

ESCAPE
SYSTEM

COMMAND
MODULE

SERVICE
MODULE

LUNAR
MODULE

INSTRUMENT
UNIT

THIRD
STAGE

SECOND
STAGE

FIRST
STAGE

363 FEET [110.64 m]

6,400,000 POUNDS [2,903,000 kg] TOTAL
6,000,000 POUNDS [2,721,600 kg] OF FUEL

Fig. 27-10. A diagram of the *Apollo 15.*

a great deal. The first stage of the *Apollo,* for example, weighs 288,750 pounds [130,979 kilograms] without fuel. That is a tremendous amount of weight. It would take a great deal of fuel to continue to accelerate this much

weight and to put it into orbit or to send it on its way out into space. In fact, it would not be possible for the first-stage booster to carry enough fuel to do this. So, as soon as the booster has done its job, it is dropped. The procedure is called **staging.**

§ 27-13 STAGING

The theory of staging is simple. You simply divide the total amounts of fuel and oxidizer needed into two or three packages, and you discard each package, one after the other, as the fuel in that package is used up.

Here is the way it works (Fig. 27-11). The big booster goes on the bottom. It is called the first-stage booster. A second-stage booster, much smaller, sits on top of it. For many missions, a third-stage, and even a fourth-stage, booster are added on. Topping all this off is the payload that is to be boosted into orbit, or into space. To start with, at launch, the first-stage booster fires and pushes the whole assembly skyward. After it has exhausted its fuel, it detaches and falls to earth. The second-stage booster fires and adds its upward thrust to the assembly above it. Notice that when the first-stage booster is detached, the assembly has lost a great amount of weight. Much less power is now required to further accelerate the upper stages with the payload.

When the second stage has burned out, it, too, detaches. The third stage fires, adding its thrust to the payload and then detaching. If there is a fourth stage, it goes through the same performance. As the stages detach, they fall earthward. Meteorlike, they flame to destruction as the friction of the air heats them to vaporizing temperatures. But sometimes, the last stage reaches sufficient speed to go into orbit instead of falling earthward, and when this happens it trails along behind the payload.

To give you an idea of what staging means in the size of payload that can be carried, think of this. A number of years ago, scientists figured that a one-stage rocket weighing 500,000 pounds [226,800 kilograms] could put a pay-

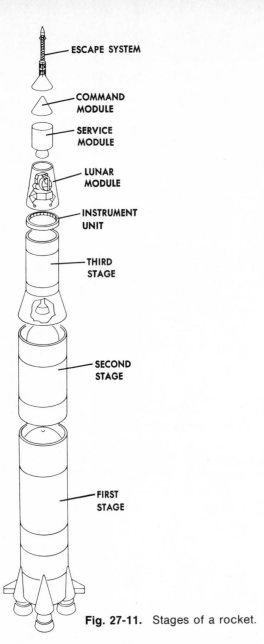

Fig. 27-11. Stages of a rocket.

ESCAPE SYSTEM

COMMAND MODULE

SERVICE MODULE

LUNAR MODULE

INSTRUMENT UNIT

THIRD STAGE

SECOND STAGE

FIRST STAGE

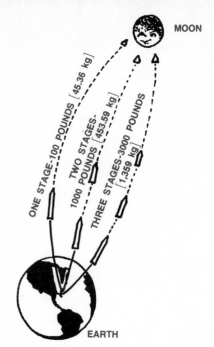

MOON

ONE STAGE-100 POUNDS [45.36 kg]

TWO STAGES-1000 POUNDS [453.59 kg]

THREE STAGES-3000 POUNDS [1,359 kg]

EARTH

Fig. 27-12. Staging enables the payload to be greatly increased.

load of 100 pounds [45.36 kilograms] on the moon (Fig. 27-12). If the fuel were divided into two packages—that is, if there were two stages—then 1,000 pounds [453.59 kilograms] could be put on the moon. But if there were three stages—with the total (three stages plus payload) weighing the same 500,000 pounds [226,800 kilograms]—then 3,000 pounds [1,359 kilograms] could be put on the moon. So by staging, the payload can be increased from 100 to 3,000 pounds [45.36 to 1,359 kilograms].

Recent developments have improved greatly on these figures. These early calculations showed that it would take about 170 pounds [77.11 kilograms] total weight to get 1 pound [0.45 kilogram] to the moon. The *Apollo,* which weighed 6,400,000 pounds [2,903,000 kilograms], put a payload of 96,000 pounds [43,545 kilograms] in orbit around the moon (Fig. 27-13). This payload included the service module, the command module, and the lunar module. This means that the *Apollo* required only about 67 pounds [30.39 kilograms] to put 1 pound [0.45 kilogram] of payload around the moon. This indicates the improved efficiency of the rockets developed in recent years. But it should not be taken as evidence that rocket efficiency will continue to improve. Actually, scientists say that the present-day rockets, using chemical fuels for combustion as the propelling force, have been made about as efficient as possible. But there are other types of propulsion engines that may be developed in the future. We will look at these later.

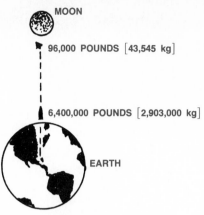

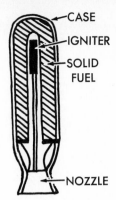

Fig. 27-14. A solid-fuel booster.

Fig. 27-13. The *Apollo* put a payload of 96,000 pounds [43,545 kilograms] in orbit around the moon.

Another thing should be said about staging. The system gets more and more complicated with each added stage. Each stage must have its own fuel system, controls, and ignition mechanisms. Also, there must be provision for detaching a stage once it has burned out or has used all its fuel. This is one reason scientists want to get along with as few stages as possible. On the other hand, there are some jobs that need more than one stage. Going to the moon or to Mars are some examples. The *Apollo* moonshots require three stages, as we explained previously.

§ 27-14 SOLID-FUEL BOOSTERS

Solid-fuel boosters are somewhat like firecrackers (Fig. 27-14). They are made up mainly of a container stuffed with a highly flammable solid. Once the fuse is lit, the solid burns rapidly and fiercely to provide a powerful thrust. In many ways, the solid-fuel rocket is simpler than the liquid-fuel rocket. It does not require separate tanks, pumps, and valves for fuel and oxidizer, or a separate combustion chamber. There is one big difference between a liquid-fuel and a solid-fuel rocket. Many liquid-fuel rockets can be turned on and off. But once the solid-fuel rocket is fired, it will continue to burn until all its fuel is burned up. Some solid-fuel rockets do have "blowout" devices that can be set off to blow an extra opening in the end

so that the fuel can burn out without producing added thrust. But once this happens, the rocket cannot be started again.

Both types of rockets have been useful in space exploration, but most of the work has been done with the liquid-fuel rockets. There are many engineers who feel that solid-propellant rockets should be developed for some of the tasks now performed by liquid-propellant rockets. Eventually, a three-stage solid-propellant rocket should be developed for space missions of the type that now use liquid-propellant rockets. The two great advantages of solid-propellant rockets over those using liquid propellants are simplicity and reliability. The fact that liquid-propellant rockets can be controlled more readily has so far been the deciding factor in their use for space missions. Remember, once a solid-propellant rocket has been ignited, it burns until the propellant is exhausted. A liquid-propellant rocket, on the other hand, can be stopped and restarted, and the rate of combustion can be speeded up or slowed down.

Solid-propellant rockets are now used in a variety of auxiliary applications. They are attached to aircraft to give additional power at takeoff. This is called **jet-assisted takeoff,** or JATO (it is also sometimes called **rocket-assisted takeoff:** RATO). Solid strap-on rockets are often used to increase the thrust of other rockets. Small solid-propellant rockets are used for small-thrust control on larger rockets and such tiny applications as jet packs for propelling individuals in space. Figure 27-15

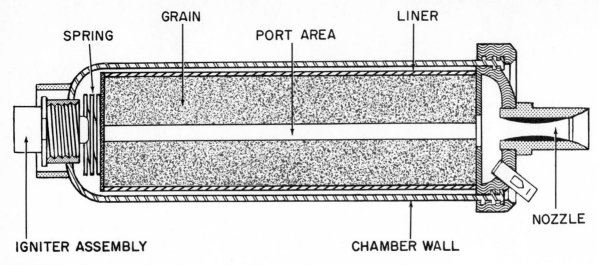

SPRING   GRAIN   PORT AREA   LINER

IGNITER ASSEMBLY   CHAMBER WALL   NOZZLE

Fig. 27-15. A cross section of a typical solid-propellant rocket engine.

shows, in sectional view, a typical solid-fuel rocket engine.

§ 27-15 LIQUID-FUEL ROCKETS

Liquid-fuel rocket engines are more complicated than those using solid fuel. Liquid-fuel-rocket power plants are designed for either **bipropellant units** or for **monopropellant units.** Bipropellant units store the fuel and oxidizer in separate tanks and then mix them for combustion (Fig. 27-9). Monopropellant units use a fuel that contains enough oxygen in its chemical makeup for combustion (Fig. 27-15).

There are numerous designs for liquid-propellant rockets, but all of them have certain basic features in common. The unit shown in Fig. 27-16 and described below is a bipropellant system.

§ 27-16 PROPELLANT TANKS

The propellant tanks (Fig. 27-16), which take up by far the most room in a liquid-propellant rocket, are used for the storage of the fuel and the oxidizer. The propellant must be delivered to the combustion chamber under pressure. Therefore, it is necessary either to place the fuel and oxidizer under pressure in their tanks

or to provide pressure pumps in the lines between the tanks and the combustion chamber.

Applications employing inert-gas pressure systems are usually used in JATO units. For extended missions, especially space flight, in which the weight of the rocket is a critical factor in achieving escape velocity, turbopump systems are more effective.

In turbopump systems, a gas turbine operates separate propellant pumps, one for fuel and one for oxidizer. The turbine must be supplied with hot gases for operation. One system, for example, uses hydrogen peroxide and a catalyst of the chemical potassium permanganate to power the turbine. The catalyst helps break down the hydrogen peroxide into water and oxygen. The water, in the form of steam, is used to drive the turbine. The turbopump then sends the propellant to the combustion chamber under pressure. The pressure in the combustion chamber is a vital factor in determining how complete the propellant combustion will be. Figure 27-17 shows a turbopump system for a bipropellant powerplant.

§ 27-17 VALVES

The throttle valves employed in liquid-propellant rockets are remote-controlled. In the turbopump unit described above, such a throttle

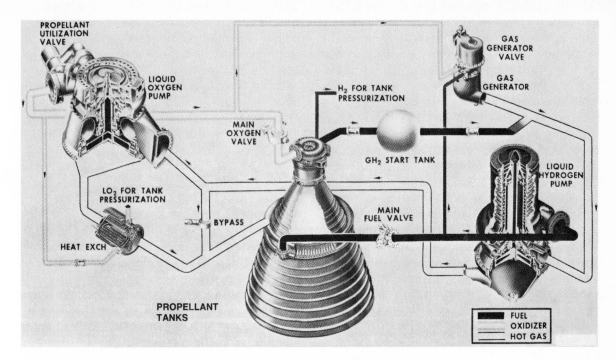

Fig. 27-16. A basic schematic for a liquid-propellant rocket engine, Rocketdyne's J-2. (*Rocketdyne*)

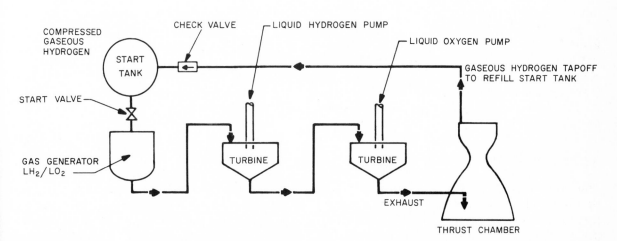

Fig. 27-17. The turbine power cycle of the J-2, showing gas spin starting system for the turbines. (*Rocketdyne*)

valve is placed in the line between the hydrogen peroxide tank and the catalyst container. This valve exerts control over the amount of steam sent to the turbine by controlling the amount of hydrogen peroxide allowed to pass into the catalyst chamber. Turbopump units include throttle valves in the line between the fuel and oxidizer tanks and the combustion chamber. Also, throttle or control valves are placed between the fuel and oxidizer tanks to control propellant flow into the combustion chamber.

Some turbopump units have a common throttle valve for fuel and oxidizer. If the fuel and the oxidizer must be delivered to the combustion chamber at different pump speeds, gears are used in the turbine shaft so that the fuel and oxidizer pumps operate at different speeds.

§ 27-18 LIQUID FUELS

Three major types of oxidizers are used in bipropellant power plants. They are liquid oxygen (LOX), fuming nitric acid (red and white), and hydrogen peroxide. The fuels used with these oxidizers are listed below.

1 Fuels for use with liquid oxygen. Ethyl alcohol, liquid hydrogen, hydrazine, ammonia, gasoline, and kerosene.

2 Fuels for use with hydrogen peroxide. Hydrazine, ethylene diamine, and ethyl alcohol.

3 Fuels for use with fuming nitric acid. Kerosene, furfuryl alcohol, gasoline, hydrazine, and aniline.

Hydrogen peroxide was used by the Germans in World War II and is used in at least one application by the British. As a general rule, hydrogen peroxide is not employed as often as either LOX or fuming nitric acid. LOX presents numerous problems. It has an extremely low temperature ($-298°$F [$-185°$C]). It is very likely to explode on impact with hydrocarbons. And, also, it has a high vapor pressure. Nitric acid also presents great difficulties. It is very corrosive. As a result, everything it comes in contact with (tanks, valves, etc.) must be constructed of either stainless steel or particular aluminum alloys.

Some successful liquid-fuel power plants have been developed that use liquid monopropellants. These are single liquid propellants that do not need a separate oxidizer.

§ 27-19 THROTTLE CONTROL AND THRUST RANGE

Although it is possible to throttle a liquid-propellant rocket, such units are not efficient over a wide thrust range. Rockets are therefore said to be **narrow-range power plants.** In order to achieve a wide thrust range and yet eliminate much throttling, multiple combustion chambers are sometimes used. The X-15, a liquid-propellant rocket aircraft, for example, has a multichamber power plant.

A variation of the multiple-chamber idea is rocket-engine clustering (Fig. 27-18). Many of the engine clusters are used to increase total thrust. They are not merely an answer to narrow thrust range. Clustering is used for rockets intended for use as boosters for spacecraft. The first stage of the *Saturn V,* for example, has a cluster of five engines. Together, they give the *Saturn V* a thrust of about $7\frac{1}{2}$ million pounds [3,401,900 kilograms].

§ 27-20 COOLING SYSTEMS

For combustion periods of more than 30 seconds, liquid-propellant rocket powerplants must be provided with some form of cooling for the combustion chamber. Several solutions have been found for such cooling. One is to circulate cooling fluid between the walls of the combustion chamber and the outer rocket walls. In some systems, however, the injector may force the cooling fluid along the inner combustion-chamber walls. Or the fluid may be injected at a number of points along the chamber walls, forming a protective coating on the walls. Or the walls themselves may be con-

Fig. 27-18. The first stage of the *Apollo Saturn 203* launch vehicle undergoes final checks before leaving the assembly facility. The 80-foot- [24.40-meter-] long stage has eight Rocketdyne H-1 engines, which produce a total thrust of 1.6 million pounds [720,000 kilograms]. (*NASA*)

structed in such a way that the coolant can come in through them.

To increase engine efficiency, the coolant used is often either the fuel or the oxidizer. When the coolant passes around the combustion chamber or over the inside walls, it picks up heat from the chamber. The coolant, fuel, or oxidizer, as the case may be, is then re-injected into the combustion chamber under pressure for combustion.

§ 27-21 IGNITION

Rocket-engine ignition is accomplished in different ways. In one unit, an electric spark is used to ignite the gas generator for the turbopump. Then, a fluid is forced into the combustion chamber. Next, the fuel and oxidizer valves open, and the LOX ignites on contact with the fluid already in the combustion chamber. Chemical elements or compounds that ignite automatically on contact with certain other elements or compounds are called **hypergolic.**

§ 27-22 ROCKET-ENGINE AIRCRAFT

There are several basic types of rocket-propelled craft: ballistic missiles, space rockets, and rocket aircraft. Rockets are also used as boosters for more conventional aircraft. Two of the aircraft which use rocket propulsion are the M2-F2 lifting body and the X-15.

§ 27-23 THE X-15

The X-15 (Fig. 26-13) is one of the fastest, highest flying aircraft ever developed. This aircraft does not take off, but is released from the parent plane at over 30,000 feet [9,144 meters] above the earth. The X-15 can fly at more than 150,000 feet [45,720 meters] and can reach speeds of about 4,000 mph [6,437 km/h]. It has an overall length of 50 feet [15.24 meters], a wingspan of 22 feet [6.71 meters], and an overall height of 13 feet [3.96 meters].

The X-15 will carry one person, 1,300 pounds [589.7 kilograms] of instrumentation, and 18,304 pounds [8,302.89 kilograms] of propellants. The aircraft weight (empty) is 12,971 pounds [5,883.43 kilograms]. Fully loaded, its weight is 31,275 pounds [14,185.74 kilograms]. Its outer shell is constructed of Inconel-X (a nickel and steel alloy), which will withstand the very high temperatures generated. The inner structure is of titanium and stainless steel.

For flight within the atmosphere, a conventional control stick is used. However, two other controls are included. With one, the pilot can control the aircraft with slight effort and is for use when the pilot is under extreme physical stress. The other control is for the ballistic control rockets and is for use when the craft is at very high altitudes.

§ 27-24 THE X-15 ENGINE

The rocket engine (Fig. 27-19) used for the main power plant in the X-15 develops more than 60,000 pounds [27,216 kilograms] of thrust (about 800,000 horsepower [596,800 kilowatts]). The propellant is composed of liquid oxygen (LOX) and anhydrous ammonia. The engine has a propellant flow rate of 10,000 pounds per minute [4,536 kilograms per minute]. Therefore, the engine can be operated at full thrust for only about 90 seconds.

The propellants are housed under pressure in tanks in the craft's fuselage. Decomposing hydrogen peroxide powers a turbine pump, through the production of high-temperature steam (Fig. 27-20). This pump sends the propellants to the engine. No propellants must be allowed to collect in the combustion chamber without combustion. If as much as 1 pint [0.47 liter] collects there, it could cause an explosion that would destroy the aircraft.

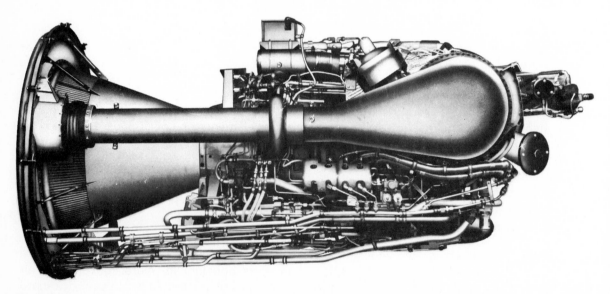

Fig. 27-19. The XLR 99 turborocket engine, which provides the power for the X-15.

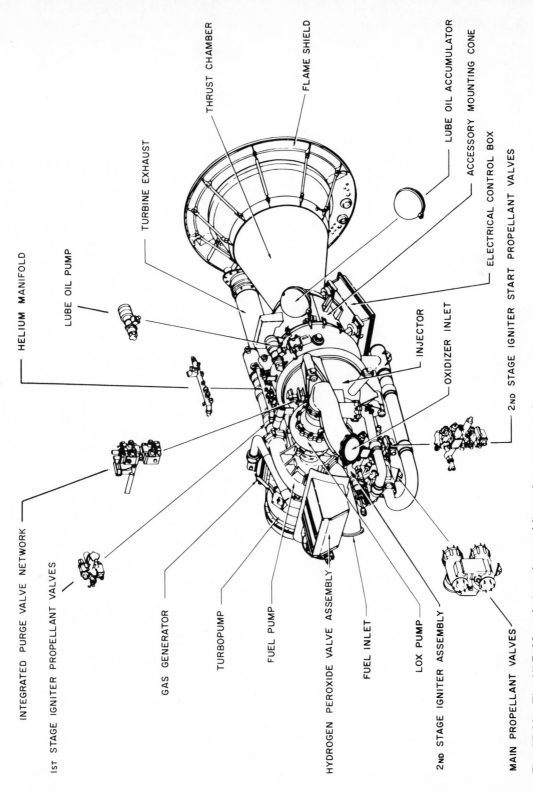

HELIUM MANIFOLD

LUBE OIL PUMP

TURBINE EXHAUST

THRUST CHAMBER

FLAME SHIELD

LUBE OIL ACCUMULATOR

ACCESSORY MOUNTING CONE

ELECTRICAL CONTROL BOX

INJECTOR

OXIDIZER INLET

2ND STAGE IGNITER START PROPELLANT VALVES

INTEGRATED PURGE VALVE NETWORK

1ST STAGE IGNITER PROPELLANT VALVES

GAS GENERATOR

TURBOPUMP

FUEL PUMP

HYDROGEN PEROXIDE VALVE ASSEMBLY

FUEL INLET

LOX PUMP

2ND STAGE IGNITER ASSEMBLY

MAIN PROPELLANT VALVES

Fig. 27-20. The XLR 99 rocket engine and its major components.

§ 27-25 AUXILIARY POWER

Auxiliary power in the X-15, used to operate atmospheric control surfaces and instruments, is provided by two small units operating on superheated steam provided by hydrogen peroxide. They are not connected with the main engine. These small accessories have many jobs in the aircraft, from powering the inertial guidance system to heating the pilot's suit. Two auxiliary units are provided in addition as a safety feature. If one fails to operate, not all the accessories are inoperative.

§ 27-26 PRESSURE AND TEMPERATURE CONTROL

Nitrogen is used in the X-15 to provide both pressure and cooling. Because of the high altitudes at which it must operate, the X-15 carries its own atmosphere with it. Liquid nitrogen is circulated about the cockpit to protect it from the friction of the aircraft's passage through the atmosphere. This same nitrogen provides a reasonable pressure within the cockpit. An air-nitrogen spray circulated in the pressure suit cools the pilot.

§ 27-27 X-15 CONTROL SYSTEMS

The X-15 has two types of controls: those for use within the atmosphere and those for use above the atmosphere. The controls for use within the atmosphere are, of course, aerodynamic in nature. They differ from the rudders, ailerons, elevators, flaps, etc., found in most conventional aircraft, however.

All but one of the control surfaces for the X-15 are located in the tail section (Fig. 26-13). The tiny, thin wings of the X-15 contain only a set of flaps. These flaps are used only when the craft is landing, and then only to slow it.

For aircraft control above the atmosphere, the X-15 relies upon ballistic nozzles. Some of these nozzles are around the nose of the aircraft, and some are at the wingtips. To move the craft's nose to the left, the pilot ejects superheated steam (obtained from decomposing hydrogen peroxide) from the nozzles to the right of the nose, and so on.

§ 27-28 M2-F2 LIFTING BODY

The M2-F2 lifting body (Fig. 26-14) was designed for exploration of flight in the transonic flight region and for research into the problems of piloted re-entry into the atmosphere from space. Figure 27-21 shows the M2-F2 in a cutaway view.

The unusual, half-cone shape of the M2-F2 gives this aircraft good flight characteristics at high speeds, but it is not nearly as good a performer at low speeds. The shape of the entire aircraft is intended to make it behave like an airfoil.

§ 27-29 LANDING ASSIST ROCKETS

In addition to its main engine, the M2-F2 carries four rocket motors, which use hydrogen peroxide to provide thrust if needed by the pilot. These rockets are throttleable and can produce thrust over a range of 100 to 400 pounds [45.36 to 181.44 kilograms].

§ 27-30 PILOTED SPACECRAFT FLIGHT

As we have already mentioned, many astronauts have ridden into the sky on a long tail of flame spouting from rocket boosters. There have been more than forty piloted spacecraft put into orbit around the earth or sent to the moon. They have varied greatly in size. The smallest was the American *Mercury* spacecraft, weighing 2,900 pounds [1,315.4 kilograms] that carried one astronaut around the earth three times and then came back to earth. The largest to date have been the *Apollo* spacecraft, which have gone to the moon and back. These weigh 96,000 pounds [43,545 kilograms].

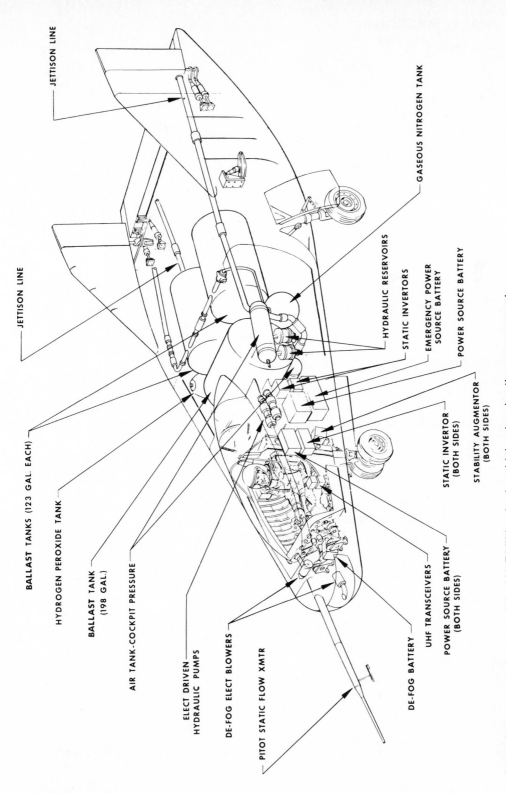

Fig. 27-21. An interior view of the M2-F2 lifting-body vehicle, showing the arrangement of components and subsystems. *(Northrop Norair)*

§ 27-31 SPACE MISSIONS OF TOMORROW

Human beings have taken their first steps into space. Neil Armstrong was the first person ever to set foot on the moon. That historic mission was but the beginning, many scientists say. They look to human exploration of Mars and other planets. Detailed plans are being drawn for space stations and for special shuttle planes that can fly back and forth between earth and the space stations (Fig. 27-22). Various types of propulsion engines are being developed and tested for use on long flights to the more distant planets and beyond. These propulsion engines would not use chemical fuels such as are now employed in present-day rockets. Instead, they would use such propulsive devices as nuclear engines or electrical thrusters.

SPACE SHUTTLE

Fig. 27-22. The space shuttle would be able to travel back and forth between earth and future space stations.

§ 27-32 POSSIBLE FUTURE SPACECRAFT POWER PLANTS

The idea behind the **nuclear engine** is that a uranium-fueled atomic reactor would superheat a working fuel such as hydrogen. The fuel would then be ejected through a nozzle, and this would impart thrust to the spacecraft, just as in chemical-fuel engines.

A scheme for **electrical propulsion** is to accelerate material to high speeds by various electrical means. These systems, if properly developed, would work fine far out in space. They would not work for launching and for getting away from any earthly body such as the earth or the moon. They would provide very low thrust, but a small push out in space continued for a long period of time would result in high spacecraft speed. Solar sails have been proposed for long space flights away from the sun (Fig. 27-23). A **solar sail** is a huge spread of material that is turned to catch the full force of the sunlight. This sunlight puts pressure on the sail. True, the pressure, or weight, of sunlight is very small, but a spacecraft with large enough sails could ultimately gain very high

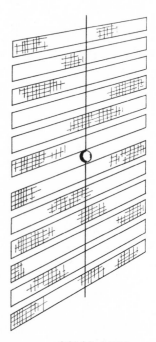

SOLAR SAILS
SPREAD TO
CATCH SUNLIGHT

Fig. 27-23. Solar sails.

speed just from the weight of the sunlight falling on the sails. Of course, it would take a long time—perhaps years—to reach astronomical speeds of, say, 100,000 miles per second [160,930 kilometers per second]. But in interplanetary and outer space travel, it will take years to reach the objective and return to earth.

Review questions

1. Robert H. Goddard is sometimes referred to as the father of rocket engineering in this country. Why?
2. What is the meaning of the expression escape velocity?
3. What causes a rocket to move?
4. Why are rocket engines necessary for use in space?
5. What are the two types of rocket engine?
6. What is a booster rocket?
7. What is meant by the term staging? How does staging help in sending rockets into orbit?
8. List the advantages and disadvantages of the solid-propellant over the liquid-propellant rocket.
9. What takes up the most room in a liquid-propellant rocket?
10. What type of power plants might someday be used to power future space missions to the more distant planets?

Study problems

1. Find out all you can about the pioneering work of Dr. Robert H. Goddard. Write a short paper on his accomplishments and his frustrations.
2. Explore the history of World War II for information about the German V-2 rocket. Prepare a brief report.
3. On October 4, 1957, the Russians put the first satellite, *Sputnik I,* into orbit. What effect did this have upon the United States?
4. Write a report on space flights.

28 MECHANICAL POWER TRANSMISSION

§ 28-1 GETTING THE POWER TO THE WHEELS

There are many ways of getting the power from the engine to the wheels. In steam locomotives, the pistons of the steam engine are linked directly to the locomotive wheels by connecting rods. As a piston moves back and forth in the cylinder, the connecting rod moves back and forth and, through linkage, causes the wheels to rotate.

Some small vehicles, such as golf carts or self-powered lawn mowers, use belts and pulleys to turn the wheels. Figure 28-1 shows a riding power mower in which the wheels are turned by belts. The belts and pulleys are the V type. As the belt wraps around the pulley, it wedges tightly into the V of the pulley so that the belt and pulley must move together.

In the automobile, the power from the engine must go through a much more complicated path in order to reach the wheels. This path is called the **power train.** It consists of several mechanisms: the transmission, the drive shaft, and the rear end, including the differential and the axle (Fig. 28-2). All these will be explained in later sections, as will gearing, shafts, torque converters, and the other parts that make up the power train. The power train must carry the power from the engine to the wheels. In addition to this, it must disconnect the wheels from the engine when the car is not moving—for instance, when the engine is being started. Also, the power train must provide several gear ratios.

Fig. 28-1. The power is carried from the engine to the wheels by belts and pulleys.

§ 28-2 VARYING GEAR RATIOS

There is a good reason for varying the gear ratios. The automobile engine will not produce very much power when it is turning at low speed. It must be running at 2,000 or 3,000 revolutions per minute (rpm) to produce anywhere near its maximum power. It takes considerable power to start a tractor or automobile moving and to achieve normal operating speed. Therefore, varying gear ratios are provided in the power train to permit the engine to turn at high speeds while the car wheels turn at low speed. This allows the engine to produce considerable power for starting the vehicle in motion and for **acceleration.** (Acceleration is an increase in speed.)

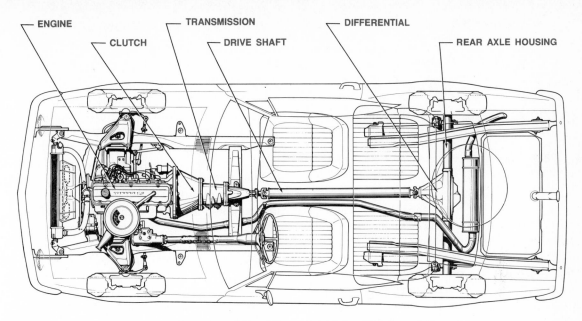

ENGINE — CLUTCH — TRANSMISSION — DRIVE SHAFT — DIFFERENTIAL — REAR AXLE HOUSING

Fig. 28-2. The power train in an automobile. (*Chevrolet Motor Division of General Motors Corporation*)

After the vehicle is moving, it takes less power to keep it in steady motion. The power train therefore changes the gear ratio between the engine and the wheels so that the engine must turn fewer revolutions for each wheel revolution. We discuss more about gear ratios and how they are varied in § 28-12. As already noted, the power train in the automobile consists of the clutch (on some cars), the transmission, the drive line, the differential, and the wheel axles (Fig. 28-2). In this chapter, we will describe the purposes and some types of these devices. Let us begin by discussing the clutch.

§ 28-3 AUTOMOTIVE CLUTCHES

The clutch is the assembly that allows power to flow from the engine to the transmission. At one time, almost all cars had clutches. But now, most cars are made with automatic transmissions and do not need clutches. However, many cars on the road today are equipped with manually shifted transmissions and clutches, so you should know about them.

§ 28-4 PURPOSES OF THE CLUTCH

The clutch is used on cars with transmissions that are shifted by hand. Its purpose is to allow the driver to couple the engine to or uncouple the engine from the transmission. The driver operates the clutch with the left foot. When the clutch is applied (in the normal running position), the power from the engine can flow through the clutch and enter the transmission. When the clutch is released by foot pressure of the driver, the engine is uncoupled from the transmission. Now, no power can flow through. It is necessary to interrupt the flow of power—to uncouple the engine—in order to shift gears.

§ 28-5 LOCATION OF THE CLUTCH

The clutch is located just behind the engine, between the engine and the transmission (Fig. 28-3). Figure 28-4 shows the parts of one type of clutch, detached from the engine. When the parts are assembled, the flywheel is bolted to

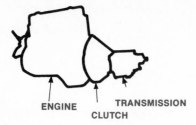

Fig. 28-3. The clutch is located between the engine and the transmission.

the crankshaft. The friction disk (also called the driven plate) is installed next. It is mounted on the end of the transmission shaft, which sticks through the clutch housing. The pressure-plate-and-cover assembly is then bolted to the flywheel. Then, the other parts are installed, as we will explain next.

§ 28-6 CONSTRUCTION OF THE CLUTCH

There are different kinds of clutches, such as the coil-spring clutch and the diaphragm-spring clutch. We start our discussion of clutch construction by looking at one of the most common types, the coil-spring clutch. This clutch has a series of coil springs set in a circle. Figures 28-5 and 28-6 show sectional and cutaway views of this clutch. The friction disk (or driven plate) is about 1 foot [0.31 meter] in diameter, and it is mounted on the transmission input shaft. The disk has splines in its hub that match the splines on the input shaft. These splines consist of two sets of teeth. The internal teeth in the hub of the friction disk match the external teeth on the shaft. When the friction disk is driven, it turns the transmission input shaft.

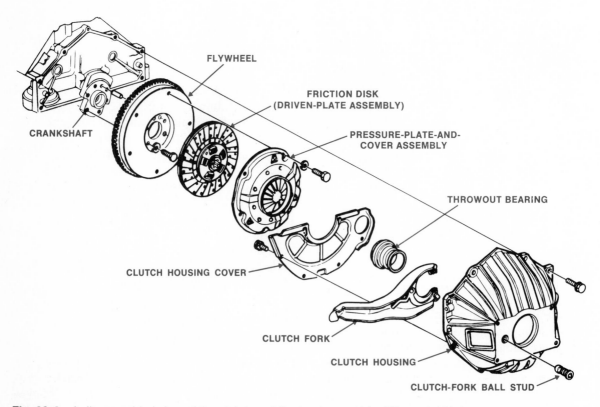

Fig. 28-4. A disassembled view of the clutch and flywheel assembly. (*Chevrolet Motor Division of General Motors Corporation*)

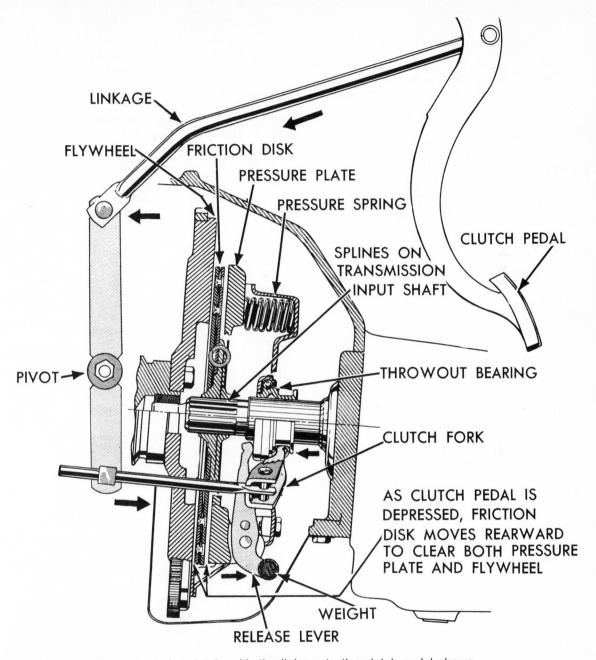

LINKAGE

FLYWHEEL   FRICTION DISK

PRESSURE PLATE

PRESSURE SPRING

CLUTCH PEDAL

SPLINES ON
TRANSMISSION
INPUT SHAFT

PIVOT

THROWOUT BEARING

CLUTCH FORK

AS CLUTCH PEDAL IS
DEPRESSED, FRICTION
DISK MOVES REARWARD
TO CLEAR BOTH PRESSURE
PLATE AND FLYWHEEL

WEIGHT

RELEASE LEVER

Fig. 28-5. Sectional view of a clutch, with the linkage to the clutch pedal shown schematically. (*Buick Motor Division of General Motors Corporation*)

The clutch also has a pressure-plate assembly, which includes a series of coil springs. The pressure-plate-and-cover assembly is attached to the engine flywheel (Fig. 28-4). The springs provide the pressure to hold the friction disk against the flywheel. Then, when the flywheel turns, the pressure plate and the friction disk also turn. When the clutch is released, the spring pressure is relieved so that the friction disk and the flywheel can rotate separately.

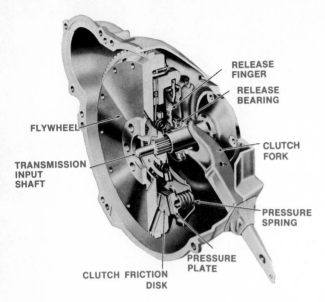

FLYWHEEL

TRANSMISSION
INPUT
SHAFT

RELEASE
FINGER

RELEASE
BEARING

CLUTCH
FORK

PRESSURE
SPRING

CLUTCH FRICTION
DISK

PRESSURE
PLATE

Fig. 28-6. A partial cutaway view of a typical clutch. (*Ford Motor Company*)

§ 28-7 OPERATION OF THE CLUTCH

Let's look at the operation of the coil-spring clutch. Figure 28-7 shows, at the left, a sectional view of the clutch. The major parts are

shown disassembled at the right. In the assembly, nine springs are used, although only three are shown at the right. Look at the assembled view at the left. Note that the springs are held between the clutch cover and the pressure plate. In the condition shown in Fig. 28-7, the springs are clamping the friction disk (driven plate) tightly between the flywheel and the pressure plate. As we mentioned, this forces the friction disk to rotate with the flywheel. In other words, the clutch is engaged.

Now, look at what happens when the driver operates the clutch pedal to disengage, or release, the clutch (see Fig. 28-8). When the clutch is released, the linkage from the pedal forces the release bearing inward (to the left in Fig. 28-8). Note that the release bearing is also called the throwout bearing. We will get to the linkage in § 28-8. First, let's see what happens when the release bearing is forced to the left.

As the release bearing moves to the left, it pushes against the inner ends of three release levers (also called release fingers). The release levers are pivoted on eyebolts, as shown in Figs. 28-5, 28-6, and 28-8. When the inner ends of the release levers are pushed in by the release bearing, the outer ends are moved to the

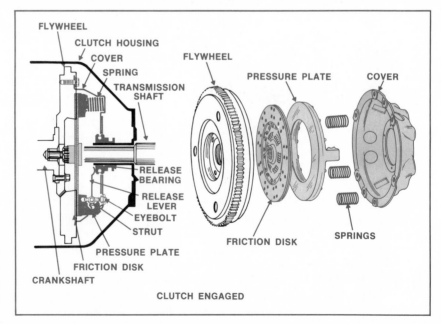

FLYWHEEL

CLUTCH HOUSING

COVER

SPRING

TRANSMISSION
SHAFT

FLYWHEEL

PRESSURE PLATE

COVER

RELEASE
BEARING

RELEASE
LEVER

EYEBOLT

STRUT

PRESSURE PLATE

FRICTION DISK

CRANKSHAFT

FRICTION DISK

SPRINGS

CLUTCH ENGAGED

Fig. 28-7. A sectional view of the clutch in the engaged position at the left. Major clutch parts are shown to the right.

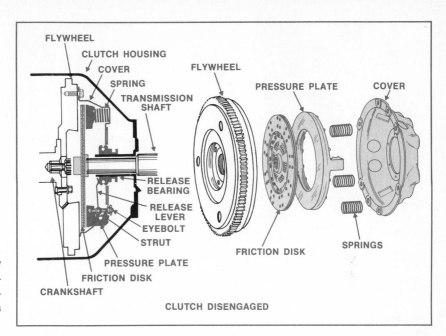

Fig. 28-8. A sectional view of the clutch in the disengaged, or released, position at the left. Major parts are shown to the right.

Labels (left view): FLYWHEEL, CLUTCH HOUSING, COVER, SPRING, TRANSMISSION SHAFT, RELEASE BEARING, RELEASE LEVER, EYEBOLT, STRUT, PRESSURE PLATE, FRICTION DISK, CRANKSHAFT

Labels (right view): FLYWHEEL, PRESSURE PLATE, COVER, FRICTION DISK, SPRINGS, CLUTCH DISENGAGED

right. This motion is carried by struts to the pressure plate (see Fig. 28-9). The pressure plate is thus moved to the right (in Fig. 28-8), and the springs are compressed. With the spring pressure off the friction disk, space appears between the disk, the flywheel, and the pressure plate. Now, the clutch is released, and the flywheel can rotate without sending power through the friction disk.

Releasing the clutch pedal takes the pressure off the release bearing. The springs push the pressure plate to the left (in Fig. 28-7). The friction disk is again clamped tightly between the flywheel and the pressure plate. The friction disk must again rotate with the flywheel. In other words, the clutch is engaged. Figure 28-10 shows, in simplified view, the applied and released position of the friction disk and pressure plate.

§ 28-8 CLUTCH LINKAGE

The only purpose of the clutch linkage is to carry the movement of the clutch pedal to the release bearing. A variety of clutch linkages are used. One of the simplest arrangements is shown in Fig. 28-5. Here, pushing down on the clutch pedal pushes the linkage, as shown by the arrows. This action causes the clutch fork to pivot and to force the release (throwout) bearing in (at the left in Fig. 28-5). A typical clutch-fork assembly, with the release (throwout) bearing, is shown in Fig. 28-11. Note that the clutch fork has a spring-held dust seal to prevent dust from entering the clutch.

§ 28-9 FRICTION DISK

The friction disk, or driven plate, is shown partly cut away in Fig. 28-12. It consists of a hub and a plate, with facings attached to the plate. The friction disk has cushion springs and dampening springs. The cushion springs are waved, or curled, slightly. The cushion springs are attached to the plate, and the friction facings are attached to the springs. When the clutch is engaged, the springs compress slightly to take up the shock of engagement. The dampening springs are spiral springs set in a circle around the hub. The hub is driven through these springs. They help to smooth out the power pulses from the engine, the result of which is that the power flow to the transmission is smooth.

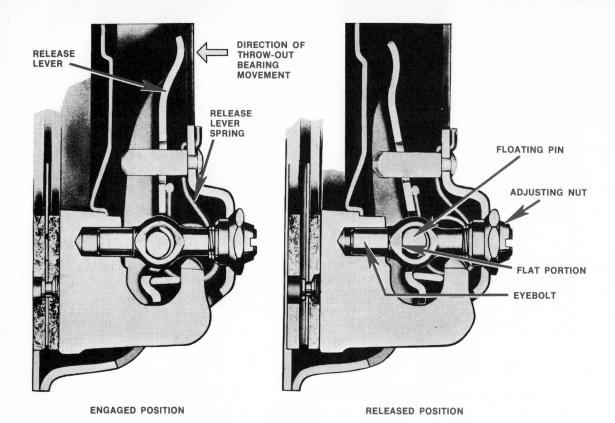

ENGAGED POSITION

RELEASED POSITION

Fig. 28-9. The two limiting positions of the pressure plate and the release lever. (*Oldsmobile Division of General Motors Corporation*)

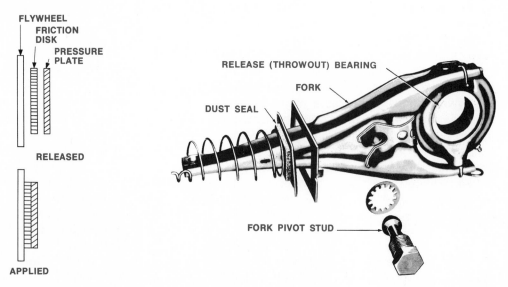

Fig. 28-10. Applied and released positions of friction disk and pressure plate.

Fig. 28-11. A clutch-fork assembly with release (throwout) bearing.

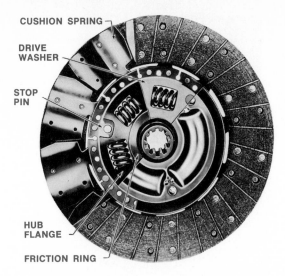

Fig. 28-12. A friction disk, or driven plate. Facings and drive washer have been cut away to show springs. (*Buick Motor Division of General Motors Corporation*)

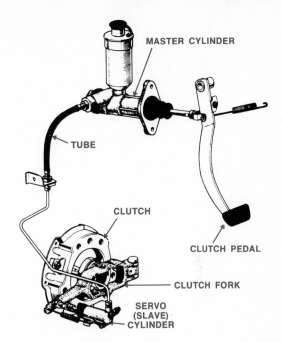

Fig. 28-13. A hydraulically operated coil-spring clutch. (*Toyota Motor Sales, Limited*)

§ 28-10 HYDRAULIC CLUTCH

The hydraulic clutch is used in vehicles in which the clutch is located far from the foot pedal. Instead of trying to install a complicated linkage between the clutch pedal and the clutch, car manufacturers use a hydraulic device. Figure 28-13 shows the clutch, the hydraulic cylinders, and the clutch pedal. This arrangement is also used on heavy-duty clutches, with high clutch-spring pressure. High clutch-spring pressure means high clutch-pedal pressure. To reduce this pedal pressure, the hydraulic clutch can be used.

Now, let's see how the hydraulic clutch works. When the driver pushes down on the clutch pedal, a pushrod is forced into a master cylinder. As the pushrod moves into the master cylinder, the rod forces a piston into the cylinder. This action puts pressure on the fluid in the cylinder, and some of the fluid is forced out. The fluid flows through a tube and into a servo cylinder at the clutch. The servo cylinder also has a piston. The fluid, flowing into the servo cylinder from the master cylinder, forces the piston in the servo cylinder

to move. This movement is carried through a pushrod to the release lever, thus releasing the clutch.

§ 28-11 MANUAL TRANSMISSIONS

Manual transmissions are transmissions that are shifted by hand. As you know, a car needs some sort of transmission. The engine should be turning fairly fast and producing considerable power when the car is first starting out. Therefore, the engine must be turning fast when the car wheels are turning slowly. Later, when the car is out on the highway, the engine is turning fast, but the wheels are also turning fast. The transmission takes care of this changing gear or speed ratio, as we will explain on following pages. As we mentioned, the manual transmission is shifted by hand. That is, the driver selects the gear ratio by hand. Most cars now come equipped with automatic transmissions. These transmissions do the job automatically. We describe automatics in Chapter 29.

Fig. 28-14. A spur gear.

Fig. 28-16. Two meshing spur gears of different sizes.

§ 28-12 GEARS

Before we get into transmissions, let's look at gears. There are many types of gears used on the automobile. All are basically similar. They all have teeth of one sort or another that mesh to carry motion from one gear to another. The simplest gear is the spur gear (Fig. 28-14). It is like a wheel with teeth. Two spur gears are shown meshed in Fig. 28-15. By "meshed" we mean that the teeth of one gear are fitted into the teeth of the other gear. When one gear rotates, the other gear also rotates.

The sizes of the two gears determine the relative speed with which they turn. The big gear turns more slowly than the small gear. But the smaller gear has greater turning force. This turning force is called **torque.** When speed is lost through gears, torque is gained.

Suppose, as a simple example, that the car is in low gear and the engine crankshaft is turning 10 times to make the car wheels turn once. This gear reduction is achieved both in the transmission and at the rear axle, in the differential (the driving mechanism at the rear

wheels). We explain the differential in later sections of this chapter. With the 10-to-1 gear reduction, there is a great increase in torque. This increase in torque gives the car the capability it needs to pull away from the curb and to accelerate rapidly.

Figure 28-16 shows two meshing spur gears, one with 12 teeth and the other with 24. The larger gear turns only half as fast as the smaller gear. In other words, while the large gear is making one complete revolution, the smaller gear is making two revolutions. If the larger gear is driving the smaller gear, there is a speed increase. There is also a torque reduction. The smaller gear turns faster, but it has less torque. If the smaller gear is driving the larger gear, there is a speed reduction but a torque increase. Let's see how these characteristics of gears are used in transmissions.

§ 28-13 FUNCTION OF THE TRANSMISSION

The typical manual transmission used on passenger cars has three forward gear ratios between the engine and the car wheels. The crankshaft must revolve about 12, 8, or 4 times to turn the car wheels once.

Fig. 28-15. Two spur gears with their teeth meshed.

In low gear, the crankshaft turns about 12 times for each car-wheel rotation. This ratio allows the engine to develop a great deal of power and torque to get the car moving. Then, in second gear, the crankshaft turns about eight times to turn the car wheels once. The car is moving, and less torque is needed for acceleration. In second gear the engine turns at a fairly high speed to produce a car speed of up to about 30 mph (miles per hour) [48.28 km/h].

In high gear, the engine crankshaft turns about four times to turn the car wheels once. While the car is in motion and in gear, normally it is not necessary to shift to lower gears. However, if additional torque is required, as when climbing a steep hill, the transmission can be shifted to a lower gear to get the higher torque.

§ 28-14 TYPES OF MANUAL TRANSMISSIONS

The most common manual transmission is the three-forward-speed unit, mentioned above and described in detail later. Many cars use a four-forward-speed transmission. In addition, heavy-duty trucks use transmissions with additional forward-speed gear positions. Some trucks have as many as 10 forward speeds.

§ 28-15 MANUAL-TRANSMISSION OPERATION

Let's use a simplified version of a standard three-speed transmission to discuss transmission gears and their operation (Fig. 28-17). In the drawing, the transmission is represented by three shafts and eight gears of various sizes. Only the moving parts are shown. The transmission housing and bearings are not shown.

Four of the gears are rigidly connected to the countershaft (Fig. 28-17). These are the driven gear, second-speed gear, first-speed gear, and reverse gear. When the clutch is engaged and the engine is running, the clutch-shaft gear drives the countershaft

driven gear. This turns the countershaft and the other gears on the countershaft. The countershaft rotates in a direction opposite, or counter, to the rotation of the clutch-shaft gear. With the gears in neutral, as shown in Fig. 28-17, and the car stationary, the transmission main shaft is not turning.

The transmission main shaft is mechanically connected by shafts and gears in the differential to the car wheels. The two gears on the transmission main shaft may be shifted back and forth along the splines on the shaft. This is done by operation of the gearshift lever in the driving compartment. The splines are matching internal and external teeth. These permit endwise movement of the gears but cause the gears and shaft to rotate together. Note in the illustrations that a floor-type shift lever is shown. This type of lever is shown since it illustrates more clearly the lever action in shifting gears. The transmission action is the same, regardless of whether a floor-type shift lever or a steering-column shift lever is used.

First gear. When the gearshift lever is operated to place the gears in *first* (Fig. 28-18), the large gear on the transmission main shaft is moved along the shaft until the large gear meshes with the small gear on the countershaft. The clutch is disengaged for this operation so that the clutch shaft and the countershaft stop rotating. When the clutch is again engaged, the transmission main shaft rotates as the driving gear on the clutch shaft drives it through the countershaft. Since the countershaft is turning more slowly than the clutch shaft, and since the small countershaft gear is engaged with the large transmission mainshaft gear, a gear reduction of approximately $3:1$ is achieved. That is, the clutch shaft turns three times for each revolution of the transmission main shaft. Further gear reduction in the differential at the rear wheels produces a still higher gear ratio (approximately $12:1$) between the engine crankshaft and the wheels.

Second gear. When the clutch is operated and the gearshift lever is moved to *second*

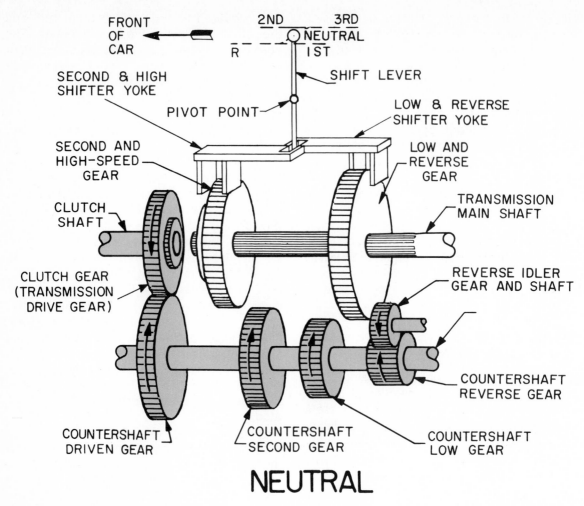

FRONT OF CAR

2ND 3RD
R NEUTRAL
1ST

SHIFT LEVER

SECOND & HIGH SHIFTER YOKE

PIVOT POINT

LOW & REVERSE SHIFTER YOKE

SECOND AND HIGH-SPEED GEAR

CLUTCH SHAFT

LOW AND REVERSE GEAR

TRANSMISSION MAIN SHAFT

CLUTCH GEAR (TRANSMISSION DRIVE GEAR)

REVERSE IDLER GEAR AND SHAFT

COUNTERSHAFT REVERSE GEAR

COUNTERSHAFT DRIVEN GEAR

COUNTERSHAFT SECOND GEAR

COUNTERSHAFT LOW GEAR

NEUTRAL

Fig. 28-17. Transmission: gears in neutral.

(Fig. 28-19), the large gear on the transmission main shaft demeshes from the small first-speed countershaft gear. The smaller transmission main-shaft gear is slid into mesh with the large second-speed countershaft gear. This meshing of the transmission main-shaft gear and the second-speed countershaft gear provides a somewhat reduced gear ratio. Now, the engine crankshaft turns only about twice to the transmission main shaft's once. The differential gear reduction increases this gear ratio to approximately 8:1.

Third gear. When the gears are shifted into *third* (Fig. 28-20), the two gears on the trans-

mission main shaft are demeshed from the countershaft gears. The second-and-third-speed gear is forced against the clutch-shaft gear. External teeth on the clutch-shaft gear mesh with internal teeth in the second-and-third-speed gear. Now, the transmission main shaft turns with the clutch shaft, and a ratio of 1:1 is obtained. Further gear reduction in the differential at the rear wheels produces a gear ratio of about 4:1 between the engine crankshaft and the wheels.

Reverse gear. When the gears are placed in *reverse* (Fig. 28-21), the larger of the transmission main-shaft gears is meshed with the

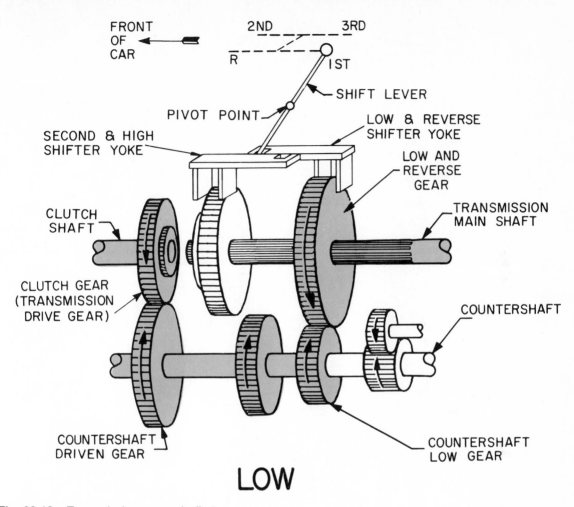

Fig. 28-18. Transmission: gears in first.

reverse idler gear. This reverse idler gear is always in mesh with the small reverse gear on the end of the countershaft. Placing the idler gear between the countershaft reverse gear and the transmission main-shaft gear causes the transmission shaft to be rotated in the opposite direction, or in the same direction as the countershaft. This reverses the rotation of the wheels so that the car backs up.

While the description above outlines the basic principles of all transmissions, somewhat more complex transmissions are used on modern cars. These include helical gears and synchromesh devices that serve to syn-

chronize the rotation of gears that are about to be meshed. This synchronization of gears eliminates clashing of the gears and makes gear shifting easier.

§ 28-16 SYNCHRONIZERS

Whenever the car is rolling, the transmission main shaft is turning and the clutch gear is spinning. Even though the clutch is disengaged, the clutch gear continues to spin until friction slows it down and stops it. Thus, when the transmission is shifted into second or third, the driver is trying to mesh gears that may be

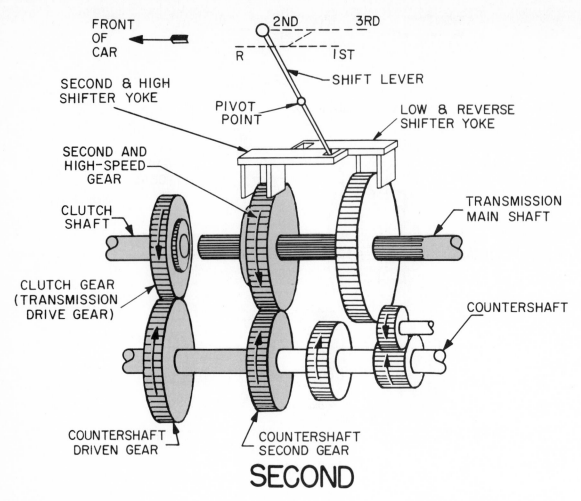

FRONT
OF
CAR

2ND   3RD

R   1ST

SHIFT LEVER

SECOND & HIGH
SHIFTER YOKE

PIVOT
POINT

LOW & REVERSE
SHIFTER YOKE

SECOND AND
HIGH-SPEED
GEAR

CLUTCH
SHAFT

TRANSMISSION
MAIN SHAFT

CLUTCH GEAR
(TRANSMISSION
DRIVE GEAR)

COUNTERSHAFT

COUNTERSHAFT
DRIVEN GEAR

COUNTERSHAFT
SECOND GEAR

SECOND

Fig. 28-19. Transmission: gears in second.

moving at different speeds. To avoid broken or damaged teeth and to make shifting easy, synchronizing devices are used in transmissions. These devices make the gears rotate at the same speed when they are about to mesh.

One type of synchronizer, shown in sectional view in Fig. 28-22, has a pair of synchronizing cones. One is an outside cone on the gear, and the other is an inside cone on the sliding sleeve. The sliding sleeve has splines that engage the splines on the gear to produce meshing. The picture at the left, labeled "neutral position," shows the general construction of the synchronizing device. The picture at the upper right shows what happens at the mo-

ment the drum, or sliding clutch sleeve, touches the gear. That is the moment when the two cones come into contact.

When contact is made, the gear and the drum are brought into synchronization. That is, they revolve at the same speed. Now, further movement of the shift lever moves the sliding sleeve into mesh with the gear. The splines on the sliding sleeve and the splines on the gear engage, as shown at the lower right. Lockup is completed. When the clutch is engaged, power can flow through the sliding sleeve and the gear.

There are other kinds of synchronizing devices. But they all work in the same general

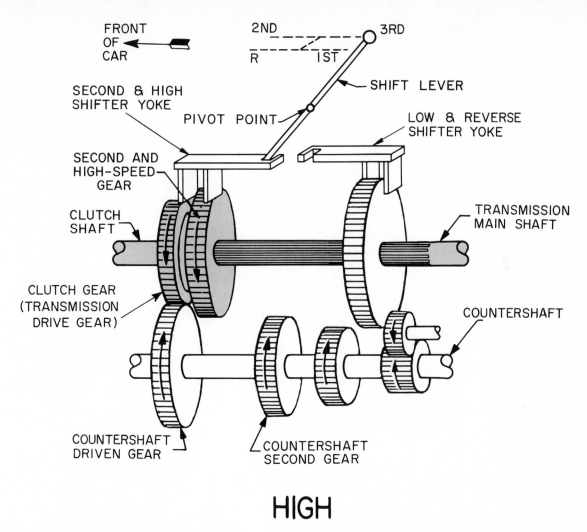

FRONT OF CAR ←

2ND — — — — 3RD
R   1ST

SHIFT LEVER

SECOND & HIGH
SHIFTER YOKE

PIVOT POINT

LOW & REVERSE
SHIFTER YOKE

SECOND AND
HIGH-SPEED
GEAR

CLUTCH
SHAFT

TRANSMISSION
MAIN SHAFT

CLUTCH GEAR
(TRANSMISSION
DRIVE GEAR)

COUNTERSHAFT

COUNTERSHAFT
DRIVEN GEAR

COUNTERSHAFT
SECOND GEAR

HIGH

Fig. 28-20. Transmission: gears in third.

manner. Whenever two gears are about to be meshed, the synchronizer brings them into synchronization. The teeth, or splines, that are about to engage are moving at the same speed, so no clash can occur.

§ 28-17 THREE-SPEED, FULLY SYNCHRONIZED TRANSMISSION

The three-speed, fully synchronized transmission has full synchronization in first and reverse as well as in second and high. Figure 28-23 shows the gears in the fully synchronized transmission in neutral and in the various gear positions. Notice that the fully synchronized transmission has a sliding clutch sleeve for the first-and-reverse gear positions. It has a second sliding clutch sleeve for the second-and-third gear positions. Study this illustration to trace the flow of power through the transmission in the four gear positions. The sliding clutch sleeve is called a synchronizer assembly in this picture.

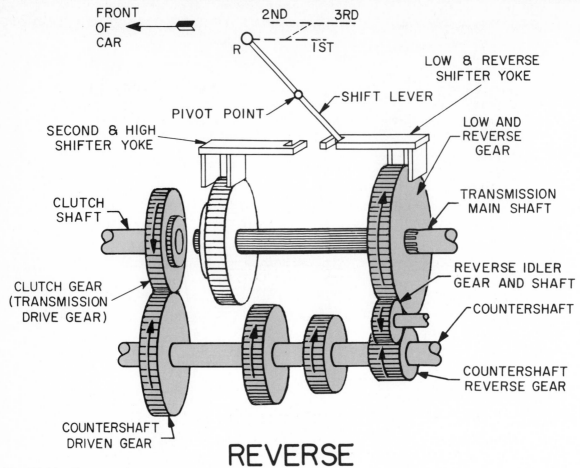

FRONT OF CAR ←

2ND   3RD
R   1ST

LOW & REVERSE SHIFTER YOKE

SHIFT LEVER

PIVOT POINT

LOW AND REVERSE GEAR

SECOND & HIGH SHIFTER YOKE

CLUTCH SHAFT

TRANSMISSION MAIN SHAFT

CLUTCH GEAR (TRANSMISSION DRIVE GEAR)

REVERSE IDLER GEAR AND SHAFT

COUNTERSHAFT

COUNTERSHAFT DRIVEN GEAR

COUNTERSHAFT REVERSE GEAR

REVERSE

Fig. 28-21. Transmission: gears in reverse.

§ 28-18 FOUR-SPEED AND FIVE-SPEED TRANSMISSIONS

Many cars are now equipped with four-forward-speed or five-forward-speed transmissions. The additional forward speeds give the car more flexibility. A popular arrangement is a four-speed transmission with a floor-mounted shift lever. This is called **four on the floor.**

In the five-speed transmissions, fifth gear is overdrive. That is, the transmission mainshift turns faster than, or overdrives, the clutch shaft. This allows the car to go faster while the engine turns slower. As a result, improved mileage is claimed.

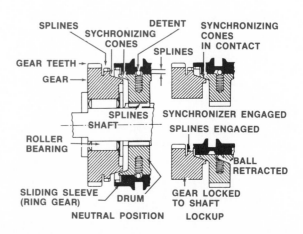

SPLINES
SYCHRONIZING CONES
DETENT
SYNCHRONIZING CONES IN CONTACT

GEAR TEETH
GEAR
SPLINES

SPLINES
SHAFT
SYNCHRONIZER ENGAGED

ROLLER BEARING
SPLINES ENGAGED

BALL RETRACTED

SLIDING SLEEVE (RING GEAR)
DRUM
GEAR LOCKED TO SHAFT

NEUTRAL POSITION
LOCKUP

Fig. 28-22. The operation of a transmission synchronizer using cones.

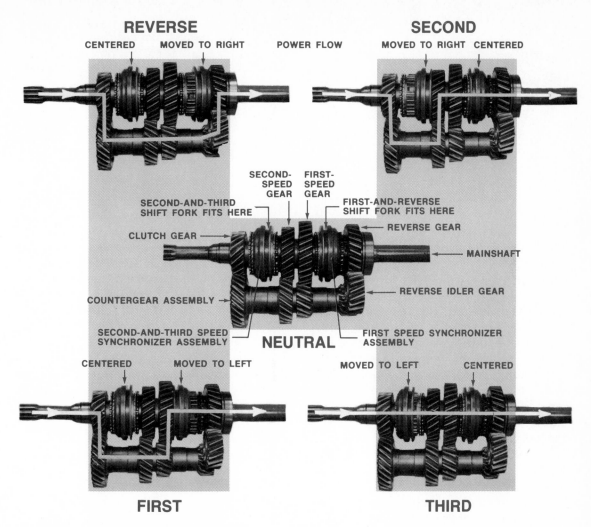

Fig. 28-23. Positions of synchronizer assemblies and power flow through a three-speed, fully synchronized transmission.

§ 28-19 GEARSHIFT LEVERS AND LINKAGES

Gearshift levers for transmissions are located either on the steering column or on the floor. Figure 28-24 shows the shifting patterns for the two arrangements. Let's take a closer look at the column shift.

To shift into first, the driver pushes down on the clutch pedal to disconnect the transmission momentarily from the engine. Then the driver lifts the shift lever and moves it back for first gear. When the lever is lifted, it pivots on

its mounting pin and pushes down on the linkage rod in the steering column. This downward movement pushes the crossover blade at the bottom of the steering column. You can see this crossover blade in Fig. 28-25. When the crossover blade is pushed down, a slot in the blade engages a pin in the first-and-reverse shift lever (Fig. 28-25). Now, when the shift lever is moved into first, the first-and-reverse lever is rotated.

To see what happens next, look at Fig. 28-26. The movement of the first-and-reverse lever on the steering column is carried to the trans-

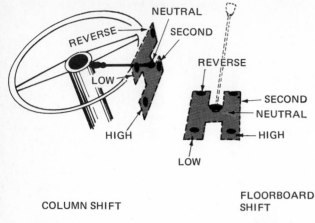

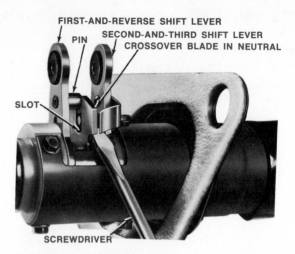

FIRST-AND-REVERSE SHIFT LEVER
SECOND-AND-THIRD SHIFT LEVER
PIN
CROSSOVER BLADE IN NEUTRAL
SLOT
SCREWDRIVER

NEUTRAL
REVERSE
SECOND
REVERSE
LOW
SECOND
NEUTRAL
HIGH
HIGH
LOW
COLUMN SHIFT
FLOORBOARD SHIFT

Fig. 28-24. Gearshift patterns for steering-column and floorboard shift levers.

Fig. 28-25. Shift levers and crossover blade at the bottom of the steering column. The screwdriver is shown holding the crossover blade in neutral for an adjustment check. (*Chrysler Corporation*)

mission by a linkage rod. The first-and-reverse lever on the transmission is rotated. This moves the first-and-reverse shift fork inside the transmission so that the first-and-reverse gear, or synchronizing drum, is moved. Figure 28-27 shows the first-and-reverse shift fork and the second-and-third shift fork.

If the shift is being made into second or third, the second-and-third shift lever at the bottom of the steering column is moved. This motion moves the second-and-third lever on the transmission. The second-and-third shift fork then moves the sliding clutch sleeve, or synchronizing drum, to shift into second or high.

Study Figs. 28-24, 28-25, and 28-26 to see how the linkage works. There are various kinds and arrangements of linkages in different cars. The floor-mounted gearshift lever, for instance, has the type of linkage shown in Fig. 28-28.

§ 28-20 DRIVE LINES

Drive lines, also called propeller shafts, carry the power from the engine and transmission to the car wheels. In most automobiles, the engine is at the front, and the rear wheels are driven. This setup means that a long drive line is required to carry the power from the engine to the rear wheels. However, when the engine is at the front and the front wheels are driven

(front-drive cars), only a short drive shaft is required to carry the power from the engine to the front wheels. Cars that have engines at the rear with rear drive (rear-drive cars) require only a short drive line to carry the power from the engine to the rear wheels.

In the following sections, we will look mainly at the standard arrangement—front-mounted engine and rear wheels driven.

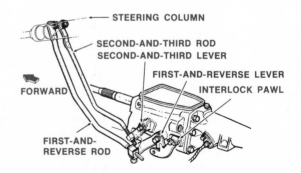

STEERING COLUMN
SECOND-AND-THIRD ROD
SECOND-AND-THIRD LEVER
FIRST-AND-REVERSE LEVER
INTERLOCK PAWL
FORWARD
FIRST-AND-REVERSE ROD

Fig. 28-26. The gearshift linkage between the shift levers at the bottom of the steering column and the transmission lever on the side of the transmission. (*Chrysler Corporation*)

370

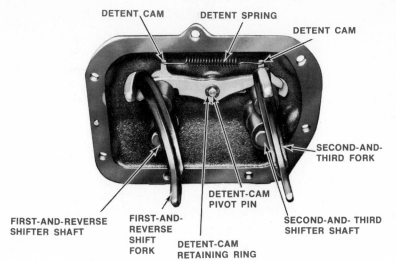

DETENT CAM   DETENT SPRING   DETENT CAM

FIRST-AND-REVERSE
SHIFTER SHAFT

FIRST-AND-
REVERSE
SHIFT
FORK

DETENT-CAM
RETAINING RING

DETENT-CAM
PIVOT PIN

SECOND-AND-
THIRD FORK

SECOND-AND- THIRD
SHIFTER SHAFT

Fig. 28-27. A transmission side cover, viewed from inside the transmission. The shift forks are mounted on the ends of levers attached to shafts. The shafts can rotate in the side cover. The detent cams and springs prevent more than one of the shift forks from moving at any one time. (*Chevrolet Motor Division of General Motors Corporation*)

Fig. 28-28. The gearshift linkage arrangement of a transmission with a floor-mounted shift lever. (*Pontiac Motor Division of General Motors Corporation*)

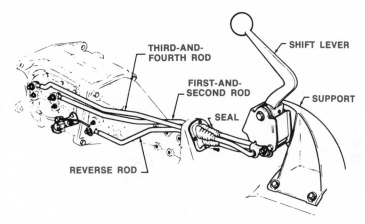

THIRD-AND-
FOURTH ROD

FIRST-AND-
SECOND ROD

SEAL

REVERSE ROD

SHIFT LEVER

SUPPORT

§28-21 CONSTRUCTION OF THE DRIVE LINE

The drive line connects the transmission output shaft to the differential at the wheel axles. The transmission and the engine are more or less rigidly attached to the car frame. But the rear wheels are attached to the car frame by suspension springs. Thus, the rear wheels move up and down, which means the following two things:

■ The drive line must change length as the wheels move up and down.

■ The angle of drive must change as the wheels move up and down.

Figure 28-29 shows how the length of the drive line and the angle of drive change as the wheels move up and down. In the top part of the picture, the wheels and differential are in the up position, which means that the drive angle is small. Also, the drive line is at its maximum length. In the bottom part of the picture, the differential and wheels are in the low position. This is their position when the wheels drop into a depression in the road. In this position, the drive angle is increased. Also, the

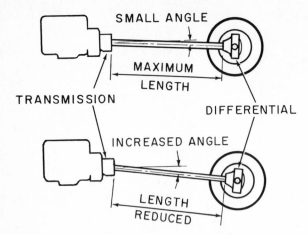

Fig. 28-29. As a rear-axle housing, with the differential and wheels, moves up and down, the angle between the transmission output shaft changes, and the length of the drive shaft also changes. The reason the drive shaft shortens as the angle increases is that the rear axle and differential move in a shorter arc than the drive shaft.

drive-line length is reduced, because as the rear wheels and differential swing down, they also move forward. The rear wheels and differential must move this way because they are attached to the springs.

Most propeller shafts are hollow tubes, with two or more universal joints and a slip joint. (We will describe these joints a little later.) Figure 28-30 shows one common type of propeller shaft. Some propeller shafts are the two-piece type and have a support at the center. Figure 28-31 shows one type of rear suspension and propeller shaft.

§ 28-22 UNIVERSAL JOINT

The universal joint allows driving power to be carried through two shafts that are at an angle to each other. Figure 28-32 shows a simple universal joint. It is a double-hinged joint, consisting of two Y-shaped yokes and a cross-shaped member. The cross-shaped member is called the spider. The four arms of the spider are assembled into bearings in the ends of the two yokes.

In operation, the driving shaft causes one of the yokes to rotate. This causes the spider to rotate. The spider then causes the driven yoke and shaft to rotate. When the driving and driven shafts are at an angle, the yokes swing around in the bearings on the ends of the spider arms.

A cross-and-two-yoke universal joint is shown in Fig. 28-33. It is almost the same as the simple universal joint shown in Fig. 28-32, except that the four bearings on the ends of the spider arms are needle bearings.

With the cross-and-two-yoke universal joint, there is some change in speed when the drive shaft and the driven shaft are at an angle to each other. The change in speed occurs because the driven yoke and driven shaft speed up and then slow down twice with every revolution of the drive line. The greater the angle between the drive and driven shafts, the greater the speedup-and-slowdown action. This type of action is not desirable because it results in increased wear of the affected parts of the drive line. To eliminate the speedup-and-slowdown action, constant-velocity universal joints are used on many cars.

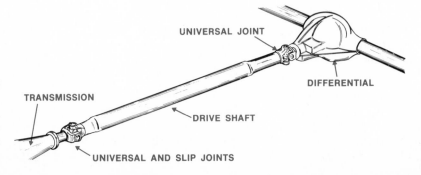

Fig. 28-30. The drive shaft connects the transmission with the differential. This is a one-piece shaft with two universal joints and one slip joint.

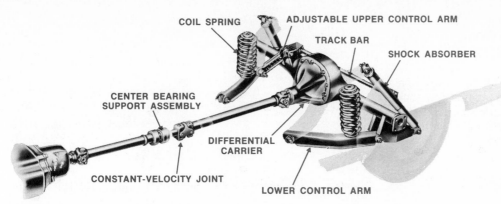

COIL SPRING   ADJUSTABLE UPPER CONTROL ARM

TRACK BAR

SHOCK ABSORBER

CENTER BEARING
SUPPORT ASSEMBLY

DIFFERENTIAL
CARRIER

CONSTANT-VELOCITY JOINT

LOWER CONTROL ARM

Fig. 28-31. One type of rear suspension and propeller shaft. (*Buick Motor Division of General Motors Corporation*)

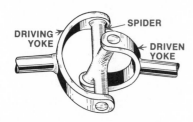

DRIVING
YOKE   SPIDER

DRIVEN
YOKE

Fig. 28-32. A simple universal joint.

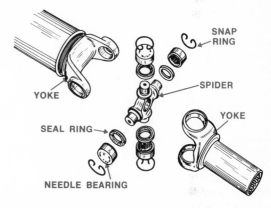

SNAP
RING

YOKE

SPIDER

YOKE

SEAL RING

NEEDLE BEARING

Fig. 28-33. A cross-and-two-yoke universal joint in disassembled view.

A constant-velocity universal joint is shown in Fig. 28-31 and in partial sectional view in Fig. 28-34. The constant-velocity universal joint consists of two simple universal joints linked by a ball and socket. The ball and socket splits the angle of the drive and driven shafts between the two universal joints of the constant-velocity unit. Because the two universal joints operate at the same angle, the speedup-and-slowdown action, which could result from the use of a single universal joint, is canceled out. In other words, the speedup resulting at any instant from the action of one universal joint is canceled out by the slowdown of the other. Thus there is no speed change through the two joints.

§ 28-23 SLIP JOINT

The change in drive-line length is taken care of by a slip joint. A slip joint is shown in Fig. 28-35. The slip joint consists of external splines

on the end of one shaft and matching internal splines on the mating hollow shaft. The splines cause the two shafts to rotate together, but permit the two to slip back and forth inside the hollow shaft. This movement allows the effective length of the propeller shaft to change as the rear axles move toward or away from the car frame. A slip joint can be seen in partial sectional view in Fig. 28-34.

§ 28-24 REAR AXLES AND DIFFERENTIALS

Power from the engine flows through the drive line to the differential. When the power arrives at the differential, it has to be split and sent

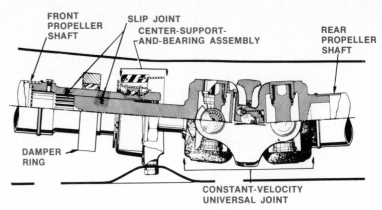

Fig. 28-34. A sectional view of a constant-velocity universal joint. (*Buick Motor Division of General Motors Corporation*)

FRONT PROPELLER SHAFT

SLIP JOINT

CENTER-SUPPORT-AND-BEARING ASSEMBLY

REAR PROPELLER SHAFT

DAMPER RING

CONSTANT-VELOCITY UNIVERSAL JOINT

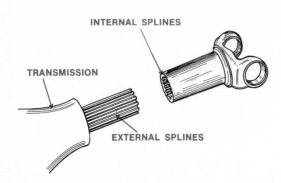

INTERNAL SPLINES

TRANSMISSION

EXTERNAL SPLINES

Fig. 28-35. The slip joint uses matching external and internal splines.

to the two rear wheels. If the car is moving in a straight line, both rear wheels travel at the same speed. But if the car is making a turn, the outer rear wheel must travel farther and faster than the inner rear wheel. The differential makes this possible.

§ 28-25 PURPOSE OF THE DIFFERENTIAL

When the car rounds a turn, the outer wheel must travel farther than the inner wheel. For example, suppose the car makes a sharp turn to the left, as shown in Fig. 28-36. The inner rear wheel, turning on a 20-foot [6.10-meter] radius, travels 31 feet [9.45 meters] during the 90-degree turn. The outer rear wheel, being nearly 5 feet [1.52 meters] from the inner wheel, turns on a $24\frac{2}{3}$-foot [7.53-meter] radius (in the

car shown), and it travels 39 feet [11.89 meters].

If the propeller shaft (drive shaft) were geared rigidly to both rear wheels, each wheel would have to skid an average of 4 feet [1.22 meters] to make the turn. On this basis, the tires would not last very long. But what is worse is that the car could not be controlled during turns. Thus, the job of the differential is to avoid these troubles by allowing one rear wheel to turn faster than the other when the car goes around a curve.

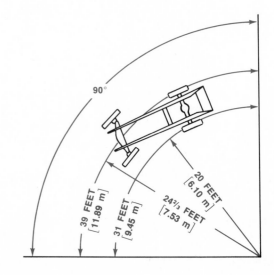

90°

39 FEET [11.89 m]

31 FEET [9.45 m]

$24\frac{2}{3}$ FEET [7.53 m]

20 FEET [6.10 m]

Fig. 28-36. The difference of wheel travel as car makes a 90-degree turn with the inner wheel turning on a 20-foot [6.10-meter] radius.

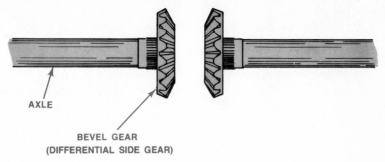

Fig. 28-37. Inner ends of the rear axles with bevel gears installed on them.

AXLE

BEVEL GEAR
(DIFFERENTIAL SIDE GEAR)

Fig. 28-38. Two meshing bevel gears.

§ 28-26 CONSTRUCTION OF THE DIFFERENTIAL

To study differential construction and operation, we will build up, gear by gear, a simple differential. The two rear wheels are mounted on axles. On the inner ends of the axles are bevel gears (see Fig. 28-37). When two bevel gears are put together so that their teeth match, one shaft can be driven by the shaft that is at a 90° angle (Fig. 28-38).

Now, let's get back to the differential. Figure 28-39 shows all the essential parts of a differential. The parts are separated so that they can be seen clearly. Keep referring to this picture as we put the parts of the differential together.

First, in Fig. 28-40, we show the addition of the differential case to the two wheel axles and bevel gears. In our discussion of differentials, we will refer to the bevel gears as differential side gears. The differential case has bearings that permit it to rotate on the two axles. Next, we add the two pinion gears and the support-

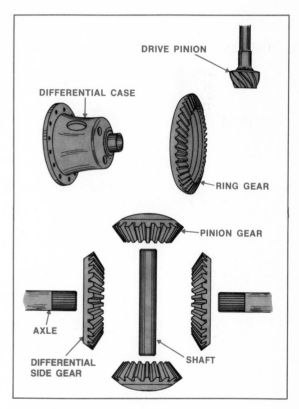

DRIVE PINION

DIFFERENTIAL CASE

RING GEAR

PINION GEAR

AXLE

DIFFERENTIAL
SIDE GEAR

SHAFT

Fig. 28-39. The basic parts of a differential.

ing shaft (Fig. 28-41). The shaft is part of the differential case. The two pinion gears are meshed with the differential side gears.

Now, we add the ring gear (Fig. 28-42). The ring gear is bolted to the flange on the differential case. Finally, we add the drive pinion (Fig. 28-43). The drive pinion is at the end of the propeller shaft (drive shaft). When the propeller shaft rotates, the drive pinion rotates, and the drive pinion rotates the ring gear.

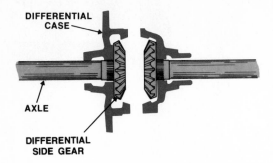

Fig. 28-40. Here, we add the differential case.

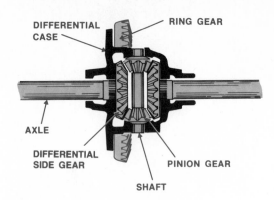

Fig. 28-42. Here, we add the ring gear.

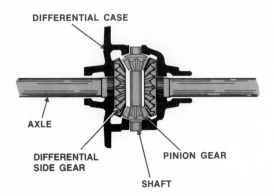

Fig. 28-41. Now, we add the two pinion gears and supporting shaft.

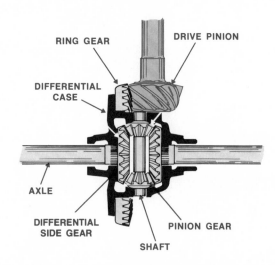

Fig. 28-43. To complete the basic differential, we now add the drive pinion. The drive pinion is meshed with the ring gear.

§ 28-27 OPERATION OF THE DIFFERENTIAL

The drive pinion on the end of the propeller shaft drives the ring gear. The rotation of the ring gear causes the differential case to rotate. When the differential case rotates, the two pinion gears and their shaft move around in a circle with the differential case. Because the two differential side gears are meshed with the pinion gears, the differential side gears must rotate, thus causing the rear axles to rotate. The wheels turn, and the car moves.

Suppose one rear wheel is held stationary. If the differential case rotates, the pinion gears would have to rotate on their shaft. The pinion gears would walk around the stationary differential side gear. As the pinion gears rotate on their shaft, they carry the rotary motion to the other differential side gear, causing it to rotate. The other wheel that is not held stationary will then rotate.

You can now see how the differential can allow one rear wheel to turn faster than the other. Whenever the car goes around a turn, the outer rear wheel travels a greater distance than the inner rear wheel. The two pinion gears rotate on their shaft and send more rotary motion to the outer wheel.

When the car is moving down a straight road, the pinion gears do not rotate on their

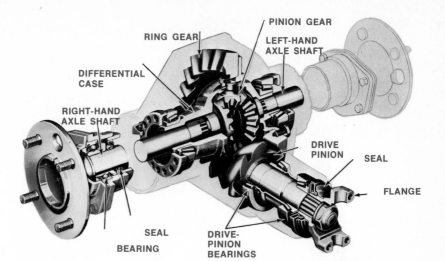

Fig. 28-44. A cutaway view of a differential and rear axle. (*Ford Motor Company*)

Labels on figure: PINION GEAR, RING GEAR, LEFT-HAND AXLE SHAFT, DIFFERENTIAL CASE, RIGHT-HAND AXLE SHAFT, DRIVE PINION, SEAL, FLANGE, SEAL, BEARING, DRIVE-PINION BEARINGS

shaft, but apply equal torque to the differential side gears. Therefore, both rear wheels rotate at the same speed.

§ 28-28 DIFFERENTIAL CONSTRUCTION

Figure 28-44 is a partial cutaway view of a differential and rear-axle assembly. As you will note, the pinions and gears are all of heavy construction to carry the power from the drive line. Figure 28-45 is a disassembled view of another differential. Notice that in Fig. 28-45, different names are used for the different gears. For instance, the drive pinion and the ring gear are called ''drive gear and pinion.'' The bevel gears are called ''side gears.'' The pinion gears are called simply ''pinions.'' Different manufacturers often use different names for the same parts in their cars. You must watch out for this in automotive work.

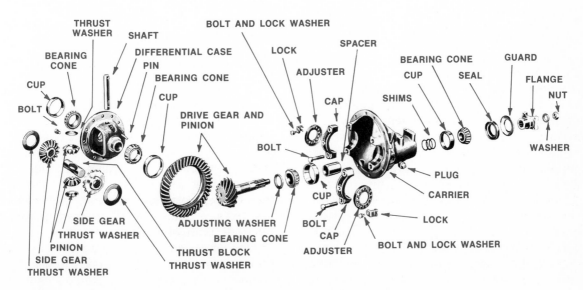

Labels on figure: THRUST WASHER, SHAFT, BOLT AND LOCK WASHER, SPACER, BEARING CONE, GUARD, BEARING CONE, DIFFERENTIAL CASE, LOCK, CUP, SEAL, FLANGE, CUP, PIN, ADJUSTER, SHIMS, NUT, BOLT, BEARING CONE, CAP, DRIVE GEAR AND PINION, BOLT, PLUG, CUP, CARRIER, SIDE GEAR, THRUST WASHER, PINION, BOLT, LOCK, SIDE GEAR THRUST WASHER, ADJUSTING WASHER, BEARING CONE, CAP, BOLT AND LOCK WASHER, THRUST BLOCK, THRUST WASHER, ADJUSTER, WASHER

Fig. 28-45. A disassembled view of a differential. (*Chrysler Corporation*)

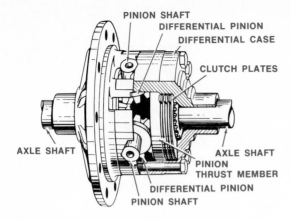

PINION SHAFT
DIFFERENTIAL PINION
DIFFERENTIAL CASE

CLUTCH PLATES

AXLE SHAFT

AXLE SHAFT
PINION
THRUST MEMBER
DIFFERENTIAL PINION
PINION SHAFT

Fig. 28-46. A cutaway view of a nonslip differential. (*Chrysler Corporation*)

Fig. 28-47. A sectional view of a nonslip differential. (*Chrysler Corporation*)

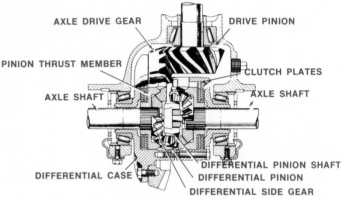

AXLE DRIVE GEAR   DRIVE PINION

PINION THRUST MEMBER   CLUTCH PLATES

AXLE SHAFT   AXLE SHAFT

DIFFERENTIAL CASE   DIFFERENTIAL PINION SHAFT
DIFFERENTIAL PINION
DIFFERENTIAL SIDE GEAR

§ 28-29 NONSLIP DIFFERENTIAL

The differential we have just studied delivers the same amount of torque to each rear wheel when both wheels have equal traction. When one wheel has less traction than the other, for example, when one wheel is slipping on ice, the other wheel cannot deliver torque. That is, all the turning effort goes to the slipping wheel. To provide good traction even though one wheel is slipping, a nonslip differential is used in many cars. It is very similar to the standard unit but has some means of preventing wheel spin and loss of traction.

One type of nonslip differential is shown in Fig. 28-46. It has two sets of clutch plates. Also, the ends of the pinion-gear shafts lie rather loosely in notches in the two halves of the differential case. Figure 28-47 is a sectional view of the nonslip differential. During normal straight-road driving, the power flow is as shown in Fig. 28-48.

Note that in Figs. 28-47, 28-48, and 28-49, the ring gear is called the "axle drive gear." The bevel gear is called the "differential side gear." The pinion gears are called the "differential pinions."

Notice that the rotating differential case carries the pinion-gear shafts around with it. Since there is considerable side thrust, the pinion shafts tend to slide up the sides of the notches in the two halves of the differential case. As the pinion shafts slide up, they are forced outward. This force is carried to the two sets of clutch plates. The clutch plates thus lock the axle shafts to the differential case. Therefore, both wheels turn.

If one wheel encounters a patch of ice or snow and loses traction, or tends to slip, the pressure is released on the clutch plates feed-

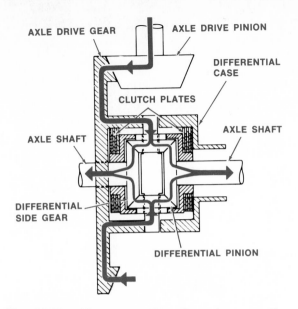

Fig. 28-48. The power flow through a nonslip differential on a straightaway.

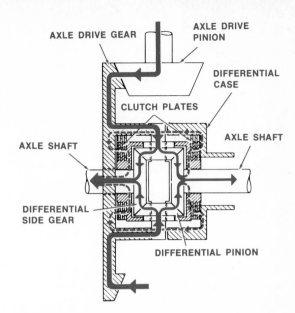

Fig. 28-49. The power flow through a nonslip differential when rounding a turn. (*Chrysler Corporation*)

ing power to that wheel. Thus the torque goes to the other wheel, and the wheel on the ice does not slip.

During normal driving, if the car rounds a turn, pressure is released on the clutch feeding the inner wheel just enough to permit some slipping. Figure 28-49 shows the action. This release of pressure permits the outer wheel to turn faster than the inner wheel.

Review questions

1. What is the name of the path through which power must go to reach the wheels?
2. Where is the pressure-plate-and-cover assembly attached?
3. When the clutch is engaged, between which parts is the friction disk clamped?
4. When the release levers are pushed in by the release bearing, what is happening to the outer ends of the release levers?
5. In the hydraulic clutch, is there direct mechanical linkage between the clutch pedal and the clutch?
6. One gear with 10 teeth is meshed with a gear with 30 teeth. Which gear will rotate faster, and how much faster will it turn?
7. If a small gear drives a big gear, is torque gained or lost?
8. What is the name of the device that allows gears to be shifted without gear clash?
9. What are the two things that the gearshift lever does when it is moved?
10. At what two places are gearshift levers located?
11. Between what two parts is the drive line connected?
12. What is the basic purpose of the differential?
13. When the car is rounding a curve, do the pinion gears turn on the shaft?
14. Does the nonslip differential have one set or two sets of clutch plates?

Study problems

1. With a car raised on a hoist, locate and name the parts of the automotive power train. On a lubrication chart, identify which of the parts of the power train require periodic lubrication.
2. On a manual transmission with the side cover removed, identify each of the gears and shift levers. Then, shift the transmission into each gear position, noting the parts that are moved by the shift lever for each gear position.
3. Obtain a drive shaft with the universal joints attached. Identify the slip joint and note how it works to compensate for changes in drive-line length. Disassemble the universal joint and identify each of the parts. Note how the universal joint is lubricated. Look for wear on any of the parts.
4. Jack up the rear end of a car and place safety stands under it. Be sure the engine is off, the transmission is in neutral, and the parking brake is released. Then try to figure out the rear axle ratio by turning the drive shaft with your hand.

29 AUTOMATIC TRANSMISSIONS

§ 29-1 HYDRAULIC TRANSMISSION OF POWER

After power is produced in an engine or other power device, it must be sent, or transmitted, to the place where it is to be used. In the automobile, for example, the power from the engine is carried through gears and shafts in the transmission, drive shaft, differential, and wheel axles to the wheels. Power in the form of electric current is transmitted from the power plant to the home, factory, office, or store where it is needed. In some special applications, power is transmitted through the flow of a fluid, as for instance in a hydroelectric power plant or in the automatic transmission, power-steering system, and hydraulic brakes used in automobiles. Actually, the automobile can be used to demonstrate nearly every kind of power transmission.

A fluid is any substance that flows easily. Therefore, it may be a gas or a liquid. Normally, a nonfoaming type of refined oil makes up the basic ingredient in a liquid fluid-power system. Fluid power then may be defined as a fluid, under pressure, transmitting power, energy, or force used to perform an operation.

In the typical hydroelectric power plant, a dam is built to hold back a big lake of water. Part of this water is then allowed to flow down through huge pipes to water turbines (Fig. 29-1). These turbines are a little like the steam and gas turbines described in earlier chapters. That is, they have blades set on a rotor so that when the water strikes the blades, it causes the rotor to spin. This rotary motion is then carried to an electric generator that produces electricity. Usually, there are several of these hydroelectric generator assemblies in the plant. Figure 29-2 shows how one of these looks on the inside.

Many of us will never see the inside of a hydroelectric power plant. However, we are all familiar with the insides of the automobile. On the car, there are three hydraulic devices that we will discuss. They are the brakes, power steering, and the torque converter used in automatic transmissions. But before we talk about these, let us look at a few fundamentals of hydraulics.

§ 29-2 HYDRAULICS

Hydraulics is the science of liquids, such as water or oil. When we study hydraulics, we learn that pressure can be exerted through liquids. We also learn that this pressure can transmit motion from one place to another. The reason for this is that liquids are incompressible. That is, they cannot be compressed to a smaller volume.

1 Incompressibility of liquids. If a gas, such as air, is put under pressure, it can be compressed into a smaller volume (Fig. 29-3). However, applying pressure to a liquid will not cause it to compress. Even under pressure, a liquid stays at its original volume.

Fig. 29-1. Modern hydroelectric power plant. (*New York Power Authority Niagara Power Project*)

2 Transmitting motion by liquid. Since liquid is not compressible, motion may be transmitted by liquid. For example, Fig. 29-4 shows two pistons in a cylinder with a liquid between them. When the applying piston is moved into the cylinder 8 inches [203.2 millimeters], as shown, the output piston will be pushed along the cylinder the same distance. In the illustration, you could substitute a solid connecting rod between piston *A* and piston *B* and get the same result. But the advantage of such a system is that you can transmit motion between cylinders by a tube (Fig. 29-5). In Fig. 29-5, as the applying piston is moved, liquid is forced out of cylinder *A,* through the tube, and into cylinder *B*. This causes the output piston to move in its cylinder.

3 Transmitting pressure by liquid. The pressure applied to a liquid is transmitted by the liquid in all directions and to every part of the liquid. For example (Fig. 29-6), when a piston with 1 square inch of area applies a force of 100 pounds on a liquid, the pressure on the liquid is 100 psi (pounds per square inch). This pressure will be registered throughout the entire hydraulic system. If the area of the piston is 2 square inches and the piston applies a force of 100 pounds, then the pressure is only 50 psi (Fig. 29-7).

In the metric system, we measure pressure in kilograms per square centimeter [kg/cm²]. For example, a pressure of 50 psi in our customary system is 3.52 kg/cm² in the metric system.

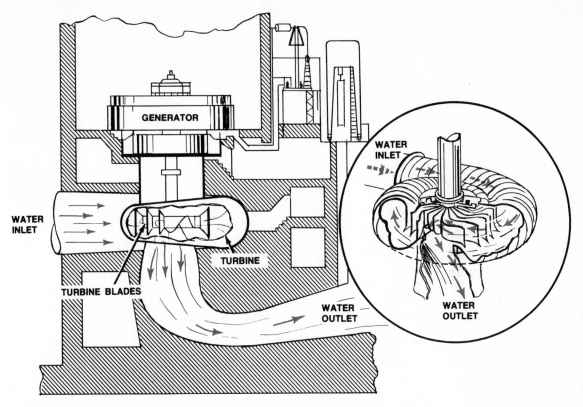

Fig. 29-2. How a hydroelectric turbine looks from the inside.

With an input-output system (Fig. 29-8), we can determine the force applied to any output piston. The force is found by multiplying the pressure in pounds per square inch by the area of the output piston in square inches. For example, the pressure shown in Fig. 29-8 is 100 psi. The output piston to the left has an area of 0.5 square inch. Thus, the output force on this piston is 100 times 0.5, or 50 pounds. The center piston has an area of 1 square inch, and its output force is therefore 100 pounds. The right-hand piston has an area of 2 square inches, and its output force is therefore 200 pounds (100 × 2). The bigger the output piston, the greater the output force. If the area of the piston were 100 square inches, for example, the output force would be 10,000 pounds. Likewise, the higher the hydraulic pressure, the greater the output force. If the hydraulic pressure on the 2-square-inch piston

GAS CAN BE COMPRESSED

LIQUID CANNOT BE COMPRESSED

Fig. 29-3. Gas can be compressed when pressure is applied; liquid, however, cannot. (*Pontiac Division of General Motors Corporation*)

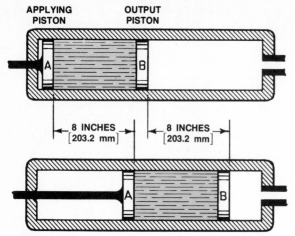

Fig. 29-4. Motion can be transmitted by liquids. When the applying piston A is moved 8 inches [203.2 millimeters], then the output piston is also moved 8 inches [203.2 millimeters]. (*Pontiac Motor Division of General Motors Corporation*)

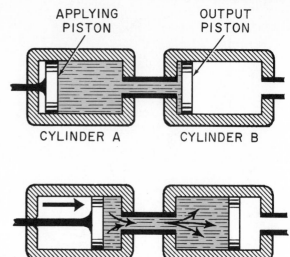

Fig. 29-5. Motion may be transmitted through a tube from one cylinder to another by hydraulic pressure. (*Pontiac Motor Division of General Motors Corporation*)

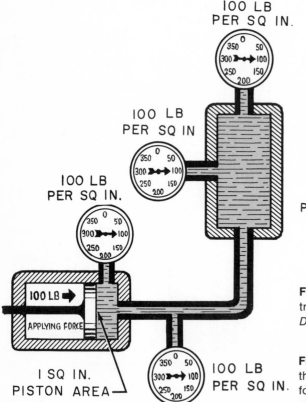

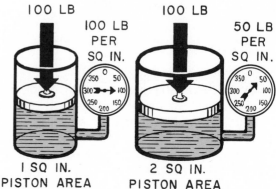

Fig. 29-6. (*Left*) Pressure applied to a liquid is transmitted equally in all directions. (*Pontiac Motor Division of General Motors Corporation*)

Fig. 29-7. (*Above*) The hydraulic pressure (psi) in the system is determined by dividing the applying force (pounds) by the area (square inches) of the applying piston. (*Pontiac Motor Division of General Motors Corporation*)

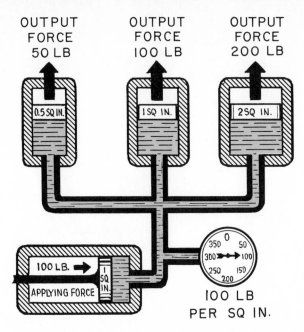

OUTPUT FORCE 50 LB   OUTPUT FORCE 100 LB   OUTPUT FORCE 200 LB

0.5 SQ IN.   1 SQ IN.   2 SQ IN.

100 LB.

APPLYING FORCE

1 SQ IN.

100 LB PER SQ IN.

Fig. 29-8. The force (pounds) applied to the output piston is the pressure in the system (psi) times the area (square inches) of the output piston. (*Pontiac Motor Division of General Motors Corporation*)

went up to 1,000 psi, then the output force on the piston would be 2,000 pounds.

In all the illustrations above, a piston-cylinder arrangement was the means of producing the pressure, and this is the method used in hydraulic brakes. However, any sort of pump or pressure-producing device can be used. Several types of pumps (gear, rotor, vane) have been used in automatic transmissions and power steering.

§ 29-3 AUTOMATIC TRANSMISSIONS

Automatic transmissions do the job of shifting gears without any assistance from the driver. They start out in low as the car pulls away from the curb. Then, the automatic transmission shifts from low gear into intermediate and then into high gear as the car picks up speed. The system is operated hydraulically, that is, with oil pressure.

There are two basic parts to the automatic transmission: the torque converter and the "gear box." The torque converter passes the power from the engine to the gear box. And that's where the shifting action takes place. In this chapter, we will look at torque converters, gears, and the methods used to produce the shifting. Let's start with a discussion of the simple fluid coupling.

§ 29-4 FLUID COUPLING

The **fluid coupling** is a special sort of clutch. It uses oil to carry the power from the driving part to the driven part. A simple version is shown in Fig. 29-9. The assembly is like a hollow doughnut, sliced in two. Each hollow half has a series of semicircular plates, called **vanes.** The two halves are enclosed in an outer cover that is attached to the flywheel. Figure 29-10 shows the arrangement. The driving half of the doughnut, called the **pump** or **impeller,** is attached to the crankshaft. The driven half, called the **turbine,** is attached to the transmission shaft.

Fig. 29-9. A simplified version of the two members of a fluid coupling, viewed from inside. (*Chevrolet Motor Division of General Motors Corporation*)

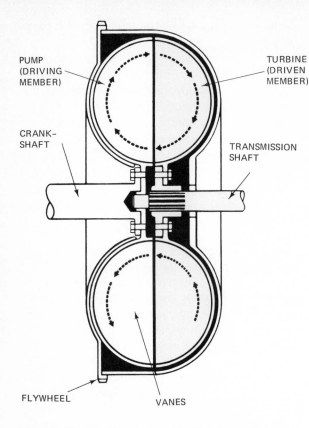

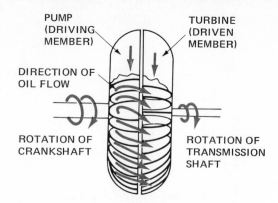

Fig. 29-11. Fluid coupling in action. Oil is throwr from the driving into the driven member. The outer casings have been cut away so that vanes can be seen.

§ 29-5 CURVED VANES

To direct the oil better, the vanes of the pump and the turbine are curved, as shown in Fig. 29-12. Also, a smaller doughnut, called a split guide ring, is centered in the vanes of the pump and the turbine. Its purpose is to help guide the oil in its path from the pump to the turbine and produce higher oil pressure against the vanes of the turbine. Figure 29-12 shows the location of the split guide ring.

§ 29-6 THE STATOR

To make the fluid coupling more effective, a third member is required. This member is called the reaction member, or **stator.** Once this third member, or stator, is added, the fluid coupling becomes a torque converter. The purpose of the stator is to change the direction of the oil coming off the turbine (driven member) vanes. To see why this change of direction is necessary, let's follow the oil in its travel between the pump and the turbine.

The oil is thrown from the pump into the turbine by centrifugal force, as we said. The oil then flows from the outer area of the turbine to the inner area and passes back into the pump. The circular arrows in Fig. 29-10 show the path the oil travels.

Fig. 29-10. Sectional view of a fluid coupling.

There is no direct mechanical connection between the pump (driving member) and the turbine (driven member). If no oil were used in the assembly, the two members could rotate independently of each other. However, filling the doughnut and cover with oil makes the difference. When the pump rotates, it throws oil into the turbine. The oil between the vanes of the pump is thrown out by centrifugal force. (**Centrifugal force** is the force that pushes things outward from the center around which they revolve.) The oil caught between the vanes of the pump has no place to go except into the turbine. The oil is thrown into the turbine with great force and hits the vanes of the turbine at an angle. In other words, the moving oil applies pressure to the vanes of the turbine, and the turbine is forced to turn. This action is shown in Fig. 29-11.

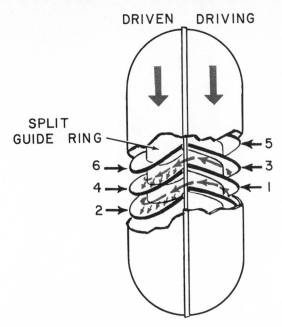

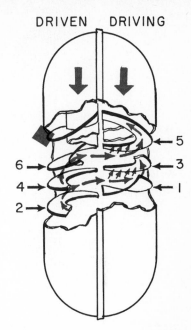

Fig. 29-12. Simplified cutaway view of two members of torque converter, showing, with heavy arrows, how oil circulates between driving- and driven-member vanes. In operation, the oil is forced by vane 1 downward toward vane 2, and thus it pushes downward against vane 2, as shown by small arrows. Oil then passes around behind the split guide ring and into the driving member again, or between vanes 1 and 3. Then, it is thrown against vane 4 and continues this circulatory pattern, passing continuously from one member to the other.

Fig. 29-13. This illustration shows what would happen if the vanes in the previous illustration were continuous. The inner ends of the vanes are not as shown here, but are as pictured in the following illustrations. Here, the split guide and the outer ends of the vanes have been cut away. If the vanes were as shown here, oil leaving the trailing edges of the driven member would be thrown upward against the forward faces of driving-member vanes, thus opposing the driving force. This effect, shown by the small arrows, would cause wasted effort and loss of torque.

Figure 29-13 shows the inner ends of the vanes and the direction in which the oil is traveling when it leaves the turbine. The oil is going in the opposite direction to the pump vanes. This opposition would waste power. The pump would have to overcome this opposing push and get the oil moving in the right direction.

§29-7 TORQUE CONVERTER

This problem is solved by putting additional stationary vanes between the inner ends of the turbine and pump vanes. Figure 29-14 is a cutaway view of the three members—pump, turbine, and stator. The pump and the turbine have been partly cut away so that you can see the vanes on the stator. Notice that the oil coming off the turbine vanes hits the stator vanes, and they change the direction of the oil from a hindering direction to a helping direction.

Figure 29-15 shows a disassembled torque converter so that you can see the shapes of the three members—turbine, stator, and pump. The stator vanes produce an effect called torque multiplication. The torque is increased, or multiplied, when the stator vanes change the direction of the oil, as shown in Fig. 29-14. Figure 29-16 shows you why.

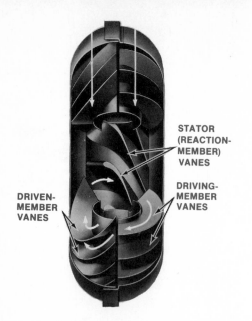

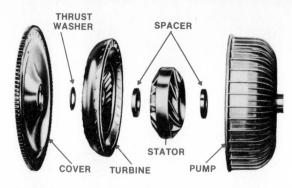

Fig. 29-15. The turbine, stator, and pump (impeller) used in a torque converter. (*Ford Motor Company*)

Fig. 29-14. A cutaway view of a torque converter with three members. The third member, the stator, serves as a reaction member. It changes the direction of oil flow, as shown by the curved arrows, under certain operating conditions. Compare this illustration with Figs. 29-12 and 29-13. (*Chevrolet Motor Division of General Motors Corporation*)

At the left in Fig. 29-16, a jet of oil is hitting a round bucket attached to a wheel. The oil pushes on the bucket and tries to turn the wheel. But the push is not great, because the oil does not give up much of its energy of motion during the one pass through the bucket.

However, when a curved vane is added, as shown at the right, things are different. Now, when the oil leaves the bucket, it hits the curved vane and is directed back into the bucket. In this way, the oil gives the bucket another push. Actually, the oil could make several circuits between the bucket and the vane. Each time the oil enters the bucket, it gives the bucket another push. This effect is what we mean by torque multiplication.

Do you see how torque multiplication applies to the torque converter? As the oil leaves the

turbine, it hits the stator vanes and is redirected into the pump in a helping direction. The pump then throws the oil back into the turbine. This is a continuous action. The repeated pushes of the oil on the turbine vanes increase the torque on the turbine. In many torque converters, the torque is more than doubled. For each 1 pound-foot [0.14 kg-m] of torque entering the pump, the turbine delivers more than 2 pound-feet [0.28 kg-m] of torque to the transmission shaft. This is torque multiplication.

However, torque multiplication takes place only when the pump is turning considerably faster than the turbine. This happens during startup, for example, when the car pulls away from the curb.

§ 29-8 STATOR ACTION

Now, let's take a look at the complete stator action, from the time the car pulls away from the curb until it is cruising down the highway. To start with, there is torque multiplication. The pump is turning fast while the turbine is turning slowly. This is the same as saying that the engine is turning fast while the car wheels are turning slowly. What happens here is the same as what happens when the manual transmission is in low gear: Speed is reduced and torque is increased.

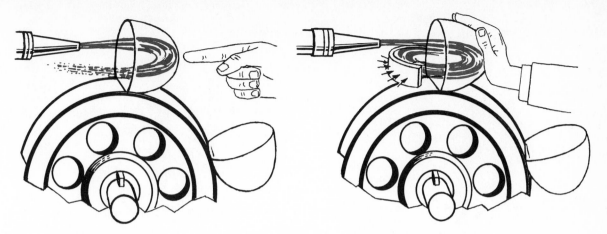

Fig. 29-16. Effect of a jet of oil on a bucket attached to a wheel. If the oil enters and leaves as at the left, the push on the bucket is small. But if the oil jet is redirected into the bucket by a curved vane as at the right, the push is increased. (*Chrysler Corporation*)

However, as the car comes up to speed, the turbine begins to "catch up" with the pump. The effect is the same as if the driver had up-shifted into high gear. Now, the oil leaving the turbine is no longer moving so forcefully in a hindering direction. The reason is that the turbine is taking up most of the energy caused by the motion of the oil. Therefore, the oil could pass directly into the pump in a helping direction. The stator is no longer needed. Actually, the stator vanes are now in the way. To get the vanes out of the way, the stator is mounted on a freewheeling mechanism.

The **freewheeling mechanism** is a one-way clutch that allows the stator to revolve freely, or "freewheel," in one direction. The mechanism locks up the stator if the stator tries to turn in the other direction. The freewheeling mechanism uses an overrunning clutch. Figure 29-17 shows how the stator is mounted on the freewheeling, or overrunning, clutch.

Figure 29-18 shows the details of the clutch. It consists of a hub, an outer ring that is part of the stator, and a series of rollers. The rollers are located in the notches in the outer ring. The outer ring is called the overrunning-clutch cam. The notches are smaller at one end than

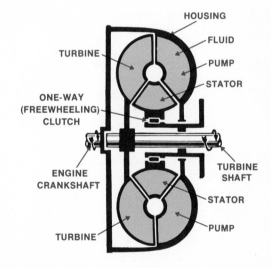

Fig. 29-17. A simplified sectional view of a torque converter, showing locations of turbine, stator, pump, and one-way clutch. (*Ford Motor Company*)

at the other. The rollers have springs behind them. When there is a push on the stator vanes from the oil leaving the turbine, the stator attempts to rotate backward. This causes the rollers to roll into the smaller ends of the notches. There, they jam and lock the stator to the hub. Thus, the stator cannot turn back-

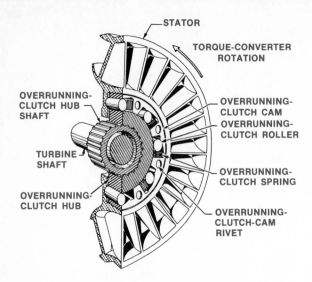

Fig. 29-18. Details of an overrunning, or one-way, clutch used to support a stator in a torque converter. (*Chrysler Corporation*)

ward. However, as the turbine speed approaches the pump speed, the direction of the oil no longer has to be changed as it leaves the turbine. It now begins to hit the other side of the stator vanes. The stator starts to revolve in a forward direction. The rollers roll out of the smaller ends of the notches and into the larger ends. There, they cannot jam, and the stator is able to run freely—to freewheel.

§ 29-9 PLANETARY GEARS

Automatic transmissions have two or more **planetary-gear** sets. The simple-looking planetary-gear assembly shown in Fig. 29-19 can do many things. It can increase the speed, reduce the speed, reverse the speed, or act as a solid shaft. The planetary-gear set consists of an internal gear (also called a ring gear), a sun gear, and a pair of planet pinions on a carrier and shaft. Before we describe how the planetary gears do their job, let's review gears.

§ 29-10 GEAR COMBINATIONS

When two gears are in mesh, as shown in Fig. 29-20, they turn in opposite directions. But if

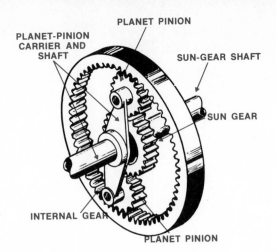

Fig. 29-19. A planetary-gear system. The planet pinions rotate on shafts that are mounted on a planet-pinion carrier. The planet-pinion carrier is attached to a shaft that is exactly aligned with the sun-gear shaft. These shafts are exactly centered in the internal gear.

Fig. 29-20. Two meshing spur gears with the same number of teeth.

another gear is put into the gear train, as shown in Fig. 29-21, the two outside gears turn in the same direction. Note that the middle gear is called an idler gear. It doesn't do any work—it is idle.

To get a combination of two gears to rotate in the same direction, you can use one internal gear. The internal gear, or ring gear, has teeth on the inside. When the spur gear and the internal gear rotate, they both rotate in the same direction (see Fig. 29-22).

Now, if another spur gear is added in the center, meshed with the small spur gear, the

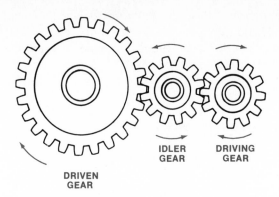

Fig. 29-21. The idler gear causes the driven gear to turn in the same direction as the driving gear.

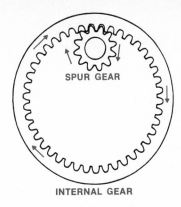

Fig. 29-22. If one internal gear is used with one external gear, both the driven and the driving gears will turn in the same direction.

combination is a simple planetary-gear system (see Fig. 29-23). The center gear is called the sun gear because the other gears revolve around it. This is similar to the way the planets in our solar system revolve around the sun. The spur gear between the sun gear and the internal gear is called the planet pinion. The reason it is so called is that it revolves around the sun gear, just as planets revolve around the sun. Now let's study the planetary-gear system.

§ 29-11 PLANETARY-GEAR OPERATION

Let's complete the planetary-gear set by adding another planet pinion. This gives us the combination shown in Fig. 29-19.

The planetary-gear sets used in automatic transmissions usually have three planet pinions. But there are various other combinations.

The two planet pinions rotate on shafts that are a part of a planet-pinion carrier (Fig. 29-19). Each of the two gears and the planet-pinion carrier are called members. The internal gear, the sun gear, and the planet-pinion-and-carrier assembly are all members. If one member is stationary and another turns, there will be either a speed increase, a speed reduction, or a speed reversal. The result depends on which member is stationary and which turns. Let's see how this works.

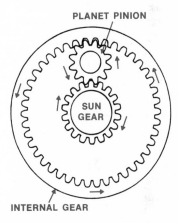

Fig. 29-23. If a sun gear is added to the arrangement shown in the previous illustration, the result is a simple planetary-gear system.

1 Speed increase 1. Suppose the sun gear is stationary and the planet-pinion carrier turns. There would be a speed increase. Why? Follow this closely! When the carrier revolves, it carries the planet pinions around with it. This movement makes the planet pinions rotate on their shafts. As the pinions rotate, they cause the internal gear to rotate also (Fig. 29-24). Note the conditions. The sun gear is stationary. The planet-pinion carrier is moving, carrying the pinions around with it. The planet pinions "walk around" the sun gear, which means that they rotate on their shafts. The inside-pinion tooth, meshed with the sun gear, is stationary

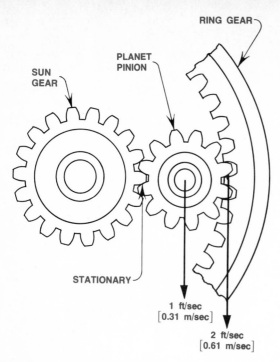

SUN GEAR

PLANET PINION

RING GEAR

STATIONARY

1 ft/sec [0.31 m/sec]

2 ft/sec [0.61 m/sec]

Fig. 29-24. If the sun gear is stationary and the planet-pinion cage is turned, the ring gear will turn faster than the cage. The planet pinion pivots about stationary teeth. If the center of the pinion shaft is moving at 1 foot [0.31 meter] per second, the tooth opposite the stationary tooth must move at 2 feet [0.61 meter] per second (it is twice as far away from the stationary tooth as the center of the shaft).

because the sun gear is stationary. This means that the outside-pinion tooth, meshed with the internal gear, is moving twice as fast as the shaft on which the planet pinion is turning. If the planet-pinion shaft is moving at 1 foot per second [0.31 meter per second (m/sec)], the outer tooth is moving at 2 feet per second [0.61 m/sec]. In other words, speed increases.

2 Speed increase 2. Another combination is to hold the internal gear stationary and turn the planet-pinion carrier. In this case, the sun gear is forced to rotate faster than the planet-pinion carrier, and there is a speed increase just as in case 1.

3 Speed reduction 1. If the internal gear turns while the sun gear is held stationary, the plan-et-pinion carrier turns more slowly than the internal gear. This is just the opposite of what we described as speed increase 1. With the internal gear turning the planet-pinion carrier, the planetary-gear set acts as a speed-reducing system.

4 Speed reduction 2. If the internal gear is held stationary and the sun gear turns, there is speed reduction. The planet pinions must rotate on their shafts. They must also walk around the internal gear, since they are in mesh with it. As the pinions rotate, the planet carrier rotates. But it rotates at a slower speed than that at which the sun gear is turning.

5 Reverse 1. To get reverse, the planet-pinion carrier can be held stationary, and the internal gear can be turned. In this case, the planet pinions act as idlers and cause the sun gear to turn in the reverse direction. Here, the system acts as a direction-reversing system, with the sun gear turning faster than the internal gear.

6 Reverse 2. A second way to get reverse is to hold the planet-pinion carrier stationary and turn the sun gear. The internal gear turns in a reverse direction, but slower than the sun gear.

7 Direct drive. If any two members are locked together, then the entire planetary-gear system acts as a solid shaft. Locking two members together locks up the system.

All the conditions discussed above are listed in the chart in Fig. 29-25. The three conditions that you will find in automatic transmissions are listed in columns 3, 4, and 6. The plane-tary-gear sets used in transmissions are designed to give direct drive, speed reduction, and reverse.

§ 29-12 HYDRAULIC SHIFT CONTROLS

We have seen how the power flows into the transmission through the torque converter and

CONDITIONS	1	2	3	4	5	6
RING GEAR	D	H	T	H	T	D
CARRIER	T	T	D	D	H	H
SUN GEAR	H	D	H	T	D	T
SPEED	I	I	L	L	IR	LR

D—DRIVEN
H—HOLD (STATIONARY)
I—INCREASING OF SPEED
L—REDUCTION OF SPEED
T—TURN OR DRIVE

Fig. 29-25. Various conditions that are possible in the planetary-gear system if one member is held and another is turned.

how planetary gears operate. Figure 29-26 shows, in simplified form, how the power flows through the torque converter and into the planetary gear. Now let's find out how the planetary gears are controlled.

There are two controls, a band and a clutch. The band consists of a brake band that surrounds a metal drum. When the band is tightened on the drum, the drum is held stationary. The clutch consists of a series of clutch plates. Half of the plates are splined to an outer ring, and the other half are splined to a hub. When oil pressure forces the two sets of clutch plates together, the clutch is engaged. When the oil pressure is released, the clutch is released. When the clutch is released, the two sets of clutch plates can rotate independently of each other.

§ 29-13 BAND AND CLUTCH

Let's put the two controls—the clutch and the band—onto the planetary-gear set. Figure 29-27 is a sectional view of a planetary-gear set with a band and a clutch. We explain later how the band and the clutch are applied. First, we want to see what happens when they are applied.

The band is shown in Fig. 29-28. Note that the band is positioned around the sun-gear drum. When the band is applied, the sun gear is held stationary. This means that the planetary-gear set acts as a speed reducer. The internal gear is turning—it is mounted on the input shaft. This arrangement forces the planet pinions to rotate. They walk around the stationary sun gear and carry the pinion carrier

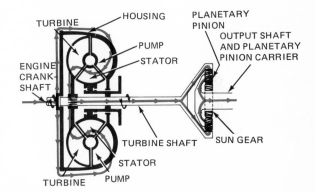

Fig. 29-26. Power flow through torque converter to planetary gears.

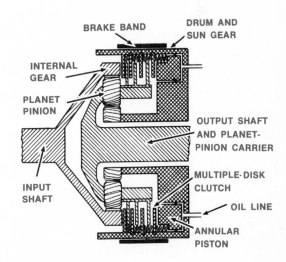

Fig. 29-27. A sectional view showing the two controlling mechanisms used in the front planetary set in an automatic transmission. One controlling mechanism consists of a brake drum and brake band; the other is a multiple-disk clutch.

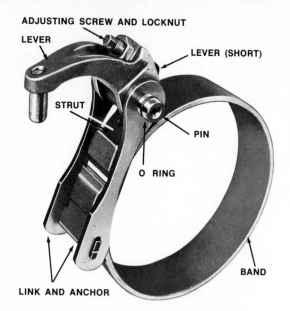

Fig. 29-28. Transmission band. (*Chrysler Corporation*)

around with them. The carrier rotates at a slower speed than the internal gear.

Now, let's see what happens if the clutch is applied instead of the band. Oil pressure that enters through the oil line causes the clutch to apply. The oil pressure forces the piston in the sun-gear drum to the left (in Fig. 29-27). The clutch plates are pushed together so that the clutch is engaged. In this situation, the planet-pinion carrier and the sun gear are locked together. The planetary-gear set is now in direct drive. In other words, the system is

locked up. Figure 29-29 shows the clutch plates. Note that they are alternately splined to the drum and to the sun gear.

The arrangement shown in Fig. 29-27 is only one of several arrangements used in automatic transmissions. In some transmissions, when the band is applied, it holds the internal gear or the planet-pinion carrier stationary. Many transmissions lock different members together when the clutch is applied. The principle is the same in all transmissions, however. There is a gear reduction when the band is applied, and there is direct drive when the clutch is applied.

§ 29-14 HYDRAULIC CIRCUITS

Figure 29-30 is a simplified diagram of a hydraulic control circuit for a single planetary-gear set in an automatic transmission. Later, we will look at the circuits for automatic transmissions that use two or more planetary-gear sets. As we have said previously, automatic transmissions use more than one planetary-gear set.

The major purpose of the hydraulic circuit is to control the shift from gear reduction to direct drive. The shift must take place at the right time, and this depends on car speed and throttle opening. These two factors produce two varying oil pressures that work against the two ends of the shift valve.

The **shift valve** is a valve spool inside a bore, or hole, in the valve body. Figure 29-31 shows

Fig. 29-29. Parts of a clutch. (*Chevrolet Motor Division of General Motors Corporation*)

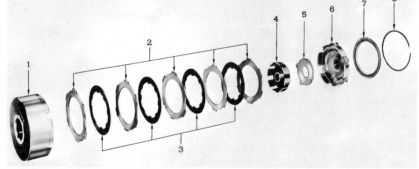

1 CLUTCH-DRUM ASSEMBLY
2 CLUTCH DRIVEN PLATE
3 CLUTCH DRIVE PLATE
4 CLUTCH HUB
5 CLUTCH-HUB THRUST WASHER
6 LOW-SUN-GEAR-AND-CLUTCH-FLANGE ASSEMBLY
7 CLUTCH-FLANGE RETAINER
8 RETAINER SNAP RING

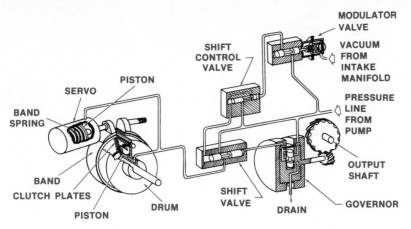

Fig. 29-30. A schematic diagram, showing the hydraulic control system for the brake-band servo and the clutch. In the system shown, the band is normally on and the clutch off; this arrangement produces gear reduction. But when the shift valve is moved, pressure from the oil pump is admitted to the front of the brake-band piston and to the clutch piston. This movement causes the brake to release and the clutch to apply. Now, with the clutch locking two planetary members together, the planetary system goes into direct drive.

Fig. 29-31. A spool valve for a shift valve.

what the spool looks like. Pressure at one end of the spool comes from the governor. Note that a governor is a device that controls, or governs, another device. In the hydraulic circuit, the governor controls pressure on one end of the shift valve.

Pressure at the other end of the spool changes as vacuum in the intake manifold changes. Let us explain. First, the governor pressure changes with car speed. The governor is driven by the output shaft from the transmission. As output-shaft speed and car speed go up, the governor pressure increases. This pressure works against the right end of the shift valve, as shown in Fig. 29-30.

Working on the left end of the shift valve is a pressure that changes as intake-manifold vacuum changes. Line pressure enters the modulator valve at the upper right in Fig. 29-30. The modulator valve contains a spool valve attached to a spring-loaded diaphragm. Vac-

uum increases in the intake manifold when the throttle is partly closed. This vacuum pulls the diaphragm in and moves the modulator spool valve to the right. The motion cuts off the line pressure going to the shift control valve. When this happens, the shift control valve moves to the right, cutting off pressure from the left end of the shift valve. This means that the shift valve is pushed to the left by governor pressure. As a result, line pressure can pass through the shift valve. Therefore, line pressure is applied to the clutch and the servo at the planetary-gear set. With this condition, the band is released and the clutch is applied. This puts the planetary-gear set into direct drive.

§ 29-15 THE FORD C6 AUTOMATIC TRANSMISSION

You cannot become an automatic-transmission expert by reading a few pages of a book. But you should know what the transmission

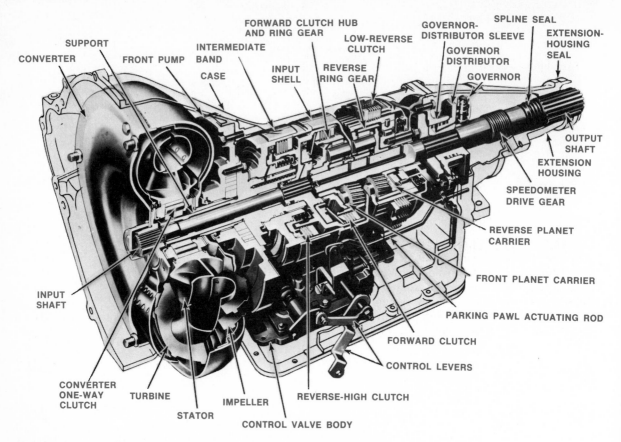

CONVERTER

SUPPORT

FRONT PUMP

INTERMEDIATE BAND

CASE

FORWARD CLUTCH HUB AND RING GEAR

INPUT SHELL

LOW-REVERSE CLUTCH

REVERSE RING GEAR

GOVERNOR-DISTRIBUTOR SLEEVE

GOVERNOR DISTRIBUTOR

GOVERNOR

SPLINE SEAL

EXTENSION-HOUSING SEAL

OUTPUT SHAFT

EXTENSION HOUSING

SPEEDOMETER DRIVE GEAR

REVERSE PLANET CARRIER

FRONT PLANET CARRIER

PARKING PAWL ACTUATING ROD

FORWARD CLUTCH

CONTROL LEVERS

REVERSE-HIGH CLUTCH

CONTROL VALVE BODY

CONVERTER ONE-WAY CLUTCH

TURBINE

STATOR

IMPELLER

INPUT SHAFT

Fig. 29-32. A cutaway view of the Ford C6 automatic transmission. (*Ford Motor Company*)

looks like inside and how it does its job. Most automatic transmissions have three forward speeds and one reverse. All have torque converters, of course. To give you an idea of what a modern automatic transmission looks like, we will study the Ford C6 transmission. This transmission is shown in cutaway view in Fig. 29-32. The complete hydraulic circuit of this transmission is shown in Fig. 29-33. All this looks pretty complicated, but if you read carefully, you will have a good idea of how this transmission works.

1 The manual shift valve. The action starts at the manual shift valve, shown at the bottom in Fig. 29-33. After starting the engine, the driver operates the selector lever on the steering column to select the driving range wanted.

The driver can put it in R, or reverse, to back up the car, or put it in D, or drive, for normal operation. In D, the transmission will automatically shift up from first to second to third. If the driver does not want to shift up to third, the shift lever can be moved to 2 or 1. In 2, the transmission will shift only from first to second. In 1, the transmission will remain in first. The driver might need 1 or 2 for pulling a heavy load, going up a steep hill, or coming down a long hill.

2 Upshifting. Let's assume that the driver selects D. Starting out, the transmission is in low. Then, as car speed increases, the shift is made from low through second to third. Shift points depend on both car speed and throttle opening, as we have already explained.

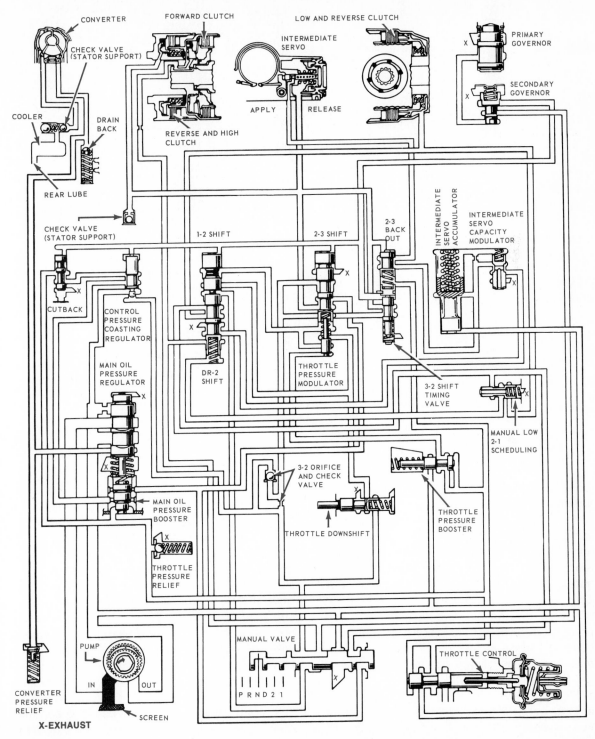

Fig. 29-33. The complete hydraulic system of the Ford C6 automatic transmission. (*Ford Motor Company*)

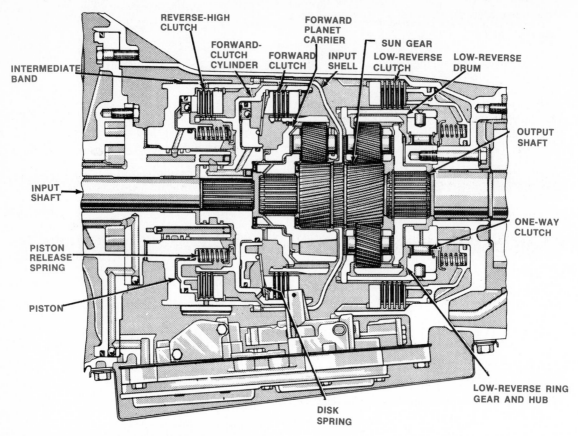

Fig. 29-34. A sectional view of the planetary-gear train in the Ford C6 automatic transmission. (*Ford Motor Company*)

3 The planetary-gear sets. The Ford C6 transmission has two interconnected planetary-gear sets, as shown in Fig. 29-34. They are interconnected by one sun gear, which is common to both. In other words, there is only one sun gear, and it serves both planetary-gear sets. By controlling the two sets, we can get the three forward and one reverse speeds.

We have to be careful here not to get confused, because in this transmission, a clutch takes the place of one of the bands. We will get to that in a moment.

4 First gear. Figure 29-35 shows the conditions in the transmission in first gear. The forward clutch is applied. This locks the front planetary ring gear to the input shaft, and it

turns with the input shaft. As the ring gear rotates, it drives the planet pinions, and they, in turn, drive the sun gear. This produces a gear reduction through the front planetary-gear set. As the sun gear turns, it drives the rear planet pinions. The rear planet pinions drive the ring gear of the rear planetary-gear set. The ring gear is splined to the output shaft, so the shaft turns. There is also gear reduction in the rear planetary-gear set. With gear reduction in both sets, the transmission is in low, or first, gear.

5 Second gear. When the upshift to second gear takes place, the hydraulic system applies the intermediate band. This band holds the sun gear and the reverse-and-third clutch drum

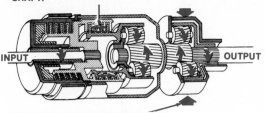

THE FORWARD CLUTCH IS APPLIED.
THE FRONT PLANETARY UNIT RING
GEAR IS LOCKED TO THE INPUT
SHAFT.

INPUT   OUTPUT

THE LOW-AND-REVERSE CLUTCH
(LOW RANGE) OR THE ONE-WAY
CLUTCH (DI RANGE) IS HOLDING THE
REVERSE UNIT PLANET CARRIER
STATIONARY.

FIRST GEAR

Fig. 29-35. A Ford C6 automatic transmission in first gear. (*Ford Motor Company*)

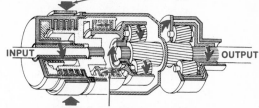

THE INTERMEDIATE BAND IS APPLIED.
THE REVERSE-AND-HIGH CLUTCH
DRUM, THE INPUT SHELL AND THE
SUN GEAR ARE HELD STATIONARY.

INPUT   OUTPUT

THE FORWARD CLUTCH IS APPLIED.
THE FRONT PLANETARY UNIT RING
GEAR IS LOCKED TO THE INPUT SHAFT.

SECOND GEAR

Fig. 29-36. A Ford C6 automatic transmission in second gear. (*Ford Motor Company*)

stationary. Now, as the arrows in Fig. 29-36 show, there is gear reduction in the front gear set only. So the transmission is in second.

6 Third, or direct, gear. When the hydraulic system sends the signals to produce the shift into third, the conditions shown in Fig. 29-37 occur. Both the forward clutch and the reverse-and-third clutch are applied. This locks the planetary-gear sets so that there is direct drive through both. The transmission is in third gear.

7 Reverse. Figure 29-38 shows the conditions in the transmission in reverse gear. Note that the reverse-and-third clutch and the low-and-reverse clutch are both applied. This condition causes gear reduction through both planetary-gear sets. Also, the direction of rotation is reversed in the rear set.

§ 29-16 OTHER AUTOMATIC TRANSMISSIONS

A variety of automatic transmissions are used in automobiles, but they all work in about the same way. All have planetary-gear sets that are controlled by bands and clutches. If you really want to study automatic transmissions, you should read other books to give you the com-

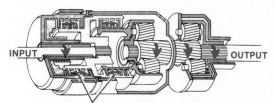

INPUT   OUTPUT

BOTH THE FORWARD AND THE
REVERSE-AND-HIGH CLUTCH ARE
APPLIED.
ALL PLANETARY GEAR MEMBERS ARE
LOCKED TO EACH OTHER AND ARE
LOCKED TO THE OUTPUT SHAFT.

HIGH GEAR

Fig. 29-37. A Ford C6 automatic transmission in third gear. (*Ford Motor Company*)

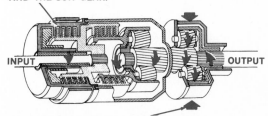

THE REVERSE-AND-HIGH CLUTCH IS APPLIED.
THE INPUT SHAFT IS LOCKED TO THE REVERSE-
AND-HIGH CLUTCH DRUM, THE INPUT SHELL,
AND THE SUN GEAR.

INPUT   OUTPUT

THE LOW-AND-REVERSE CLUTCH IS APPLIED.
THE REVERSE UNIT PLANET CARRIER
IS HELD STATIONARY.

REVERSE

Fig. 29-38. A Ford C6 automatic transmission in reverse gear. (*Ford Motor Company*)

plete picture. For example, *Automotive Transmissions and Power Trains,* a book in the *McGraw-Hill Automotive Technology Series,* covers in detail both the operation and servicing of the various types of automatic transmissions.

Review questions

1. What is used in the torque converter to carry power from the pump to the turbine?
2. What is the stator doing when the pump is turning much faster than the turbine?
3. What are the purposes of a planetary-gear set in an automatic transmission?
4. When a spur gear meshes with an internal gear, do both revolve in the same direction or in opposite directions?
5. What is the name of the center gear in the planetary-gear set?
6. What is the name of the gears between the center gear and the ring gear in the planetary-gear set?
7. If the sun gear is stationary and the ring gear turns, does the planet-pinion carrier turn faster or slower than the ring gear?
8. What are the two controls used in automatic transmissions to control the planetary-gear sets?
9. In the simplified version of the planetary-gear set and its controls, which control device puts the gear set into reduction?
10. Which control device puts the set into direct drive?

Study problems

1. Using one of the shop fluid couplings and one of the shop torque converters, show and explain their differences.
2. The next time you take a ride in a car that has an automatic transmission, try to "feel" each of the shifts that the transmission makes. Determine if it is a two-speed automatic transmission (only one upshift), a three-speed (two upshifts), or a four-speed (three upshifts).

30 | FLUID POWER TRANSMISSION

§ 30-1 POWER FOR STOPPING AND STEERING

As we mentioned earlier, many different ways of transmitting power are used in the automobile. Power is not always used to make the car go faster. For example, power is used to slow the car. This is called brake action. Power must also be used to steer the car. If the driver supplies all the power for steering, the car is said to have manual steering. But suppose that by turning the steering wheel, the driver controls a separate power supply that furnishes most of the power needed to steer the car. Then, the vehicle is said to have power steering. In the automobile, both the brake system and the power-steering system are familiar examples of using fluid to transmit power.

§ 30-2 HYDRAULIC BRAKES

There are two general kinds of brake assemblies used at the wheels of automobiles, the kind with shoes and drums, and the kind with disks, or rotors, and shoes. Although the two are differently constructed, they act in a similar manner. They apply strong frictional pressure to a brake drum or to a disk attached to the rotating wheel so that the wheel is slowed down or stopped.

The pressure comes from the hydraulic part of the brake system. Figure 30-1 is a simplified view of a hydraulic brake system. The hydraulic components consist of a master cylinder, tubing or brake lines, and cylinders at each wheel.

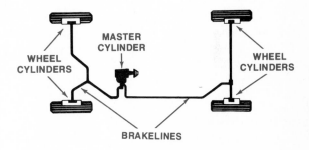

Fig. 30-1. View of a hydraulic brake system. (*Ford Motor Company*)

§ 30-3 BRAKE OPERATION

Braking action starts at the brake pedal (Fig. 30-2). When the pedal is pushed down, brake fluid is sent from the master cylinder to the wheels. At the wheels, the fluid pushes brake shoes against revolving drums or disks. The friction between the stationary shoes and the revolving drums or disks slows and stops them. This slows or stops the revolving wheels, which, in turn, slow or stop the car.

Figure 30-3 shows the brake lines, or tubes, through which the fluid flows. You will notice that there are two chambers and two pistons in the master cylinder. One chamber is connected to the front-wheel brakes. The other chamber is connected to the rear-wheel brakes. This braking arrangement is called a dual braking system. Dual braking systems are

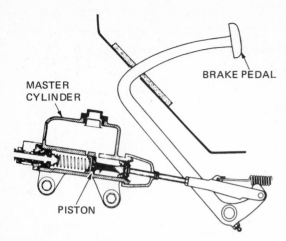

Fig. 30-2. Relationship between the brake pedal and the master cylinder. (*Pontiac Motor Division of General Motors Corporation*)

used on modern cars. There is a reason for splitting the system into two parts. If one part should fail, the other can still work and provide braking until the failed part can be fixed.

In earlier braking systems, there was only one chamber in the master cylinder, which was connected to all four wheel brakes. If one part failed, the whole system failed. The dual braking system provides added protection, because the rear and front sections seldom fail at the same time.

§ 30-4 BRAKE SHOES

The brake shoes (Fig. 30-4) are lined with a tough asbestos material that can withstand the heat and dragging effect imposed when they are forced against the drum. During hard braking, the shoe may be pressed against the drum with a pressure as great as 1,000 psi [70.31 kg/cm²]. Since friction increases as the pressure increases, this produces a strong frictional drag on the brake drum and a strong braking effect on the wheel. Disk brakes operate in a very similar way, with the brake shoe pressing against the disk. We discuss more about disk brakes in § 30-10.

A great deal of heat is also produced by the frictional effect between the brake shoes and drum. When you rub your hands together vigorously, they become warm. In like manner, when the drum rubs against the shoe, the drum and shoe get warm. In fact, under extreme braking conditions, temperatures may reach 500°F [260°C] or higher. Some of this heat goes through the brake linings to the shoes and backing plate, where it is radiated to the surrounding air. But most of it is absorbed by the brake drum. Some brake drums have cooling fins to provide additional radiating surface for getting rid of the heat more quickly. Excessive temperatures are not good for brakes, since they may char the brake linings. Also, with the linings and drums hot, less effective braking action results. This is the reason that brakes "fade" when they are used continuously for relatively long periods, as, for instance, in coming down a mountain or a long hill.

Some special-performance vehicles, such as racing cars, are equipped with metallic brakes. Instead of linings of asbestos material, these brakes have a series of metallic pads attached

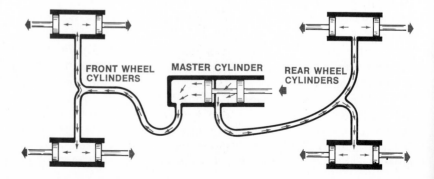

Fig. 30-3. The flow of brake fluid to the four wheel cylinders when the piston is pushed into the master cylinder.

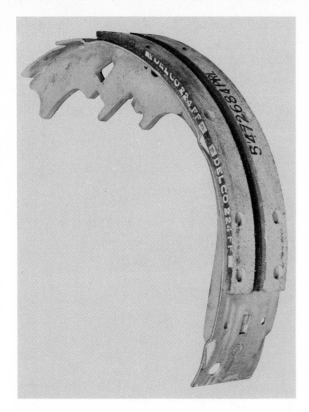

Fig. 30-4. Brake shoe of the type used on drum brakes.

to the brake shoes. These brakes can withstand more severe braking and higher temperatures and have less tendency to fade.

§30-5 BRAKE LINES

The brake fluid is carried by steel pipes, called the **brake lines,** from the master cylinder to connecting points. These points are located on the car frame near the four wheels. From these connecting points, a special flexible hose is used to connect the brake lines to the brake mechanisms at the wheels.

§30-6 BRAKE FLUID

The brake fluid is *not oil!* Brake fluid is a very special fluid that is little affected by high or low temperatures. Also, brake fluid does not damage the metal and rubber parts in the braking

system. Ordinary oil will damage these parts. For this reason, only the brake fluid recommended by the manufacturer should be put into the brake system.

◆ CAUTION ◆

Never put ordinary oil in a brake system. Ordinary oil will cause rubber parts in the system, such as the piston cups, to swell and go to pieces. This could cause complete brake failure and lead to a fatal accident. Never use anything but the brake fluid recommended by the car manufacturer!

---

§30-7 MASTER CYLINDER

In Figs. 30-5 to 30-8, we will build a master cylinder for a dual braking system. Look at Fig. 30-5 first. Here we see, at the bottom, a sectional view of the master cylinder without the internal parts. At the top we see the cylinder with the diaphragm, cover, and retainer separated from one another.

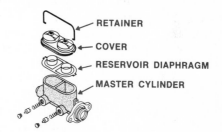

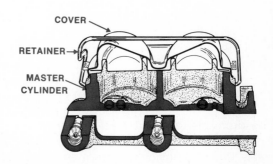

Fig. 30-5. A master cylinder with its reservoir diaphragm, cover, and retainer.

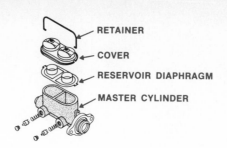

RETAINER

COVER

RESERVOIR DIAPHRAGM

MASTER CYLINDER

Fig. 30-6. Here, we have added the pushrod and boot.

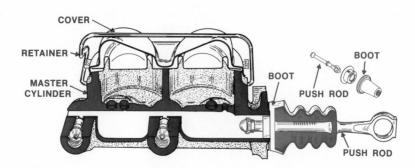

COVER

RETAINER

MASTER CYLINDER

BOOT

PUSH ROD

BOOT

PUSH ROD

Next, in Fig. 30-6, we add the boot and pushrod. In the car, the pushrod is connected by linkage to the brake pedal. When the brake pedal is pushed down, the pushrod is pushed into the master cylinder.

In Fig. 30-7, we add the primary piston parts. When the pushrod is pushed into the master cylinder, the primary piston is moved to the left and into the cylinder. The fluid ahead of the piston is therefore pushed out of the cylinder. The fluid flows through the brake lines to the rear-wheel brakes.

We complete the building of the master cylinder by adding the secondary piston, as shown in Fig. 30-8. The secondary piston, like the primary piston, is pushed to the left (in Fig. 30-8) by the pushrod. The movement of the secondary piston sends brake fluid to the front-wheel brakes.

§ 30-8 DRUM BRAKES

The drum brake has a steel or iron drum to which the wheel is bolted. The drum and wheel rotate together. Inside the drum and attached to the steering knuckle or axle housing is the brake mechanism. The brake mechanism at the front wheels is attached to the steering knuckle. The brake mechanism at the rear wheels is attached to the axle housing.

Figure 30-9 shows the moving parts of the drum brake. There are two brake shoes at each wheel. The bottoms of the shoes are held apart by an adjusting screw—also known as a star wheel. The tops of the shoes are held apart by a wheel cylinder. The shoes are made of metal. Riveted to each shoe is a facing of friction material. The facing is called the **brake lining.** These brake linings are made of tough asbestos material that can hold up under the rubbing pressure and the heat produced during braking.

Figure 30-10 shows the addition of the brake backing plate and the wheel spindle. The brake backing plate is bolted to the steering axle at the back wheels, or to the steering knuckle at the front wheels. At the right in Fig. 30-10, you can see the brake drum in place over the brake shoes. The drum has been partly cut away so that you can see one of the shoes.

The shoes are attached at one point to the brake backing plate. The attachment is loose so that the shoes can move around a little. We have now described all the internal brake parts. Now let's see how the parts work.

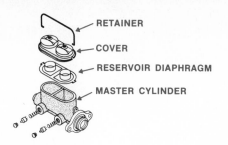

RETAINER
COVER
RESERVOIR DIAPHRAGM
MASTER CYLINDER

Fig. 30-7. Continuing to add parts, here, we show the primary piston parts.

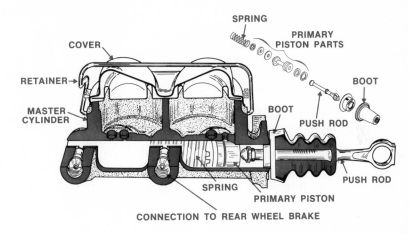

SPRING
PRIMARY PISTON PARTS
COVER
RETAINER→
MASTER CYLINDER
BOOT
BOOT
PUSH ROD
PUSH ROD
SPRING
PRIMARY PISTON
CONNECTION TO REAR WHEEL BRAKE

Fig. 30-8. Here are the complete assembly and all parts for the master cylinder.

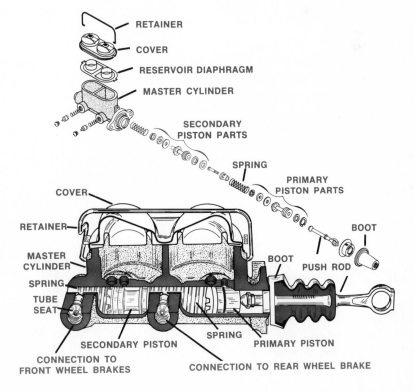

RETAINER
COVER
RESERVOIR DIAPHRAGM
MASTER CYLINDER
SECONDARY PISTON PARTS
SPRING
PRIMARY PISTON PARTS
COVER
RETAINER→
MASTER CYLINDER
BOOT
SPRING
PUSH ROD
TUBE SEAT
BOOT
SPRING
PRIMARY PISTON
SECONDARY PISTON
CONNECTION TO FRONT WHEEL BRAKES
CONNECTION TO REAR WHEEL BRAKE

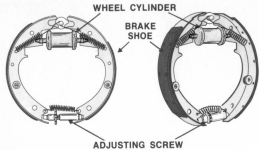

WHEEL CYLINDER

Fig. 30-9. These are the operating components of drum brakes for the rear (*left*) and front (*right*) wheels.

§ 30-9 OPERATION OF THE DRUM BRAKE

When brake fluid is forced from the master cylinder, it flows through the brake lines into the wheel cylinder (Fig. 30-11). A wheel cylinder is shown disassembled at the top and in sectional view at the bottom of Fig. 30-12. You will notice that there are two pistons, with piston cups, inside the wheel cylinder. When the brake fluid is forced into the wheel cylinder, it pushes the pistons apart. This action pushes the brake-shoe actuating pins out. The brake shoes are therefore forced tightly against the rotating brake drum. The friction between the brake linings and the drum slows or stops the rotation of the drum and the wheel.

§ 30-10 DISK BRAKES

The disk brake (also spelled disc) has a metal disk instead of a drum. Figure 30-13 shows a

disk-brake assembly of the floating caliper type. There is a flat shoe, or pad, located on each side of the disk. In operation, these two flat shoes are forced tightly against the rotating disk. The shoes grip the disk, just as you would squeeze a piece of paper between your finger and thumb as you picked it up. Figure 30-14 shows how a fixed-caliper type of disk brake works. Fluid pressure from the master cylinder forces the pistons to move in. This action pushes the friction pads of the brake shoes tightly against the disk. The friction between the shoes and the disk slows and stops the disk, thus providing the braking action.

§ 30-11 POWER BRAKES

Many cars now have power brakes. With power brakes, only a relatively light pressure is required to brake the car. When the brake pedal is pushed down, a vacuum-operated device takes over and does most of the job of pushing the pistons into the master cylinder. The vacuum comes from the intake manifold of the car. Figure 30-15 is a simplified drawing that shows how the system works. The system includes a cylinder in which a tight-fitting piston can move. When vacuum is applied to one side of the piston, the atmospheric pressure causes the piston to be pushed to the right, as shown. This movement pushes the piston rod into the hydraulic cylinder (called the master cylinder in the actual power-brake system).

In the power-brake system, the brake pedal does not directly work on the master cylinder. Instead, the brake pedal works a vacuum valve,

Fig. 30-10. Here, we have added the brake backing plates to both brakes and the wheel spindle to the front brake. To the right, we show a brake drum partly cut away so that you can see the shoe inside.

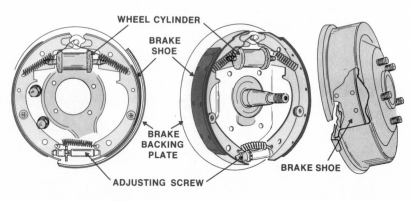

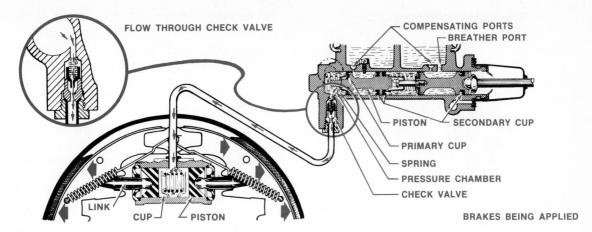

FLOW THROUGH CHECK VALVE

COMPENSATING PORTS
BREATHER PORT

PISTON
SECONDARY CUP
PRIMARY CUP
SPRING
PRESSURE CHAMBER
CHECK VALVE

LINK
CUP
PISTON

BRAKES BEING APPLIED

Fig. 30-11. Conditions in a drum-type brake system with the brakes applied. Brake fluid flows from the master cylinder to the wheel cylinder, as shown, causing the wheel-cylinder pistons to move outward and thereby apply the brakes.

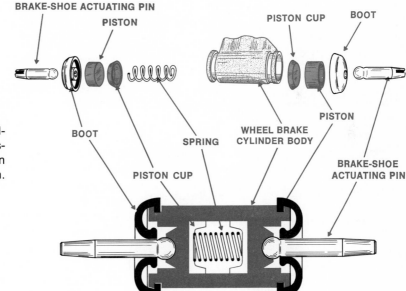

BRAKE-SHOE ACTUATING PIN
PISTON
PISTON CUP
BOOT
PISTON
BOOT
SPRING
WHEEL BRAKE CYLINDER BODY
BRAKE-SHOE ACTUATING PIN
PISTON CUP

Fig. 30-12. The wheel cylinder for a drum brake, disassembled at the top and in sectional view at the bottom.

which then admits vacuum to the power cylinder. Figure 30-16 is a sectional view of one widely used power-brake unit.

§ 30-12 PNEUMATICS

Before we discuss air brakes, let's review a few basic principles. Gases are composed of widely spaced molecules that move about freely. For this reason, gases may be compressed and reduced in volume. Gases must be compressed in volume in order to develop pressures greater than normal. By crowding the gas molecules into a space much smaller than originally occupied, the molecules will exert pressure on the container walls. If the pressure normally placed upon a volume of gas was doubled, the gas would occupy one-

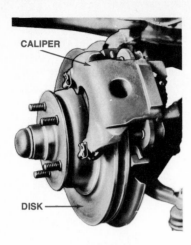

Fig. 30-13. A disk-brake assembly of the floating-caliper type. (*Ford Motor Company*)

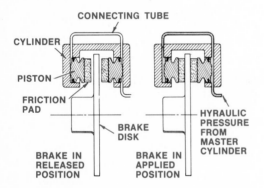

Fig. 30-14. Sectional views showing how hydraulic pressure forces friction pads (shoes) inward against the brake disk to produce the braking action.

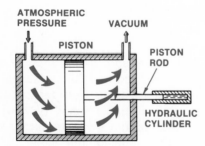

Fig. 30-15. If there is a vacuum on one side of the piston and atmospheric pressure on the other side, the piston must move toward the vacuum side.

half its original space. The gas thus confined can then be used to exert a force as it tries to assume its initial volume and space. Gases under pressure are used to provide force for pneumatic hammers, air brakes, and pneumatic pistons. Compression of gases is accomplished by using pistons as in a tire pump and air compressor. Gases are also compressed by using a rotary-type compressor such as the automobile- and diesel-engine supercharger, which we discussed in earlier chapters.

§ 30-13 AIR BRAKES

The air-brake system used on big trucks and railroad cars is a well-known example of fluid power using a gas. In this type of system, pressurized air is stored in tanks and made available for brake application. To simplify our discussion, a truck air-brake system is shown in Fig. 30-17.

Various devices are necessarily included in a vehicle air-brake system, not only to provide the braking force, but also to supply built-in safety factors. In a typical air-brake system, the following sequence normally takes place.

Compressed air from the compressor is sent to the storage tanks, or air reservoir tanks. The compressor governor controls the output by cutting out the compressor when the pressure reaches 100 to 105 psi [7.03 to 7.38 kg/cm²]. Should the pressure fall below 80 to 85 psi [5.62 to 5.98 kg/cm²], the governor allows the compressor to cut in and restore the pressure to its preset maximum.

Air from the compressor is carried off through two lines. One line leads directly to the reservoir tanks. The other line carries the air through the compressor governor to an air gage, where a reading may be taken of the pressure in the system. All air-brake systems employ a low-air-pressure switch that will set up a warning buzz or flash a red light should the pressure fall below a safe operating level. From the pressure gage line, the air is carried to an outlet from the second tank and then piped directly to the brake control valve. The

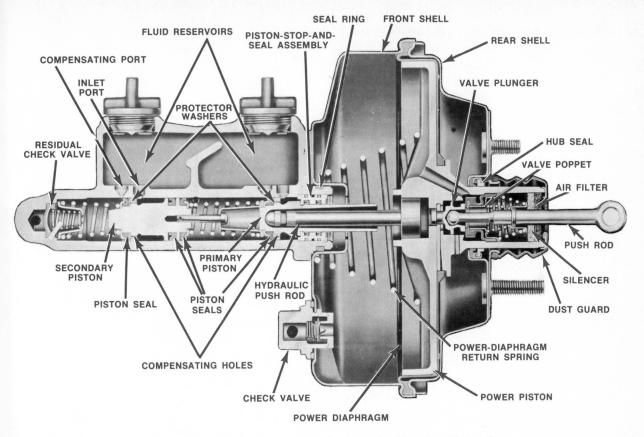

Fig. 30-16. (*Above*) A sectional view of a power-brake unit. (*Cadillac Motor Car Division of General Motors Corporation*)

Fig. 30-17. A truck air-brake system.

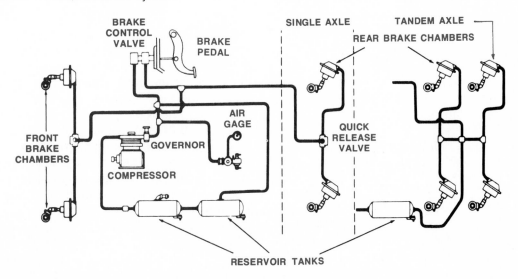

pressure at the brake control valve is the same as the reading on the pressure gage.

When the brake pedal is depressed, the brake control valve is opened, and full pressure is allowed to pass to the front brake chambers. Another smaller line carries full tank pressure to the rear brake chambers.

An air brake chamber is used at each wheel. The brake chamber converts the energy of the compressed air into the mechanical force and motion required to apply the brakes. Each chamber contains a diaphragm with a rod attached to it. When air pressure is applied to the diaphragm, the rod is pushed outward. This motion of the rod applies force to the connecting linkage to apply the brakes.

When the brake pedal is released, the pressure is exhausted quickly from the brake chambers through a quick-release valve.

§ 30-14 POWER STEERING

The steering system permits the front wheels to be pivoted on their supports to the right or to the left so that the car can be steered. Figure 30-18 is a simplified drawing of a steering system. The steering wheel is mounted on a steering shaft that extends into the steering gear. The bottom end of the shaft has a worm gear that rotates as the wheel is turned. A gear sector is meshed with the worm gear. Rotation of the worm gear causes the gear sector to rotate. This movement causes the pitman arm, attached to the sector, to swing to the right or to the left. This action, in turn, pushes or pulls on the tie rods attached to the pitman arm. The steering-knuckle arms, attached to the front wheels, are therefore forced to swing the wheels to the right or to the left on their pivots.

On more than half the cars manufactured in the United States, a hydraulic system is used to assist the driver in steering the car. Such hydraulic systems are called **power-steering systems.** The principle of power steering is very simple. A booster arrangement is provided that is set into operation when the steering-wheel shaft is turned. The booster then takes over and does most of the work of steering.

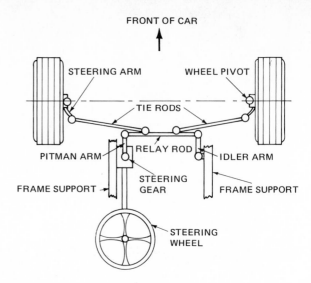

Fig. 30-18. Simplified drawing of a steering system.

Power steering has used compressed air, electrical mechanisms, and hydraulic pressure. The latter is now the only type used for automobile power-steering mechanisms.

In the power-steering system, an oil pump is used to send a fluid under pressure into the steering gear. This high-pressure fluid then does most of the work of steering. Figure 30-19 shows the power-steering system for a front-wheel-drive car. The power-steering system looks almost like the manual system, except that the steering gear is a little larger. The pump is not shown. Figure 30-20 shows the mounting arrangement for the steering gear and pump in one Chevrolet model. The pump is driven by a belt from the crankshaft pulley. In operation, the pump produces a high pressure on the power-steering fluid. This fluid is a special form of oil.

Figure 30-21 shows the working parts of a typical power-steering pump. The rotor rotates, and the vanes move into and out of the slots in the pump rotor. As the vanes move out of the slots, the space between the vanes increases. Fluid is sucked into the space. Then, with further rotation, the vanes are pushed back into the slots. This decreases the space between the vanes. The fluid is therefore forced out under pressure. The pump has a

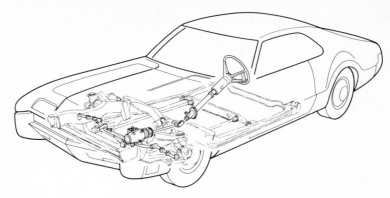

Fig. 30-19. The power-steering system for a front-wheel-drive car. (*Oldsmobile Division of General Motors Corporation*)

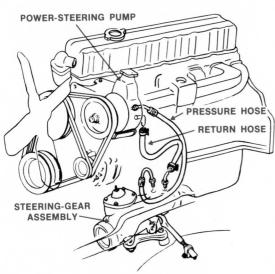

POWER-STEERING PUMP

PRESSURE HOSE

RETURN HOSE

STEERING-GEAR ASSEMBLY

Fig. 30-20. The power-steering system for a six-cylinder car.

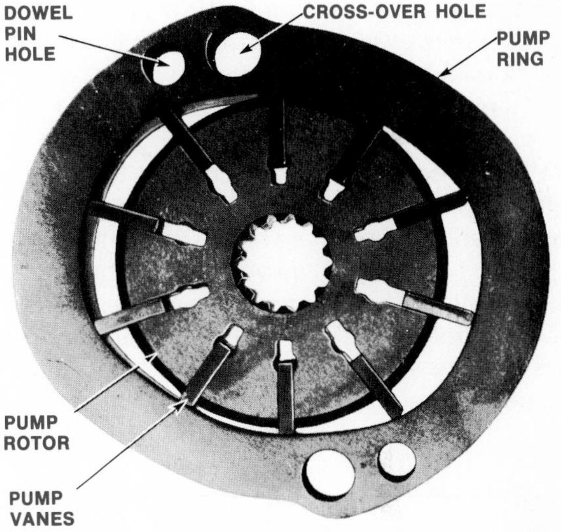

DOWEL PIN HOLE

CROSS-OVER HOLE

PUMP RING

PUMP ROTOR

PUMP VANES

Fig. 30-21. Internal working parts of a vane-type power-steering pump. The vanes follow the oval shape of the housing and move in and out of the slots in the rotor. (*Chevrolet Motor Division of General Motors Corporation*)

pressure-relief valve, which opens if the pressure goes too high.

§ 30-15 POWER-STEERING GEAR

A typical power-steering gear is shown in cutaway views in Fig. 30-22. It is a recirculating-ball steering gear, with a power piston and a control valve added. The piston has a rack on it, as shown in Fig. 30-22. This rack is meshed with the gear sector on the pitman shaft.

The steering gear uses a small torsion bar that twists as the steering wheel is turned. This actuates the control valve. In the straight-ahead position, shown in Fig. 30-23, the control valve is centered. Equal fluid pressure is applied to both ends of the piston. However, note what happens when a right turn is made (Fig. 30-24). When the driver applies turning effort to the steering wheel, the front wheels resist the effort to swing them away from straight ahead. As a result, the torsion bar is twisted. This twisting moves the control valve off center. Now, the fluid flows through the control valve, as shown at the lower right in Fig. 30-24. Pressure is released on the right-hand side of the piston. But pressure is applied to the left-hand side of the piston. The fluid pressure therefore pushes the piston to the right, and this causes the gear sector on the pitman shaft to turn. The pitman arm swings and causes the front wheels to swing to the right.

To make a left turn, pressure is relieved from the left-hand side of the piston and applied to the right-hand side. The piston moves to the left, causing the gear sector to turn to the left.

The amount of pressure applied to the piston depends on the conditions at the front wheels. If it takes only a little pressure on the steering arms to swing the wheels, only a little pressure is applied to the piston. But if it takes a lot of pressure to swing the wheels, then more pressure is applied to the piston.

Here's how the pressure is controlled. The amount of pressure at the steering arms at the wheels determines how much the torsion bar will be twisted. If only a little push or pull is required to swing the wheels, then the torsion bar will twist only a little. As a result, the control valve will not turn as much. Less pressure will be applied to the piston. But if there is great resistance at the wheels, for instance, during parking, the torsion bar will be twisted more.

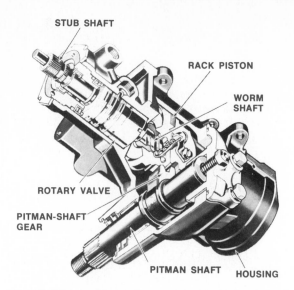

Fig. 30-22. A rotary-valve power-steering gear, cut away to show internal parts. (*Cadillac Motor Car Division of General Motors Corporation*)

This twisting motion turns the control valve more so that more pressure is applied to the piston.

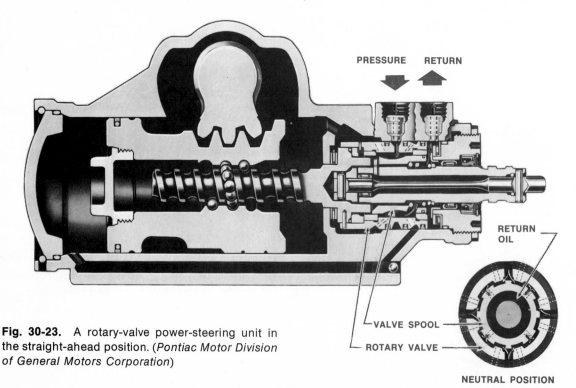

Fig. 30-23. A rotary-valve power-steering unit in the straight-ahead position. (*Pontiac Motor Division of General Motors Corporation*)

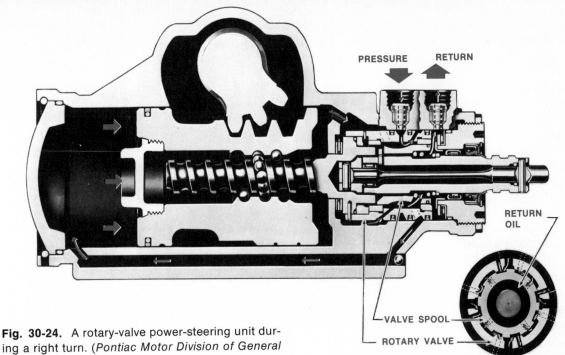

PRESSURE **RETURN**

RETURN OIL

VALVE SPOOL
ROTARY VALVE

RIGHT-TURN POSITION

Fig. 30-24. A rotary-valve power-steering unit during a right turn. (*Pontiac Motor Division of General Motors Corporation*)

There are other types of power-steering gears. But all operate on the same general principles. As the driver applies turning effort to the steering wheel, the power-steering gear takes over and provides most of the turning effort.

Review questions

1. What are the two general kinds of brake assemblies used at the wheels of automobiles?
2. What carries the brake fluid from the master cylinder to the wheel cylinders?
3. How many pistons are there in the master cylinder of the dual braking system?
4. What do the brake shoes push against when the brakes are applied?
5. Why should oil never be put into the brake system?
6. Where does the pressure that operates the power-steering gear come from?

Study problems

1. The hydroelectric power plant is an example of power transmitted through the flow of a fluid. Give examples of other devices using the flow of fluid to create power.
2. Obtain information as to the reason for using air brakes on heavy duty vehicles.
3. Obtain the component parts of a power-steering device and build an operating system. Check fluid pressure at necessary points.

31 | SHOP SAFETY

§ 31-1 SAFETY IN THE SHOP

Shopwork is varied and interesting. The shop is where the action is. It is where all the power-mechanics service jobs are carried out. These jobs include grinding valves, replacing bearings, honing engine cylinders, adjusting carburetors, and so forth. These and many other jobs are discussed in Chapter 39.

Before you start working in the shop, you should know about safety. Safety in the shop means protecting yourself and your fellow workers from possible danger or injury. In this chapter, you will learn how to protect yourself and others by safely working with and handling tools.

§ 31-2 SAFETY IS YOUR JOB

Yes, safety is your job. When working in the shop, you are being "safe" if you are protecting your eyes, your fingers, your hands— yourself—from danger at all times. And, just as important, safety means looking out for the safety of those around you.

On the following pages there are safety guidelines that you should follow when you work in the shop. These guidelines are based on common sense. If you pay attention to the guidelines, you can prevent accidents. Remember, safety is your job.

§ 31-3 LAYOUT OF THE SHOP

The first thing to do when you start working in the shop is to find out where everything is located. This includes the different power tools as well as the car lifts and the work areas.Many shops mark off work areas with painted lines on the floor (Fig. 31-1). The lines warn customers and workers away from danger zones in which machinery is operating. The lines also remind workers to keep their tools and equipment within the work areas.

Notice the warning signs posted around the shop. These signs are there to remind you about safety. Read them carefully and follow the instructions at all times. Failure to follow instructions is the most common cause of accidents in the shop.

§ 31-4 EMERGENCIES

If there is an accident and someone gets hurt, notify your instructor at once! Your instructor will know what to do—call the school nurse, a doctor, or an ambulance. If there is a fire, get help at once. The quicker you get at a fire, the easier it is to control.

§ 31-5 FIRE PREVENTION

Locate the fire extinguishers in the shop, and learn how to operate them (Fig. 31-2). Gasoline is such a familiar item in the shop that people often forget that it can be extremely dangerous. A spark or lighted match in a closed place filled with gasoline vapor can cause an explosion. Even the spark from a light switch can set off gasoline vapors. There have been cases in which employees washed the shop floor with gasoline—with the doors closed—and then

Fig. 31-1. The layout of a shop. (*Mercer County Area Vocational-Technical School*)

Fig. 31-2. A fire extinguisher.

turned off the lights. The spark from the light switch set off a terrible explosion of gasoline vapor that not only destroyed the service station but also injured or killed the employees.

Remember that it is illegal and dangerous to pour gasoline down floor drains. Gasoline can form vapors in the sewer line, and these vapors could be set off by a lighted match or cigarette thrown down a drain.

To prevent explosions, keep the doors open or the ventilator system going if there is gasoline vapor around. Wipe up spilled gasoline at once and put the rags outside to dry. Never light or smoke cigarettes around gasoline. If you are working on an engine with a leaky carburetor, fuel line, or pump, catch the leaking gasoline in a container or with rags, and put the rags outside as soon as possible. Fix the leak right away. Be very careful to avoid sparks around the car. Store gasoline in an approved safety container (Fig. 31-3). Never— NEVER—keep gasoline in a glass jug. The jug

Fig. 31-3. Store all inflammable liquids in safety cans.

could break and cause a terrible explosion or fire.

Oily rags are another possible source of fire, because the oil on the rags might cause so much heat to develop that the rags ignite spontaneously, or catch fire. This is called **spontaneous combustion,** and it results from a chemical action that produces heat and fire. Oily rags and waste should be put into special closed metal containers where they can do no harm (Fig. 31-4).

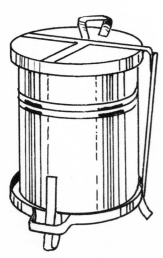

Fig. 31-4. Type of safety can used for storage of oily rags.

§ 31-6 TAKING CARE OF YOURSELF IN THE SHOP

Some people say, "Accidents will happen!" Safety experts do not agree. They say, "Accidents are caused"—caused by carelessness, by inattention to the job at hand, by the use of damaged or incorrect tools, and sometimes by just plain stupidity. To keep accidents from happening, follow these simple safety guidelines:

1. Work quietly, and give the job your undivided attention.
2. Keep your tools and equipment under control (Fig. 31-5).
3. Keep jack handles out of the way. Stand creepers against the wall when they are not in use (Fig. 31-6).

Fig. 31-5. Tools should not be scattered; always keep them within easy reach.

4. Never indulge in horseplay or other foolish activities. You could cause someone to get seriously hurt.

5. Don't put sharp objects, such as screwdrivers, in your pocket. You could cut yourself or get stabbed. Or you could ruin the upholstery in a car.

6. Make sure that your clothes are suitable for the job (Fig. 31-7). Dangling sleeves, scarves, or ties can get caught in machinery and cause serious injuries. Do not wear sandals or open-toe shoes. Wear full leather shoes with nonskid rubber heels and soles. Steel-toe safety shoes are best for shop work.

7. Wipe excess oil and grease off your hands and tools so that you can get a good grip on tools or parts.

8. If you spill oil or grease, or any liquid, on the floor, clean it up so that no one will slip and fall.

9. Never use compressed air to blow dirt from your clothes and never point a compressed-air hose at another person. Flying particles could cause serious eye injuries.

10. Always wear goggles or a face shield on *any* job where there is danger from flying particles (Fig. 31-8).

11. Watch out for flying sparks from grinding wheels or welding operations. The sparks can set your clothes on fire.

12. To protect your eyes, wear goggles when using chemicals such as solvents. If you get a chemical in your eyes, quickly flush them out with lots of water, and then see the school nurse or a doctor at once.

13. When using a car jack, make sure it is centered so that it won't slip. And *never*—NEVER—jack up a car while someone is working under it! People have been killed when the jack slipped and the car fell down on them! Always use car stands, or supports, properly placed, before going under a car (Fig. 31-9).

14. Always use the right tool for the job. The wrong tool could damage the part being worked on and could also cause injury to you or to those around you.

Fig. 31-6. When jacking up a car, always finish with the handle pointing up so that no one can trip over the handle. When creepers are not in use, stand them up against the wall where no one can stumble over them. (*Mercer County Area Vocational-Technical School*)

Fig. 31-7. A face shield and proper clothing are important accident-prevention safeguards.

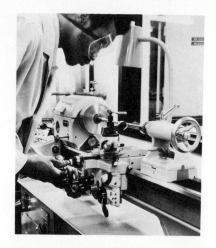

Fig. 31-8. Always wear goggles or a face shield when using a machine that can throw chips or particles.

Fig. 31-9. Properly located car stands will hold the car safely.

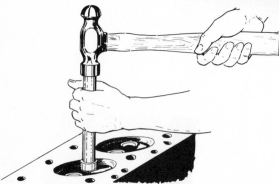

Fig. 31-10. Use brass or plastic drift between hammer and hardened surface.

§ 31-7 TAKING CARE OF TOOLS

Keep tools clean and in good condition. Always use the proper tool for the job. Do not use a hardened hammer or punch on a hardened surface (Fig. 31-10). Hardened steel is brittle and could shatter under your heavy blows. Slivers could then enter your hand, or worse, your eye. Use a soft hammer on hardened steel parts.

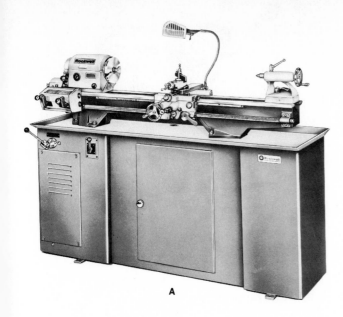

A

C

Fig. 31-11. Common power tools found in the shop: (A) bench lathe; (B) drill press; (C) pedestal grinder.

B

§ 31-8 SAFETY WITH POWER TOOLS

A great variety of power tools are found in modern shops (Fig. 31-11). They all have electric motors to turn either the cutting tool or the workpiece quickly and easily. Observe certain cautions when working with power tools in order to work safely.

1. Read and understand the instructions that explain how to operate the power tool you are about to use.
2. Have your instructor check you out and give you permission to operate the tool.
3. Keep hands and clothing away from moving parts.
4. Make sure that the work is securely held so that nothing can come loose when the tool is turned on.
5. Be ready to turn the tool off quickly if something seems wrong. Never take a chance. Be sure everything is in good order before proceeding.
6. When using portable power tools such as drills (Fig. 31-12), do not drag the tool around by the electric cord.
7. Always make sure that portable power tools are grounded. Make sure the cord is not in someone's way.

Fig. 31-12. An electric drill—a hand power tool.

8. After finishing with a portable tool, clean it and return it to the bench or tool crib, where it belongs.
9. Never attempt to adjust or oil moving machinery unless the instructions specifically state that this should be done.

Review questions

1. What is the first thing you should do if there is an accident in the shop?
2. Why should you never use gasoline when the garage doors are closed?
3. How can oily rags be dangerous?
4. What is wrong with operating a car in a garage when the garage doors are closed?
5. What are the precautions fo observe when you are using a power tool?

Study problems

1. Make a list of the important safety features of your shop. First, identify and write down the locations of the fire extinguishers. Next, note the location of the fire alarm and telephone. Then, locate the first aid kit.
2. Carefully inspect each fire extinguisher in the shop. Be sure you known how to operate it and which types of fires it can be used against. Check the inspection tag to make sure that the fire extinguisher has been serviced recently. If the fire extinguisher is equipped with a pressure gage, be sure that the fire extinguisher is still properly charged. If it has been discharged, notify your instructor immediately.

32 | HAND AND POWER TOOLS

§ 32-1 MODERN POWER MECHANICS

In the modern automotive and transportation industry, new types of transmissions, engines, steering systems, brakes, and other components are constantly being designed and tested. If a new design is accepted for production, engineers who specialize in servicing and repair procedures are called in to study the new design. They work out the proper methods of repairing the new device. If the standard tools available in the average service garage will not do the job, the engineers design new kinds of tools that will. In other chapters of this book, there are pictures and descriptions of these special tools. For instance, a special dial indicator should be used to measure the diameter of an engine cylinder. A complete mechanic's tool set is shown in Fig. 32-1.

Despite the fact that there are probably thousands of different *kinds* of tools, there are only a few basic *types* of tools, including tools for measuring, for hammering, for cutting, and for turning bolts and nuts. When working on any engine or mechanical device, it may be necessary to use measuring tools of some kind to determine whether parts are worn or damaged. For repairing or replacing worn parts and for removing bolts or nuts, wrenches or screwdrivers are needed. Hammering or cutting tools are used as new parts are fitted into machines.

Some tools are operated by compressed air or by electric motors. These are called **power tools.** The more common hand and power tools are described in this chapter.

§ 32-2 TURNING TOOLS

Most machines, including engines, are put together with bolts, screws, and nuts. To take an engine apart, or disassemble it, these bolts, screws, and nuts must be turned and taken off so that the parts can be separated. The most common tools to perform this job are wrenches and screwdrivers.

§ 32-3 SCREWDRIVERS

The typical **screwdriver** (Fig. 32-2) has a handle and a shank, the end of which has been flattened to form a blade. The blade tip should be flat on the end so that it fits squarely into the screw slot (Fig. 32-3). If it is tapered, it will tend to rise out of the screw slot when it is turned. This makes the removal or installation of a screw more difficult.

Some screws have what resembles two screw slots at right angles. The **Phillips screw** is one of these. It requires a special screwdriver with a tip to fit the crossed slots (Fig. 32-4).

The **offset screwdriver** is a variation of the standard screwdriver (Fig. 32-5). With the

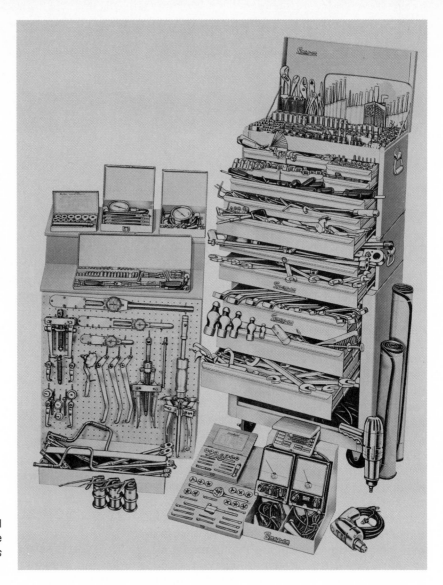

Fig. 32-1. A complete tool set used by an automobile mechanic. (*Snap-on Tools Corp.*)

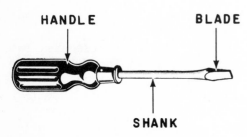

HANDLE   BLADE

SHANK

Fig. 32-2. A screwdriver.

Fig. 32-3. The screwdriver tip should fit the screw slot precisely.

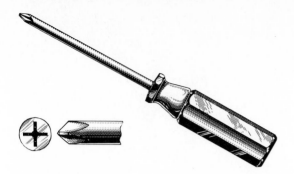

Fig. 32-4. Phillips screwdriver and the slotted head of a Phillips screw.

Fig. 32-5. An offset screwdriver.

offset screwdriver, a screw that is set in an awkward place can be reached and turned. The blade tips are set at right angles so that first one end of the screwdriver can be used and then the other to keep the screw turning.

The **Allen wrench** (Fig. 32-6) is a special form of turning tool. It has a hexagonal (six-sided) shape and fits into a hexagonal hole in the head of a screw.

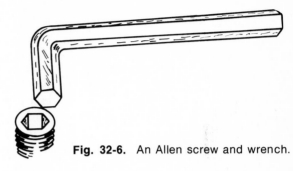

Fig. 32-6. An Allen screw and wrench.

§ 32-4 WRENCHES

Bolt heads and nuts commonly used in machines and engines are of the **hexagonal type.** That is, they have six flats around the outer surface (Fig. 32-7). These flats permit the use of a wrench to turn the nut or bolt.

A typical wrench of the **open-end type** is shown in Fig. 32-8. Most have the open end at an angle of 15° with the handle, as shown, although a few have a 90° angle for special purposes. The open-end wrench is used by placing the open end over the flats of the bolt head or nut. The wrench handle provides considerable leverage so that the bolt or nut can be tightened or loosened easily.

Box wrenches (Fig. 32-9) serve the same purpose as open-end wrenches. However, they are more useful in many cases, because the box fits down over the nut or bolt head and will not slip off. The most widely used box wrench is the **12 point,** which means the box has 12 notches in it, spaced 30° apart. One disadvantage of the box wrench is that it must

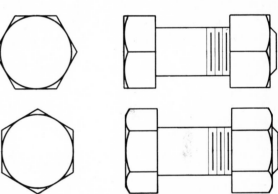

Fig. 32-7. A hex-head bolt and nut.

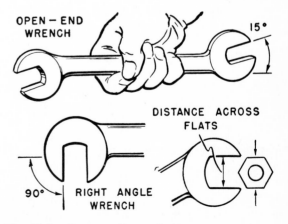

Fig. 32-8. Open-end wrenches.

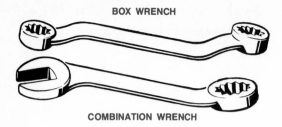

BOX WRENCH

COMBINATION WRENCH

Fig. 32-9. Box wrench and combination wrench.

be lifted clear off the nut and put on in a different position with each swing of the wrench when working in a restricted space. Here is where the open-end wrench is more convenient to use; it will slip onto the nut more easily. Thus, many wrenches are made as combination open-end and box (Fig. 32-9). The box is used to break the nut or bolt loose, and then the tool is reversed so the open-end wrench can be used for backing the nut off or the bolt out.

When using a wrench, it is usually better to pull on it than to push. Sometimes, when the wrench is pushed, the nut or bolt gives way suddenly, and you can bang your hand against the machinery before you can catch yourself. This will not happen if you pull the wrench. Even when pulling, however, be ready for the nut or bolt to break loose suddenly. If it does this and you are not expecting it, you might pull the wrench off and hit yourself with it before you can stop.

§ 32-5 ADJUSTABLE WRENCH

The **adjustable wrench** (Fig. 32-10) has an adjustable jaw that can be moved back and forth to narrow or widen the distance between the jaws. Thus, the adjustable wrench can be made to fit many sizes of nut or bolt heads. This tool is not intended to take the place of the regular wrench but is handy to have for special jobs or odd sizes of nuts or bolts. When using this wrench, make sure that it is properly tightened on the nut or bolt head. Attach the wrench so that the adjustable jaw will be on the inside of the turning motion, as shown in

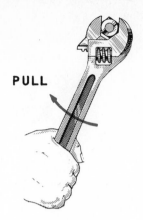

PULL

Fig. 32-10. Adjustable open-end wrench.

Fig. 32-10. In this way, the pulling pressure will keep the adjustable jaw tight against the nut or bolt head.

§ 32-6 SOCKET WRENCHES

The **socket wrench** is similar to the box wrench, except that the sockets are detachable from the handle so that a single handle can be used for many sizes of sockets. Figure 32-11 shows typical sockets. The drive end of a socket fits into the handle, and the socket fits down over the nut or bolt head.

A variety of handles are used with sockets (Fig. 32-12). The ratchet handle permits the

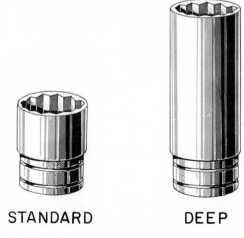

STANDARD   DEEP

Fig. 32-11. Sockets.

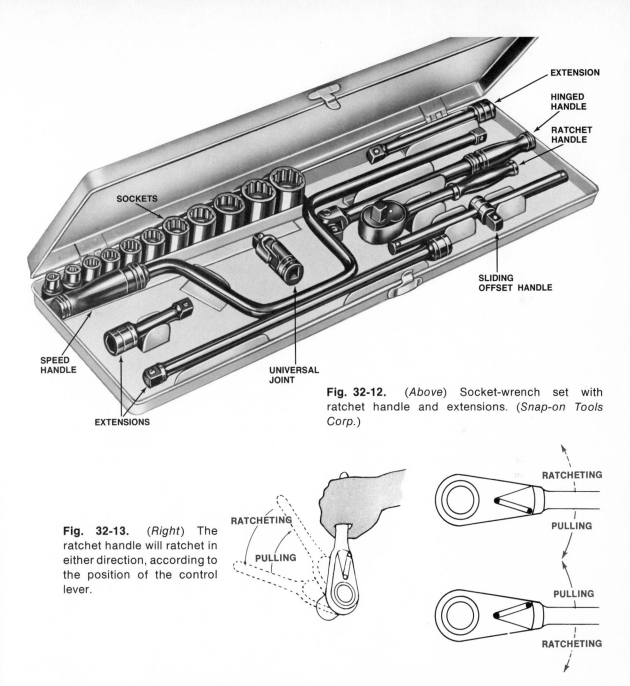

EXTENSION

HINGED HANDLE

RATCHET HANDLE

SOCKETS

SLIDING OFFSET HANDLE

SPEED HANDLE

UNIVERSAL JOINT

EXTENSIONS

Fig. 32-12. (*Above*) Socket-wrench set with ratchet handle and extensions. (*Snap-on Tools Corp.*)

RATCHETING

PULLING

RATCHETING

PULLING

PULLING

RATCHETING

Fig. 32-13. (*Right*) The ratchet handle will ratchet in either direction, according to the position of the control lever.

handle to be swung back and forth without lifting the socket from the nut or bolt head with each swing (Fig. 32-13). The ratchet can be switched from one side to the other so that the handle will either loosen a nut, ratcheting on the back swing, or tighten a nut, ratcheting in the opposite direction.

Another type of handle has a hinge that permits working at an angle on a bolt or nut in a difficult place. There is also a universal joint that can be attached between the handle and the socket. The sliding offset handle permits the application of heavy leverage if necessary. All these tools are identified in Fig. 32-12.

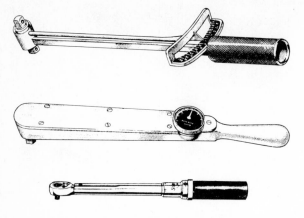

§ 32-7 TORQUE WRENCH

The **torque wrench** (Fig. 32-14) is a special form of wrench that has a measuring device to indicate the amount of **torque** (twist) being applied to a bolt or nut. In much of today's machinery, engineers have carefully worked out the amount that bolts or nuts should be tightened. If they are not tightened enough, they may loosen in service. If they are tightened too much, then too much strain may be put on the bolts or machine parts and they might break. In either case, the result could be severe damage to the machine. To prevent this, nuts and bolts must be tightened just the right amount. The torque wrench does this job. As the handle is pulled, the amount of torque is registered on the dial. With other types of torque wrenches, the specification is set by turning the handle. A loud click occurs when that setting is reached. Use of the torque wrench permits very accurate tightening of nuts and bolts.

§ 32-8 PLIERS

Pliers (Fig. 32-15) are a special form of adjustable wrench. The jaws are adjustable as the two legs of the handle are moved on the pivot.

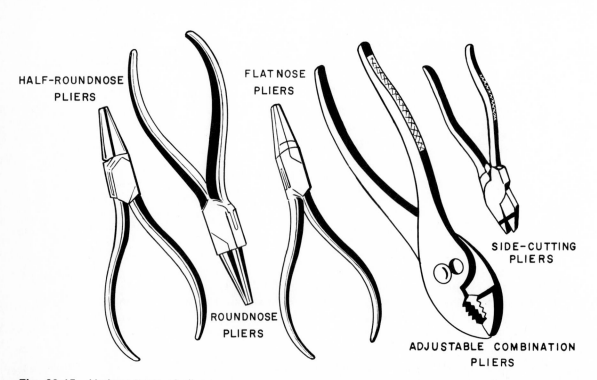

HALF-ROUNDNOSE PLIERS

FLAT NOSE PLIERS

ROUNDNOSE PLIERS

SIDE-CUTTING PLIERS

ADJUSTABLE COMBINATION PLIERS

Fig. 32-15. Various types of pliers.

FLAT CHISEL

CAPE CHISEL

ROUNDNOSE CAPE CHISEL

DIAMOND-POINT CHISEL

Fig. 32-16. Various types of chisels.

Thus, the jaws can be widened to grip an object and to hold or turn it. Pliers should not be used on nuts or bolt heads, because they will round off the edges of the hex and roughen the flats so that a wrench will no longer fit properly.

Some pliers have a side cutter that can be used to cut wire. There are also regular **nippers** that have cutting edges instead of jaws as in the pliers. These can be used to cut wire, thin sheet metal, small bolts, and so on.

§ 32-9 CUTTING TOOLS

Cutting tools have two basic functions: removing metal by chipping away small bits of the workpiece with a sharp edge (or many sharp edges), or severing, that is, slicing off pieces of metal by the shearing process. Cutting hand tools used in the metal shop include hacksaws, drills, chisels, and files.

§ 32-10 CHISELS

The **chisel** has a single cutting edge and is driven with a hammer to cut metal. Several different shapes of chisels are shown in Fig. 32-16. Each shape has its special purpose, but the chisel most commonly used is the plain flat chisel. The chisel should be held loosely in the left hand, with the right hand swinging the hammer (or the reverse if you are left-handed). The reason for holding the chisel loosely is that if the hammer does not strike square or misses,

the left hand will give with the blow and will be less subject to injury. Goggles should be worn when chipping with a chisel, since metal chips may be propelled from the workpiece or chisel and fly into the eyes. A chisel that has mushroomed on the end because of repeated hammer blows should be dressed on a grinding wheel (Fig. 32-17) so that the turned-over metal is removed. Otherwise, this metal could break off and fly away as the hammer strikes it. Not only is it hard to be accurate with a chisel in this condition, but the flying chips might cut your hand or hurt someone else. Notice also how the cutting edge of the chisel should be dressed on the grinding wheel (Fig. 32-18).

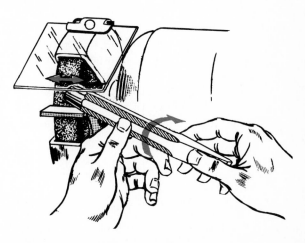

Fig. 32-17. Grinding the mushroom from the chisel head.

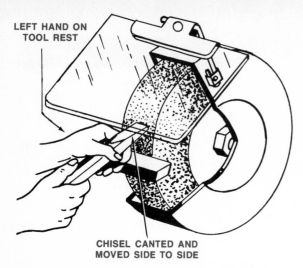

LEFT HAND ON TOOL REST

CHISEL CANTED AND MOVED SIDE TO SIDE

Fig. 32-18. Dressing, or grinding, a chisel. (*General Motors Corporation*)

§ 32-11 FILES

The **file** is a little like a series of tiny chisels each with a sharp cutting edge. Files have many sizes, shapes, and uses. A typical file with the parts named is shown in Fig. 32-19. As the face of the file is moved across a metal piece, the cutting edges or teeth remove shavings from the metal. The coarseness of the file determines how thick the shavings will be. The term **cut** is used to designate the coarseness or fineness of the file. For instance, a **coarse-cut file** has relatively large teeth that are comparatively far apart. A **fine-cut file** has many more teeth, much smaller and closer together. Such a file is called a **smooth,** or **dead-smooth, file.** Terms used to indicate coarseness or fineness are, in order: **rough, coarse, bastard, second-cut, smooth-cut** and **dead-smooth.**

When only one series of teeth has been cut across the face of the file, with all cuts parallel to each other, the file is known as a **single-cut file,** regardless of its coarseness. When the file has two series of cuts across its face in two different directions, it is known as a **double-cut file** (Fig. 32-20).

In addition to the classification by cut, files are also classified according to shape. Files may be **flat, triangular, half-round,** or **round,** and may or may not have taper (Fig. 32-21). The selection of a file depends upon the type of job it will be called upon to do.

There are a number of cautions to be observed in using a file. First, never attempt to use a file without putting a handle on the tang. Otherwise your hand might slip and you would drive the tang into your hand. To install the handle, put the file tang into the hole in the handle, and then tap the butt end of the handle on the bench. This drives the file firmly into the handle. Never try to hammer the file into the handle. The file is brittle and hammering on it could cause it to shatter.

Second, when starting a filing job, be sure that the part to be filed is fastened down or clamped firmly in a vise. Use soft jaws in the vise if the part needs to be protected from scarring (Fig. 32-22).

Third, in using the file, make the forward, or cutting, stroke smooth and firm, using the proper amount of downward pressure to get a cut. A heavy pressure tends to overload the teeth so that the cut is uneven. A light pressure will not allow the teeth to cut properly and may even tend to dull the teeth.

Fourth, on the back, or return, stroke, lift the file clear of the work. Dragging the file teeth back over the work wears the cutting edges of the teeth. The exception to this is when the file is being used on soft metal, which tends

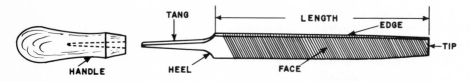

TANG **LENGTH** **EDGE** **TIP** **HANDLE** **HEEL** **FACE**

Fig. 32-19. A typical file with parts indicated. (*General Motors Corporation*)

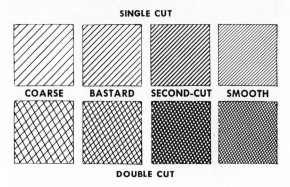

SINGLE CUT

COARSE BASTARD SECOND-CUT SMOOTH

DOUBLE CUT

Fig. 32-20. File cuts. (*General Motors Corporation*)

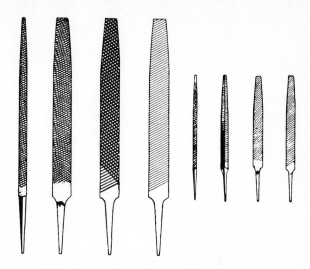

Fig. 32-21. Various types of files.

to clog the teeth. In this case, dragging the teeth on the back stroke will help to clean the teeth. The teeth can also be cleaned with a file card. The file card is a wire-bristled brush that is moved across the face of the file in line with the teeth. Tapping the handle on the bench every few strokes will also help to keep the teeth clean.

Fifth, when a file is not in use, it should be put away carefully. If it is thrown into a drawer with other files or tools, the teeth will be chipped and dulled. Files are subject to rusting and should be kept in a dry place.

▶ CAUTION ◀

Never attempt to use a file as a pry bar and never hammer on it. The file is brittle and could shatter in a dangerous manner.

---

§ 32-12 PUNCHES

Punches (Fig. 32-23) are used to knock out rivets or pins and are also used to align parts for assembly and to mark the locations of holes to be drilled. They come in several shapes, each having its own special use. The **starting punch** is used with a hammer to loosen a pin or shaft. The **pin punch,** which has a long thin shank, can then be used to drive the pin out (Fig. 32-24).

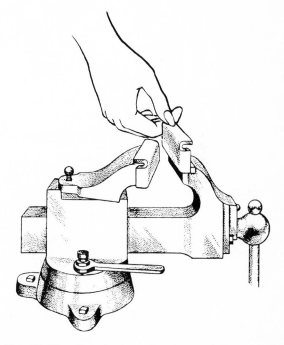

Fig. 32-22. Bench vise, showing soft jaws being put into place on vise jaws to protect against scarring the workpiece.

The **center punch** (Fig. 32-23) is used for marking hole locations for drilling. It is also used for marking parts before disassembly so

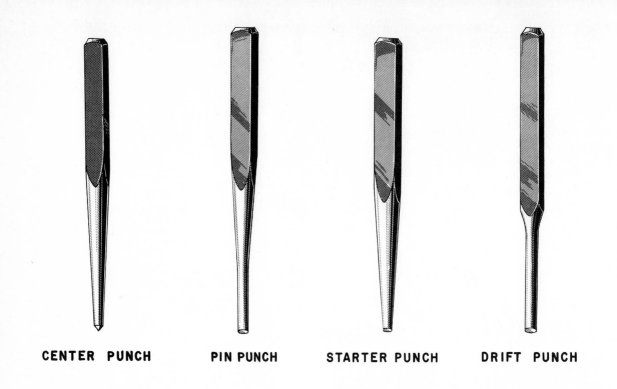

CENTER PUNCH   PIN PUNCH   STARTER PUNCH   DRIFT PUNCH

LINING-UP PUNCH

Fig. 32-23. Various types of punches.

that they can be reassembled in the same relative positions. For example, it might be possible to put a cover plate on a housing in more than one position. But if, before disassembly, a punch mark is made on the edge of the cover plate, and a punch mark then made next to it on the housing, the two punch marks can be aligned on reassembly so that the cover plate may be put back on in its proper position.

§ 32-13 DRILLS

Drills are tools for making holes. The type of material into which the hole is to be drilled determines the type of drill to be used. The typical drill, or **twist drill,** as it is called, is a

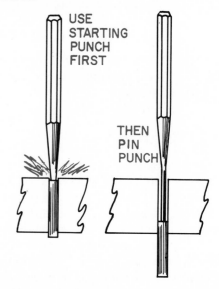

USE STARTING PUNCH FIRST

THEN PIN PUNCH

Fig. 32-24. Using starting and pin punches. (*General Motors Corporation*)

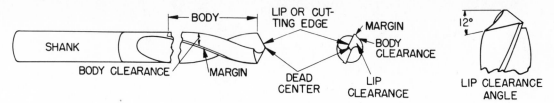

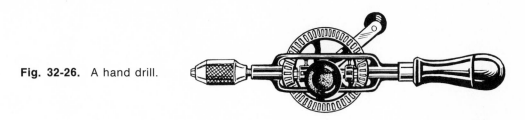

Fig. 32-25. The parts of a twist drill.

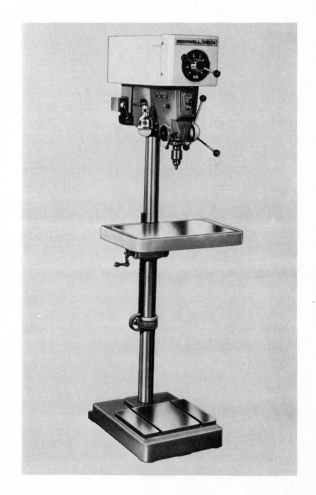

Fig. 32-26. A hand drill.

cylindrical bar in which helical grooves have been cut (Fig. 32-25). Note that the drill has two cutting edges cut into the material. The resulting chips pass up the helical grooves and away from the working surface.

While twist drills are often used with hand-operated drills (Fig. 32-26), most drills for the shop are powered by an electric motor. These use a set of reduction gears that turn the twist drill at a relatively slow speed. For more accurate or for larger hole drilling, a **drill press** is used (Fig. 32-27). This has a heavy upright to support the drill motor and drill. It also has a platform on which the work can be placed and attached, if necessary. A handle is provided that turns a gear and moves a rack to bring the drill down to the work.

Cautions to observe in the use of the electric drill (and other power tools) are outlined in Chapter 31, ''Shop Safety.''

The center punch should always be used to mark the location of a hole to be drilled in a metal piece. The center-punch mark locates the position of the hole accurately and keeps the drill from wandering (Fig. 32-28).

§ 32-14 HACKSAW

The **hacksaw** is the saw commonly used for cutting metals. The blades are replaceable and are held, for sawing, in a metal adjustable

Fig. 32-27. A drill press.

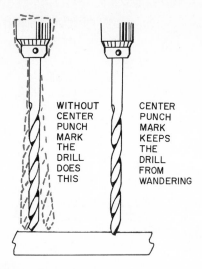

Fig. 32-28. Center-punching a hole location will keep the drill from wandering. (*General Motors Corporation*)

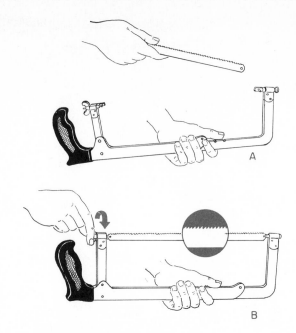

Fig. 32-29. (*A*) Installing the blade on a hacksaw. (*B*) Tightening the blade. Note that teeth point *away* from the handle.

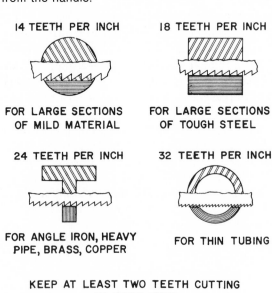

14 TEETH PER INCH

FOR LARGE SECTIONS OF MILD MATERIAL

18 TEETH PER INCH

FOR LARGE SECTIONS OF TOUGH STEEL

24 TEETH PER INCH

FOR ANGLE IRON, HEAVY PIPE, BRASS, COPPER

32 TEETH PER INCH

FOR THIN TUBING

KEEP AT LEAST TWO TEETH CUTTING TO AVOID THIS

Fig. 32-30. Select the proper blade for each material.

frame (Fig. 32-29). Blades are made with from 14 to 32 teeth per inch. Using a blade with the wrong number of teeth for the job will make the job more difficult and could also cause the blade to break.

Figure 32-30 illustrates correct and incorrect blades for various jobs. After the proper blade is selected, it should be placed in the hacksaw frame and tightened to the proper tension (Fig. 32-29). Teeth should point away from the handle so that the cut is made on the push (forward) stroke.

When using the hacksaw (Fig. 32-31), take smooth, even, forward strokes and use steady pressure to get a good cut. On the return stroke, lift the blade slightly so that the back edges of the teeth will not rub. Rubbing could dull the teeth.

In sawing sheet metal, it may be found that even with a 32-tooth blade, only one tooth will be in contact with the work at one time. For such work, clamp the sheet metal in the vise between two blocks of wood and cut through both the wood and metal at the same time.

Take good care of blades. Do not throw them carelessly into the drawer with other tools, because this will dull the teeth. Instead, wipe

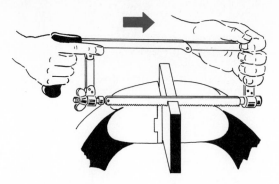

Fig. 32-31. The proper way to hold a hacksaw.

them with a lightly oiled cloth and lay them away in a drawer where they will not be damaged.

§ 32-15 TAPS AND DIES

Taps and **dies** are used to cut screw threads. A **tap** can be run down into a hole to cut internal threads. A **die** is run over the outside of a rod to make external threads.

Taps are made in several styles (Fig. 32-32). To tap a hole that has been drilled, the proper tap is put into a **tap wrench** (Fig. 32-33). The tap selected is based on the size of the hole and the type of thread to be cut. Also considered is whether the hole is to be tapped all the way through or not. For a hole that goes completely through a piece of metal, use the **taper tap.** To thread a hole only part way, use the **plug tap.** To thread a hole all the way to the bottom when it does not go all the way through a piece of metal, use the **bottoming tap.**

To tap the hole: Apply a lubricant such as soluble oil to the tap. Start the tap squarely in the hole, and then turn the wrench smoothly and evenly so that the tap cuts the thread. Every couple of turns the tap should be backed off about a quarter turn so that chips can be cleared and more lubricant applied.

Dies cut outside threads (Fig. 32-34). The die is held in a **die stock** (Fig. 32-35). To thread a rod, the die is fastened into the stock. The end of the rod should be chamfered so that the die will start easily. Then the die should be

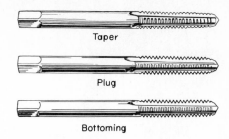

Fig. 32-32. Hand taps. (*Greenfield Tap and Die Corporation*)

Fig. 32-33. Hand-tap wrenches. (*Greenfield Tap and Die Corporation*)

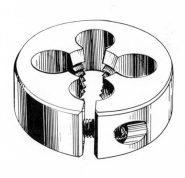

Fig. 32-34. Dies. (*Greenfield Tap and Die Corporation*)

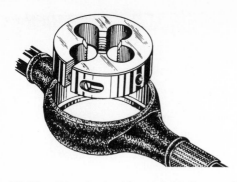

Fig. 32-35. Die stock. (*Greenfield Tap and Die Corporation*)

lubricated and turned onto the rod, smoothly and evenly. Every couple of turns, the die should be backed off a part turn so that more lubricant can be applied.

§ 32-16 GRINDERS

The cutting element of the grinder is the grinding wheel. **Grinding wheels** (Fig. 32-36) are made of abrasive material. They are made in many sizes and with many different kinds and grades of abrasive. For coarse work, the abrasive particles are relatively large and cut the material very rapidly. An example might be the grinding wheels used to take the rough edges off metal castings. For fine work, such as sharpening small tools, a relatively fine grinding wheel would be used.

Great care must be taken when the grinding wheel is used. The wheel is driven by an electric motor and revolves at fairly high speed. The wheel could be broken by heavy blows, heavy pressure, or excessive tightening of the spindle nut. During the grinding of many metals, showers of hot sparks are thrown off by the wheel. These sparks can be seen in Fig. 32-36. Figure 32-37 shows a grinding wheel being used to sharpen a screwdriver. When finished, the screwdriver should look as shown in Fig. 32-3.

Goggles should always be worn when the grinding wheel is used, and the safety shield should be in place (Fig. 32-37). When using the

Fig. 32-36. A shower of sparks is produced by the grinding wheel cutting the workpiece.

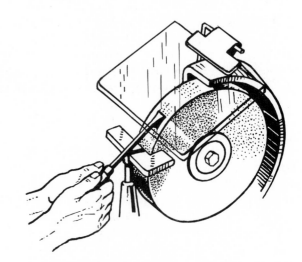

Fig. 32-37. Grinding wheel used to sharpen a screwdriver.

wheel, do not allow the metal being ground to get overheated. This is particularly important when grinding a tool such as a screwdriver, chisel, or drill. Overheating of the tool will draw the temper of the steel and make the tool so

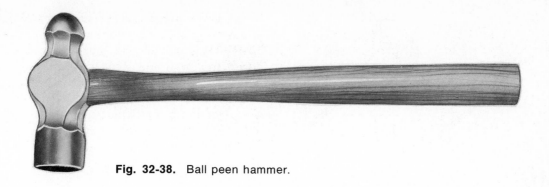

Fig. 32-38. Ball peen hammer.

soft that the edge would not last long. To prevent overheating, dip the tool in water repeatedly during the grinding operation.

▶ CAUTION ◀

Do not use the grinder, or any other powered tool, until your instructor has explained to you how it operates and has told you to go ahead. These tools are easy and safe to use if they are used properly. But they can cause injuries if they are improperly used.

---

§ 32-17 HAMMERS

A variety of **hammers** are used in the shop. You are probably familiar with the claw hammer, which has a claw opposite the driving head for pulling nails. This is a carpenter's tool and is not of much use around the machine shop. Instead, the **ball peen hammer** is used (Fig. 32-38). When using this hammer, grip the handle firmly near its end and swing it so that the face strikes the object squarely (Fig. 32-39).

If the work requiring the use of a hammer is apt to be dented or otherwise damaged by the hard face of the ball peen hammer, then a **soft-faced hammer** should be used (Fig. 32-40).

Check the head of the hammer occasionally to make sure that it is firmly in place on the handle. A wedge or screw is used to spread

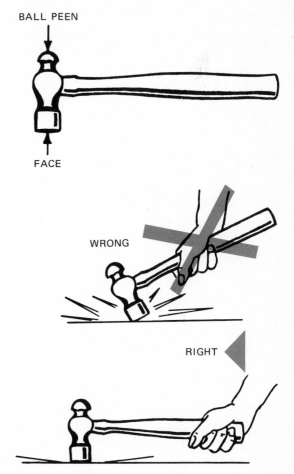

Fig. 32-39. Correct striking action. (*General Motors Corporation*)

the handle and tighten it in the eye of the hammerhead. If the wedge or screw has come loose, it should be driven tight. Someone could

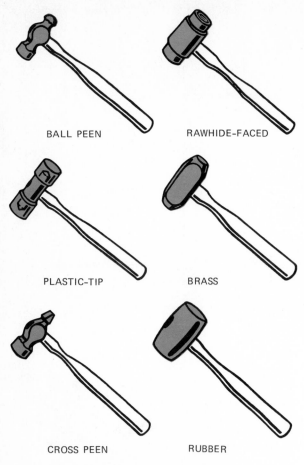

BALL PEEN

RAWHIDE-FACED

PLASTIC-TIP

BRASS

CROSS PEEN

RUBBER

Fig. 32-40. Various types of hammers.

The **arbor press** has a handle that rotates a gear which is meshed with a rack (Fig. 32-41). This causes the rack to move up or down. The lower end of the rack holds the tool. A considerable amount of pressure can be exerted through the arbor press and tool on work placed on the platform. The arbor press can therefore be used to press bearings, pins, and other tightly held parts into or out of an assembly.

be very seriously hurt if the hammerhead should happen to fly off when the hammer is swung.

§32-18 OTHER COMMON TOOLS

The **bench vise** is used to hold the object being worked on (see Fig. 32-22). When the handle is turned, a screw in the base of the vise shifts the movable jaw toward or away from the stationary jaw. When objects that are easily marred are put into the vise, the soft jaws should be used. These soft jaws are caps of a soft metal (such as copper) that are much less apt to scratch or dent an object gripped in the vise jaws.

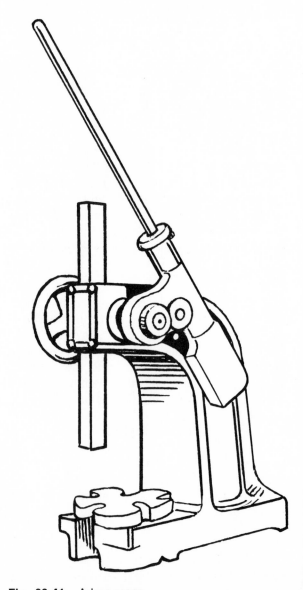

Fig. 32-41. Arbor press.

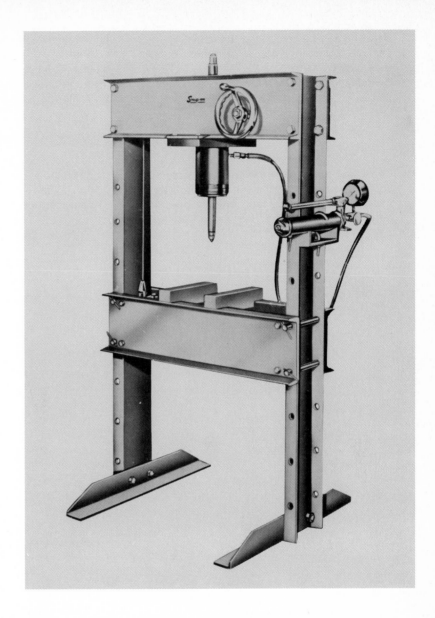

Fig. 32-42. Hydraulic press.

Larger presses of this type are operated hydraulically and can exert thousands of pounds of pressure (Fig. 32-42). Such presses are sometimes used to straighten car frames or to bend or straighten heavy metal parts.

Review questions

1. List at least 10 basic hand tools that should be in the toolbox of every engine mechanic.
2. What are the advantages and disadvantages of box wrenches?

3. List the types of socket-wrench handles and describe how each is used.
4. Why is a torque wrench necessary in power mechanics?

5. Describe how to select and use a file.
6. List the common types of punches. Explain their uses.
7. Describe the procedure for drilling and tapping a hole.
8. Why is a center punch used before drilling a hole?
9. List the safety rules to follow when using grinders in the power-mechanics shop.
10. Name the hammer used in the machine shop.

Study problem

1. Check out a complete toolbox with tools from your instructor. Empty the contents onto a workbench. Pick up each tool, one by one, and return it to the toolbox. As you do so, write down the name and most common use of each tool. When you finish, check your list against your instructor's inventory sheet.

33 | MEASUREMENTS AND MEASURING TOOLS

§33-1 MEASUREMENTS

Almost any service job performed by a mechanic requires the taking of measurements. In some cases, thickness, diameter, or length must be measured. In other cases engine-power output, vacuum, or speed may need to be measured. Taking some sort of measurement is often the first step in any testing procedure or repair job. Many measurements are made in inches or fractions of an inch. However, as the United States begins to use the metric system, the meter and its divisions will become familiar to the mechanic. We will study more about the metric system of measurement in Chapter 34. Now, let's discuss some of the frequently used measuring tools.

§33-2 RULER

The common ruler or steel scale (Fig. 33-1) is marked off in inches and fractions of an inch. Sometimes these markings are as small as $\frac{1}{64}$ inch. Other rulers or steel scales are marked with both inches and centimeters, or millimeters (Fig. 33-2). The metric ruler has only metric markings.

However, the ruler is not of great use to the mechanic. It has two serious disadvantages. One, it is not accurate enough to measure many machine parts. Two, it cannot be used to take certain measurements because of its shape.

§33-3 FEELER GAGES

Feeler gages are strips or blades of hardened steel or other metal, ground or rolled with extreme accuracy to the proper thickness. They are generally supplied in sets (Fig. 33-3), with each blade being marked with its thickness in thousandths of an inch, thousandths of a millimeter, or both. For example, in Fig. 33-3, the gage marked 0.003 inch is also marked in its metric equivalent of 0.08 millimeter.

§33-4 STEPPED FEELER GAGES

Some feeler gages have two steps, or thicknesses; these are **stepped feeler gages** (Fig. 33-4). The tip of the blade of a stepped feeler gage is thinner than the rest of the blade (Fig. 33-5). The blade marked 4–6 in Fig. 33-4 is

Fig. 33-1. A ruler, or steel scale.

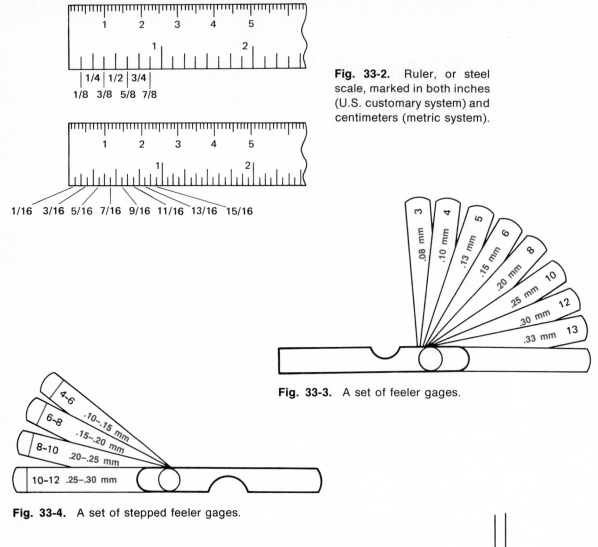

Fig. 33-2. Ruler, or steel scale, marked in both inches (U.S. customary system) and centimeters (metric system).

Fig. 33-3. A set of feeler gages.

Fig. 33-4. A set of stepped feeler gages.

STEPPED FEELER GAGE

0.012 IN.

0.012—READ AS TWELVE THOUSANDTHS
0.010—READ AS TEN THOUSANDTHS
(OR ONE HUNDREDTH)

0.010 IN.

Fig. 33-5. The blade of a stepped feeler gage. The tip is thinner than the rest of the blade.

440

0.004 inch [0.10 millimeter] thick at the tip and 0.006 inch [0.15 millimeter] thick on the thicker portion that starts about $\frac{1}{2}$ inch back from the end of the blade. This type of feeler gage is handy on certain jobs where the specifications might call, for example, for a clearance of 0.005 inch [0.13 millimeter]. By using the 0.004–0.006 inch [0.10–0.15 millimeter] gage, the adjustment can be made so that the 0.004-inch [0.10-millimeter] portion will fit, and the 0.006-inch [0.15-millimeter] portion will not fit.

§ 33-5 WIRE FEELER GAGES

Wire feeler gages (Fig. 33-6) are similar to the flat feeler gages, except that they are made of carefully calibrated steel wire of the proper thickness. They are useful in checking spark-plug gaps and similar dimensions. Metric wire feeler gages are also available. When specifications are given in the metric system, and you do not have metric gages, refer to Chapter 34, "Using The Metric System."

§ 33-6 CALIPERS

Calipers can be used with a scale to make fairly accurate measurements of many parts. Figure 33-7 shows typical outside and inside calipers. The **outside caliper** can be used to measure the diameter of a shaft. First, it is adjusted by turning the thumb nut until the legs will just slip over the shaft with a slight drag (Fig. 33-8). Then, the distance between the two legs can be measured with a scale (Fig. 33-9).

In a similar manner, the **inside caliper** can be used to measure the diameter of a hole (Fig. 33-10). It is adjusted until the legs can be moved in the hole with a slight drag. Then, the distance between the legs is checked with a scale (Fig. 33-11).

A **dial caliper** is shown in Fig. 33-12. With this precision measuring tool, outside and inside measurements can be read directly. Although a dial caliper is a relatively expensive tool, it is a very quick and accurate tool to use. The dial caliper is read in about the same way

Fig. 33-6. A set of wire feeler gages.

Fig. 33-7. *Top:* Outside calipers. *Bottom:* Inside calipers.

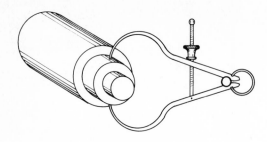

Fig. 33-8. Using outside calipers to measure the diameter of a shaft.

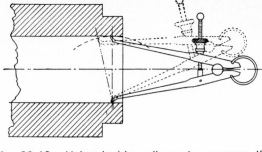

Fig. 33-10. Using inside calipers to measure the diameter of a hole.

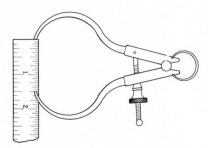

Fig. 33-9. Measuring the distance between the legs of the caliper.

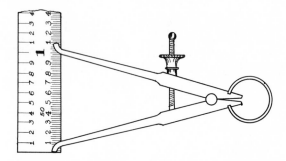

Fig. 33-11. Measuring the distance between the legs of the caliper.

as a dial indicator. This is discussed in § 33-12. There is more about calipers in §§ 33-13 and 33-14.

§ 33-7 PRECISION MEASURING INSTRUMENTS

Measurements made with a caliper and scale are usually not accurate enough for engine service and repair work. In such work, measurements must be taken in much finer detail than $\frac{1}{64}$ inch [0.397 millimeter]. Thus, for such measurements, more precise instruments are used, such as the micrometer, the dial indicator, and the vernier caliper. Let's begin our study of precision measuring instruments by reviewing how to read the micrometer.

§ 33-8 MICROMETERS

One special form of caliper is called a **micrometer,** or **mike** (Figs. 33-13 and 33-14). The mi-

crometer measures much more accurately than an ordinary caliper. In addition, the measurements can be read directly from the micrometer itself without the use of a separate ruler or steel scale.

Instead of reading in fractions of an inch like many rulers, the micrometer measures in tenths, hundredths, thousandths, and sometimes ten-thousandths of an inch. That is, it uses the decimal system. To make it easy for you to convert fractions of an inch into decimals, and vice versa, Fig. 33-15 can be used. For instance, $\frac{1}{4}$ inch is equal to 0.250 inch, and $\frac{15}{16}$ inch equals 0.9375 inch.

§ 33-9 USING THE MICROMETER

In the shop, the micrometer is usually called a **mike,** and measuring something with the mike is called **miking.** To use the micrometer to mike the diameter of a round shaft, or stock,

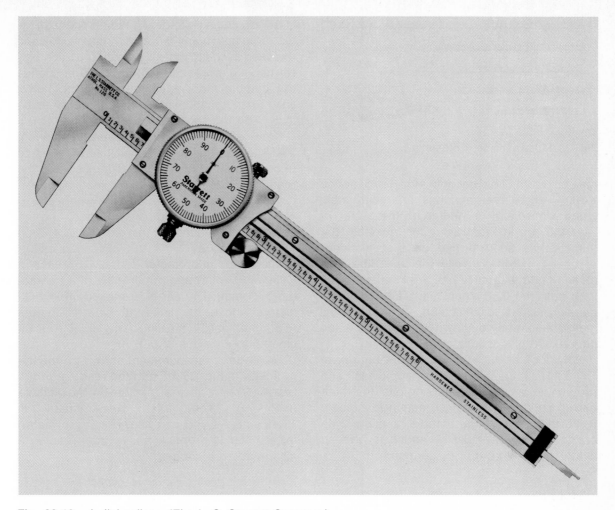

Fig. 33-12. A dial caliper. (*The L. S. Starrett Company*)

it is held as shown in Fig. 33-16. The thimble is turned until the piece being measured is a light-drag fit between the anvil and the end of the spindle. Then, the measurement is read off the barrel and thimble. The barrel markings each indicate 0.025 inch, and four of them make 0.1 inch. Thus, if three of the figures are visible, as shown at the top in Fig. 33-17, then the thimble will have uncovered 12 markings and the measurement will be *at least* 0.300 inch.

Notice we said *at least.* Noting the figures on the barrel is only the first step in reading the

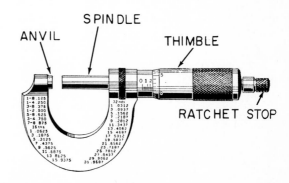

Fig. 33-13. A micrometer.

Fig. 33-14. An inside micrometer.

measurement. The second step is to notice exactly where the markings on the thimble stand with regard to the line on the barrel.

At the top in Fig. 33-17, for instance, the thimble has been turned so that marking-number 4 on it lines up with the barrel line. This means that the thimble has been turned 0.004 inch past the 0.300-inch mark. Each thimble marking is equal to 0.001 inch. Thus, the actual measurement being shown at the top is 0.304 inch (0.300 plus 0.004).

Now, look at the middle and the bottom of Fig. 33-17. In the middle, the thimble has been turned just one thimble marking past the ninth barrel marking. The barrel markings indicate 0.225 inch. The single thimble marking, indicating 0.001 inch, must be added to give an actual measurement of 0.226 inch.

As you can see, to get a reading on the mike, first note how many barrel markings have been exposed as the thimble is backed off. Each barrel marking represents 0.025 inch. Then, notice the thimble marking aligned with the barrel line. Each thimble marking represents 0.001 inch. Add the two to get the total.

Now, notice the bottom measurement in Fig. 33-17. Here, the thimble has been turned to uncover only eight barrel markings so that the reading is 0.200 inch. But the thimble has been turned until its marking 24 aligns with the barrel line. This figure is thus 0.024 inch, and the total measurement is therefore 0.224 inch (0.200 plus 0.024).

The **metric micrometer** (Fig. 33-18) reads in millimeters (mm) and hundredths of a millimeter (0.01 mm). The barrel is marked off in millimeters above the line and half millimeters (0.50 mm) below the line. The thimble is divided into divisions of 0.01 mm.

To read the measurement shown on the micrometer in Fig. 33-18, add the reading on the barrel (11 mm) to the reading on the thimble (0.45 mm). The total measurement is 11.45 mm. As you can see, reading a metric micrometer is probably easier than reading a mike marked in inches.

$\frac{1}{64}$	.0156	$\frac{17}{64}$	.2656	$\frac{33}{64}$	.5156	$\frac{49}{64}$	.7656
$\frac{1}{32}$	.0312	$\frac{9}{32}$	.2812	$\frac{17}{32}$	.5312	$\frac{25}{32}$	.7812
$\frac{3}{64}$	.0468	$\frac{19}{64}$	.2969	$\frac{35}{64}$	.5469	$\frac{51}{64}$	.7969
$\frac{1}{16}$	.0625	$\frac{5}{16}$	.3125	$\frac{9}{16}$	.5625	$\frac{13}{16}$	.8125
$\frac{5}{64}$	.0781	$\frac{21}{64}$	.3281	$\frac{37}{64}$	.5781	$\frac{53}{64}$	.8281
$\frac{3}{32}$	.0937	$\frac{11}{32}$	.3437	$\frac{19}{32}$	.5937	$\frac{27}{32}$	.8437
$\frac{7}{64}$	.1094	$\frac{23}{64}$	.3594	$\frac{39}{64}$	.6094	$\frac{55}{64}$	.8594
$\frac{1}{8}$	.125	$\frac{3}{8}$	.375	$\frac{5}{8}$	.625	$\frac{7}{8}$	.875
$\frac{9}{64}$	.1406	$\frac{25}{64}$	.3906	$\frac{41}{64}$	.6406	$\frac{57}{64}$	.8906
$\frac{5}{32}$	.1562	$\frac{13}{32}$	.4062	$\frac{21}{32}$	.6562	$\frac{29}{32}$	.9062
$\frac{11}{64}$	.1719	$\frac{27}{64}$	.4219	$\frac{43}{64}$	.6719	$\frac{59}{64}$	.9219
$\frac{3}{16}$	.1875	$\frac{7}{16}$	.4375	$\frac{11}{16}$	.6875	$\frac{15}{16}$	.9375
$\frac{13}{64}$	.2031	$\frac{29}{64}$	.4531	$\frac{45}{64}$	.7031	$\frac{61}{64}$	.9531
$\frac{7}{32}$	.2187	$\frac{15}{32}$	.4687	$\frac{23}{32}$	.7187	$\frac{31}{32}$	.9687
$\frac{15}{64}$	.2344	$\frac{31}{64}$	.4844	$\frac{47}{64}$	.7344	$\frac{63}{64}$	.9843
$\frac{1}{4}$	.25	$\frac{1}{2}$	.5	$\frac{3}{4}$	.75	1	1.0

Fig. 33-15. Decimal equivalents.

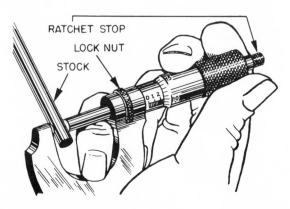

Fig. 33-16. Using a micrometer to measure the diameter of a rod.

RATCHET STOP

LOCK NUT

STOCK

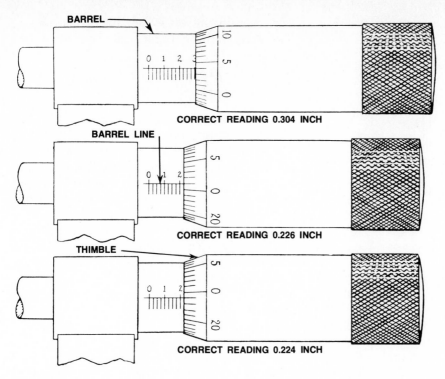

CORRECT READING 0.304 INCH

CORRECT READING 0.226 INCH

CORRECT READING 0.224 INCH

Fig. 33-17. Reading a micrometer.

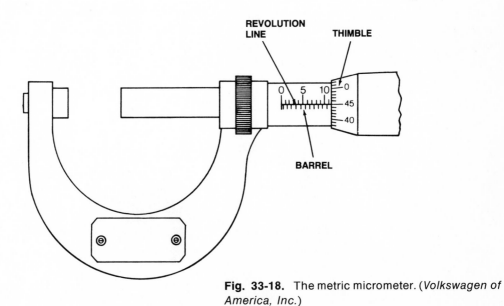

Fig. 33-18. The metric micrometer. (*Volkswagen of America, Inc.*)

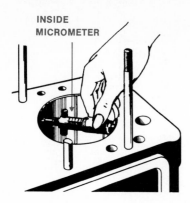

Fig. 33-19. Using an inside micrometer to measure the diameter, or bore, of an engine cylinder. (*Austin Healey*)

Many variations of the micrometer are used in shop work. For example, there are inside micrometers (Fig. 33-14) for measuring hole diameters (Fig. 33-19). In addition, much of the precision power machinery in the shop has micrometer adjustments. Such tools as cylinder honing or boring equipment, lathes, and grinders have micrometer adjustments so that the machining operations can be carefully controlled.

A recent development is the digital micrometer (Fig. 33-20). This micrometer makes mike reading faster and easier because the actual measurement appears automatically in the "counter" on the frame. Figure 33-20 shows a digital micrometer that reads in the metric system, with the measurement in millimeters appearing in the counter.

§ 33-10 CAUTIONS IN USING MICROMETERS

The micrometer has very precisely cut screw threads in it, and *rough treatment will ruin it.* The micrometer should not be thrown about carelessly on the work bench. It should be kept in a special drawer or container where it is protected from dirt and danger of other tools being dropped on top of it. It should be wiped clean after every use.

Never clamp it on the piece being measured. The thimble should be tightened only enough to cause the micrometer to drag slightly as it slides over the piece. Clamping will distort the threads and frame and could ruin the micrometer.

Never try to measure a piece being turned in a lathe with the piece revolving. The micrometer might tighten on the piece and be torn out of your hand. This could not only ruin the work and the micrometer, but you could also be injured.

§ 33-11 DIAL INDICATORS

The dial indicator is a gage that uses a dial face and a needle to register measurements (Fig. 33-21). The dial indicator has a movable

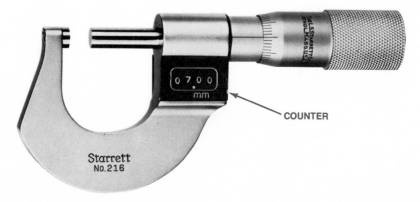

Fig. 33-20. A digital micrometer that reads in the metric system. (*The L. S. Starrett Company*)

COUNTER

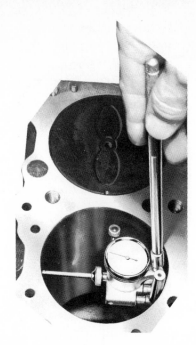

Fig. 33-21. A dial indicator being used to detect wear in an engine cylinder. (*Pontiac Motor Division of General Motors Corporation*)

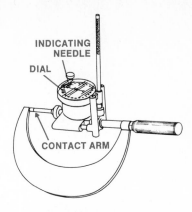

Fig. 33-22. A dial indicator can be used to measure the diameter, or bore, of an engine cylinder. Once the reading is taken, the dial is set to zero and the reading is measured with a micrometer. (*Pontiac Motor Division of General Motors Corporation*)

contact arm. When the arm is moved, the needle rotates on the dial face to show movements in thousandths of an inch. The dial indicator is used to measure end play in shafts or gears, movement of contact points, cylinder bores in engine blocks, and so on. Figure 33-21 shows a dial indicator being used to measure the bore, or diameter, of a cylinder. As the dial indicator is moved up and down, any difference in the diameter will cause the needle to move. Differences in the cylinder diameter at various points indicate cylinder wear.

To use the dial indicator to find the actual diameter of a cylinder, insert the indicator in the cylinder and note the position of the needle. Then, remove the dial indicator and use a micrometer, as shown in Fig. 33-22. You must adjust the mike until the needle is in the same position as it was in the cylinder. Then, read the setting of the mike to get the actual dimension in thousandths of an inch.

§ 33-12 HOW TO READ A DIAL INDICATOR

There are two scales on some dial indicators (Fig. 33-23). The outer scale is usually marked in measurements of 1/100 of an inch (0.01 inch). If an inner scale is used, it frequently records or counts the number of revolutions made by the large indicator needle.

On the metric dial indicator (Fig. 33-24), the outer scale is divided into 100 divisions, each division representing 1/100 of a millimeter (0.01 mm). The inner scale is divided into 10 divisions. Each division is 1 millimeter. So, every complete revolution of the outer scale represents 1 millimeter.

Now, let's read the measurement shown on the dial indicator in Fig. 33-24. The needle on the inner scale has passed 4 but has not quite reached 5, making a reading of 4 mm. To find out how much past 4 we've gone, we must add the reading on the outer dial. The dial on the outer scale determines the fractional (1/100-mm) reading. In Fig. 33-24, the outer needle is pointing to 98, which is read and written as 0.98. Added together, the total reading is 4.98 mm.

Fig. 33-23. A dial indicator. (*The L. S. Starrett Company*)

§ 33-13 VERNIER CALIPER

The main disadvantage of using ordinary calipers, such as we discussed in § 33-6, is that they do not give a direct reading of the measurement taken. To find the actual measurement, the setting of the caliper must be measured with a steel scale or ruler.

In 1631, Pierre Vernier, a French mathematician, invented an accurate, direct-reading scale for use in making linear measurements. When Vernier's scale was combined with an ordinary caliper, the result was a measuring instrument called a **vernier caliper.**

Basically, the vernier caliper is made of two graduated steel rules (Fig. 33-25). One rule is fixed and is called the fixed rule or frame. To one end of the frame is attached the fixed jaw (Fig. 33-25). The second rule is moveable and slides along the frame. One end of the sliding rule has the other measuring jaw attached (Fig.

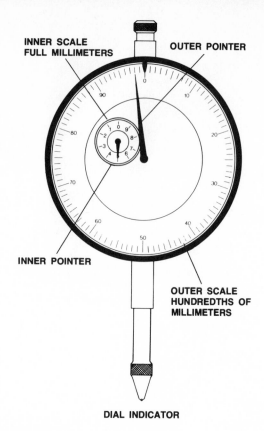

DIAL INDICATOR

Fig. 33-24. A metric dial indicator. (*Volkswagen of America, Inc.*)

33-25). This jaw is called the adjustable, or sliding, jaw.

§ 33-14 HOW TO READ A VERNIER CALIPER

To measure an outside diameter with a vernier caliper, place the object snugly between the jaws of the caliper (Fig. 33-26). The number of inches is read on the fixed scale on the frame. To this is added the number of tenths that are seen between the last inch reading and the 0 on the vernier scale. Then the number of 0.025-inch marks seen between the last tenth reading and the 0 on the vernier scale are added. Finally, we read the number of lines from 0 on the reverse scale to the point where

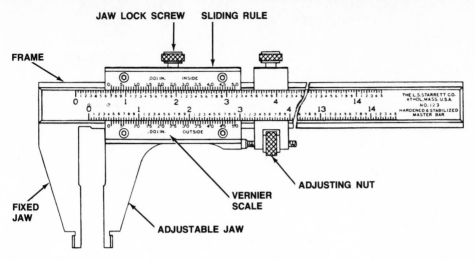

Fig. 33-25. The vernier caliper. (*The L. S. Starrett Company*)

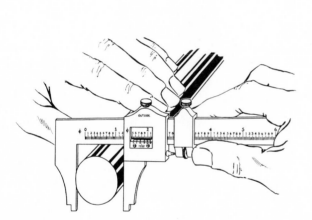

Fig. 33-26. Using the vernier caliper to measure an outside diameter.

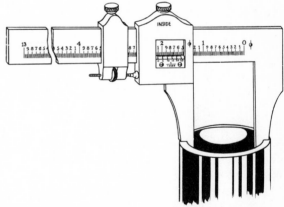

Fig. 33-27. The vernier caliper can also be used to measure an inside diameter.

the line on the vernier scale coincides exactly with a line on the fixed scale. Each of these lines represents $\frac{1}{1,000}$ of an inch (0.001 inch).

Figure 33-27 shows the use of a vernier caliper to measure an inside diameter. Some vernier calipers have slightly different scales on one side to be used for reading inside diameters. Other vernier calipers have different 0 marks to be used when reading inside diameters.

Figure 33-28 shows a simplified metric vernier caliper. On this caliper, the fixed scale is divided into 1-millimeter sections. The moveable vernier scale is divided into 10 lines, each representing $\frac{1}{10}$ of a millimeter (0.10 mm). To read the metric vernier caliper, count the number of millimeters on the fixed scale that precede the vernier scale 0. Then, look at the vernier scale until a line is located that coincides exactly with a line on the fixed scale.

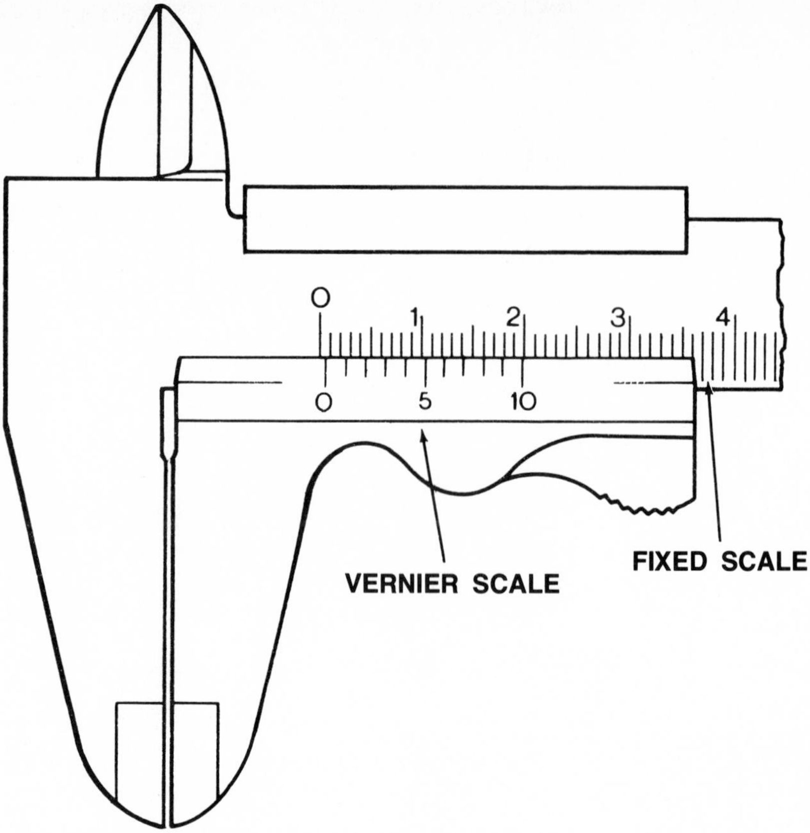

Fig. 33-28. A simplified metric vernier caliper. (*Volkswagen of America, Inc.*)

VERNIER SCALE

FIXED SCALE

Count each line on the vernier scale from 0 to the point where the two lines coincide. Since each line is 0.10 of a millimeter, be sure to count by tens. This is the decimal portion of the reading, which must be added to the first reading to obtain the complete measurement.

Review questions

1. What is the other name for a ruler?
2. How thick is a stepped feeler gage marked "006–008"?
3. What jobs are best performed using wire feeler gages?
4. When ordinary calipers are used to measure a part, how is the actual measurement read?
5. Name three precision measuring instruments.
6. What is the name of the part of the mike that is turned?
7. What type of mike is used to measure the diameter of a cylinder?
8. Describe the difference between reading an inch-scale micrometer and a metric-scale micrometer.
9. What cautions should be observed in using micrometers?
10. Name several measurements in an engine that can be made with a dial indicator.

Study problems

1. Make two large vernier scales from wood or cardboard for use in the classroom. Mark one scale in the U.S. customary system and the other scale in the metric system.
2. Select the correct size micrometer, measure the diameters of the different shafts assigned to you, and record the measurements.

3. Obtain six spark plugs and get a reading of the gap setting with a flat feeler gage. Check the setting with a wire feeler gage. Was there a difference? Why?

34 USING THE METRIC SYSTEM

§ 34-1 SYSTEMS OF MEASUREMENTS

The two most widely used measuring systems are the U.S. customary system and the metric system. The U.S. customary system (*USCS*) uses inches, feet, miles, pints, quarts, gallons, and so on. The United States is the only major country in the world still using the USCS. Most countries already use the metric system or are in the process of switching to it.

Cars, motorcycles, and machinery imported into the United States are dimensioned in the metric system. This means that all measurements, all nuts and bolts, are in metric units. If you work on cars and other machinery from foreign countries, you will need an extra set of tools—metric tools. In the meantime, the United States is gradually switching to the metric system. This is resulting primarily from its adoption by individual industries, trade organizations, professional societies, and the like. Let's take a look at the U.S. customary system. We will discuss the metric system in later sections.

§ 34-2 HISTORY OF THE U.S. CUSTOMARY SYSTEM

Our familiar USCS measurements originated hundreds of years ago in England. They were brought to this country by the first settlers.

England had developed a system of measurement that was well suited to that country's commercial needs. The system had evolved from contact with many different countries and cultures through hundreds of years of trade. Babylonian, Egyptian, Anglo-Saxon, and Norman French influences had helped shape the English measurement system. But perhaps the greatest influence was that of the Roman Empire. They used the number 12 as a base for their number system. So, from the Romans came a familiar measurement, the division of 1 foot into 12 inches (Fig. 34-1).

While many colonies in America followed the English system of measurement, some did not. Different standards among the colonies made trade difficult. The colonial leaders recognized the problem and did something about it. The Articles of Confederation in 1781 and the U.S. Constitution in 1790 gave Congress the power to fix uniform standards for weights and measures. These uniform standards were to be used by all colonies. The effect of this action was to create what we know today as the U.S. customary system. It is almost the same today as it was when first used by the early colonists.

Measurements in the U.S. customary system include:

Length	Liquid
12 inches = 1 foot	16 ounces = 1 pint
3 feet = 1 yard	2 pints = 1 quart
5,280 feet = 1 mile	4 quarts = 1 gallon

Weight
16 ounces = 1 pound
2,000 pounds = 1 ton

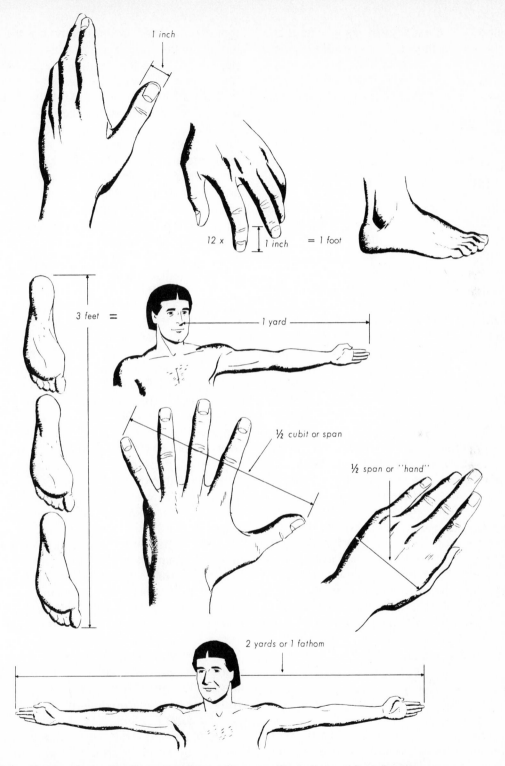

Fig. 34-1. Some of our familiar measurements come from the ancient Romans. (*General Motors Corporation*)

Looking at all these figures, as well as at the different units of measure, you could become a little confused, especially if you were not already familiar with them. Yet these are only some of the figures and units of measure a person must learn to work with the U.S. customary system.

§ 34-3 HISTORY OF THE METRIC SYSTEM

Not all the world was satisfied with the English system of measure. Many people thought it was too complicated because the units bore little relation to each other. Many other people did not use the English system at all. They preferred units that were peculiar to their own country or locality.

In 1670, a Frenchman named Gabriel Mouton recognized the advantages of a worldwide uniform measurement system. He proposed a decimal measurement system. It would be based on the length of 1 minute of arc of a great circle of the earth. Few people listened to Mouton.

In 1671, a French astronomer named Jean Picard suggested that the length of a pendulum beating seconds should be the basic unit of length for a new measurement system. Again, few people listened.

However, during the next 100 years, many people gradually became aware of the need for a standard measurement system for all countries. During the French Revolution in 1790, the National Assembly of France asked the French Academy of Sciences to "deduce an invariable standard for all the measures and all the weights." A commission appointed by the Academy created what we know today as the metric system of measurement.

§ 34-4 UNITS OF THE METRIC SYSTEM

In use, the metric system proved to be both scientific and simple. Through the years, some changes have been made, simplifying the system even further. In the beginning, the unit of length in the metric system was based on a portion of the earth's circumference. Measures for volume and weight were determined from the unit of length. This meant that the basic units of the system were related to each other as well as to nature. The simplicity of the system lay in the fact that the larger and smaller elements of each unit could be obtained by multiplying or dividing the basic units by 10 and its multiples. Multiplication or division could be performed simply by moving the decimal point. Thus the metric system is a base-10, or decimal, system.

The unit of length was originally called a **metre,** which we now spell **meter** (Fig. 34-2). The name came from the Greek word "metron," meaning "a measure." An actual standard representing the meter was to be constructed. It was to equal one ten-millionth of the distance from the North Pole to the equator, measured along a certain meridian of the earth that ran near Dunkirk, France, and Barcelona, Spain.

The metric unit of mass, or weight, was called the **gram** (Fig. 34-3). It was defined as the mass of 1 cubic centimeter of water at its temperature of maximum density. A cubic centimeter is a cube that is $\frac{1}{100}$ of a meter on each side. The cubic decimeter (a cube $\frac{1}{10}$ of a meter on each side) was chosen as the unit of fluid capacity. This measure was given the name **liter** (Fig. 34-3).

Slowly, other countries began to accept the metric system after it became compulsory in France in 1840. In 1866, the U.S. Congress passed an act making it "lawful throughout the United States of America to employ the weights and measures of the metric system in all contracts, dealings or court proceedings."

With great increases in industry and technology, even better metric standards were needed to keep up with scientific advances. Thus, the Metric Convention, a treaty signed by 17 countries, including the United States, came into being. The Convention set up well-defined metric standards of length and mass. New metric standards were constructed and

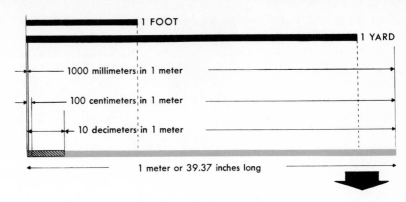

1 FOOT

1 YARD

1000 millimeters in 1 meter

100 centimeters in 1 meter

10 decimeters in 1 meter

1 meter or 39.37 inches long

Fig. 34-2. The meter is the metric unit of length. (*General Motors Corporation*)

10 meters in 1 dekameter

100 meters in 1 hectometer

1000 meters in 1 kilometer

1 mile

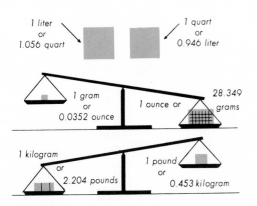

1 liter
or
1.056 quart

1 quart
or
0.946 liter

1 gram
or
0.0352 ounce

1 ounce or

28.349
grams

1 kilogram
or
2.204 pounds

1 pound
or
0.453 kilogram

Fig. 34-3. The gram is the metric unit of mass, and the liter is the metric unit of fluid capacity (volume). (*General Motors Corporation*)

distributed to each country signing the treaty. By 1900, 35 countries, including most of the nations of continental Europe and most of South America, had officially accepted the metric system.

In Sèvres, France, the International Bureau of Weights and Measures was set up. Its job was to act as a clearing house for all information about the practical use and refinement of the metric system. Because more precise and practical ways of using the measurement units are constantly being developed, the General Conference of Weights and Measures was established. This is a diplomatic organization that meets periodically to ratify and adopt improvements in the system and standards as they develop.

In 1960, the General Conference adopted an extensive revision and simplification of the metric system. The name Le Système International d'Unités (International System of Units), commonly referred to as SI, was adopted for this modernized metric system. Since that time, further improvements have been made to SI by the General Conference. The meter is now defined as 1,650,763.73 wavelengths, measured in a vacuum, of the orange-red line of the light waves from the element krypton-86 (Fig. 34-4).

Today, with the exception of the United States and a few small countries, all the other nations of the world are either using the metric system or are committed to metric usage (Fig. 34-5). In 1971, the Secretary of Commerce transmitted to Congress the results of a 3-year study authorized by the Metric Study Act of

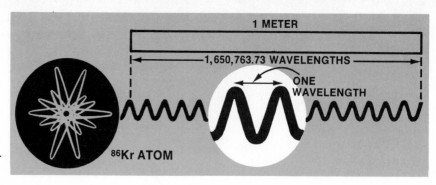

Fig. 34-4. The precise definition of the meter.

1968. At that time, the recommendation was made that the United States change to predominant use of the metric system. It was recommended that this be done through a coordinated national program to help minimize expense and confusion. Subsequently, bills making this recommendation law have been introduced in Congress. However, none of the bills has passed. The United States remains the only industrialized nation in the world without national policy guidelines to implement the use of the metric system.

§ 34-5 HOW THE METRIC SYSTEM WORKS

As you can see, from our brief discussions of the U.S. customary system and the metric system, the USCS just grew for several hundred years. It did not grow logically and, therefore, did not lend itself to scientific and industrial use.

In contrast, the metric system is scientific (Fig. 34-6). It is based on units that are divided into multiples of 10, just as our monetary system is based on dividing the dollar into 100 parts. In the metric system, for example, 1 meter (which is 39.37 inches) is 10 decimeters, or 100 centimeters, or 1,000 millimeters. Volume, in the metric system, is measured in liters. One liter is equal to 1.057 quarts.

As you can see in Fig. 34-6, in the metric system, three units of measure replace many different units used in the U.S. customary

system. Weight is measured in grams, length in meters, and volume in liters. By changing to the metric system, the United States would not have to make any change in its money system. That is already based on the decimal system. Also, time is measured in the same units of hours, minutes, and seconds used in the U.S. customary system. Now, let's look more closely at some basic metric measurements.

Length

1 millimeter = 0.039 inch
10 millimeters = 1 centimeter (0.394 inch)
100 centimeters = 1 meter (39.37 inches, or a little more than 1 yard)
1,000 millimeters = 1 meter
1,000 meters = 1 kilometer (0.62 mile)

Volume (liquid)

1,000 milliliters = 1 liter (1.057 quarts)

Weight

1,000 milligrams = 1 gram (0.035 ounce, or the approximate weight of a paper clip)
1,000 grams = 1 kilogram (2.2 pounds)
1,000,000 milligrams = 1 kilogram
1,000 kilograms = 1 metric ton

You have probably noticed that the names of metric units sometimes include prefixes (milli, centi, kilo, etc.) as in milliliter, centimeter, and kilogram. These prefixes indicate multiples

Fig. 34-5. The map shows that most of the world uses the metric system.

457

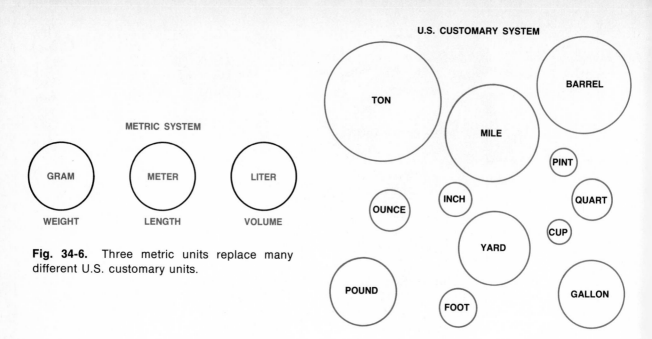

Fig. 34-6. Three metric units replace many different U.S. customary units.

LENGTH

1 in. (inch)	= 25.4 mm (millimeters)	or 0.0254 m (meter)
1 cm	= 0.39 in.	or 0.03281 ft (foot)
1 mm (millimeter)	= 0.039 in.	or 0.003281 ft
1 ft	= 30.48 cm	or 0.3048 m
1 mi (mile)	= 1.609 km (kilometers)	or 1,609 m
1 km	= 0.62 mi	or 3,281 ft

VOLUME/CAPACITY

1 cu in. (cubic inch)	= 16.39 cu cm (cubic centimeters)	or 0.01 l (liter)
1 cu cm	= 0.061 cu in.	or 0.001 l
1 l	= 61.02 cu in.	or 1.057 qt (quarts)
1 gal (gallon)	= 4 qt	or 3.780 l

WEIGHT

1 kg (kilogram)	= 2.2 lb (pounds)	= 35.2 oz (ounces)
1 lb	= 0.454 kg	

Here are the metric measurements, taken from the complete metric system table, that you will work with most often.

LENGTH

1 km	= 1,000 m	= 100,000 cm
1 m	= 100 cm	= 1,000 mm (millimeters)

VOLUME/CAPACITY

1 kl (kiloliter)	= 1,000 l	= 100,000 cl (centiliters)
1 l	= 1.000 cu cm	= 1,000 ml (milliliters)

WEIGHT

1 kg (kilogram)	= 1,000 g (grams)	= 100,000 cg (centigrams)

Fig. 34-7. Conversion table: the U.S. customary system versus the metric system.

or submultiples of the units. The most commonly used prefixes, and the multiplication factors they indicate, are given below.

Prefix	Multiplication factor
kilo	1,000 (one thousand)
centi	0.01 (one hundredth)
milli	0.001 (one thousandth)

Thus, the term kilometer means 1,000 meters; a centimeter is $\frac{1}{100}$ of a meter; and a millimeter is $\frac{1}{1,000}$ of a meter.

The conversion table (Fig. 34-7) gives you the measurement equivalents you will need to know to work on imported cars and motorcycles. With this table, you can convert USCS measurements to metric measurements, or vice versa.

In most service manuals for imported cars and motorcycles, the manufacturers give specifications in both the U.S. customary measurements and in metric measurements. Thus, the piston stroke of a certain engine would be given as 73.7 millimeters (2.90 inches). Specified torque for tightening main-bearing cap bolts would be given as 4.5–5.5 kg-m (32.5–39.5 pound-feet).

If you are going to work on imported cars and motorcycles, you need a set of metric wrenches and sockets. There are some nuts and bolts on imports that domestic tools will not fit. Study Fig. 34-8 to familiarize yourself with the difference between metric and standard sizes of sockets.

Many mechanics work with the metric system regularly. They memorize the frequently used measurements. Others find the metric equivalent by working a separate arithmetic problem for each specification. Such a process is slow, and mistakes can be made easily.

Metric equivalents can be found much faster and more accurately by use of computed metric conversion tables such as those shown in Fig. 34-9. Study these carefully and learn how they are used. These tables can save you time in determining the metric measurements.

§ 34-6 FUTURE OF THE METRIC SYSTEM IN THE UNITED STATES

Each year since 1971, a bill dealing with metrication has been introduced in Congress. But none of these bills has passed. The bill would establish a national guideline for the process of metrication. Without this guideline, it is up to almost anyone whether to adopt the metric system or to continue with the USCS system.

The American automotive manufacturers have chosen to adopt the metric system and are now in the process of converting to total metric usage. How they are doing this is typified by the way General Motors is handling its metric program.

General Motors has a plan to familiarize car owners and service technicians with the metric system. This plan calls for introducing metric measurements into the owners' manuals and service manuals for 1976 model vehicles. According to the proposed plan, the conversion from inches to metric will take place in three consecutive years. In the 1976 service manual, a measurement of 1 inch would appear as "1.000 inch [25.4 mm]." This is the practice followed in this textbook.

In 1977, the manuals would have the metric measurement first, followed by the inch equivalent. For example, "25.4 mm [1.000 inch]." And in 1978's and all later manuals, plans tentatively call for the inch measurement to be left out altogether.

§ 34-7 GENERAL MOTORS METRIC EXPERIENCE

General Motors has gained a lot of service insight and experience in working with metric in the United States by importing the German-built Opel automobile. The car is marketed through Buick dealers. The message from the domestic dealers servicing the Opel, an all-metric car, was clear. A metric car could be serviced in conventional United States dealer-

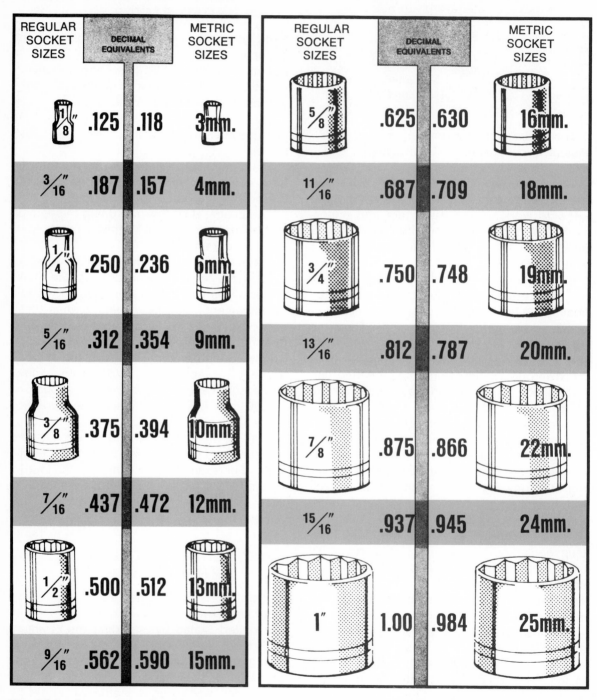

REGULAR SOCKET SIZES	DECIMAL EQUIVALENTS		METRIC SOCKET SIZES	REGULAR SOCKET SIZES	DECIMAL EQUIVALENTS		METRIC SOCKET SIZES
1/8″	.125	.118	3mm.	5/8″	.625	.630	16mm.
3/16″	.187	.157	4mm.	11/16″	.687	.709	18mm.
1/4″	.250	.236	6mm.	3/4″	.750	.748	19mm.
5/16″	.312	.354	9mm.	13/16″	.812	.787	20mm.
3/8″	.375	.394	10mm	7/8″	.875	.866	22mm.
7/16″	.437	.472	12mm.	15/16″	.937	.945	24mm.
1/2″	.500	.512	13mm.	1″	1.00	.984	25mm.
9/16″	.562	.590	15mm.				

Fig. 34-8. Comparison of metric and standard sizes of sockets. (*Dana Corporation*)

MILLIMETERS TO INCHES

mm	1	2	3	4	5	6	7	8	9	10	11	12	13
inches	0.0394	0.0787	0.1181	0.1575	0.1968	0.2362	0.2756	0.3150	0.3543	0.3937	0.4331	0.4724	0.5118
mm	14	15	16	17	18	19	20	21	22	23	24	25	26
inches	0.5512	0.5905	0.6299	0.6693	0.7087	0.7480	0.7874	0.8268	0.8661	0.9055	0.9449	0.9842	1.0236
mm	27	28	29	30	31	32	33	34	35	36	37	38	39
inches	1.0630	1.1024	1.1417	1.1811	1.2205	1.2598	1.2992	1.3386	1.3779	1.4173	1.4567	1.4961	1.5354
mm	40	41	42	43	44	45	46	47	48	49	50	51	52
inches	1.5748	1.6142	1.6535	1.6929	1.7323	1.7716	1.8110	1.8504	1.8898	1.9291	1.9685	2.0079	2.0472
mm	53	54	55	56	57	58	59	60	61	62	63	64	65
inches	2.0866	2.1260	2.1653	2.2047	2.2441	2.2835	2.3228	2.3622	2.4016	2.4409	2.4803	2.5197	2.5590
mm	66	67	68	69	70	71	72	73	74	75	76	77	78
inches	2.5984	2.6378	2.6772	2.7165	2.7559	2.7953	2.8346	2.8740	2.9134	2.9527	2.9921	3.0315	3.0709
mm	79	80	81	82	83	84	85	86	87	88	89	90	91
inches	3.1102	3.1496	3.1890	3.2283	3.2677	3.3071	3.3464	3.3858	3.4252	3.4646	3.5039	3.5433	3.5827
mm	92	93	94	95	96	97	98	99	100				
inches	3.6220	3.6614	3.7008	3.7401	3.7795	3.8189	3.8583	3.8976	3.9370				

INCHES TO MILLIMETERS

inches	1/64	1/32	3/64	1/16	5/64	3/32	7/64	1/8	9/64	5/32	11/64	3/16	13/64
mm	0.3969	0.7937	1.1906	1.5875	1.9844	2.3812	2.7781	3.1750	3.5719	3.9687	4.3656	4.7625	5.1594
inches	7/32	15/64	1/4	17/64	9/32	19/64	5/16	21/64	11/32	23/64	3/8	25/64	13/32
mm	5.5562	5.9531	6.3500	6.7469	7.1437	7.5406	7.9375	8.3344	8.7312	9.1281	9.5250	9.9219	10.3187
inches	27/64	7/16	29/64	15/32	31/64	1/2	33/64	17/32	35/64	9/16	37/64	19/32	39/64
mm	10.7156	11.1125	11.5094	11.9062	12.3031	12.7000	13.0969	13.4937	13.8906	14.2875	14.6844	15.0812	15.4781
inches	5/8	41/64	21/32	43/64	11/16	45/64	23/32	47/64	3/4	49/64	25/32	51/64	13/16
mm	15.8750	16.2719	16.6687	17.0656	17.4625	17.8594	18.2562	18.6531	19.0500	19.4469	19.8437	20.2406	20.6375
inches	53/64	27/32	55/64	7/8	57/64	29/32	59/64	15/16	61/64	31/32	63/64		
mm	21.0344	21.4312	21.8281	22.2250	22.6219	23.0187	23.4156	23.8125	24.2094	24.6062	25.0031		

FAHRENHEIT TO CELSIUS (Centigrade)

°F	-20	-15	-10	-5	0	5	10	15	20				
°C	-28.9	-26.1	-23.3	-20.6	-17.8	-15.0	-12.2	-9.4	-6.7				
°F	25	30	35	40	45	50	55	60	65	70	75	80	85
°C	-3.9	-1.1	1.7	4.4	7.2	10.0	12.8	15.6	18.3	21.1	23.9	26.7	29.4
°F	90	95	100	105	110	115	120	125	130	135	140	145	150
°C	32.2	35.0	37.8	40.6	43.3	46.1	48.9	51.7	54.4	57.2	60.0	62.8	65.6
°F	155	160	165	170	175	180	185	190	195	200	205	210	212
°C	68.3	71.1	73.9	76.7	79.4	82.2	85.0	87.8	90.6	93.8	96.1	98.9	100.0
°F	215	220	225	230	235	240	245	250	255	260	265		
°C	101.7	104.4	107.2	110.0	112.8	115.6	118.3	121.1	123.9	126.6	129.4		

Fig. 34-9. Computed metric conversion tables. (*Buick Motor Division of General Motors Corporation*)

(continued on next page)

FEET TO METERS

ft	0	1	2	3	4	5	6	7	8	9	ft
	m	m	m	m	m	m	m	m	m	m	
—		0.305	0.610	0.914	1.219	1.524	1.829	2.134	2.438	2.743	—
10	3.048	3.353	3.658	3.962	4.267	4.572	4.877	5.182	5.486	5.791	10
20	6.096	6.401	6.706	7.010	7.315	7.620	7.925	8.230	8.534	8.839	20
30	9.144	9.449	9.754	10.058	10.363	10.668	10.973	11.278	11.582	11.887	30
40	12.192	12.497	12.802	13.106	13.411	13.716	14.021	14.326	14.630	14.935	40
50	15.240	15.545	15.850	16.154	16.459	16.764	17.069	17.374	17.678	17.983	50
60	18.288	18.593	18.898	19.202	19.507	19.812	20.117	20.422	20.726	21.031	60
70	21.336	21.641	21.946	22.250	22.555	22.860	23.165	23.470	23.774	24.079	70
80	24.384	24.689	24.994	25.298	25.603	25.908	26.213	26.518	26.822	27.127	80
90	27.432	27.737	28.042	28.346	28.651	28.956	29.261	29.566	29.870	30.175	90
100	30.480	30.785	31.090	31.394	31.699	32.004	32.309	32.614	32.918	33.223	100

MILES TO KILOMETERS

mile	0	1	2	3	4	5	6	7	8	9	mile
	km	km	km	km	km	km	km	km	km	km	
—		1.609	3.219	4.828	6.437	8.047	9.656	11.265	12.875	14.484	—
10	16.093	17.703	19.312	20.921	22.531	24.140	25.750	27.359	28.968	30.578	10
20	32.187	33.796	35.406	37.015	38.624	40.234	41.843	43.452	45.062	46.671	20
30	48.280	49.890	51.499	53.108	54.718	56.327	57.936	59.546	61.155	62.764	30
40	64.374	65.983	67.593	69.202	70.811	72.421	74.030	75.639	77.249	78.858	40
50	80.467	82.077	83.686	85.295	86.905	88.514	90.123	91.733	93.342	94.951	50
60	96.561	98.170	99.779	101.39	103.00	104.61	106.22	107.83	109.44	111.04	60
70	112.65	114.26	115.87	117.48	119.09	120.70	122.31	123.92	125.53	127.14	70
80	128.75	130.36	131.97	133.58	135.19	136.79	138.40	140.01	141.62	143.23	80
90	144.84	146.45	148.06	149.67	151.28	152.89	154.50	156.11	157.72	159.33	90
100	160.93	162.54	164.15	165.76	167.37	168.98	170.59	172.20	173.81	175.42	100

SQUARE INCHES TO SQUARE CENTIMETERS

in²	0	1	2	3	4	5	6	7	8	9	in²
	cm²	cm²	cm²	cm²	cm²	cm²	cm²	cm²	cm²	cm²	
—		6.452	12.903	19.355	25.806	32.258	38.710	45.161	51.613	58.064	—
10	64.516	70.968	77.419	83.871	90.322	96.774	103.226	109.677	116.129	122.580	10
20	129.032	135.484	141.935	148.387	154.838	161.290	167.742	174.193	180.645	187.096	20
30	193.548	200.000	206.451	212.903	219.354	225.806	232.258	238.709	245.161	251.612	30
40	258.064	264.516	270.967	277.419	283.870	290.322	296.774	303.225	309.677	316.128	40
50	322.580	329.032	335.483	341.935	348.386	354.838	361.290	367.741	374.193	380.644	50
60	387.096	393.548	399.999	406.451	412.902	419.354	425.806	432.257	438.709	445.160	60
70	451.612	458.064	464.515	470.967	477.418	483.870	490.322	496.773	503.225	509.676	70
80	516.128	522.580	529.031	535.483	541.934	548.386	554.838	561.289	567.741	574.192	80
90	580.644	587.096	593.547	599.999	606.450	612.902	619.354	625.805	632.257	638.708	90
100	645.160	651.612	658.063	664.515	670.966	677.418	683.870	690.321	696.773	703.224	100

CUBIC INCHES TO CUBIC CENTIMETERS

in³	0	1	2	3	4	5	6	7	8	9	in³
	cm³	cm³	cm³	cm³	cm³	cm³	cm³	cm³	cm³	cm³	
—		16.387	32.774	49.161	65.548	81.935	98.322	114.709	131.097	147.484	—
10	163.871	180.258	196.645	213.032	229.419	245.806	262.193	278.580	294.967	311.354	10
20	327.741	344.128	360.515	376.902	393.290	409.677	426.064	442.451	458.838	475.225	20
30	491.612	507.999	524.386	540.773	557.160	573.547	589.934	606.321	622.708	639.095	30
40	655.483	671.870	688.257	704.644	721.031	737.418	753.805	770.192	786.579	802.966	40
50	819.353	835.740	852.127	868.514	884.901	901.289	917.676	934.063	950.450	966.837	50
60	983.224	999.611	1015.998	1032.385	1048.772	1065.159	1081.546	1097.933	1114.320	1130.707	60
70	1147.094	1163.482	1179.869	1196.256	1212.643	1229.030	1245.417	1261.804	1278.191	1294.578	70
80	1310.965	1327.352	1343.739	1360.126	1376.513	1392.200	1409.288	1425.675	1442.062	1458.449	80
90	1474.836	1491.223	1507.610	1523.997	1540.384	1556.771	1573.158	1589.545	1605.932	1622.319	90
100	1638.706	1655.093	1671.481	1687.863	1704.255	1720.642	1737.029	1753.416	1769.803	1786.190	100

CUBIC FEET TO CUBIC METERS

ft³	0	1	2	3	4	5	6	7	8	9	ft³
	m³	m³	m³	m³	m³	m³	m³	m³	m³	m³	
—		0.0283	0.0566	0.0850	0.1133	0.1416	0.1699	0.1982	0.2265	0.2549	—
10	0.2832	0.3115	0.3398	0.3681	0.3964	0.4248	0.4531	0.4814	0.5097	0.5380	10
20	0.5663	0.5947	0.6230	0.6513	0.6796	0.7079	0.7362	0.7646	0.7929	0.8212	20
30	0.8495	0.8778	0.9061	0.9345	0.9628	0.9911	1.0194	1.0477	1.0760	1.1044	30
40	1.1327	1.1610	1.1893	1.2176	1.2459	1.2743	1.3026	1.3309	1.3592	1.3875	40
50	1.4159	1.4442	1.4725	1.5008	1.5291	1.5574	1.5858	1.6141	1.6424	1.6707	50
60	1.6990	1.7273	1.7557	1.7840	1.8123	1.8406	1.8689	1.8972	1.9256	1.9539	60
70	1.9822	2.0105	2.0388	2.0671	2.0955	2.1238	2.1521	2.1804	2.2087	2.2370	70
80	2.2654	2.2937	2.3220	2.3503	2.3786	2.4069	2.4353	2.4636	2.4919	2.5202	80
90	2.5485	2.5768	2.6052	2.6335	2.6618	2.6901	2.7184	2.7468	2.7751	2.8034	90
100	2.8317	2.6800	2.8884	2.9167	2.9450	2.9733	3.0016	3.0300	3.0583	3.0866	100

GALLONS (U.S.) TO LITERS

U.S. gal	0	1	2	3	4	5	6	7	8	9	U.S. gal
	l	l	l	l	l	l	l	l	l	l	
—		3.7854	7.5709	11.3563	15.1417	18.9271	22.7126	26.4980	30.2834	34.0638	—
10	37.8543	41.6397	45.4251	49.2105	52.9960	56.7814	60.5668	64.3523	68.1377	71.9231	10
20	75.7085	79.4940	83.2794	87.0648	90.8502	94.6357	98.4211	102.2065	105.9920	109.7774	20
30	113.5528	117.3482	121.1337	124.9191	128.7045	132.4899	136.2754	140.0608	143.8462	147.6316	30
40	151.4171	155.2025	158.9879	162.7734	166.5588	170.3442	174.1296	177.9151	181.7005	185.4859	40
50	189.2713	193.0568	196.8422	200.6276	204.4131	208.1985	211.9839	215.7693	219.5548	223.3402	50
60	227.1256	230.9110	234.6965	238.4819	242.2673	246.0527	249.8382	253.6236	257.4090	261.1945	60
70	264.9799	268.7653	272.5507	276.3362	280.1216	283.9070	287.6924	291.4779	295.2633	299.0487	70
80	302.8342	306.6196	310.4050	314.1904	317.9759	321.7613	325.5467	329.3321	333.1176	336.9030	80
90	340.6884	344.4738	348.2593	352.0447	355.8301	359.6156	363.4010	367.1864	370.9718	374.7573	90
100	378.5427	382.3281	386.1135	389.8990	393.6844	397.4698	401.2553	405.0407	408.8261	412.6115	100

(continued on next page)

POUNDS TO KILOGRAMS

(continuation of Fig. 34-9)

lb	0	1	2	3	4	5	6	7	8	9	lb
	kg	kg	kg	kg	kg	kg	kg	kg	kg	kg	
—		0.454	0.907	1.361	1.814	2.268	2.722	3.175	3.629	4.082	—
10	4.536	4.990	5.443	5.897	6.350	6.804	7.257	7.711	8.165	8.618	10
20	9.072	9.525	9.979	10.433	10.886	11.340	11.793	12.247	12.701	13.154	20
30	13.608	14.061	14.515	14.969	15.422	15.876	16.329	16.783	17.237	17.690	30
40	18.144	18.597	19.051	19.504	19.958	20.412	20.865	21.319	21.772	22.226	40
50	22.680	23.133	23.587	24.040	24.494	24.948	25.401	25.855	26.308	26.762	50
60	27.216	27.669	28.123	28.576	29.030	29.484	29.937	30.391	30.844	31.298	60
70	31.751	32.205	32.659	33.112	33.566	34.019	34.473	34.927	35.380	35.834	70
80	36.287	36.741	37.195	37.648	38.102	38.555	39.009	39.463	39.916	40.370	80
90	40.823	41.277	41.730	42.184	42.638	43.092	43.545	43.998	44.453	44.906	90
100	45.359	45.813	46.266	46.720	47.174	47.627	48.081	48.534	48.988	49.442	100

POUNDS PER SQUARE INCH TO KILOGRAMS PER SQUARE CENTIMETER

lb/in²	0	1	2	3	4	5	6	7	8	9	lb/in²
	kg/cm²	kg/cm²	kg/cm²	kg/cm²	kg/cm²	kg/cm²	kg/cm²	kg/cm²	kg/cm²	kg/cm²	
—		0.0703	0.1406	0.2100	0.2812	0.3515	0.4218	0.4921	0.5625	0.6328	—
10	0.7031	0.7734	0.8437	0.9140	0.9843	1.0546	1.1249	1.1952	1.2655	1.3358	10
20	1.4062	1.4765	1.5468	1.6171	1.6874	1.7577	1.8280	1.8983	1.9686	2.0389	20
30	2.1092	2.1795	2.2498	2.3202	2.3905	2.4608	2.5311	2.6014	2.6717	2.7420	30
40	2.8123	2.8826	2.9529	3.0232	3.0935	3.1639	3.2342	3.3045	3.3748	3.4451	40
50	3.5154	3.5857	3.6560	3.7263	3.7966	3.8669	3.9372	4.0072	4.0779	4.1482	50
60	4.2185	4.2888	4.3591	4.4294	4.4997	4.5700	4.6403	4.7106	4.7809	4.8512	60
70	4.9216	4.9919	5.0622	5.1325	5.2028	5.2731	5.3434	5.4137	5.4840	5.5543	70
80	5.6246	5.6949	5.7652	5.8356	5.9059	5.9762	6.0465	6.1168	6.1871	6.2574	80
90	6.3277	6.3980	6.4683	6.5386	6.6089	6.6793	6.7496	6.8199	6.8902	6.9605	90
100	7.0308	7.1011	7.1714	7.2417	7.3120	7.3823	7.4526	7.5229	7.5933	7.6636	100

POUND-FEET TO KILOGRAM-METERS

lb-ft	0	1	2	3	4	5	6	7	8	9	lb-ft
	kg-m	kg-m	kg-m	kg-m	kg-m	kg-m	kg-m	kg-m	kg-m	kg-m	
—		0.138	0.276	0.415	0.553	0.691	0.829	0.967	1.106	1.244	—
10	1.382	1.520	1.658	1.796	1.934	2.073	2.211	2.349	2.487	2.625	10
20	2.764	2.902	3.040	3.178	3.316	3.455	3.593	3.731	3.869	4.007	20
30	4.146	4.284	4.422	4.560	4.698	4.837	4.975	5.113	5.251	5.389	30
40	5.528	5.666	5.804	5.942	6.080	6.219	6.357	6.495	6.633	6.771	40
50	6.910	7.048	7.186	7.324	7.462	7.601	7.739	7.877	8.015	8.153	50
60	8.292	8.430	8.568	8.706	8.844	8.983	9.121	9.259	9.397	9.535	60
70	9.674	9.812	9.950	10.088	10.227	10.365	10.503	10.641	10.779	10.918	70
80	11.056	11.194	11.332	11.470	11.609	11.747	11.885	12.023	12.161	12.300	80
90	12.438	12.576	12.714	12.855	12.991	13.129	13.267	13.405	13.544	13.682	90
100	13.820	13.958	14.096	14.235	14.373	14.511	14.649	14.787	14.925	14.064	100

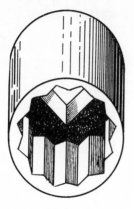
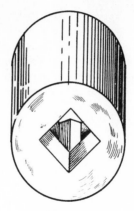

Fig. 34-10. The drive end of a socket is always made for the U.S. customary system.

SOCKET END—
MAY BE METRIC
OR USCS

DRIVE END—
ALWAYS USCS—
¼, ⅜, or ½ INCH

ship facilities. The mechanic and technician in the dealership needed only routine new-product training. The fact that the Opel is an all-metric car did not present a servicing problem.

One factor that helped ease service problems on the Opel was the almost traditional purchase by the dealer of a set of special tools for each new year and model of American car. So, buying a few special tools made to metric sizes presented nothing new to the Buick dealer carrying the Opel line, or to the service technician fixing the car.

Mechanics took the Opel in stride and bought the few metric sockets and wrenches needed to fix it. There is one little-recognized fact about the metric system as applied to hand tools. While bolt-head sizes are expressed in millimeters, socket drivers, such as speed handles and ratchets, are based on inches throughout the world. That is, the drive end of a socket is always made for the U.S. customary system ¼-, ⅜-, and ½-inch drive sizes (Fig. 34-10).

To the mechanic and service technician, this means that only the new metric-size sockets must be added to the tool box. The ratchets and handles already owned fit both metric- and inch-size sockets. Today, knowing how to work with the metric system should be just as basic to a mechanic as knowing how to check a battery.

Review questions

1. How do the U.S. customary system and the metric system differ?
2. Which system is easier to use in science and industry? Why?
3. Name the prefixes used in the metric system for one thousand, one-hundredth, and one-thousandth.
4. To work on metric cars and motorcycles, you should have certain tools in your tool box. What are they?
5. Socket drivers, such as ratchets, and the drive ends of sockets are made in three sizes. What are these sizes, and which measurement system are they part of?

Study problem

1. Study the conversion table until you understand how to convert inches to millimeters, millimeters to inches, miles to kilometers, gallons to liters, and so on. Then make the following conversions:

1 inch = __?__ millimeter(s)
1 millimeter = __?__ inch(es)
1 gallon = __?__ liter(s)
1 liter = __?__ gallon(s)
1 mile = __?__ kilometer(s)
1 kilometer = __?__ mile(s)
1 kilogram = __?__ pound(s)
1 pound = __?__ kilogram(s)

35 SMALL-ENGINE MAINTENANCE AND SERVICE

§ 35-1 COMMON SMALL-ENGINE ABUSES

Small engines are built to work hard. They have large crankshafts and bearings. The number of years that a small engine will continue to perform satisfactorily depends, in part, on the care and maintenance it gets. Some of the abuses that shorten an engine's life include:

1. Allowing dirt to get into the engine. This can result from not servicing the air cleaner and fuel strainer properly, from improperly replacing spark plugs, from contaminating the fuel, and so on.
2. Failure to feed the engine the proper oil-fuel mixture so that the engine is inadequately lubricated. This means that it will wear rapidly and fail early.
3. Overloading the engine so that it works too hard and wears out fast.
4. Running the engine too fast. Some people change the governor setting so that the engine will run faster and handle heavier loads. The engine will wear out rapidly under these conditions. Overspeeding is a sure way to shorten engine life.
5. Failure to store the engine properly during the off season. Many engines operate machines that are in use only part of the year. When they are not to be used for several weeks or months, engines should be prepared for the idle period. Not preparing the

engine properly can lead to early engine failure.

§ 35-2 WINTER STORAGE

For winter storage, drain the fuel tank, and run the engine to use up the fuel in the carburetor. Fuel left in the carburetor is apt to form gum that will clog fuel passages. Remove the spark plug and pour a tablespoon or so of heavy engine oil into the combustion chamber. Turn the engine over a few times to distribute the oil over the engine parts. Replace the plug. Store the machine in a warm, dry place.

◆ CAUTION ◆

When you are working on the "business" side of equipment (under the mower, where the blade is, on the chain saw, etc.), always make sure that the engine cannot start by turning the engine off and by disconnecting the spark-plug wire.

---

§ 35-3 TROUBLESHOOTING TWO-CYCLE ENGINES

When trouble occurs with an engine, a definite procedure of checking or **troubleshooting** the engine to locate the trouble should be used. With a definite procedure, the cause of trouble can be quickly found. Trouble will be minimized if the engine is properly maintained. Trouble-shooting and maintenance are described below.

§ 35-4 TWO-CYCLE-ENGINE TROUBLES

The two most common complaints about two-cycle engines are failure to start and lack of power. In addition, the engine may **surge** (repeatedly increase in speed and then slow down), it may gradually lose power as it is operated, or it may fire irregularly.

One point must be carefully checked before any work is started on the engine itself, especially when the complaint is lack of power. This is to make sure that the machine itself is not at fault. For example, dull blades on a lawn mower work the engine much harder. A heavy stand of grass makes hard work for a mower engine if the blades are dull. Similarly, a power saw with dull saw teeth will work the engine harder and might cause the operator to say that the engine lacks power.

If you are fairly sure that the engine itself is to blame for the trouble, then follow the procedures given below for the different kinds of trouble.

§ 35-5 ENGINE WILL NOT START

Failure of the engine to start could be due to lack of fuel, fuel not feeding to the carburetor, carburetor not feeding fuel to the air passing through the air horn to the engine, clogged air filter, clogged exhaust ports, defective ignition system, or internal engine damage. To check out the engine and locate the trouble, proceed as follows:

1. Make sure that there is clean gasoline in the fuel tank.
2. Be sure that the vent in the fuel-tank cap is clear. If it is clogged, the engine may start, but soon stop. This is because the clogged vent does not permit gasoline to flow rapidly enough from the fuel tank to the carburetor.

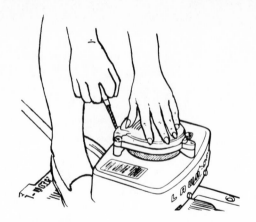

Fig. 35-1. Check engine compression by slowly pulling the engine through the compression stroke with the starter rope.

3. Check engine compression by slowly pulling the engine through the compression stroke with the starter (Fig. 35-1). Be sure the ON-OFF switch is *off.* If the starter is of the rope type, you can judge the compression by the feel. For instance, if the engine spins very easily, then there is little compression. This is probably due to a loose cylinder head, defective head gasket, loose spark plug, cracked head or cylinder, or broken piston rings or piston. First, check the spark plug. If it is tight, then you should look for the other causes of trouble, disassembling the engine as necessary.

If the engine uses a windup starter or an electric starter, you will have to judge the compression by the way the engine acts when it is cranked. With the windup starter, if release of the spring turns the engine over unusually fast or long, you can suspect loss of compression. The same thing can be said about the electric starter. But be careful not to blame the engine if the battery is run down or if the starting motor is at fault.

When checking compression, listen for unusual squeaks, squeals, or scraping or knocking sounds. Any of these could mean worn bear-

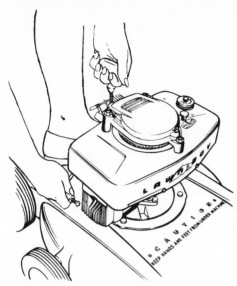

Fig. 35-2. Check the ignition system by disconnecting the high-voltage lead from the plug and putting a bolt into it to get metal contact. Hold the lead with the bolt in it about $\frac{3}{16}$ inch [4.76 millimeters] from the cylinder head while cranking the engine.

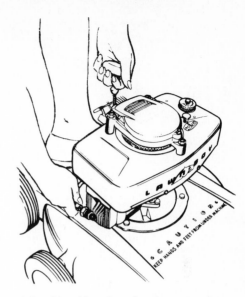

Fig. 35-3. Checking the plug for a spark.

ings, scored cylinder walls or pistons, or broken rings or other broken parts. If you hear such noises, do not try to start the engine before carefully checking engine parts, disassembling the engine as necessary to examine them.

If the engine has normal compression, it will resist the pull of the rope (or act normally when starting is attempted with the windup or electric starter). Another sign of good compression is a sucking sound when the engine is spun fast, followed by a sort of cough as the engine stops after the spin, indicating that the engine is taking in air normally.

Try to start by choking the engine, making sure that the ON-OFF switch is turned on, and then cranking the engine. If the engine shows normal compression but will not start, then the ignition system or the carburetor is probably at fault. Check the ignition system first by disconnecting the high-voltage lead from the spark plug. Pull back the rubber hood to expose the lead clip, or put a bolt into it to get metal contact. Hold the clip or bolt about $\frac{3}{16}$

inch [4.76 mm] from the cylinder head and crank the engine (Fig. 35-2). If a strong spark jumps to the cylinder head, then the ignition system is probably okay.

If no spark occurs, then the ignition system is probably at fault, and it should be checked. Causes of trouble could be dirty or worn contact points, points out of adjustment, or a defective capacitor, high-voltage lead, ON-OFF switch, or magneto coil.

If a spark does occur, examine the spark plug to see if it can deliver the spark to the engine cylinder. Remove the plug, reattach the high-voltage lead to it, and lay it or hold it against the cylinder head (Fig. 35-3). Crank the engine. If no spark jumps the gap, the spark plug is probably at fault. Examine it for cracks, black sooty deposits on the porcelain or electrodes, burned electrodes, or wide gap (Fig. 35-4). Any of these could prevent a good spark.

There is one other point to notice when checking a spark plug just removed from an engine that has been cranked and has failed to start. If the end is wet with gasoline, then chances are that fuel is getting to the engine. Put your finger over the spark-plug hole in the head and crank the engine with the choke on

Fig. 35-4. Defective spark plug, showing cracks, sooty deposits, etc.

Fig. 35-5. Using your finger to feel if fuel is entering the cylinder head.

(Fig. 35-5). If your finger gets wet, it is added evidence that fuel is getting through.

If the end of the plug, or your finger, does not get wet with gasoline, then the carburetor is at fault. The trouble could be due to clogged lines or nozzle, incorrect adjustment, or a defective float system.

Try to adjust the carburetor. If this fails, then the carburetor will have to be removed for disassembly and repair. A typical adjustment is as follows: Turn the adjusting knob, shown in Fig. 35-6, down (clockwise) to bring the main nozzle needle down on its seat. Do not turn it down tight, since this might damage the seat or the needle, and they would require replacement. Back off the knob two full turns. Close the choke and crank to see if gasoline appears in the engine (using your finger on the plug hole to check). If gasoline now appears, replace the plug and try to start. If the engine starts, open the choke in a normal manner as the engine runs and warms up. If the engine runs roughly, it may be getting too much gasoline. Turn the needle knob in to produce a leaner mixture. After the engine is warmed up, turn the needle knob in until the engine begins to die from an excessively lean mixture. Then, back out about $\frac{1}{4}$ turn. This should be the best adjustment.

§ 35-6 ENGINE STARTS BUT LACKS POWER

A common cause of this trouble in two-cycle engines is clogged exhaust ports. Carbon that forms as a result of the combustion action in the cylinder often cakes up around the exhaust ports (Fig. 35-7). As this buildup continues, the engine is less and less able to exhaust burned gases. Thus, less fresh charge can enter the engine cylinder. This means that engine power is lost. After some time, if the accumulations are not removed, the engine will barely run. To

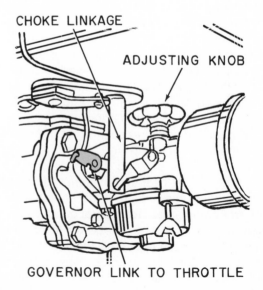

Fig. 35-6. Location of adjusting knob and choke linkage.

CHOKE LINKAGE

ADJUSTING KNOB

GOVERNOR LINK TO THROTTLE

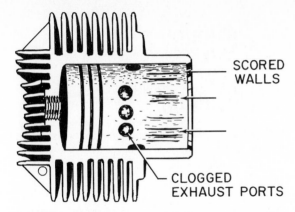

Fig. 35-7. Cutaway view of cylinder head, showing scored walls caused by the carbon deposits on the exhaust ports.

remove the accumulations, take off the exhaust muffler. Turn the engine so that the piston covers the exhaust ports. Then, use a screwdriver or a hardwood scraper (Fig. 35-8) to carefully scrape away the carbon accumulations. The piston will keep particles from falling into the cylinder, where they could cause trouble. Be extremely careful to avoid scratching the piston. Blow out all loose particles from the ports.

If clogged exhaust ports are not the cause of lack of power, then check and adjust the carburetor. Chances are that the carburetor is supplying an overrich or overlean mixture.

If the lack of power is not due to clogged exhaust ports or faulty carburetor action, then the trouble probably is in the engine itself. It could be due to worn pistons or cylinders or to worn or broken rings. One other possible cause should be considered, and this is a defective reed valve in the crankcase. If this valve is not seating properly, it may not hold compression in the crankcase. The result is that not enough air-fuel mixture will be retained in the crankcase. The charge going to the combustion chamber on intake will not be enough for the engine to develop full power. If the reed valve is warped or bent so that it does not lie flat against the inlet holes, it should be replaced.

Fig. 35-8. Use a hardwood scraper or a screwdriver to remove carbon from the exhaust ports.

§ 35-7 ENGINE SURGES

If the engine **surges,** that is, repeatedly speeds up and slows down, the trouble probably is in either the carburetor or the governor. Try readjusting the carburetor as already explained. If this does not cure the trouble, then check the governor. Things to look for in the governor are binding of the linkage between the governor and the throttle valve, a weak or damaged spring, or worn or binding governor parts.

If engine speed is not correct, it can be adjusted on some models by bending the linkage between the governor and the throttle valve. On other models, adjustment is made by changing governor springs. *Do not attempt to change speed by stretching a spring.* Chances are the spring will not hold its new set and engine operation will be unsteady.

§ 35-8 ENGINE LOSES POWER

If the engine starts off okay but gradually loses speed as it warms up, the most likely cause is in the fuel system. For example, the vent in the fuel-tank cap might be clogged, or the needle in the float bowl might be stuck. In either case, too little gasoline gets through to the carburetor, and the engine slows down because it is fuel-starved.

Lack of lubrication in the engine, as, for instance, from failure to put oil in the gasoline, might cause loss of power as the engine warms up. Chances are that this would soon cause complete engine failure from seized bearings or scored cylinder walls or pistons (Fig. 35-9).

§ 35-9 IRREGULAR FIRING

If the engine fires irregularly, it could be due to a weak spark or to poor carburetion. Check the spark as already described. Replace the coil or capacitor; clean and adjust or replace wires as necessary. Check and adjust the carburetor as already described.

§ 35-10 MAINTENANCE OF TWO-CYCLE ENGINES

There are only a few steps of maintenance on the engine, but these must be taken regularly if you expect the engine to give you long and trouble-free service.

1. Clean and re-oil the carburetor air filter regularly. You will usually find instructions on the engine on when and how to do this job. To clean the filter, the element must be removed and washed in clean gasoline. Then it must be re-oiled. Usually, the recommendation calls for doing this every 10 hours of engine operation. If operating conditions are especially dusty, clean the filter every 5 or 6 hours.

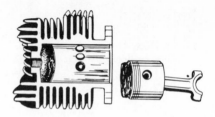

Fig. 35-9. Scored piston and cylinder walls caused by lack of lubrication.

2. Check the tightness of all bolts and nuts on the engine and on the whole machine periodically. They sometimes loosen up in service, and if not retightened, parts may become damaged or lost.

3. Lubricate all bearings outside the engine, as, for instance, the wheel bearings on a power mower. Make sure oil reservoirs are filled (as, for instance, that on the chainsaw lubricator).

4. Make sure that the blades (or saw teeth) are sharp and that the rest of the assembly is in good condition.

5. Keep the fuel tank filled. If it is allowed to sit around only partly filled, air will enter and leave the tank as the temperature changes. This will introduce moisture into the tank. The moisture will condense and ultimately cause severe rusting of the tank (metal tanks). Not only will this damage the tank, but rust particles may get into the carburetor and cause clogging of the fuel passages.

6. Keep the machine clean. Wipe it off periodically to remove oil, grass clippings, dust, and so on. Remember that collecting of such trash around the engine will act as a blanket so that the engine may overheat. On mowers, clean off the accumulations of grass clippings from the inside of the housing.

7. Clean the fuel filter or strainer periodically. Some engines have a bowl-type of fuel strainer, as shown in Fig. 35-10. On these, the bowl is removed so that the strainer

can be taken out and washed in clean gasoline. Other engines have a strainer inside the fuel tank. Some of these can be removed for cleaning. Others are permanently mounted in the tank and can be cleaned only by removing the fuel tank and washing it out several times with solvent.

8. For winter storage, drain the fuel tank, and run the engine to use up the fuel in the carburetor. Fuel left in the carburetor is apt to form gum that will clog fuel passages. Remove the spark plug and pour a tablespoon or so of heavy engine oil into the combustion chamber. Turn the engine over a few times to distribute the oil over the engine parts. Replace the plug. Store the machine in a warm, dry place.

◆ CAUTION ◆

When working on the "business" side of equipment (under the mower, where the blade is, on the chain saw, etc.), always make sure that the engine cannot start by turning the engine off and by disconnecting the spark-plug wire.

---

§35-11 OPERATING TWO-CYCLE ENGINES

There are several things to remember about operating a two-cycle engine. These include how to lubricate it, how to start it, how to run it, how to stop it, and how to store it.

§35-12 LUBRICATING TWO-CYCLE ENGINES

You lubricate many two-cycle engines by adding oil in the recommended amount to the fuel. The oil-fuel mixture enters the crankcase, as already explained. The fuel, in vapor form, passes on up to the combustion chamber as a part of the air-fuel mixture. Part of the oil, in mist form, is retained in the crankcase, where it lubricates the piston, rings, and crankshaft bearings. Some of the oil does get

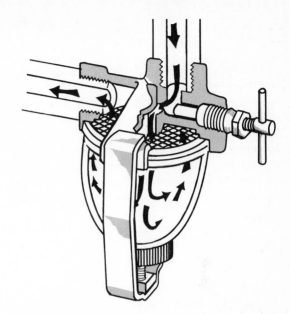

Fig. 35-10. Sediment-bowl type of fuel strainer. Arrows show direction of fuel flow.

up into the combustion chamber, where it is burned along with the air-fuel mixture.

It is critical that the right amount of oil be mixed with the fuel. The manufacturer's recommendations—usually on a metal plate attached to the engine—should be carefully followed. Too much oil will cause the exhaust ports to clog up very quickly. Too little oil will cause rapid wear of the piston, rings, cylinder wall, and bearings.

To mix the oil and fuel, use an approved metal gasoline can. *Never*—NEVER—use glass jugs. Jugs can break and flood the area with gasoline. The slightest spark could then set off a disastrous explosion and fire. Some containers have an oil-measuring cup at the filler opening. This makes it easy to measure out the required amount of oil. First, fill the container about half full with gasoline. Then, measure out and add the oil. Close the container and shake it vigorously. Add the rest of the gasoline, close the container, and shake it vigorously again. Make sure the oil and gasoline are thoroughly mixed.

§ 35-13 STORING GASOLINE

There are local and state laws about storing gasoline. There is also common sense. Remember that it is difficult to close any container tight enough to keep it from leaking gasoline fumes. And these fumes are *very*—VERY—explosive. So it is just common sense *never* to store gasoline containers inside a closed room such as the basement. Houses have been destroyed by gasoline vapors leaking from a container in the basement. The vapors can be set off by the pilot light of the furnace or water heater, or by an electric spark from turning a switch on or off.

Gasoline deteriorates if it is stored for any length of time. This is the reason that an engine that is to be stored for any length of time should have the fuel tank and carburetor drained. Otherwise, the stale gasoline can deposit gum and varnish on critical parts, and this could cause poor engine performance. A carburetor overhaul would then be needed.

Fig. 35-11. Make sure the mower is level and that you have it under control by holding it or by having a foot on it before using the rope-wind starter.

For the same reason, you should not store gasoline for a long period of time and expect to use it.

§ 35-14 STARTING THE ENGINE

There are more complaints about hard starting than about anything else on small engines. You can start small engines more easily if you follow the correct procedure.

If the engine uses a rope-wind starter, make sure the equipment is level so it will not tip over. If you can, put one foot on the equipment to hold it steady (Fig. 35-11).

If the equipment has brakes, apply them before starting the engine. If it has a clutch, disengage it if possible so that the machine will not start to move when the engine starts. On riding equipment, operate the controls from the driver's seat. Then, you can quickly stop the engine if something goes wrong.

Make sure the shutoff valve to the carburetor is open. Adjust the throttle and choke, or use the primer. Then, turn on the ignition and crank the engine.

♦ CAUTION ♦

When starting a chain saw, put it on the ground or brace it so that it will not get out of control when you crank it. If you don't have full control, the saw could get away from you, with disastrous results.

If you are using an electric starter, avoid cranking for long periods, because this can damage the starter.

If the engine does not start right away, open the choke valve part way and try again. It may be that the engine has flooded; that is, it has gotten too much gasoline.

Once the engine has started, allow it to run for a minute or two so that it can warm up. Never gun a cold engine or try to take full power from it. Give it a chance to warm up first.

§ 35-15 OPERATING THE ENGINE

Overloading and overspeeding are the two most common causes of small-engine trouble and short engine life. We have already mentioned these abuses in § 35-1.

If the engine is new or rebuilt, work it only lightly for the first few hours to give it a chance to get broken in. Adjust the carburetor for a fairly rich mixture for the first 10 hours.

§ 35-16 STOPPING THE ENGINE

Remove any load from the engine before stopping it. It is a good idea to slow the engine to idle for a minute or two before turning it off. This cools the engine more gradually.

After stopping the engine, close the fuel-tank shutoff valve if the engine has one. This relieves the pressure on the carburetor float system or diaphragm and prevents fuel leaks.

§ 35-17 STORING THE ENGINE

If you do not plan to use the engine for a month or so, put it in storage, as outlined in § 35-2.

§ 35-18 SERVICING A TWO-CYCLE ENGINE

Servicing a two-cycle engine is much simpler than servicing an automotive four-cycle engine. There are many fewer parts. Nevertheless, you have to follow the same basic rules about cleanliness, using the right tools, and following the recommended procedures exactly.

When trouble occurs in a small engine, it may require anything from a minor adjustment to a complete disassembly to fix the trouble. However, remember that these engines are relatively inexpensive and that repair time and parts are relatively costly. Thus, if an engine is in bad shape, and a new piston, rings, connecting rod, and bearings are required, it would probably be cheaper to buy a new engine. However, the old engine would be useful to students to practice on. That is, they could practice tearing it down and building it up.

Disassembly of a small engine is relatively simple. Figure 35-12 shows a cutaway view of a small engine. In the paragraphs that follow, we will touch briefly on the major steps in servicing the small engine.

§ 35-19 DISASSEMBLY

Disconnect the gas line from the carburetor. Remove the choke rod from the choke shaft.

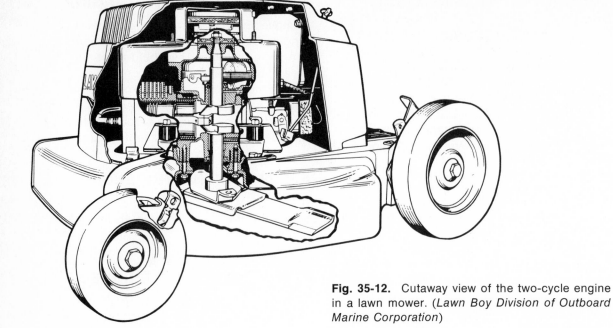

Fig. 35-12. Cutaway view of the two-cycle engine in a lawn mower. (*Lawn Boy Division of Outboard Marine Corporation*)

Take out attaching screws and remove the complete engine shroud, gasoline tank, and recoil starter as an assembly (Fig. 35-13).

Remove the flywheel nut and washer. To hold the flywheel while the nut is being loosened, a piston stop can be used on some models (Fig. 35-14). This device is installed in place of the spark plug and has a long rod that comes into contact with the piston head. Now, when the flywheel nut is turned, the piston moves up to the stop, and the stop keeps the piston, crankshaft, and flywheel from moving.

Lift the starter pulley, plate, screen, pin, and spring off the flywheel (Fig. 35-15). Remove the flywheel. If it sticks, put the nut back on the end of the crankshaft, turning it down until it is nearly flush with the end of the crankshaft. Now, use a soft metal hammer, lift up on the flywheel, and tap gently on the nut (Fig. 35-16). The nut protects the end of the crankshaft so that the screw threads will not become battered.

Lift off the governor yoke, arms, and collar as an assembly and set it to one side (Fig. 35-17). Remove the governor lever and wear-

block assembly (Fig. 35-18). Remove three screws, and lift off the magneto plate (Fig. 35-19).

Take out the screws and remove the carburetor, air filter, and reed-plate assembly as a unit (Fig. 35-20). Bend down the tangs of the

Fig. 35-13. Removing engine shroud, gas tank, and starter as an assembly. (*Lawn Boy Division of Outboard Marine Corporation*)

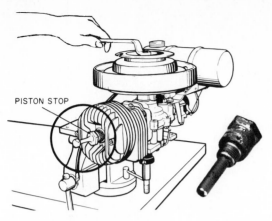

Fig. 35-14. Using piston stop to hold the piston while the flywheel nut is being removed. (*Lawn Boy Division of Outboard Marine Corporation*)

Fig. 35-16. Hammering on crankshaft nut to loosen the flywheel. (*Lawn Boy Division of Outboard Marine Corporation*)

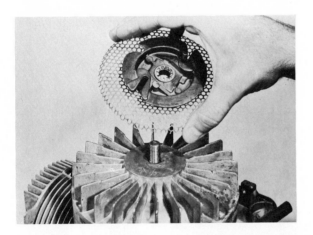

Fig. 35-15. Removing starter pulley, plate, and screen. (*Lawn Boy Division of Outboard Marine Corporation*)

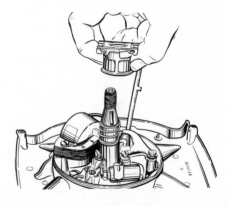

Fig. 35-17. Removing the governor yoke. (*Lawn Boy Division of Outboard Marine Corporation*)

lock plates and remove the two screws attaching the connecting-rod cap to the rod (Fig. 35-21), using either a screwdriver or a box wrench as required by the type of screw.

Remove the cylinder bolts and separate the crankcase from the cylinder (Fig. 35-22). Now, you can slip the piston-and-rod assembly out of the cylinder (Fig. 35-23). Before you slip the piston clear out of the cylinder, move it back and forth a few times to check for binding. If the piston moves hard in the cylinder, then there is probably a scuffed condition that will require replacement of the piston and rings and honing of the cylinder wall. Examine the piston, after removing it, for scratches, scuff marks, or scores. Any of these indicate undue wear that may require replacement of the piston. Examine the piston rings for the same conditions.

If the cylinder wall shows signs of wear so that it will have to be refinished, discard the piston and rings. A new, oversized piston will be required to fit the enlarged cylinder. Cylinder service is discussed below.

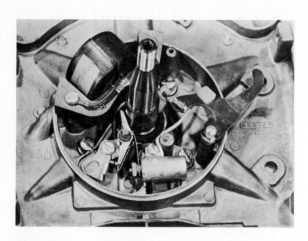

Fig. 35-18. Location of the governor lever that must be removed. (*Lawn Boy Division of Outboard Marine Corporation*)

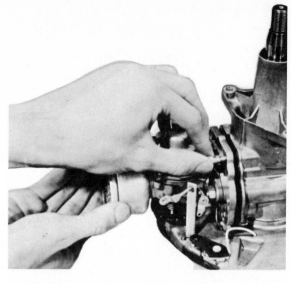

Fig. 35-20. Removing the carburetor and reed-plate assembly. (*Lawn Boy Division of Outboard Marine Corporation*)

Fig. 35-19. Removing screws so that the magneto plate can be removed. (*Lawn Boy Division of Outboard Marine Corporation*)

Fig. 35-21. Removing the connecting-rod-cap screws. (*Lawn Boy Division of Outboard Marine Corporation*)

Continue disassembly by removing the crankshaft from the crankcase through the top, as shown in Fig. 35-24. You have now disassembled the engine, as shown in Fig. 35-25.

Fig. 35-22. Detaching the cylinder head by removing the head bolts. (*Lawn Boy Division of Outboard Marine Corporation*)

Fig. 35-24. Removing the crankshaft from the crankcase. (*Lawn Boy Division of Outboard Marine Corporation*)

Fig. 35-23. Taking the piston-and-rod assembly from the cylinder. (*Lawn Boy Division of Outboard Marine Corporation*)

§ 35-20 SERVICING THE CYLINDER

Examine the cylinder for cracks, stripped threads in the bolt holes, broken fins, or scores or other damage in the cylinder bore. Any of these requires replacement of the cylinder. Next, if the cylinder appears in good condition, use an inside micrometer or a telescoping gage to check the bore for an out-of-round condition (Fig. 35-26). If the cylinder bore is scored or worn out of round, then it must be honed to a larger size so that a larger size piston and rings can be installed. Pistons are supplied in standard oversizes, for example, 0.010, 0.020, and 0.030 inch [0.25, 0.51, and

Fig. 35-25. Major components of the engine. (*Lawn Boy Division of Outboard Marine Corporation*)

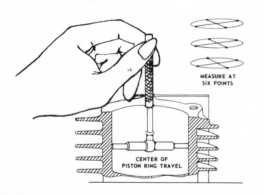

Fig. 35-26. Checking the cylinder bore with a telescoping gage. (*Briggs & Stratton Corporation*)

0.76 millimeter] oversize. The cylinder must be enlarged to take one of these oversize pistons.

Figure 35-27 shows the honing procedure to be used in the cylinder of a four-cycle, one-cylinder engine. The procedure is similar for the cylinder of the two-cycle engine, except that it must be done from the crankcase end of the cylinder. When the honing job is fin-

ished, the cylinder wall should have a cross-hatch pattern like the one shown in Fig. 35-28. This finish requires that the hone be moved up and down at the right speed while the hone is rotating at the right speed.

◗ CAUTION ◗

The cylinder must be cleaned very carefully after the honing job is finished. First, wash the cylinder in kerosene. Then, use a stiff brush, soap, and water to wash out the cylinder bore.

---

§ 35-21 SERVICING PISTON, RINGS, AND CONNECTING ROD

Rods are attached to pistons in several ways. In one, the piston pin is a press fit in the rod and must be driven out with a special punch (Fig. 35-29). Another design uses lock clips to hold the piston pin in place (Fig. 35-30). On these, needle-nose pliers must be used to re-

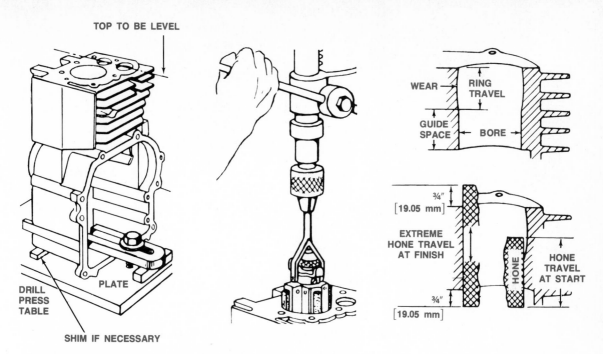

Fig. 35-27. Honing the cylinder. (*Briggs & Stratton Corporation*)

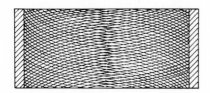

Fig. 35-28. Crosshatch appearance of a properly honed cylinder bore. (*Briggs & Stratton Corporation*)

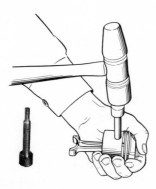

Fig. 35-29. Removing the piston pin with a special punch. (*Lawn Boy Division of Outboard Marine Corporation*)

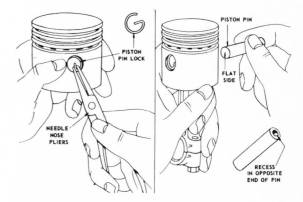

Fig. 35-30. Removing the piston-pin lock. (*Briggs & Stratton Corporation*)

move the clips so that the piston pin can be slipped out. With the piston pin removed, the rod and piston may be separated.

Remove the piston rings, one at a time, using a ring remover, as shown in Fig. 35-31, to spread the rings so that they can be slipped off over the head of the piston.

481

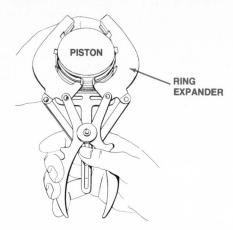

Fig. 35-31. Using the ring expander to remove the piston rings from the piston. (*Briggs & Stratton Corporation*)

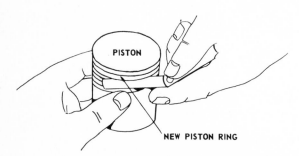

Fig. 35-32. Checking the side clearance of the piston ring in the ring groove. (*Briggs & Stratton Corporation*)

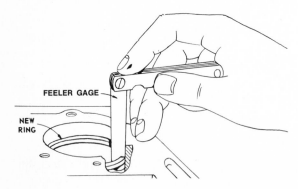

Fig. 35-33. Checking the ring gap with the ring inserted in the cylinder bore. (*Briggs & Stratton Corporation*)

Discard the piston if it is scored, shows wear spots, or is collapsed to an out-of-round condition. Some engine manufacturers recommend that the fit of the piston to the cylinder bore be checked by inserting the piston into the bore with a feeler gage along its side. This will tell you whether or not there is too much clearance. Excessive wear will cause this condition. The remedy is to refinish the cylinder and install a new oversize piston. Always discard the old piston if the cylinder is refinished, because a new, oversize piston must be installed.

If the piston appears in good condition and is to be reused, check for ring groove wear by cleaning all carbon from the top ring groove. Put a new ring of the proper size and kind in the top ring groove. Then, check the clearance between the ring and the side of the groove with a feeler gage (Fig. 35-32). If the clearance is excessive (see manufacturer's specifications), discard the piston and install a new one.

To check rings, clean all carbon from them, especially their ends. Checking one at a time, insert the rings in the cylinder bore, pushing them down 1 inch [25.4 millimeters] into the bore. Check the ring gap with a feeler gage (Fig. 35-33). If it is excessive, throw the ring away.

Some connecting rods have sleeve bearings in the big end; others use needle bearings. On rods with sleeve bearings, new rods must be used if the bearings are worn. (Larger engines have separate bearing shells that can be replaced by themselves so that the rods can be reused.) Needle bearings can be replaced. New sets come in strips (Fig. 35-34). To install a new set of needles, lay the strip on your forefinger, as shown in Fig. 35-34, and carefully strip off the backing. Then, curl the finger around the crankpin so that the needles transfer from your finger to the crankpin. The grease on the needles will hold them in place.

Figure 35-35 shows a set of rings properly installed on a piston. Note that the rings on this piston are of three types and that the oil ring goes on the bottom. New rings are installed

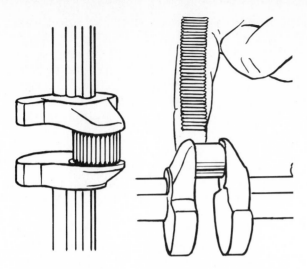

Fig. 35-34. Needles in place around the crankpin; hold them on your finger to apply them to the crankpin. (*Lawn Boy Division of Outboard Marine Corporation*)

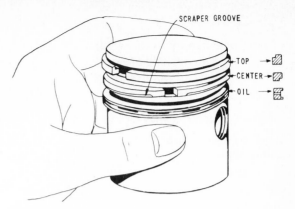

Fig. 35-35. Position of the piston rings on the piston. (*Briggs & Stratton Corporation*)

with a device called the ring expander (Fig. 35-31).

After the connecting rod has been re-attached to the piston-and-ring assembly, then the rings must be compressed into the piston ring grooves before the piston can be inserted into the cylinder bore. A ring compressor is required for this job. Figure 35-36 shows how the ring compressor is used. The right-hand picture in this illustration shows a piston being installed in a four-cycle engine, but the principle of installation in the two-cycle engine is similar. There is one difference, however: In the two-cycle engine, the piston is pushed up into the cylinder from the crankcase side.

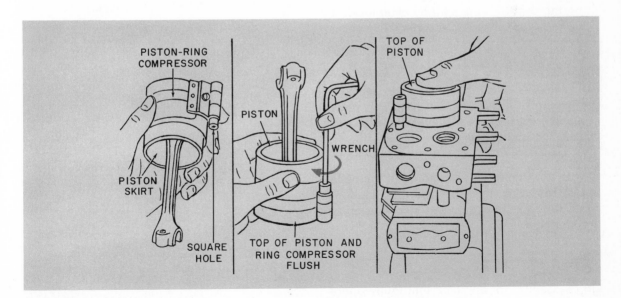

Fig. 35-36. Using the ring compressor to install the piston-and-ring assembly. (*Briggs & Stratton Corporation*)

§ 35-22 GASKETS

Always use new gaskets when reassembling the engine. Old gaskets are probably hard and brittle and will not hold compression.

§ 35-23 REED VALVE

Make sure the reeds are not bent or damaged and that they seal tightly when they close. Replace the assembly if the reeds are not in good condition.

§ 35-24 SERVICING BEARINGS

If ball bearings are in good condition, leave them in place in the crankcase. But if they require replacement, be sure to use the special tool supplied by the manufacturer to remove the old bearing and install the new one (Fig. 35-37).

§ 35-25 RECOIL STARTER

If the recoil starter needs service, refer to Fig. 35-38 (which shows a typical unit in exploded view). To disassemble it, take out the screws and lift off the cap, with the spring coiled inside it.

▶ CAUTION ◀

The spring is coiled in the cap under tension, so be careful that it does not jump free when you remove the cap. Leave the spring in the cap unless the spring requires replacement.

Note how the rope is anchored in the pulley before removing it so that you can anchor it again in the same way when you reassemble the starter.

§ 35-26 IGNITION

On reassembly, or any time that faulty operation indicates the ignition system should be checked, note the condition of the breaker points. If they are burned or pitted, replace

Fig. 35-37. Using a special punch to install a new crankshaft bearing. (*Lawn Boy Division of Outboard Marine Corporation*)

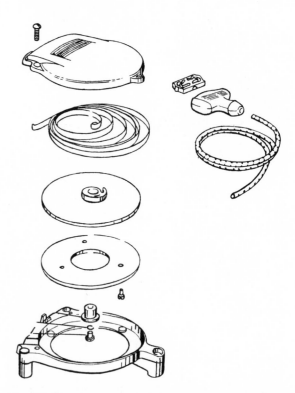

Fig. 35-38. Exploded view of a recoil starter. (*Lawn Boy Division of Outboard Marine Corporation*)

them. Check the point gap with a feeler gage of the correct thickness as specified by the manufacturer (Fig. 35-39). Adjust by loosening

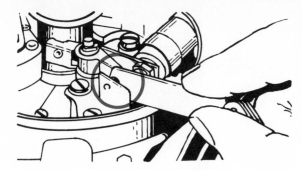

Fig. 35-39. Checking the point gap in the magneto. (*Lawn Boy Division of Outboard Marine Corporation*)

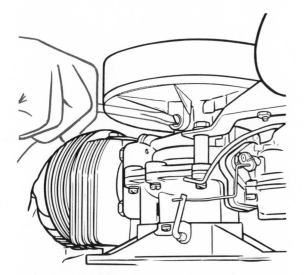

Fig. 35-40. Adjusting the ignition timing. (*Lawn Boy Division of Outboard Marine Corporation*)

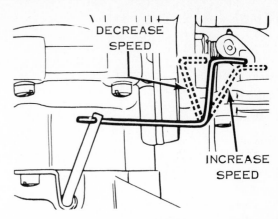

Fig. 35-41. Adjusting the governor speed on one type of engine. (*Lawn Boy Division of Outboard Marine Corporation*)

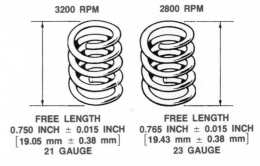

Fig. 35-42. On some engines, the governor spring must be changed to change the governed engine speed. (*Lawn Boy Division of Outboard Marine Corporation*)

the locking screw and turning the pivot to get the proper gap. Then, tighten the locking screw and recheck the gap. The rubbing block of the breaker arm must be on the high point of the cam when the gap is checked. Then draw a piece of clean white paper between the points to remove any grease or dirt.

On some models, the ignition may be timed by loosening a locking screw (Fig. 35-40) and then shifting the magneto plate one way or the other until the engine is running smoothly. Tighten the screw. The engine must be warmed up and running at medium speed when this adjustment is made.

§ 35-27 GOVERNOR

A variety of governors have been used on the two-cycle engine. If the engine is not being governed at the correct speed, the governor should be adjusted. On some models, this is done by bending the link between the throttle and the governor (Fig. 35-41). On others, the governor spring is changed (Fig. 35-42).

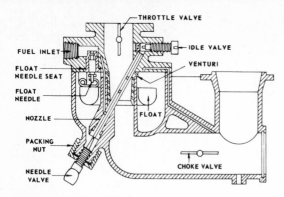

Fig. 35-43. Sectional view of a carburetor. (*Briggs & Stratton Corporation*)

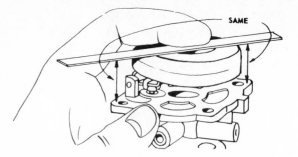

Fig. 35-44. Checking the float level of the carburetor. (*Briggs & Stratton Corporation*)

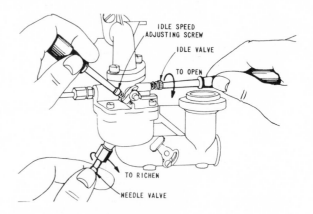

Fig. 35-45. Final adjustments on one model of carburetor. (*Briggs & Stratton Corporation*)

▶ **CAUTION** ◀

Do not attempt to adjust governor speed by stretching the spring!

---

§ 35-28 CARBURETOR

Many different models of carburetors have been used. A sectional view of one type is shown in Fig. 35-43. If carburetor repair is needed, then either the carburetor should be replaced with a new one, or else it must be disassembled so that new parts can be installed. There are carburetor repair kits available which contain all necessary new parts.

▶ **CAUTION** ◀

Always install new gaskets when repairing a carburetor. The old gaskets are probably hardened and will not provide a good seal so that leakage would occur if they were used again.

It is important to adjust the float so that the gasoline will be maintained at the proper height in the float bowl. The procedure of checking the float on one model is shown in Fig. 35-44. The float should be parallel to the body mounting surface with the body gasket in place and the float valve and float installed. Bend the tang on the float, if necessary, to bring the float to parallel.

Final adjustment on one model of carburetor is shown in Fig. 35-45. Turn in the idle-speed adjusting screw and the needle valve until they just seat.

▶ **CAUTION** ◀

Do not tighten, because this could damage the seats!

Back off the needle $1\frac{1}{2}$ turns. Back off the idle valve $\frac{3}{4}$ turn. Start the engine; allow it to warm up. Turn the needle valve in until the engine misses from a lean mixture. Back out until the engine begins to run unevenly from an overly rich mixture. Turn the needle valve to the midway point between lean and rich so that the engine runs smoothly. Then, hold the throttle at the idle position and set the idle speed to proper specifications by turning the adjusting screw.

Review questions

1. What are causes of failure of a two-cycle engine to start?
2. What effect does a mixture of gasoline and air that is too rich or too lean have upon engine operation?
3. If your engine starts easily, but loses speed as it warms up, what is the likely cause of the problem?

4. What are the eight important steps in the maintenance of a two-cycle engine?
5. How is the two-cycle engine lubricated?
6. Should old gasoline be used in a small engine?

Study problems

1. You are about to mow your lawn and you find that your two-cycle engine will not start. List the steps you would take to locate the source of trouble.
2. Remove, clean, and adjust the spark plugs of a two-cycle engine in accordance with the manufacturer's manual.

3. Completely disassemble, examine the parts of, reassemble, adjust, and test operate a two-cycle engine in your school shop. Have your instructor check each step.
4. While the engine in Problem 3 is disassembled, identify each part. Use the exploded view in the manufacturer's manual.

36 AUTOMOTIVE-ENGINE TROUBLESHOOTING

§ 36-1 PROCEDURE FOR TROUBLESHOOTING

Troubleshooting is the procedure that is used to locate causes of trouble. We start with the specific complaint and make a series of tests that will pin down the trouble to a specific engine component. In this unit, the common complaints such as engine will not turn over, or engine will not run, or engine misses, etc., are covered, and the tests needed to track down the trouble causes are described.

§ 36-2 ENGINE WILL NOT TURN OVER

If the engine will not turn over when starting is attempted, chances are that the battery is run down. Make a quick check by turning on the car lights and trying to start. If the lights come on bright and stay bright, there is an open circuit in the starting motor or in the starting-motor wiring circuit or control circuit.

If the lights dim considerably, the battery is run down. However, mechanical trouble in the engine or starting motor could also cause high current draw, which would dim the lights.

If the lights dim only slightly, listen for the sound of the starting motor running. It is possible that the starting motor is running, but for some reason the drive pinion is not engaging with the flywheel.

If the lights go out when cranking is attempted, the chances are that there is a bad connection at one of the battery terminals.

If the lights burn dimly or not at all when they are turned on, then the battery is probably run down.

§ 36-3 ENGINE TURNS OVER SLOWLY BUT DOES NOT START

This could be due to a run-down battery, a defective starting motor, bad connections in the starting-motor wiring circuit, or mechanical trouble in the engine. Remember that in cold weather, especially if a heavy oil is used, even an engine in good condition will not turn over very rapidly.

§ 36-4 ENGINE TURNS OVER AT NORMAL SPEED BUT DOES NOT START

With this condition, the battery and starting motor are in normal condition, and the cause of trouble is probably in the ignition or fuel system. First, consider the possibility that the fuel tank is empty or that the engine has been overchoked and is flooded. That is, the choke valve remained closed too long so that too much fuel has entered the intake manifold and engine cylinders. This can be cleared by cranking the engine with the throttle held wide open. If this does not start the engine, check for an ignition spark. Disconnect the high-voltage lead from the center terminal of the

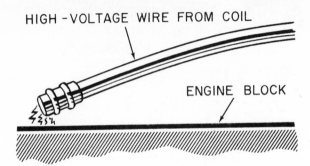

Fig. 36-1. Testing for a spark by holding the high-tension lead $\frac{1}{4}$ inch [6.35 millimeters] from the engine block.

ignition-distributor cap and hold the lead clip about $\frac{1}{4}$ inch [6.35 millimeters] from the engine block (Fig. 36-1). Crank the engine to see if a good spark occurs. If it does not, the trouble is in the ignition system. If a spark does occur, the trouble is probably in the fuel system (although ignition timing could be off to prevent starting).

Troubles in the fuel system that could prevent starting include a faulty fuel pump that does not deliver fuel to the carburetor, or troubles in the carburetor such as clogged jets or systems. Air leakage into the intake manifold, due to a defective manifold or carburetor-mounting gasket, could also prevent starting. In addition, if the failure to start occurs with a hot engine, then consider the possibility that the automatic choke is stuck closed and is causing flooding of the engine. Take off the air cleaner to check the position of the choke valve (Fig. 36-2).

◆ CAUTION ◆

Never try to start the engine, or run the engine, with the air cleaner off. The engine could backfire and cause you to get burned. Also, the backfire could ignite any gasoline vapors around the engine and cause a terrible fire.

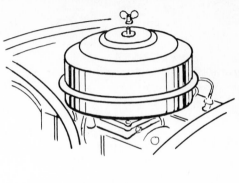

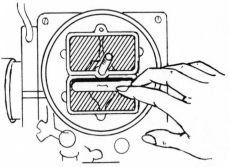

Fig. 36-2. Remove the air cleaner by unscrewing the retaining wingnut. If the engine is hot, the choke valve should be open.

§ 36-5 ENGINE RUNS BUT MISSES

A missing engine is a rough engine that runs unevenly and lacks power. Missing can be caused by many things, from a faulty spark plug to defective valves or piston rings. One way to track down a miss is to short out each spark plug in turn with a screwdriver (Fig. 36-3). Hold on to the insulated handle so that you will not be shocked. Put the screwdriver between the spark-plug terminal and the engine block or head. This prevents a spark from occurring so that the cylinder will miss. If the engine speed or rhythm changes, then the cylinder has been delivering power. But if the speed or rhythm does not change, then the cylinder was already missing before the plug was shorted out. Check all cylinders in this way. If you find a missing cylinder, put in a new spark plug and try again. If the cylinder still

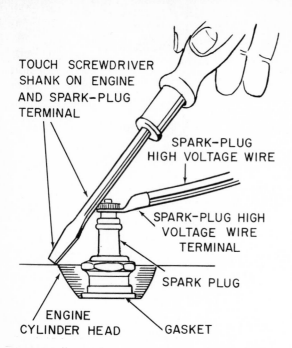

Fig. 36-3. Use a screwdriver to short out each spark plug in turn.

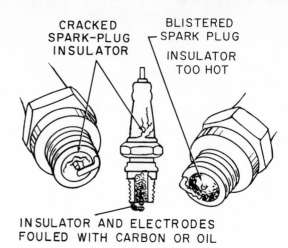

Fig. 36-4. Cracked, worn, or dirty spark plugs may cause the engine to lack power.

misses, chances are that the trouble is in the engine, caused by defective valves or rings, stuck manifold heat-control valve, clogged exhaust, and so on.

On many engines, you cannot get to the spark plugs to short them out as described above. On these engines, remove the spark-plug lead from the spark plug or from the distributor cap. Note any changes in rpm.

Sometimes the miss will seem to jump around from one cylinder to another and is hard to locate. This sort of trouble is most likely due to a faulty fuel system that is not supplying the proper ratio of air-fuel mixture.

Performing an engine tune-up (Chapter 37) will usually locate the cause of trouble so it can be eliminated.

§ 36-6 ENGINE LACKS POWER

This trouble could be caused by many things. The best procedure is to tune up the engine so that all conditions that could cause the trouble can be checked and corrections made where necessary. Some of the causes of lack of power include:

1. Fuel system faulty because of fuel-pump defects, lines or filters clogged, air leaks into the intake manifold, carburetor jets or circuits worn or clogged, malfunctioning positive crankcase ventilation (PCV) valve, misadjusted throttle linkage between the accelerator pedal and the carburetor, a faulty choke, the high-speed system or accelerator pump not operating properly, and so on.
2. Ignition system at fault due to incorrect timing, a weak ignition coil, worn or dirty spark plugs (Fig. 36-4), bad wiring, contact points burned or out of adjustment, and so on.
3. In the engine, numerous conditions could cause loss of power, including worn or burned valves or valve seats, worn pistons, rings, or cylinder walls, rings clogged with carbon, and defective bearings.
4. A clogged exhaust, from a bent or collapsed exhaust pipe, tail pipe, or muffler, could create back pressure that would reduce engine power.

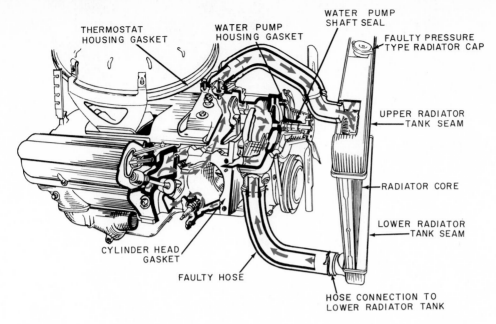

Fig. 36-5. V-8 engine cooling system, showing areas where coolant leakage may occur. (*Pontiac Motor Division of General Motors Corporation*)

5. An overheated engine will lose power. High temperatures may also cause vapor locks in the fuel pump and fuel lines. This condition results when the fuel vaporizes in the pump or line and prevents the pump from delivering fuel to the carburetor.

6. Friction or slippage in the transmission or drive line will put an extra burden on the engine so that it loses power.

§ 36-7 ENGINE OVERHEATS

When the engine overheats, the first thing you think of is that the cooling system has lost coolant. If coolant has been lost, there may be a leak in the cooling system that should be corrected. Leaks can occur in the hoses, at the hose connections, in the radiator, or at the cylinder-head gasket (Fig. 36-5). Also, block or head cracks may allow leakage.

Other possible causes of engine overheating include a loose or broken fan belt, a defective radiator hose, or a defective thermostat that fails to open properly.

Conditions outside of the cooling system that might cause engine overheating include late ignition or valve timing, lack of engine oil, or high-altitude or hot-climate operation. Also, freezing of the coolant could restrict circulation and cause hot spots to develop in the cooling system that would result in boiling of the coolant in those spots.

§ 36-8 ENGINE IDLES ROUGHLY

This could be due to an improperly set idle speed or mixture. But it could also be due to most of the conditions that cause lack of engine power, described in § 36-6.

§ 36-9 ENGINE STALLS

If the engine starts and then stalls, note whether the stalling takes place as the engine warms up, after idling or slow-speed driving, or after high-speed driving.

If the engine stalls as it warms up, the choke valve may be stuck closed, the manifold heat-

In a six-cylinder engine the PCV valve is located in:
1. Rocker arm cover
2. Base of carburetor
3. Hose

Fig. 36-6. Location of PCV valve in a six-cylinder engine.

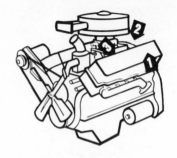

In a V-8 engine the PCV valve is located in:
1. Rocker arm cover
2. Rear of engine
3. Carburetor base

Fig. 36-7. Location of PCV valve in a V-8 engine.

control valve may be stuck, or the hot-idle speed may be set too low. Also, stalling may be due to vapor lock caused by overheating of the engine.

If the positive crankcase ventilation (PCV) valve is sticking, it will cause poor idling and stalling. Figures 36-6 and 36-7 show the locations of the PCV valve in six-cylinder and V-8 engines.

If the engine stalls after idling or low-speed driving, it could be due to a defective fuel pump that cannot deliver enough fuel at low speed. Also, the carburetor idle speed or mixture adjustment could be incorrect, or the float level in the carburetor set too high.

When the car is driven slowly, or sits with the engine running (as for instance, in a traffic jam), the engine is apt to overheat. This, in turn, causes vapor lock and engine stalling.

If the engine stalls after high-speed driving, it could be due to temporary engine overheating that produces a vapor lock.

§ 36-10 ENGINE BACKFIRES

Backfiring could be due to overheating of the engine, excessive carbon in the cylinders and on the spark plugs, hot or sticking valves, an excessively rich or lean mixture due to carburetor troubles, or late ignition timing.

§ 36-11 EXCESSIVE OIL CONSUMPTION

Oil is lost from the engine in three ways: by burning in the combustion chambers, by leakage in liquid form, and by passing out of the crankcase through the ventilating system in the form of mist.

External leaks can often be found by inspecting the seals around the oil pan, valve cover plate, timing-gear housing, and oil-line and filter connections (Fig. 36-8).

Burning of oil in the combustion chambers gives the exhaust gas a bluish tinge. Oil can enter the combustion chambers in three ways: past clearance between intake-valve stems and guides, past piston rings, and through the PCV system.

If the clearance between the intake-valve stems and guides is excessive, then oil will seep through this clearance and into the cylinder on every intake stroke (Fig. 36-9). The remedy here is to install a valve seal or a new valve guide, and possibly a valve.

Probably the most common cause of excessive oil consumption is the passage of oil into the combustion chambers between the piston

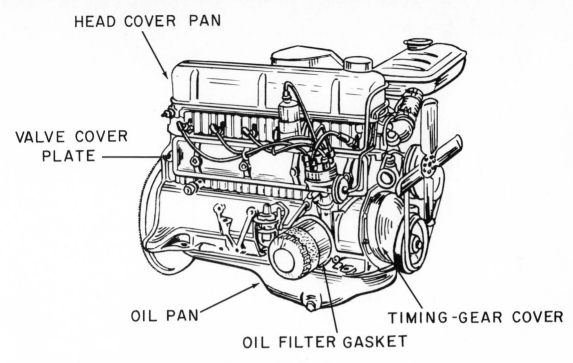

HEAD COVER PAN

VALVE COVER PLATE

OIL PAN

OIL FILTER GASKET

TIMING-GEAR COVER

Fig. 36-8. Six-cylinder engine, showing possible locations where oil leakage may occur. (*Ford Motor Company*)

rings and cylinder walls. This is often called **oil pumping** and is due to worn, tapered, or out-of-round cylinder walls or to worn or clogged oil control rings. Also, worn engine bearings could make the situation worse. Worn bearings will allow more oil to leak past the bearings and be thrown on the cylinder walls. High speed can also increase oil consumption. Oil gets hotter and thinner at high speed. Also, the oil pump pumps more oil to the bearings at high speed. This additional and thinner oil puts an added load on the oil control rings. An engine uses several times as much oil at 60 mph (miles per hour) [96.56 km/h] as it uses at 30 mph [48.28 km/h].

§36-12 LOW OIL PRESSURE

If the oil pressure is low, it often means that engine bearings are worn. The bearings can pass so much oil that the oil pump cannot keep

the pressure up. Also, the chances are that the bearings farthest away from the pump will be oil-starved because the nearer bearings will drain off most of the oil.

Other causes of low oil pressure include a worn oil pump, a broken or cracked oil line, a weak pressure-relief-valve spring, or a clogged oil line to the pump. Oil dilution, foaming, sludge, or insufficient or thin oil may cause low oil pressure.

§36-13 EXCESSIVE FUEL CONSUMPTION

The best procedure to locate the cause of excessive fuel consumption is to tune up the engine so that any defects can be found and eliminated (Chapter 37). Remember, however, that the operating conditions play an important part in fuel economy. Fuel consumption will be high if the engine is used only for short trips

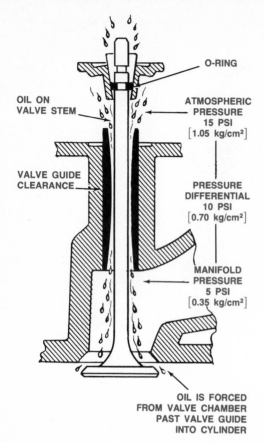

Fig. 36-9. Worn valve guide shows pressure differences causing oil to flow into combustion chamber.

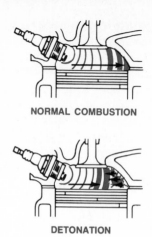

NORMAL COMBUSTION

DETONATION

Fig. 36-10. *Top:* View of normal combustion pressure waves. *Bottom:* View of opposing pressure waves from the uncontrolled second explosion, called detonation. (*General Motors Corporation*)

with cooling-off periods between so that it is always operating cold or on the warm-up cycle. High-speed operation also reduces fuel economy. Best fuel economy is obtained at consistent, medium-speed operation.

§ 36-14 ENGINE NOISES

Some engine noises are trouble signs, others are more or less normal. Explanations of some of the important ones follow.

1 Valve and tappet noise. This is a regular clicking noise that results from the taking up of the clearance, or lash, in the valve train as the valve is closed. There will be no such noise in an engine with properly operating hydraulic valve lifters. If an engine with hydraulic valve lifters does have this noise, then the hydraulic valve lifters are sticking.

On engines with mechanical valve lifters, excessive valve and tappet noise may mean that the clearance, or lash, is excessive. A quick check can be made by inserting a feeler gage between the valve stem and rocker arm, or lifter. If the noise is reduced, then there is excessive clearance. If the noise is not reduced, the noise is probably coming from weak valve springs, worn lifter faces, lifters loose in the block, rough adjustment-screw face, or rough cams on the cam shaft. Of course, the noise might not be from the valve train at all. The noise could be coming from some other part of the engine. (See the other causes of noise described in the following paragraphs.)

2 Spark knock. This is a pinging or rattling sound that is a characteristic sound of detonation. It is most noticeable when the engine is accelerated on a hard pull. Some knock is normal. When it is excessive, it could be due to fuel having an octane rating too low for the engine (Fig. 36-10). Another cause may be excessive carbon deposits in the engine cylinders. Excessively advanced ignition timing or some of the conditions that cause engine backfiring may also cause detonation.

3 Connecting-rod noise. If a connecting-rod bearing or crankpin is worn, there will be excessive clearance, and this clearance will cause noise. The noise is most noticeable when the engine is "floating," that is, not accelerating or decelerating. Connecting-rod noise can be pinpointed by shorting out one spark plug at a time. The noise will be reduced when the cylinder that is responsible is not delivering power. The noise can also come from a misaligned rod or from inadequate oil.

4 Piston-pin noise. This noise results from a worn or loose piston pin, a worn bushing, or lack of oil. It is somewhat similar to valve and tappet noise, but it is more of a metallic double knock that is regular and especially noticeable at idle or relatively low speed. One way to track it down is to advance the ignition spark a few degrees and then short out spark plugs one at a time with the engine idling. The noise will be reduced when the noisy cylinder is shorted out.

5 Piston-ring noise. This is also similar to valve and tappet noise, but it is more of a clicking, snapping, or rattling noise. It is due to low ring tension, worn rings or cylinder walls, or broken rings. A test for this condition can be made by removing the spark plugs and pouring an ounce of heavy oil into each cylinder. Crank the engine for several revolutions to work the oil down to the rings. Then replace the plugs and start the engine. If the noise has been reduced, chances are that it is being caused by the rings.

6 Piston-slap. This is a muffled, hollow, bell-like sound that results from the rocking back and forth of the piston in the cylinder. If it occurs only when the engine is cold, then the condition is not serious. But if it also occurs with the engine hot, then further checking and possibly service are required. Causes could be inadequate oil, worn pistons or cylinder walls, collapsed piston skirts, excessive piston clearance, or misaligned connecting rods.

7 Crankshaft knock. This is a heavy, dull metallic knock that is most noticeable when the engine is under a heavy load or accelerating, especially when cold. When the noise is regular, it is probably due to worn main bearings. When the noise is irregular and sharp, it is probably due to a worn end-thrust bearing that allows excessive and sudden end movement of the crankshaft.

8 Other noises. Parts that are loosely mounted, such as the alternator, starting motor, horn, water pump, and so on, will cause a variety of rattling, grinding, or scraping noises. In addition, other automotive components such as the transmission, drive line, wheel bearings, differential, and so on, can develop noise.

Review questions

1. How would you check an engine for ignition spark?
2. What troubles in the carburetor could prevent starting?

3. What happens when an engine suffers from vapor lock? What usually causes vapor lock?
4. How would freezing of the coolant cause an engine to boil soon after starting?

5. Name three ways that oil can be lost from an engine.

Study problems

1. List in proper sequence the steps to be taken in locating and correcting the following troubles:
   a. Engine will not turn over
   b. Engine turns over at normal speed but does not start

6. List five kinds of engine noises. Explain the cause for each.

   c. Engine runs but misses
   d. Engine idles roughly

2. The instructor will place several malfunctions in all shop engines. Each student will have a time limit in which to restore the assigned engine to proper running order.

37 | AUTOMOTIVE-ENGINE TUNE-UP AND SERVICE

§ 37-1 TESTING INSTRUMENTS

When an engine is not running properly, special testing instruments can be used to pinpoint the trouble. In much the same way a doctor uses testing instruments to pinpoint the cause of a patient's illness. Neither the doctor nor the automotive technician tries to prescribe a cure until each one has a good idea of what is causing the ailment to be treated.

Various testing instruments are described in this chapter, along with explanations of how they are used.

§ 37-2 TACHOMETER

The tachometer (Fig. 37-1) is a device to measure how fast the engine is running. It reports this on a dial in **rpm** (revolutions per minute). The tachometer is connected into the ignition circuit and counts the number of times, per minute, that the ignition-distributor contact points open. Or, in the electronic ignition system, the tachometer counts the number of times per minute that the pickup coil sends a surge of current (the signal) to the electronic control unit. (This is the number of times per minute that the primary circuit is interrupted.) It translates this into rpm.

§ 37-3 CYLINDER COMPRESSION TESTER

As the miles pile up, engine valves, piston rings, pistons, and cylinder walls wear. This wear permits more and more leakage of compression from the combustion chamber in the cylinder. When an engine loses compression or cannot hold compression, it means that part of the compressed air-fuel mixture leaks past the piston rings or past the valves. Also, during the power stroke, some of the high-pressure gas leaks past these same points. This can cause serious power loss because this leakage reduces the pressure applied to the pistons. Leakage of combustion gases past the piston rings is called **blow-by.** These gases blow by the piston rings.

To use the compression tester, remove all the spark plugs from the engine. Hold the tester tightly against the spark-plug hole (Fig. 37-2) of one cylinder and crank the engine with the starting motor. Make sure the throttle is wide open so that the carburetor does not deliver fuel to the engine. Note the maximum compression as indicated by the needle on the tester dial. Check all the cylinders in this way.

If the compression does not come up to specifications, there is leakage past the valves, piston rings, or through a defective head gasket. You can pinpoint the trouble further by pouring a little heavy engine oil into the cylinder through the spark-plug hole. If the compression then increases to a more normal figure, it means that the leakage has been past the piston rings—the heavy oil has improved the seal and reduced the leakage. But if the compression does not increase, then the leakage is past the valves or through a defective head gasket.

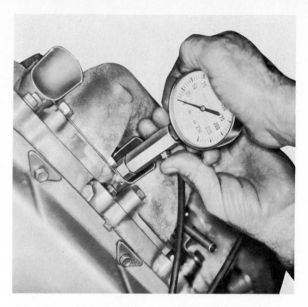

Fig. 37-2. A cylinder-compression tester in use. (*Chevrolet Motor Division of General Motors Corporation*)

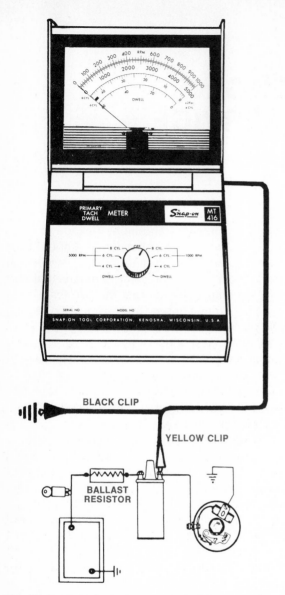

BLACK CLIP

YELLOW CLIP

BALLAST RESISTOR

Fig. 37-1. A tachometer connected to an engine. (*Snap-on Tools Corporation*)

With low compression, the cylinder head must come off so that engine parts can be inspected and serviced as necessary.

§ 37-4 CYLINDER-LEAKAGE TESTER

The cylinder-leakage tester does about the same job as the compression tester, but in a different way. It applies air pressure to the cylinder with both valves closed. The tester then shows if air is leaking out of the cylinder. If it is, listen at any of the three listening points as shown in Fig. 37-3 to determine where the air leak is.

§ 37-5 VACUUM GAGE

This tester measures the vacuum in the intake manifold. If the vacuum is low, or is erratic when the engine runs, then it indicates certain engine troubles, as explained in a following paragraph. To use the vacuum gage, connect it to the intake manifold at the place provided (Fig. 37-4). Then, with the engine at operating temperature and idling, note the vacuum reading. Here is what the different readings mean (Fig. 37-5).

1. A steady reading that is up to specifications indicates that the engine is operating normally. (There is apt to be somewhat more fluctuation of the needle on some of the high-performance engines.)

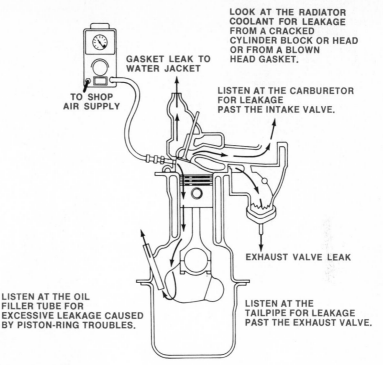

Fig. 37-3. Here is how the cylinder-leakage tester works. It applies air pressure to the cylinder through the spark-plug hole with the piston at TDC and both valves closed. Points where air is leaking can then be pinpointed, as shown. (*Sun Electric Corporation*)

LOOK AT THE RADIATOR COOLANT FOR LEAKAGE FROM A CRACKED CYLINDER BLOCK OR HEAD OR FROM A BLOWN HEAD GASKET.

GASKET LEAK TO WATER JACKET

LISTEN AT THE CARBURETOR FOR LEAKAGE PAST THE INTAKE VALVE.

TO SHOP AIR SUPPLY

EXHAUST VALVE LEAK

LISTEN AT THE OIL FILLER TUBE FOR EXCESSIVE LEAKAGE CAUSED BY PISTON-RING TROUBLES.

LISTEN AT THE TAILPIPE FOR LEAKAGE PAST THE EXHAUST VALVE.

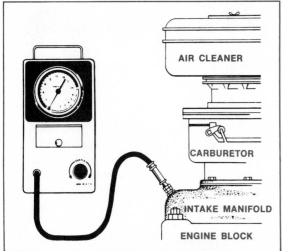

Fig. 37-4. A vacuum gage connected to an intake manifold for a manifold vacuum test. (*Sun Electric Corporation*)

2. A steady reading that is too low could mean leakage past piston rings or late valve timing.

3. A very low reading could mean air leaks through the intake-manifold gasket or the carburetor-mounting gasket, or leaks around the throttle-valve shaft.

4. Oscillations of the needle that increase as engine speed is increased could mean weak valve springs.

5. A gradual falling back of the needle indicates a clogged exhaust line.

6. Regular falling back of the needle could mean that either the spark plug in one cylinder is not firing or a valve is stuck open.

7. Irregular falling back of the needle is probably due to sticking valves that stick irregularly.

8. If the needle oscillates slowly, the probability is that the air-fuel mixture is too rich and that the carburetor will need servicing.

§ 37-6 EXHAUST-GAS ANALYZER

At one time, the major use of the exhaust-gas analyzer was to adjust the carburetor. It is still used for that purpose, but today it has the added job of checking out the emission con-

**LOW AND STEADY READING INDICATES
LOW COMPRESSION, AIR LEAKS, OR
LATE IGNITION TIMING**

**RAPID VIBRATION WHEN
ENGINE IS ACCELERATED INDICATES
WEAK VALVE SPRINGS**

**INTERMITTENT
DROP OF NEEDLE INDICATES STICKY
VALVES**

**FLOATING MOTION OF NEEDLE
INDICATES RICH MIXTURE**

Fig. 37-5. Typical vacuum gage readings.

trols on the car. The tester is now called an exhaust-emission tester (Fig. 37-6). As previously mentioned, certain amounts of hydrocarbons and carbon monoxide leave the engine in the exhaust system because of incomplete combustion. The exhaust-emission tester de-

Fig. 37-6. Exhaust-emission tester. (*Sun Electric Corporation*)

tects these and reports on how much of these pollutants there is in the exhaust gas. If the amount is over the legal limit, then corrections must be made. This could mean adjusting the carburetor or servicing the emission controls on the engine.

§ 37-7 IGNITION TIMING LIGHT

For good engine performance, the ignition system must be timed so that the sparks occur in the cylinders at the correct time toward the end of the compression strokes. Early (advanced) or late (retarded) timing will reduce engine performance. The timing light is used to set the ignition timing.

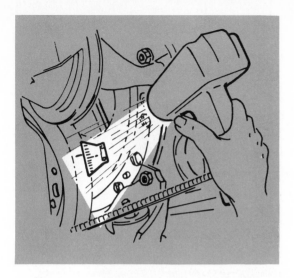

Fig. 37-7. The timing light flashes every time number 1 spark plug fires.

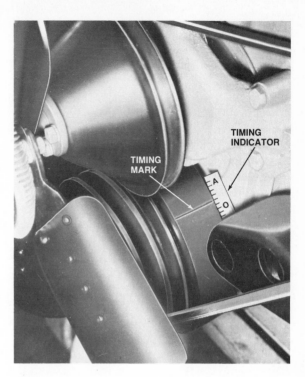

Fig. 37-8. Ignition timing marks on flywheel and on crankshaft pulley. (*Chevrolet Motor Division of General Motors Corporation*)

The timing light has a special lamp in it that flashes on and off almost instantly when it is connected to a source of electricity. When used on the engine, it is connected to number one spark plug. Every time this plug fires, the timing light flashes on. It goes off again almost instantly. Thus, when the light is pointed at the rotating crankshaft pulley, it seems to make the crankshaft pulley stand still (Fig. 37-7). That is, the crankshaft pulley is lighted only at the instant that number one spark plug fires, and that is the only time you can see it. If the ignition timing is correct, a timing mark on the crankshaft pulley will be aligned with a stationary pointer or tab on the front of the engine (Fig. 37-8). If the timing is not correct, then the ignition distributor is loosened in its mounting and turned one way or the other to bring the markings into alignment. Then the distributor-mounting clamp is tightened.

§ 37-8 OSCILLOSCOPE

The oscilloscope (Fig. 37-9) is a high-speed voltmeter. It shows on the face of a television-like picture tube curves and lines repre-

senting the voltages in the ignition system. If there is trouble, the picture will show it and indicate what is wrong.

§ 37-9 DYNAMOMETER

The dynamometer (Fig. 37-10) tests the engine output as it appears at the drive wheels of the car. The car is driven onto a pair of heavy rollers, as shown in Fig. 37-11. Then the engine is started, the car is put into gear, and the car drives the rollers. The rollers can be loaded various amounts to simulate different driving conditions. Thus, the engine performance can be tested. Various test instruments are connected during the test to pinpoint any trouble. The dynamometer can also test the transmission and differential. Also, some dynamometers are designed to test wheel alignment, suspension, brakes, and steering.

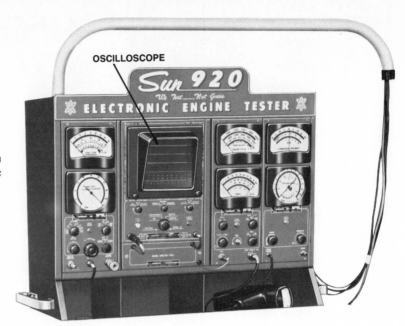

OSCILLOSCOPE

Fig. 37-9. Engine tester with oscilloscope. (*Sun Electric Corporation*)

Fig. 37-10. A chassis dynamometer of the flush-floor type. The rollers are set at floor level. (*Sun Electric Corporation*)

502

Fig. 37-11. An automobile in place on a chassis dynamometer. The rear wheels drive the dynamometer rollers, which are flush with the floor. At the same time, instruments on the test panel measure car speed, engine power output, engine vacuum, and so on. (*Sun Electric Corporation*)

§ 37-10 OTHER TESTING INSTRUMENTS

A variety of other testing instruments are used by automotive mechanics to determine the condition of the various components of the engine systems. For example, there are testers to test fuel pumps, cooling systems, ignition coils, condensers, distributor contact-point opening, alternator output, and so on.

§ 37-11 ENGINE TUNE-UP

The purpose of an engine tune-up is to check the various engine components so that anything below par can be brought up to specifications. This restores engine performance. In other words, the engine is tuned up and its performance is improved. Many service technicians use a test report such as the one shown in Fig. 37-12 as a guide and record. The test report also gives the service technician something specific to show the customer so that the customer can see what is wrong and

what corrections are necessary. A typical tune-up procedure follows. The engine should be at operating temperature.

1. Connect oscilloscope, if available, and check ignition system.
2. Remove all spark plugs and check compression of each cylinder.
3. Clean and regap all plugs, or get new plugs. Install them.
4. Check battery.
5. Check starting motor.
6. Check charging system (alternator and regulator).
7. Check drive belts and replace them if worn.
8. Check distributor rotor, cap, and spark-plug wires.
9. Check, clean, and adjust distributor contact points.
10. Check ignition timing.
11. Recheck system with oscilloscope to see if previously indicated troubles are eliminated.
12. Check manifold heat-control valve.

CUSTOMER_____ PHONE_____ DATE_____

MAKE-YEAR-MODEL_____ MILEAGE_____

SIGNATURE _____

Complete Engine Performance

TEST PROCEDURE
(USING A SUN INFRA-RED ENGINE PERFORMANCE TESTER)

CUSTOMER'S COMMENTS:

	TEST PROCEDURE	READ	SPECS.	RESULTS	GOOD	BAD
START	Cranking Voltage	Voltmeter				
	Cranking Coil Output	Scope (Display)				
	Cranking Vacuum	Vacuum Gauge				
IDLE	Idle Speed	Tachometer				
	Dwell	Dwell-Meter				
	Initial Timing	Timing Advance Unit				
	Hydrocarbons P.P.M.	Hydrocarbons Meter				
	Carbon Monoxide	Carbon Monoxide Meter				
	PCV Test	Carbon Monoxide Meter				
	Manifold Vacuum	Vacuum Gauge				
CRUISE	Dwell Variation	Dwell Meter				
	Coil Polarity	Scope (Display)				
	Spark Plug Firing Voltage	Scope (Display)				
	Maximum Coil Output	Scope (Display)				
	Secondary Circuit Insulation	Scope (Display)				
	Secondary Circuit Condition	Scope (Raster)				
	Coil and Condenser Condition	Scope (Raster)				
	Breaker Point Condition	Scope (Raster)				
	Cam Lobe Accuracy	Scope (Superimposed)				
	Hydrocarbons P. P.M.	Hydrocarbons Meter				
	Carbon Monoxide Percent	Carbon Monoxide Meter				
	Cylinder Power Balance	Tachometer				

	Record R.P.M.	1	2	3	4	5	6	7	8

	TEST PROCEDURE	READ	SPECS.	RESULTS	GOOD	BAD
ACCEL-ERATION	Spark Plugs Under Load	Scope (Display)				
	Accelerator Pump Action	Carbon Monoxide Meter				
HIGH SPEED	Timing Advance	Timing Advance Unit				
	Charging Voltage	Voltmeter				
	Hydrocarbons P.P.M.	Hydrocarbons Meter				
	Carbon Monoxide Percent	Carbon Monoxide Meter				
	Exhaust Restriction	Vacuum Gauge				

Fig. 37-12. A test report form. (*Sun Electric Corporation*)

13. Check fuel pump with a tester.
14. Clean or replace the air-cleaner filter.
15. Check choke, throttle valve, and linkage.
16. Inspect all engine vacuum fittings, hoses, and connections.
17. Clean the engine oil-filler cap, if it has a filter.
18. Check the cooling system, water hoses, connections, radiator, pump, fan clutch, and antifreeze strength.
19. Check and replace PCV valve if necessary.
20. Check condition of emission controls including air-injection parts, vapor recovery system, and so on.

21. Adjust engine valves if necessary.
22. Adjust carburetor idle speed and mixture screws.
23. Road test the car or test it on a dynamometer.
24. Check the door-jamb sticker to see if an oil and oil-filter change is due. Also, note schedule for a chassis lubrication.

§ 37-12 AUTOMOTIVE-ENGINE SERVICE

The following pages describe various services that automotive engines may require.

Regardless of the service job required, there are two things the mechanic must keep in mind. One of these is cleanliness; the other is accuracy.

The major enemy of a good service job is dirt. Even a trace of dirt left on a bearing or cylinder wall can ruin an otherwise good service job. Thus, be absolutely certain that you follow instructions. Do not leave dirt or abrasive on engine parts when you finish a service job.

Before any major engine-service work, the engine block should be cleaned. One widely used method is steam cleaning. The steam melts accumulated grease and dirt and washes it away, leaving clean metal. Before steam cleaning an engine, electrical units should be removed or covered so that steam will not damage them.

Engine-service work must be accurately done according to the specifications of the engine manufacturer. Some measurements and clearances in engines are specified in thousandths of an inch [hundredths of a millimeter]. If these specifications are not followed, engine trouble can result. For instance, a bearing clearance that is off only 0.002 inch [0.05 millimeter] could result in a costly engine failure in a few hundred miles. So you *must* be accurate, and you *must* follow instructions. If in doubt about a particular measurement or procedure, look it up in the manufacturer's manual.

§ 37-13 ENGINE VALVES

The various troubles that valves can have are the following:

1 Valve sticking. This can be caused by gum or carbon build-up on the valve stem or guide or to a warped valve stem.

2 Valve burning. This can be caused by a sticking valve, eccentric seat, weak valve spring, or deposits on the valve seat or face.

3 Valve breakage. This could be due to over-

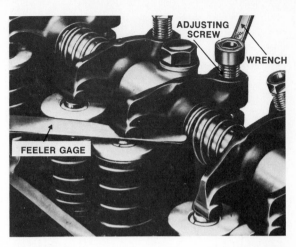

Fig. 37-13. Adjustment of valve-tappet clearance, or valve lash, on an overhead-valve engine. (*Ford Motor Company*)

heating, excessive clearance in the valve train, an eccentric valve seat, or scratches on the stem due to careless handling of the valve.

4 Valve-face wear. If the valve gets too hot, seats poorly, or if there are deposits on the valve face or seat, the face and seat will wear more rapidly and may burn.

§ 37-14 VALVE SERVICE

Valve service includes the following:

1 Adjusting clearance in valve trains (solid lifters). Figure 37-13 shows how this is done on engines using the shaft-mounted rocker arms. Figure 37-14 shows how it is done on engines using ball-stud-mounted rocker arms.

2 Adjusting valve trains with hydraulic valve lifters. On these, no adjustments are normally required. But if valves and valve seats are ground, changing the length of the valve train, then adjustment may be required. Or else new shorter push rods should be installed.

3 Valve grinding. If valve faces are worn or pitted, the valves can be ground on a valve-refacing machine (Fig. 37-15).

Fig. 37-14. Adjustment of valve-tappet clearance on an engine with the rocker arms independently mounted on ball studs. Backing the stud nut out increases clearance. (*Chevrolet Motor Division of General Motors Corporation*)

CHUCK FOR VALVE   GRINDING WHEEL

Fig. 37-15. A valve grinder, or refacing machine. (*Black and Decker Manufacturing Company*)

4 Valve-seat grinding. Valve seats should be ground at the same time that the valves are ground (Fig. 37-16). Both must be ground so that the meeting faces are at the proper angles.

5 Camshaft and bearings. If the camshaft and bearings are worn, the camshaft must be removed for checking. A worn or bent camshaft should be discarded. Worn bearings should be replaced.

§ 37-15 ENGINE BEARINGS

In addition to the camshaft bearings, there are the bearings that support the crankshaft (called the main bearings) and the bearings that attach the rods to the crankshaft (the rod bearings). New bearings are required if the old bearings are worn.

To install new rod bearings, the oil pan is removed and the rod caps are removed so that the old bearings can be taken off and new bearings installed.

To install new main bearings, the crankshaft does not have to be removed. However, ex-

perts advise removal of the crankshaft to assure the best job. The lower bearing halves can be removed by removing the rod caps (remove one at a time so that the crankshaft has support). The upper halves are removed with a roll-out tool stuck in the crankshaft oil hole (Fig. 37-17). When the crankshaft is rotated, the bearing half is forced out.

Bearings should be checked for fit with Plastigage (Figs. 37-18 and 37-19) or with shim stock.

§ 37-16 CONNECTING RODS, PISTONS, AND RINGS

To service pistons and rings, the rod-and-piston assemblies must be removed from the engine. First, the ring ridge, if present, must be removed. The ridge marks the upper limit of ring travel. If it is not cut away, it will catch on the top ring and break the ring and piston (Fig. 37-20). Then, rods can be removed from the top of the engine. The cylinder head must be removed, and rod caps taken off, before this can be done. Piston pins are then removed to separate the pistons and rods.

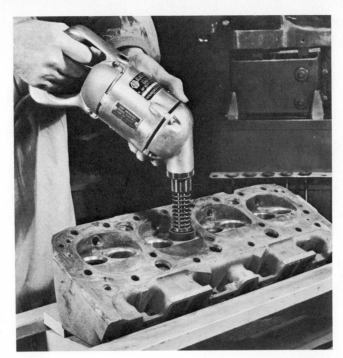

Fig. 37-16. Concentric valve-seat grinder. The stone is rotated at high speed. About once every revolution, it is automatically lifted off the valve seat so that it can throw off loosened grit and grindings. (*Black and Decker Manufacturing Company*)

◆ CAUTION ◆

Do not mix the parts. All parts that are to be reused must go back together exactly as originally found, and must be put back into the same cylinders from which they were removed.

Next, rings are removed, using a special ring remover (Fig. 37-21). Pistons should be cleaned. This includes cleaning out the ring grooves. Next, the fit of the pistons to the cylinders should be checked. If a piston is too loose, it should be discarded and a new oversize piston obtained. This means that the cylinder will have to be machined to a larger size so that it will take the oversize piston with the correct clearance.

As a rule, new piston rings should be used every time an engine is torn down. The old rings will not do their job on a rebuilt engine. The new rings should be checked for fit in the cylinder and also in the piston-ring grooves.

Fig. 37-17. Removal of the upper main bearing. The crankshaft journal is shown partly cut away so that the tool can be seen inserted into the oilhole in the journal. (*Chrysler Corporation*)

Fig. 37-18. A Plastigage strip in place in readiness for a check of the bearing clearance. (*Chevrolet Motor Division of General Motors Corporation*)

After the rings are installed on the piston and the piston is reattached to the rod with the piston pin, the assembly should be dipped in oil to cover the rings with oil. Then a ring compressor should be used to reinstall the assembly (Fig. 37-22). Rods should be reattached to the crankshaft with the rod caps. Then, other

parts that have been removed, such as cylinder head, oil pan, and so on, should be reinstalled.

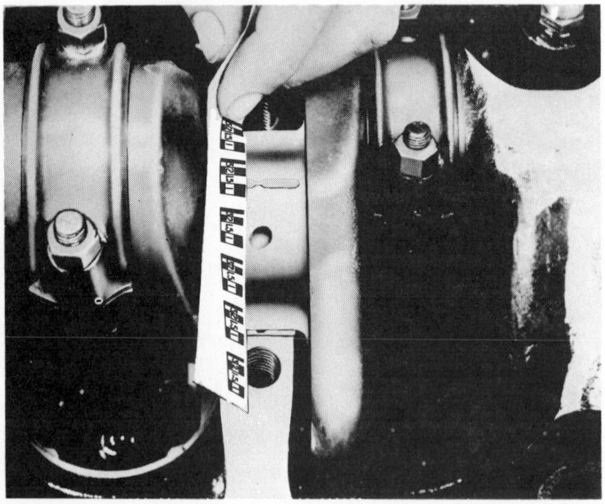

§ 37-17 CRANKSHAFT

If the crankshaft is removed, it should be cleaned and checked for alignment and for worn or scratched journals. Oil holes should be cleaned with solvent and a valve-guide brush. If the main or crank journals are worn, they can be turned down in a crankshaft lathe. Then, new undersize bearings will be required.

§ 37-18 CYLINDER BLOCK

If the cylinders are worn, they will require honing or boring. If they are only slightly worn, honing may do the job. Figure 37-23 shows a dial indicator being used to measure cylinder wear. Heavy wear requires reboring (Fig. 37-25). As a rule, if the cylinder walls require machining, they must be finished to a larger size to take oversize pistons. New pistons are coated with special metals. They cannot be machined to a smaller size, because this would remove the special metal and ruin the pistons.

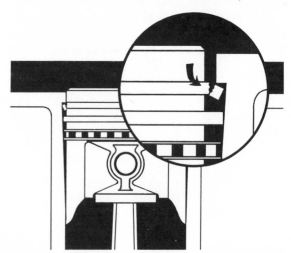

Fig. 37-20. How a ring ridge caused by cylinder wear may break the ring if the piston is withdrawn without removing the ridge. (*Sealed Power Corporation*)

Fig. 37-19. The flattening of Plastigage is checked to determine the bearing clearance. (*Chevrolet Motor Division of General Motors Corporation*)

Fig. 37-21. Using a piston-ring-expander tool to remove or install a compression ring on a piston. (*Service Parts Division of Dana Corporation*)

Therefore, the cylinders must be finished to the proper size to take the new pistons, as we mentioned.

A cylinder hone is shown in Fig. 37-24 installed in a cylinder and ready to be used. It has a set of stones that cut metal from the cylinder wall when the hone is rotated in the cylinder. The boring machine has a cutting tool that cuts metal as it revolves in the cylinder (Fig. 37-25).

After the honing or boring job, the cylinder walls must be washed down with a stiff brush and soapy water. All traces of metal cuttings must be removed. Any traces left will get onto bearings and cause serious trouble.

Fig. 37-23. Measuring cylinder diameter with a dial indicator. Once the reading is taken, the dial should be set at zero and the reading measured with a micrometer. (*Pontiac Motor Division of General Motors Corporation*)

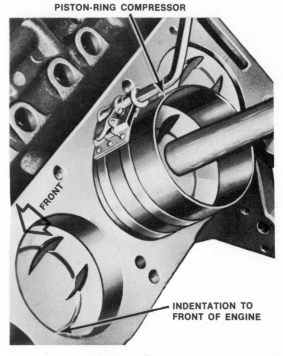

Fig. 37-22. A piston-ring compressor tool is used to install a piston with rings. (*Chevrolet Motor Division of General Motors Corporation*)

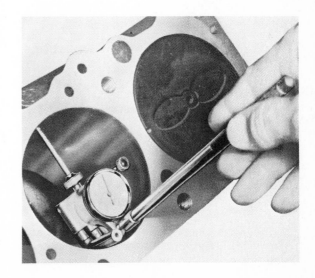

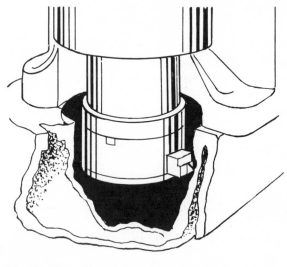

Fig. 37-24. (*Left*) A cylinder hone in place in a cylinder. In operation, the hone revolves in the cylinder, and the abrasive stones in the hone remove material from the cylinder wall.

Fig. 37-25. (*Above*) Revolving head of a cylinder-boring machine, showing the cutting tool.

Review questions

1. Name five instruments used in an engine tune-up.
2. Which instrument could most accurately detect a blown head gasket?
3. What is the purpose of the exhaust-gas analyzer?
4. What is the purpose of an engine tune-up?
5. Describe two ways to check main bearing clearance.
6. Why would machining pistons to a smaller size ruin the pistons?

Study problems

1. Review the causes of the various vacuum gage readings. Then, install a vacuum gage on an engine and diagnose the engine's condition.
2. Follow the procedure outlined in § 37-11 and perform a complete engine tune-up. Write down any abnormal conditions that you find and tell what you did to correct them.

38 | SERVICING THE AUTOMOTIVE-ENGINE SYSTEMS

§ 38-1 SERVICING THE ENGINE SYSTEMS

For the engine to run, it must have four systems: cooling, lubricating, fuel, and electrical. We have discussed all these systems earlier. Now let us see what troubles might occur in these systems and how they can be fixed.

§ 38-2 COOLING SYSTEM SERVICE

The three most common complaints about problems in the engine that might come from trouble in the cooling system are engine overheating, slow engine warm-up, and cooling system leaks.

Overheating. This could be due to the following:

1. Loss of coolant due to leakage
2. Accumulations of rust or scale that require cleaning the system
3. Collapsed hose that prevents normal coolant circulation
4. Defective thermostat
5. Defective water pump
6. Loose or worn fan belt
7. Overloading the engine, high altitude, long periods of idling or slow speed operation, and such troubles as insufficient engine oil and wrong ignition timing

Slow warm-up. This is probably due to a defective thermostat that stays open.

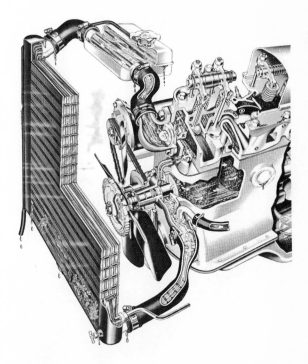

Fig. 38-1. Various places in the cooling system where leaks might develop. (*Union Carbide Corporation*)

Leaks. Figure 38-1 shows places in the cooling system where coolant might leak out.

§ 38-3 COOLING SYSTEM CHECKS

If the radiator has an expansion tank, you can tell at a glance if the system needs water. On those without expansion tanks, you must remove the radiator pressure cap to tell.

511

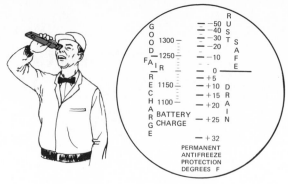

Fig. 38-3. Reading the refractometer.

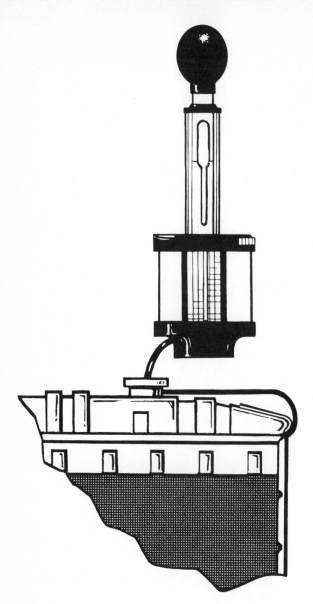

Fig. 38-2. Cooling-system hydrometer used to check coolant for amount of freeze-up protection. (*Ford Motor Company*)

◆ CAUTION ◆

Use care when removing a radiator cap! If the engine is hot, the coolant may boil as you release the pressure. Some manufacturers say that you must not remove the pressure cap from a hot engine. Instead, wait until the engine is cool.

You need a special tester to check the strength of the antifreeze. There are several types of testers. With one type, coolant is sucked up into a tube. Then you can determine the antifreeze strength by checking the height of a float inside the tube (Fig. 38-2). In another, you put a few drops on a plate and look through an eyepiece (Fig. 38-3) to get the reading.

Note that the antifreeze should be changed every 2 years, according to most automotive manufacturers. This means draining the old coolant and adding a mixture of fresh antifreeze and water. Sometimes, the system must also be flushed out to remove accumulated rust and scale.

§ 38-4 LUBRICATING SYSTEM SERVICE

The two complaints you might hear that are related to the lubricating system are that the engine uses too much oil or that the indicator light (or oil-pressure gage) is acting up. An engine that uses too much oil either has oil leaks, is worn, or is being driven at high speeds. Leaks may occur in such places as around the oil-pan gasket, valve cover, and crankshaft oil seals (Fig. 38-4). If the engine is worn, it will burn oil in the combustion chambers. The piston rings in a worn engine cannot keep the oil from getting into the combustion

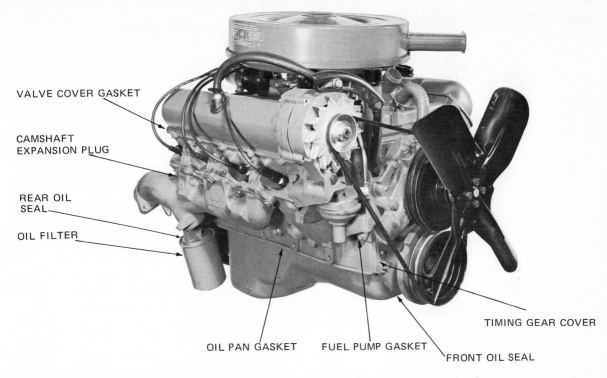

VALVE COVER GASKET

CAMSHAFT
EXPANSION PLUG

REAR OIL
SEAL

OIL FILTER

OIL PAN GASKET   FUEL PUMP GASKET

TIMING GEAR COVER

FRONT OIL SEAL

Fig. 38-4. Exterior view of an engine, showing points where oil may be lost. (*Federal-Mogul*)

chambers. The exhaust gas will have a bluish tinge due to the burned oil. High-speed operation burns more oil because the piston rings have a harder time controlling the oil.

§ 38-5 CHANGING OIL

Different manufacturers have different recommendations for the length of time between oil changes. Chrysler Corporation recommends that for a car in normal service, the oil should be changed every three months or 4,000 miles [6,437 kilometers], whichever comes first. Ford recommends changing the oil every four months or 4,000 miles [6,437 kilometers]. General Motors recommends every four months or 6,000 miles [9,656 kilometers]. For severe service, the oil should be changed at more frequent intervals. Ford says to change it every two months or 2,000 miles [3,219 km] if the car is operated in severe service—dusty condi-

tions, towing trailers, or much idling and short-trip operation.

The oil filter should be changed every other time the oil is changed (Fig. 38-5). For severe service, change it every oil change.

Fig. 38-5. An oil filter being removed. (*Chrysler Corporation*)

Fig. 38-6. Draining engine oil in preparation for adding fresh oil. (*Mobil Oil Company*)

To change the oil, the car is put on a lift and raised. Then a drain pan is put into position and the drain plug removed (Fig. 38-6). After the oil is drained, the plug is replaced, the car lowered, and the required amount of fresh oil is put into the engine.

§ 38-6 FUEL SYSTEM SERVICE

In the fuel system, the only regular services are cleaning of the air cleaner and of the fuel filter. In addition, the idle speed and idle mixture may require adjustment. Also, if troubles develop, the carburetor or the fuel pump may have to be taken off the engine for service.

§ 38-7 AIR CLEANER SERVICE

The air cleaner (Fig. 38-7) performs the very important job of keeping dirt from getting into the engine. Dirt can ruin the engine, and thus the air cleaner must be kept in good condition. The usual recommendation is that the air cleaner should be taken off the car and cleaned every time the engine oil is changed. The air cleaner is easily removed by removing

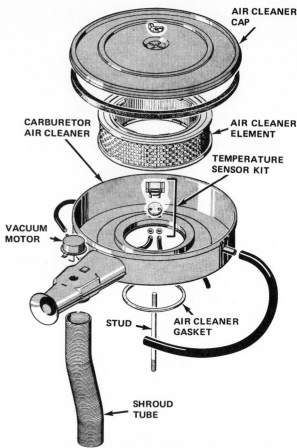

Fig. 38-7. Typical carburetor air cleaner for a late-model V-8 engine. (*American Motors Corporation*)

wing nuts and supporting clamps, if used. Also, all hoses and vacuum lines must be disconnected. Different elements require different cleaning procedures. Always refer to the manufacturer's shop manual for specific instructions.

§ 38-8 FUEL FILTER SERVICE

On some models, the fuel filter can be removed, cleaned, and put back on the engine. On other models, clogged elements can be replaced with new elements by removing the lower section (Fig. 38-8).

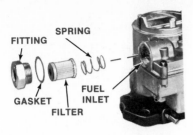

Fig. 38-8. The fuel filter located in the carburetor. (*Buick Motor Division of General Motors Corporation*)

§38-9 FUEL PUMP SERVICE

About the only time the fuel pump requires service is when it develops internal trouble and fails to deliver sufficient fuel. This will show up as engine trouble. An engine that is starved for fuel will stall and lack intermediate and high-speed performance. If the fuel pump is defective, it should be replaced with a new unit. Figure 38-9 shows the installation of a fuel pump on a six-cylinder engine.

§38-10 CARBURETOR OVERHAUL

Never attempt to disassemble a carburetor without a job sheet or manual that applies to the carburetor. Special carburetor overhaul kits are supplied for many carburetors. These have all the necessary parts, including jets, gaskets, washers, and so on, that are required, as well as specific step-by-step instructions on how to do the job.

§38-11 ELECTRICAL SYSTEM SERVICE: BATTERY SERVICE

The battery should be checked periodically to make sure it is in a charged condition so that it can perform well when called upon to crank the engine. Battery condition is more critical in cold weather: the colder an engine, the harder it is to start.

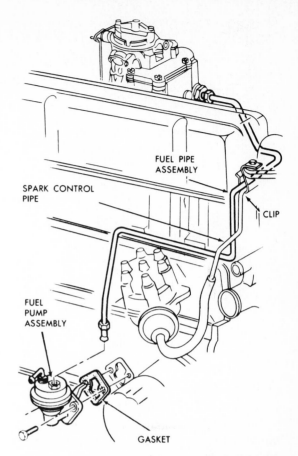

Fig. 38-9. Typical installation of a fuel pump on a six-cylinder engine. (*Chevrolet Motor Division of General Motors Corporation*)

◆ CAUTION ◆

The sulfuric acid in the battery electrolyte is very corrosive and will eat holes in cloth, shoes, and metal. Sulfuric acid can cause painful and severe burns if it gets on your skin, and it could blind you if it gets in your eyes. In case of accident, the sulfuric acid should be flushed away with quantities of water. Be sure to do this immediately. Be very careful to keep baking soda on hand at all times. The action of the baking soda will neutralize the sulfuric acid when sprinkled on it.

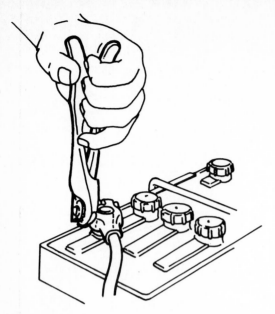

Fig. 38-10. Using battery-cable pliers to loosen the nut-and-bolt type of battery cable. (*General Motors Corporation*)

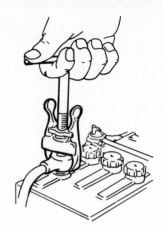

Fig. 38-11. Using a special clamp puller to pull the cable from a battery terminal. (*United Delco Division of General Motors Corporation*)

▶ CAUTION ◀

The gases that form in the tops of battery cells during charge are very explosive. These are the same gases that are used as fuel for some of the largest space rockets. Therefore, never bring an open flame near the tops of battery cells.

---

§ 38-12 BATTERY CHECKS

Whenever a battery is checked or watered, it should be looked over for corrosion, loose hold-down clamps, and dirt. A battery top can be cleaned of dirt by tightening the vent caps, brushing the top with a solution of baking soda and water, and then flushing off the top with water. Care must be used to keep soda from getting into the battery.

If the cable clamps and terminals are corroded, the clamps should be disconnected from the terminals and both should be cleaned. On the bolt-type clamp, use a box wrench or special pliers (Fig. 38-10) to loosen the nut,

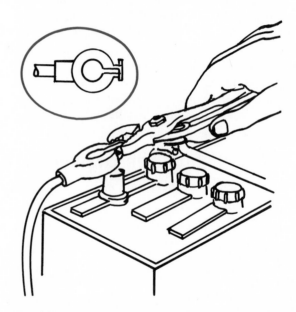

Fig. 38-12. Using pliers to loosen the spring-ring type of cable clamp from a battery terminal. (*United Delco Division of General Motors Corporation*)

and use a special clamp puller (Fig. 38-11) to pull the clamp off the terminal. To remove the spring type of clamp, use pliers to spring the ring apart (Fig. 38-12).

Fig. 38-13. Using a battery carrier.

After the clamps are off, clean them and the terminals with special wire brushes. When it is necessary to carry a battery to or from the car, always use the proper type of battery carrier

(Fig. 38-13). Several kinds of testers can be used to check the battery state of charge. One type is a hydrometer. In this type, liquid from the battery is sucked up into the glass tube of the hydrometer. You then check to see how far the float is sticking out of the liquid (Fig. 38-14). In a second type, you put a few drops of the liquid on a plate. Then you look through an eyepiece to check the state of charge (See Fig. 38-3. This same tester is used to test the cooling system antifreeze). Remember the cautions above about the sulfuric acid in the battery.

Other testers are electrical in nature. If you work with batteries, you will be told how to use these testers.

§ 38-13 BATTERY CHARGING

If a battery runs down, it can be charged from a battery charger and brought back up to good condition, if it is otherwise okay. If the battery is old or has a cracked case or other damage, then chances are you will have to discard it and get a new one. But remember, the better the battery is cared for, the longer it will last. Good care includes adding water when necessary, keeping the battery clean, and keeping it in a charged condition.

§ 38-14 BATTERY CARE

In addition to being checked periodically for state of charge, the battery should also be watered if necessary. A battery in normal

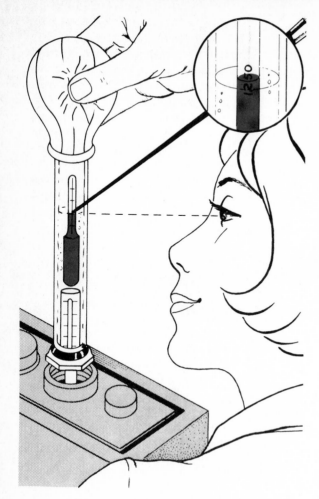

Fig. 38-14. Using a battery hydrometer to check the specific gravity of a battery cell. The reading should be taken at eye level.

service tends to lose a little water from evaporation and from the chemical actions taking place inside it. When adding water, don't overdo it. If the battery cells are overfilled, the acid will bubble out when the battery is charged and get on engine parts.

Note that if a battery requires a great deal of water, and water has to be added frequently, the chances are that the battery is being overcharged. Overcharging overheats the battery

and causes it to lose water fast. In this case, the alternator-regulator systems should be checked and adjustments made as necessary to bring the charging rate down when the battery is charged.

§ 38-15 STARTING MOTOR SERVICE

Very little in the way of service is normally required on the starting motor. The shop manuals of many manufacturers call for servicing starting motors only at the time that the engine is overhauled. Other starting motors have provision for oiling and for cover bands that can be removed so that the brushes and commutator can be inspected. These units should be oiled and inspected periodically.

In case of trouble, the starting motor should be removed from the engine for disassembly and repair. The starting motor can be tested either on or off the engine.

§ 38-16 ALTERNATOR-REGULATOR SERVICE

If the alternator-regulator system is not operating properly, then checks will be required to determine whether it is the alternator or the regulator that is at fault. Different alternators and regulators require different testing methods. Always refer to the manufacturer's shop manual covering the specific model being checked and serviced to determine the procedure to follow and the equipment to use.

§ 38-17 IGNITION SYSTEM SERVICE

To service the ignition system, a variety of testing instruments are required, including an ignition-coil tester, a condenser tester, and a distributor tester. The distributor tester should have a dwell meter to check the contact-point opening. In addition to these instruments, a timing light is necessary.

Fig. 38-15. Using the oscilloscope to pinpoint engine problems. (*Sun Electric Corporation*)

§ 38-18 OSCILLOSCOPE

The oscilloscope is a high-speed voltmeter (Fig. 37-9). It tests the ignition system secondary voltages and shows these voltages on the television-like screen (Fig. 38-15). The way that the picture looks tells the experienced automotive technician whether or not the ignition system is working properly. If the ignition system has some defect, it will change the picture so that the technician can tell what is wrong. For instance, if a spark plug is not firing so that a cylinder is missing, the voltage will be differ-

ent for that spark plug. This instantly pinpoints the trouble for the technician.

§ 38-19 DISTRIBUTOR TESTER

The distributor tester (Fig. 38-16) has a variable drive arrangement and a vacuum source that can be used to check the operation of the distributor-centrifugal and vacuum-advance mechanisms. If these are not correct, the distributor must be either repaired or replaced. On some distributors, the advance mechanisms are adjustable.

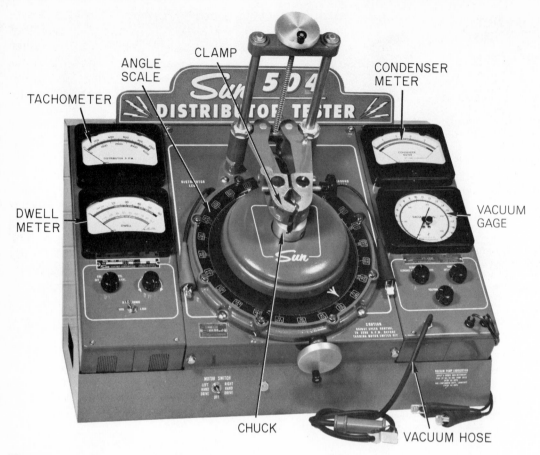

TACHOMETER

ANGLE
SCALE

CLAMP

CONDENSER
METER

DWELL
METER

VACUUM
GAGE

CHUCK

VACUUM HOSE

Fig. 38-16. Ignition distributor tester. (*Sun Electric Corporation*)

§ 38-20 IGNITION-COIL AND CONDENSER TESTERS

The condenser tester applies voltage to the condenser to determine whether or not the condenser can hold up. It tests for shorts or grounds, insulation resistance, series resistance, and capacity.

The ignition-coil tester checks the coil for such defects as high-voltage breakdown in the secondary, high primary resistance, open circuits, and shorted turns in the primary or secondary winding. If all these conditions are checked out and found within specified range, then the coil is in good condition.

§ 38-21 IGNITION TIMING

Chapter 37 describes the use of a timing light to set the ignition timing (see Figs. 37-7 and 37-8). This procedure is necessary every time the contact points have been adjusted or the distributor has been moved in its mounting. Timing is always checked and adjusted during engine tune-up.

§ 38-22 IGNITION SYSTEM TROUBLES

Troubleshooting was covered in Chapter 36. Several of the troubles listed in Chapter 36

could be caused by defects in the ignition system. It is important to remember, however, that these troubles could also be caused by other engine components, as explained in Chapter 36. Using the oscilloscope and other testing instruments will locate the source of trouble.

§ 38-23 EMISSION SYSTEM SERVICE

Servicing emission-control systems includes checking the PCV valve and several filters. These filters include the air filter in the air-injection system, the filter at the bottom of the canister in the fuel-vapor recovery system (Fig. 38-17), and the PCV filter on the inside of the air cleaner housing. These filters should be checked or replaced at the mileages specified by the manufacturer.

Fig. 38-17. Checking a filter in the emission-control system.

Review questions

1. List four possible causes of engine overheating.
2. How often should an oil filter be changed?
3. Describe how to clean a heavily corroded battery.

4. What condition is usually indicated when a battery requires water frequently?
5. When should the starting motor be overhauled?

Study problem

1. Locate each piece of test equipment in your shop that is used for checking automotive engine systems. Study the instruction manual for each tester. Then practice connecting and using each tester on an engine in the shop until you can properly use each one.

39 YOUR FUTURE CAREER IN POWER MECHANICS

§ 39-1 JOBS IN POWER MECHANICS

As you explore the world of power mechanics by studying this book, you are probably thinking about a possible future in this field. You may be wondering what sort of career you can find if you decide to take up power mechanics as your life's work. Let us look at a few possible careers.

§ 39-2 AIRCRAFT MECHANIC

Aircraft mechanics keep airplanes, from the small single-engine jobs to the biggest jets, in good operating condition. They are skilled workers who really must know their business. Every time an airplane takes off, their work is being tested. If they fail to do their job properly, a mechanical failure might lead to a major airplane disaster. Being an aircraft mechanic is a very rewarding job. It pays very well, and there is a real sense of pride in doing the job well.

Aircraft mechanics do routine preventive maintenance. That is, they change engine oil, lubricate wheels and other airplane parts, check out all systems in the aircraft, and make whatever repairs are necessary. Some mechanics are aircraft-engine specialists. They disassemble engines for inspection of parts, repair, and reassembly. Most mechanics are located in the larger cities on the main airline routes.

To become an aircraft mechanic, you must be in good health, young (to start with), and have some experience in and interest in mechanical things. If you found this course in general power mechanics interesting, you might think about becoming an aircraft mechanic. The larger airlines train apprentices in 3- to 4-year programs of instruction and work experience. There are trade schools where you can learn the basics in a much shorter time. The more you know to start with, the better chance you have of getting a job with an airline. Naturally, the airlines want good people who have had some education and experience.

§ 39-3 AUTOMOTIVE MECHANIC

It is easier to become a qualified automotive mechanic than an aircraft mechanic. This is only natural, because the airplane is, after all, much more complicated than an automobile. The automotive mechanic today is a well-paid and highly respected member of our society who performs preventive maintenance, diagnoses troubles, and makes repairs. Preventive maintenance is the periodic examination, adjustment, repair, or replacement of automotive parts. When mechanical or electrical trouble occurs, the mechanic diagnoses the trouble, finds the cause, and fixes it. The mechanic uses a variety of testing instruments, ranging from cylinder compression testers to oscilloscopes and dynamometers.

The tools required range from simple hand tools to numerous special tools designed to do specific jobs such as adjusting front alignment or removing and replacing engine pistons.

Some automotive mechanics are general mechanics. They work on a variety of automotive components. Others are specialists, such as the automatic transmission specialist or the air-conditioning specialist. Others specialize in tune-up and spend their time making the checks, adjustments, and repairs that restore the engine to prime operating condition. They often use scientific testing equipment such as oscilloscopes and dynamometers.

Front-end mechanics check front alignment and balance wheels. They use special alignment equipment and wheel-balancing machines.

Brake mechanics adjust brakes, replace brake linings, and repair brake parts such as calipers, wheel cylinders, and master cylinders.

Body mechanics specialize in repairing collision damage, straightening or replacing body panels, replacing broken glass, straightening or replacing frame components, and repairing or adjusting doors, trunk lids, and hoods.

There is a good future in this profession. Many of the operators and owners of automotive repair facilities and service stations started out as automotive mechanics. The automotive mechanic can become service manager, parts manager, car salesman, manufacturer's representative, or owner of a garage or automotive dealership.

§ 39-4 BOAT MECHANIC

Boat engines have many things in common with automotive engines. A boat mechanic makes periodic inspections of boat engines, performs engine tune-ups, makes adjustments, and repairs or replaces parts that are not up to specifications. Boat mechanics may specialize in either gasoline or diesel engines or may work on both. Other fields of specialization include outboard motor mechanic or inboard motor mechanic. Each requires a considerable degree of skill and knowledge.

The boat mechanic uses many of the same tools and testing equipment as the automotive mechanic.

§ 39-5 DIESEL MECHANIC

The diesel mechanic specializes in the inspection, testing, adjustment, and repair of diesel engines. Diesel engines come in a great variety of sizes, ranging from small units used in automobiles to huge engines used in construction equipment and locomotives. Even larger diesel engines rated at thousands of horsepower are used in power plants and in ships. Reliability is the keyword with diesel engines. A diesel-powered truck loaded with frozen food worth thousands of dollars has to arrive on time. If it breaks down on the road, the frozen food may melt and become worthless. Thus, the diesel mechanic must be a person who can keep the engines running and who can do everything necessary to prevent breakdowns. In a sense, the diesel mechanic is like the aircraft mechanic. Nothing is ever left to chance. Every factor is considered and everything is done to assure perfect functioning of the equipment.

§ 39-6 MOTORCYCLE MECHANIC

More and more motorcycles are being sold in the United States. More and more motorcycle mechanics are being hired. The motorcycle mechanic must have many of the same skills as the automotive mechanic. That is, the motorcycle mechanic must be able to use special equipment to test, adjust, and overhaul motorcycle engines and other motorcycle components such as transmissions and brakes.

§ 39-7 OTHER POSSIBILITIES

The above by no means exhausts the possible careers for the student who finishes a course in general power mechanics. The student who successfully completes the course has learned many basic facts about how power is produced, how it is transmitted, and how it is used. The student has also learned certain hands-on skills, that is, how to handle tools and make repairs and adjustments to engines. Even though the student may never go directly into repair or servicing work, these basic skills will be of great value if the student goes into some other phase of power mechanics such as sales, supervision, or management. Usually, the people at the top, the ones with the most important jobs, started at the bottom, learning the fundamentals, and moving up step by step.

And you, as a student who has completed the course in general power mechanics, have taken an important first step to a satisfying and productive career.

INDEX

A

ac, 37–39
Adjustable wrench, 424
Afterburner, 325–327
Air brakes, 408–410
Air cleaner, 194–195, 514
    servicing of, 514
Air-cooled engines, 52
Air filter, 66–67
Air-injection system, 231–232
Aircraft, 305–333, 347–351
    engines, 305–333
    gas turbines, 321–322
    propellers, 312–314
    rocket-engine, 347–351
    supersonic transport, 331–333
Airplane, 305–332
Alternating current, 37–39
Alternator, 82, 168–174, 518
    regulators, 173
    servicing of, 518
    stator, 169
Ammeter, 145
Ampère, André, 10
Antifreeze, 212
Apollo, 340–343
Arbor press, 436–437
Atlas rocket, 336
Atmospheric pressure, 47
Atomic energy, 42
Atoms, 43–44, 73
Automatic transmissions,
    381–400
Automotive charging systems,
    163–174
Automotive electrical equip-
    ment, 142–188

Automotive emission controls,
    226–234
Automotive power sources, 235

B

Battery, 142, 150–153
    construction of, 150–152
    ratings of, 152–153
    servicing of, 515–518
Bearings, engine, 61–62
    types of, 57–58
Bench vise, 436
Bendix drive, 159–161
Blow-by, 497
Boilers, steam, 25
Boosters, rocket, 340–341
    solid-fuel, 343–344
Bore, 135
Box wrenches, 423–425
Brake horsepower, 140–141
Brakes, 401–410
    air, 408–410
    disk, 406–407
    drum, 404–406
    hydraulic, 401–407
    power, 406–410

C

Calipers, 441–442
    vernier, 448–450
Camshafts, 124
    overhead, 131–132
Carburetor, 68–70, 195–201
    marine, 293–294
    overhaul of, 515

Cars, electric, 239–241
Catalytic converter, 230–234
Centrifugal advance, 182–183
Change of state, 44–45
Charging systems, automotive,
    163–174
Chisels, 427
Choke valve, 69–70
Clutch, automotive, 335–361
    hydraulic, 361
Coal, 19–21
Combustion, 45, 227
Combustion engines, 12
Compression ratio, 136
Condenser, ignition, 180–181
    tester for, 520
Conductors, 74
Connecting rods, 117–119
Cooling systems, engine,
    207–213, 511–512
    antifreeze for, 212
    expansion tank in, 212
    fan in, 208–209
    marine, 292–293
    purpose of, 207
    radiator in, 210
    radiator pressure cap in,
        210–212
    servicing of, 511–512
    temperature indicator in,
        212–213
    thermostat in, 210
    water pump in, 207–208
Crankcase ventilation, 221–223,
    227–228
Crankshaft, 49–51, 109–111
Cugnot, N. J., 7
Current, electric, 74

Current regulator, 167
Cutout relay, 167
Cylinder block, 101–105
Cylinder compression tester, 497–498
Cylinder head, 105
Cylinder-leakage tester, 498

D

dc, 37–39
Dial indicators, 446–447
Dies, 433–434
Diesel engines, 254–266
    applications, 264–266
    emissions from, 265–266
    four-cycle type, 258–259
    fuel for, 261–264
    fuel system for, 259–261
    history of, 254–255
    two-cycle type, 255–257
Differentials, 373–379
    nonslip, 378–379
Diodes, 170–172
Dipstick, 225
Direct current, 37–39
Dirt bike, 274–275
Disk brakes, 406–407
Distributor, ignition (*see* Ignition distributor)
Distributor tester, 519
Drills, 430–431
Drive lines, 370–372
Drum brakes, 404–406
Dynamometer, 501

E

Edison, Thomas Alva, 11
EGR system, 230–231
Electric cars, 239–241
Electric current, 74
Electric fuel pumps, 193–194
Electric lights, 36–37
Electric starter, 80
Electrical generating plants, 36 (*see also* Power plant, electric)
Electrical system service, automotive, 515–521
Electricity, 9–12, 142–188
Electromagnets, 10, 75, 147–148
Electronic ignition system, 184–188

Electrons, 73
Emission controls, automotive, 226–234
    servicing of, 521
Energy, 17–22
Engines, 43–72, 84–141, 235–333, 467–521
    air-cooled, 52
    aircraft, 305–333
        jet, 309, 320
        piston-type of, 314–320
        radial, 314–315
        reaction, 338–340
        turbojet, 309
    bearings in, 61–62, 109–114
    combustion, 12
    cooling system (*see* Cooling systems, engine)
    Diesel (*see* Diesel engines)
    external-combustion, 12
    fan, 208–209
    four-cycle, 84–141
        bearings in, 109–114
        construction of, 101–134
        flywheel in, 114–115
        operation of, 84–91
        types of, 91–100
        valves in, 85–86, 124–133
    internal-combustion, 43–72, 84–141, 235–333
    lubricating system (*see* Lubricating system)
    marine, 278–304
    motorcycle (*see* Motorcycle engines)
    noises in, 494–495
    off-the-road, 13
    oil for, 214
    operating cycles of, 53
    principles of, 48
    rotary (*see* Gas turbine; Wankel engine)
    servicing of, 497–510
    small (*see* Small engine)
    steam, 235–237
    Stirling, 238–239
    troubleshooting of, 488–495
    tune-up of, 503–505
    two-cycle (*see* Two-cycle engines)
    types of, 43
    Wankel (*see* Wankel engines)
Escape velocity, 337
Exhaust gas, cleaning of, 229–234
Exhaust-gas analyzer, 499–500

Exhaust gas recirculating system, 230–231
Exhaust manifold, 108
Exhaust system, 109, 202–203
Expansion tank, 212
External-combustion engines, 12

F

Faraday, Michael, 10
Feeler gage, 439–441
Files, 428–429
Fire prevention, 414–416
Fitch, John, 6
Float bowl, 67–68
Fluid coupling, 385–387
Fluid power transmission, 401–413
Flywheel, 114–115
Flywheel propulsion, 242–243
Ford, Henry, 12–13
Ford C6 automatic transmission, 395–399
Fossil fuels, 19
Four-cycle engines, 84–141
    construction of, 101–134
    I-head, 98–99, 124
    L-head, 96–98
    measurements of, 125–141
    operation of, 84–91
    overhead camshaft, 99–100
    types of, 91–100
    valves in, 85–86
Freewheeling mechanism, 389–390
Friction, 58–61
Friction disk, 359
Fuel cells, 240
Fuel filter, 191
    servicing of, 514
Fuel for power, 14–15
Fuel gage, 204
Fuel injection, 201
Fuel pump, 192–194
    servicing of, 515
Fuel system, 65–72, 189–206, 293
    air cleaner in, 194–195
    automotive, 189–206
    carburetor in, 195–201
    filter in, 191
    fuel-injection type of, 201
    fuel pump in, 192–194
    fuel tank in, 191
    marine, 293

servicing of, 514–515
two-cycle engine, 65–72
vapor-return line in, 192
vapor separator in, 192
Fuel tank, 65, 191
Fuel-vapor recovery systems, 228
Fulton, Robert, 6

G

GM Electrovair, 240
Gas, natural, 21
Gas turbine, 244–246, 321–322
aircraft, 321–322
Gaskets, 105–106
Gasoline, 65, 189–190
storing of, 474
Gear ratios, 354–355
Generators, 10, 163–168
Geothermal power plants, 21–22
Gilbert, William, 10
Glider, 305
Goddard, Robert, 335
Governors for two-cycle
engines, 70–72
Gravity, 47
overcoming of, 336–338
Grinders, 434–435
Gunpowder, 334

H

HEI, 186–187
Hacksaw, 431–433
Hammers, 435–436
Hand tools, 421–437
Heat, 44
Helicopter, 308
Hero's steam turbine, 305
High energy ignition, 186–187
Holmes, Frederick, 10
Horsepower, 2, 18–19, 137–141
brake, 140–141
Hybrid cars, 241
Hydraulic brakes, 401–407
Hydraulics, 381–385
Hydrogen fuel, 204–205

I

Ignition coil, 39–40, 175–180
testers for, 520

Ignition distributor, 175–184
contact points in, 175–177
Ignition switch, 187–188
Ignition system, 175–188
advance mechanism in,
182–184
condenser in, 180–181
distributor in, 175–184
electronic, 184–187
marine, 294–297
servicing of, 518–521
spark plugs in, 178–180
switch in, 187–188
timing of, 182
two-cycle engine, 76–80
Ignition timing, 182, 520
Ignition timing light, 500–501
Inboard engine, 278–280,
298–303
Insulation, 74–75, 146
Intake manifold, 106–108

J

Jet engines, 309, 320, 327
accessories for, 327

K

Kilowatts, 138

L

Lenoir, Jean Joseph E., 12
Lifting body, M2-F2, 350
Locomotives, 6–8
Lubricating system, 214–225
crankcase ventilation in,
221–223
dipstick in, 225
oil filter in, 220–221
oil-pressure indicators in,
223–224
oil pump in, 219–220
positive crankcase ventilation
in, 222–223
purpose of, 214
relief valve in, 220
servicing of, 512–514

M

M2-F2 lifting body, 350
Magnetism, 144–145

Magnets, 146–147
Manifold, exhaust, 108
intake, 106–108
Manifold heat control, 203
Marine engines, 278–304
Master cylinder, 403–404
Measurements, 439–450,
452–466
metric, 452–466
Measuring tools, 439–450
Metric system, 452–466
units of, 454–458
Metric to USCS conversion
tables, 456–464
Micrometers, 442–446
Minibike, 274
Molecules, 43–44
Motorboats, 280–281
Motorcycle engines, 267–277
four-cycle, 271–274
two-cycle, 267–271
types of, 267

N

Natural gas, 21
Nonslip differential, 378–379
Nuclear age, 13–14

O

Oberth, Herman, 335
Oersted, Hans Christian, 10
Off-the-road engines, 13
Ohm, George, 10
Ohm's law, 148
Oil, 19–21, 214–225
additives in, 215
service ratings of, 214–215
supply of, 20–21
viscosity of, 214–215
Oil filter, 220–221
Oil-pressure indicators, 223–224
Oil pumps, 219–220
Oil shale, 21
Oscilloscope, 501, 519
Otto, Nickolaus, 12
Outboard drives, 285
Outboard engine, 278–297
V-4 type, 286–291
Overhead-camshaft engine,
99–100, 131–132
Overrunning clutch, 161–162

P

PCV system, 222–223, 227–228
  location of, 492
Piston displacement, 135–136
Piston rings, 48, 88–89, 121–124
Piston stroke, 53, 85
Pistons, 117–120
Plane (*see* Aircraft)
Planetary gears, 390–392
Pliers, 426–427
Pneumatics, 407–408
Pollution from automobiles, 226
Positive crankcase ventilating
      system, 222–223, 227–228
Power, 1–35, 137–138
  fuel for, 14–15
  history of, 1–16
  steam (*see* Steam power)
Power brakes, 406–410
Power mechanics, jobs in,
      522–524
Power plant, electric, 36–42
Power plants, 36–42
  air pollution from, 40–41
  Edison's, 11
  solar, 22
  thermal pollution from, 41–42
  water, 22
  wind, 22
Power steering, 410–413
Power train, 354
Power transmission, mechani-
      cal, 354–379
Pressure, 45
Primer, 70
Propellant tanks, 344
Propeller, aircraft, 306–307,
      312–314
  boat, 281–285
Propeller shafts, 370–372
Protons, 73
Pulsejets, 321
Punches, 429–430

R

Radial engine, 314
Radiator, 210
Radiator pressure cap, 210–212
Ramjets, 320
Reaction engines, 338–340
Rear axles, 373–379

Regulator, alternator, 173
  generator, 167–168
Relief valve, 220
Resistance, 148
Rocker arms, 128–130
Rocket boosters, 340–341
Rocket-engine aircraft, 347–351
Rockets, 334–352
  liquid-fuel, 344
Rotary engines (*see* Gas turbine;
      Wankel engine)
Ruler, 439
Run-on, preventing, 234

S

SST, 331–333
Safety in the shop, 414–420
Satellites, 336
Saturn V rocket, 346
Savery, Thomas, 5
Screwdrivers, 421–423
Secondary wiring, 179–180
Shop safety, 414–420
Slip joint, 373
Small engines, 467–487
  abuses of, 467
  disassembly of, 475–478
  maintenance of, 467–487
  servicing of, 467–487
  troubleshooting of, 467–472
  winter storage of, 467
Snorkel tube, 229
Snowmobiles, 275–277
Socket wrenches, 424–425
  metric, 460
Solar power plants, 22
Solenoid, starting motor, 162
Solid-fuel boosters, 343–344
Space age, 15
Space missions, 352
Spacecraft, piloted, 350
  power plants, 352–353
Spark plugs, 178–180
Sputnik, 336
Staging, 341–343
Starter-generator, 80–81
Starting motor, 80–82, 153–162,
      297, 518
  construction of, 157–158
  marine, 297
  operation of, 157
  servicing of, 518
  two-cycle engine, 80–82

Stator, alternator, 169
Steam, uses of, 23–24
Steam car, 237
Steam engines, 5–9, 27–32,
      235–237
  reciprocating, 27–32
Steam generating units, 24–25
Steam power, 3–9, 23–35
  producing, 24
Steam turbine, 3–5, 32–35
Steamboats, 6
Steering, power, 410–413
Stevens, John, 6
Stirling engine, 238–239
Stroke, 85, 135
Superchargers, 317–319
Superheaters, steam, 26–27
Supersonic transport aircraft,
      331–333
Synchronizers, 365–367

T

TCS system, 231
Tachometer, 497
Taps, 433–434
Temperature indicator, 212–213
Thermactor system, 232
Thermostat, 210
Thermostatic air cleaner, 195
Thrust bearing, 113–114
Tools, hand, 421–437
  measuring, 439–450
Torque, 139–140
Torque converter, 387–390
Torque wrench, 426
Transmission-controlled spark
      system, 231
Transmissions, automatic,
      381–400
  fluid coupling in, 385–387
  hydraulic controls in,
      392–395
  hydraulics in, 381–385
  planetary gears in, 390–392
  torque converter in, 387–390
Transmissions, manual, 361–370
Trailbike, 274
Transformer, 39
Trevithick, Richard, 7
Tsiolkovsky, K. E., 334–335
Tune-up, engine, 503–505
Turbine, gas, 244–246
Turbofans, 327

Turbojet engine, 309, 323–327
Turboprops, 327
Two-cycle engines, 52–64
  exhaust valves in, 55
  fuel system for, 65–72
  governors for, 70–72
  ignition systems for, 76–80
  lubricating of, 62–63
  operation of, 54–55
  reed valves in, 55–57
  starting systems for, 63, 80–82

U

U.S. customary system (USCS),
  452–454
USCS to metric conversion
  tables, 456–464
Universal joint, 372–373

V

V-2 rocket, 335–336
Vacuum, 47–48

Vacuum advance, 183–184
Vacuum gage, 498–499
Valve guides, 126
Valve lifters, hydraulic, 130–131
Valve seats, 126–127
Valve springs, 128
Valve timing, 125
Valves, engine, 85–86
Vapor-return line, 192–193
Vapor separator, 193
Verbiest, Ferdinand, 7
Vernier caliper, 448–450
Vibration damper, 115
Volta, Alessandro, 10
Voltage, 145–146
Voltage regulator, 167–168
Voltaic pile, 10
Volumetric efficiency, 138–139

W

Wankel engine, 246–253
  cooling system in, 253
  fuel system in, 252

  ignition system in, 251–252
  operation of, 248–251
Water mills, 2–3
Water power plants, 22
Water pump, 207–208
  marine, 286
Watt, James, 506
Wind power plants, 22
Windmills, 3
Work, 17–18, 137
Wrenches, 423–426
  adjustable, 424
  socket, 424–425
  torque, 426
Wright, Orville, 305
Wright, Wilbur, 305

X

X-15, 348–350